Lecture Notes in Computer Science 16089

Founding Editors

Gerhard Goos
Juris Hartmanis

AF166227

The series Lecture Notes in Computer Science (LNCS), including its subseries Lecture Notes in Artificial Intelligence (LNAI) and Lecture Notes in Bioinformatics (LNBI), has established itself as a medium for the publication of new developments in computer science and information technology research, teaching, and education.

LNCS enjoys close cooperation with the computer science R & D community, the series counts many renowned academics among its volume editors and paper authors, and collaborates with prestigious societies. Its mission is to serve this international community by providing an invaluable service, mainly focused on the publication of conference and workshop proceedings and postproceedings. LNCS commenced publication in 1973.

Jorge Carrillo-de-Albornoz ·
Alba García Seco de Herrera · Julio Gonzalo ·
Laura Plaza · Josiane Mothe · Florina Piroi ·
Paolo Rosso · Damiano Spina ·
Guglielmo Faggioli · Nicola Ferro
Editors

Experimental IR Meets Multilinguality, Multimodality, and Interaction

16th International Conference of the CLEF Association, CLEF 2025
Madrid, Spain, September 9–12, 2025
Proceedings

 Springer

Editors
Jorge Carrillo-de-Albornoz ⓘ
Universidad Nacional de Educación
a Distancia
Madrid, Spain

Alba García Seco de Herrera ⓘ
Universidad Nacional de Educación
a Distancia
Madrid, Spain

Julio Gonzalo ⓘ
Universidad Nacional de Educación
a Distancia
Madrid, Spain

Laura Plaza ⓘ
Universidad Nacional de Educación
a Distancia
Madrid, Spain

Josiane Mothe ⓘ
IRIT Université de Toulouse
Toulouse, France

Florina Piroi ⓘ
Technische Universität Wien
Vienna, Austria

Paolo Rosso ⓘ
Universitat Politècnica de València
Valencia, Spain

Damiano Spina ⓘ
RMIT University
Melbourne, VIC, Australia

Guglielmo Faggioli ⓘ
University of Padua
Padova, Italy

Nicola Ferro ⓘ
University of Padua
Padova, Italy

ISSN 0302-9743   ISSN 1611-3349 (electronic)
Lecture Notes in Computer Science
ISBN 978-3-032-04353-5   ISBN 978-3-032-04354-2 (eBook)
https://doi.org/10.1007/978-3-032-04354-2

Preface

Since 2000, the *Conference and Labs of the Evaluation Forum* (CLEF) has played a leading role in stimulating research and innovation in the domain of multimodal and multilingual information access. Initially founded as the *Cross-Language Evaluation Forum* and running in conjunction with the *European Conference on Digital Libraries* (ECDL/TPDL), CLEF became a standalone event in 2010 combining a peer-reviewed conference with a multi-track evaluation forum. The combination of the scientific program and the track-based evaluations at the CLEF conference creates a unique platform to explore information access from different perspectives, in any modality and language.

The CLEF conference has a clear focus on experimental information retrieval (IR) as seen in evaluation forums (like the CLEF Labs, TREC, NTCIR, FIRE, MediaEval, RomIP, TAC) with special attention to the challenges of multimodality, multilinguality, and interactive search, ranging from unstructured to semi-structured and structured data. The CLEF conference invites submissions on new insights demonstrated by the use of innovative IR evaluation tasks or in the analysis of IR test collections and evaluation measures, as well as on concrete proposals to push the boundaries of the Cranfield/TREC/CLEF paradigm.

CLEF 2025[1] was organized by the Universidad Nacional de Educación a Distancia (UNED), Spain, from 9 to 12 September 2025. CLEF 2025 was the 16th year of the CLEF Conference and the 26th year of the CLEF initiative as a forum for IR Evaluation. The conference format remained the same as in past years and consisted of keynotes, contributed papers, lab sessions, and poster sessions, including reports from other benchmarking initiatives from around the world. All sessions were organized in person, but also allowed for remote participation for those who were not able to attend physically.

CLEF 2025 continued the initiative introduced in the 2019 edition, during which the *European Conference for Information Retrieval (ECIR)* and CLEF joined forces: ECIR 2025[2] hosted a special session dedicated to CLEF Labs where lab organizers presented the major outcomes of their Labs and their plans for ongoing activities. This was reflected in the ECIR 2025 proceedings, where CLEF Lab activities and results were reported as short papers. The goal was to engage the ECIR community in CLEF activities and disseminate the research results achieved during CLEF evaluation cycles through submission of papers to ECIR.

The following scholars were invited to give a keynote talk at CLEF 2025: Joanna Bryson, Full Professor of Ethics and Technology at the Hertie School, Germany, José Hernández-Orallo, Full Professor of Computer Science at the Universitat Politècnica de València, Spain, and Sameer Antani, Principal Investigator and a senior researcher in the Division of Intramural Research of the National Library of Medicine (NLM) at the National Institutes of Health (NIH), USA.

---

[1] https://clef2025.clef-initiative.eu.

[2] https://ecir2025.eu/.

CLEF 2025 received a total of 14 scientific submissions, of which a total of 6 papers (5 long and 1 poster) were accepted. Each submission was double-blindly reviewed by at least two program committee members, and the program chairs oversaw the reviewing and follow-up discussions. Several papers were a product of international collaboration. This year, researchers addressed the following important challenges in the community: selective search as first stage retrieval, detection of AI-generated biomedical texts, multilingual datasets for authorship attribution, with focus on medical disinformation detection, authorship analysis by detecting writing style changes, the use of LLMs in authorship analysis and attribution, LLM use in taxonomy generation, personalized education (K-12) with help from LLMs, and document visual question answering through a vision language model.

Like in previous editions, since 2015, CLEF 2025 invited CLEF lab organizers to nominate a "best of the labs" paper, among those submitted in the CLEF 2024 labs, which was reviewed as a full paper submission to the CLEF 2025 conference, according to the same review criteria and PC. Five full papers were accepted for this "best of the labs" section.

The conference integrated a series of workshops presenting the results of lab-based comparative evaluations. A total of 20 lab proposals were received and evaluated in peer review based on their innovation potential and the quality of the resources created. The 14 selected labs represented scientific challenges based on new datasets and real-world problems in multimodal and multilingual information access. These datasets provide unique opportunities for scientists to explore collections, develop solutions for these problems, receive feedback on the performance of their solutions, and discuss the challenges with peers at the workshops. In addition to these workshops, the labs reported results of their year-long activities in overview talks and lab sessions. Overview papers describing each of the labs are provided in this volume. The full details for each lab are contained in a separate publication, the Working Notes.[3]

The 14 labs running as part of CLEF 2025 comprised mainly labs that continued from previous editions at CLEF (BioASQ, CheckThat!, ELOQUENT, eRisk, EXIST, ImageCLEF, JOKER, LifeCLEF, LongEval, PAN, qCLEF, SimpleText, and Touché) and a new pilot/workshop activity: TalentCLEF. In the following, we give a few details for each of the labs organized at CLEF 2025 (presented in alphabetical order):

BioASQ: Large-scale biomedical semantic indexing and question answering[4] aimed to push the research frontier towards systems that use the diverse and voluminous information available online to respond directly to the information needs of biomedical scientists. This edition of BioASQ offered the following tasks: *Task 1 (13b) – Biomedical Semantic Question Answering:* Benchmark datasets of biomedical questions, in English, along with gold standard (reference) answers constructed by a team of biomedical experts. The participants have to respond with relevant articles, and snippets from designated resources, as well as exact and "ideal" answers. *Task 2 – Synergy: Question Answering for developing problems:* Biomedical experts pose unanswered questions for

---

[3] Faggioli, G., Ferro, N., Rosso, P., and Spina, D. editors (2025). *CLEF 2025 Working Notes.* CEUR Workshop Proceedings (CEUR-WS.org).

[4] https://www.bioasq.org/workshop2025.

developing problems, such as COVID-19, receive the responses provided by the participating systems, and provide feedback, together with updated questions in an iterative procedure that aims to facilitate the incremental understanding of developing problems in biomedicine and public health. *Task 3 – MultiClinSum: Multilingual Clinical Summarization:* a shared task on the automatic summarization of lengthy clinical case reports written in different languages. The organizers distribute lengthy clinical case reports written in English, Spanish, French, and Portuguese. The participants generate summaries of the clinical case reports. The evaluation is based on a comparison with manual summaries of the clinical case reports. *Task 4 – BioNNE-L: Nested Named Entity Linking in Russian and English:* A shared task on NLP challenges in entity linking, also known as medical concept normalization (MCN), for English and Russian languages. The train/dev datasets include annotated mentions of disorders, anatomical structures, and chemicals. The participants normalize the entity mentions to concept names and unique UMLS identifiers. The evaluation is based on a comparison with manual nested named entity linking annotations. *Task 5 – ElCardioCC: Clinical Coding in Cardiology:* The ELCardioCC task on automated clinical coding concerns i) the assignment of cardiology-related ICD-10 codes to discharge letters from Greek hospitals, ii) the extraction of the specific mentions of ICD-10 codes from the discharge letters. The evaluation is based on metrics, such as micro and macro F-measure for Subtask (i) and token F-measure for Subtask (ii). *Task 6 – GutBrainIE: Gut-Brain Interplay Information Extraction:* The GutBrainIE task aims to foster the development of Information Extraction (IE) systems that support experts by automatically extracting and linking knowledge from scientific literature, facilitating the understanding of gut-brain interplay and its role in neurological disease. The task is divided into two subtasks: i) extraction of named entities and linking them to concepts in a reference ontology, and ii) identifying binary relations between entity pairs.

CheckThat! Lab on Checkworthiness, Subjectivity, Persuasion, Roles, Authorities and Adversarial Robustness[5] The eighth edition of the CheckThat! lab at CLEF presented a diverse set of challenges aimed at advancing technology to support and enhance the journalistic verification process. This edition revisited core tasks in the verification pipeline while also introducing auxiliary tasks such as subjectivity identification, claim normalization, and fact-checking numerical claims, with a particular emphasis on scientific web discourse. These tasks pose complex classification and retrieval problems at the document level, including in multilingual contexts. The lab was organized into the following tasks: *Task 1 – Subjectivity:* Given a sentence from a news article, determine whether it is subjective or objective. This is a binary classification task and is offered in Arabic, English, Bulgarian, German, and Italian for mono- and multi-lingual settings. Additionally, unseen languages like French and Spanish are considered for zero-shot settings. *Task 2 – Claim Normalization:* Given a noisy, unstructured social media post, the task is to simplify it into a concise form. This is a generation task, offered in 20 languages: English, Arabic, Bengali, Czech, German, Greek, French, Hindi, Korean, Marathi, Indonesian, Dutch, Punjabi, Polish, Portuguese, Romanian, Spanish, Tamil, Telugu, and Thai. *Task 3 – Fact-Checking Numerical Claims:* This task focuses on verifying claims with numerical quantities and temporal expressions. Numerical claims

---

[5] https://checkthat.gitlab.io/clef2025/.

are defined as those requiring validation of explicit or implicit quantitative or temporal details. Participants must classify each claim as True, False, or Conflicting based on a short list of evidence. *Task 4 – Scientific Web Discourse Processing (SciWeb)*, which was further divided into two subtasks. *Subtask 4.1 – SciWeb Discourse Detection:* This task aims at classifying the different forms of science-related online discourse. Namely, given a tweet, this multilabel task aims at detecting whether a tweet contains a scientific claim or scientific reference or is referring to science contexts or entities. *Subtask 4.2 – SciWeb Claim-Source Retrieval:* Given a tweet containing a scientific claim and an informal reference to a scientific paper, this task aims at retrieving the scientific paper that serves as the source for the claim from a given pool of candidate scientific papers.

ELOQUENT lab for evaluation of generative language model quality[6] addressed high-level quality criteria through a set of open-ended shared tasks implemented to require minimal human assessment effort. It offered the following tasks: *Task 1 – Voight-Kampff:* Generate text samples for a classifier to distinguish between human-authored and machine-generated text. *Task 2 – Robustness and Consistency:* Explore how much a generative language model's output is affected by stylistic, dialectal, or other non-topical variation in the input. *Task 3 – Preference Score Prediction:* Predict human preferences between sets of LLM-generated responses collected from human assessors, and generate judgments to explain the choice made. *Task 4 – Sensemaking:* Given a set of possibly noisy texts, generate questions and answers about the topic.

eRisk: Early Risk Prediction on the Internet[7] explored the evaluation methodology, effectiveness metrics, and practical applications (particularly those related to health and safety) of early risk detection on the Internet. This year's edition of eRisk included the following tasks: *Task 1 – Search for Symptoms of Depression:* Rank sentences from users according to their relevance to each of the 21 symptoms of the BDI-II questionnaire. Training data consists of sentence-tagged datasets from 2023 and 2024, with new test data including contextual information (previous and next sentences). *Task 2 – Contextualized Early Detection of Depression:* Participants analyze full conversational interactions to classify users with signs of depression, considering the conversational context beyond isolated user writings. The test phase includes writings with full conversational dynamics, while the training phase uses isolated user submissions. *Pilot Task – Conversational Depression Detection via LLMs:* Participants interact with a persona powered by a large language model (LLM) that is fine-tuned using types of depressive and non depressive users. The objective is to detect signs of depression, with participants limited to a specified number of messages to engage with the LLM.

EXIST: sEXism Identification in Social neTworks[8] aimed to capture and categorize sexism, from explicit misogyny to other subtle behaviors, in social networks. In 2024 the EXIST campaign included multimedia content in the format of memes, stepping forward research on more robust techniques to identify sexism in social networks. Following this line, in 2025 we focused on TikTok videos in the challenge, thus including in the dataset the three most important sources of sexism spreading: text, images, and videos. Consequently, it is essential to develop automated multimodal tools capable

---

[6] https://eloquent-lab.github.io/.

[7] https://erisk.irlab.org/.

[8] https://nlp.uned.es/exist2025/.

of detecting sexism in text, images, and videos to raise alarms or automatically remove such content from social networks, because platforms' algorithms often amplify content that perpetuates gender stereotypes and internalized misogyny. This lab contributed to the creation of applications that identify sexist content in social media across all three formats. This task was divided into three tasks, each split into three subtasks. *Task 1 – Sexism Identification and Characterization in Tweets Subtask 1.1 – Sexism Identification in Tweets:* The first subtask is a binary classification. The systems have to decide whether or not a given tweet contains or describes sexist expressions or behaviors (i.e., it is sexist itself, describes a sexist situation, or criticizes a sexist behavior). *Subtask 1.2 – Source Intention in Tweets:* This subtask aims to categorize sexist messages according to the intention of the author in one of the following categories: (i) direct sexist message, (ii) reported sexist message, and (iii) judgemental message. *Subtask 1.3 – Sexism Categorization in Tweets:* The third subtask is a multiclass task that aims to categorize sexist messages according to the type or types of sexism they contain (according to a categorization proposed by experts that takes into account the different facets of women that are undermined): (i) ideological and inequality, (ii) stereotyping and dominance, (iii) objectification, (iv) sexual violence, and (v) misogyny and non-sexual violence. *Task 2 – Sexism Identification and Characterization in Memes Subtask 2.1 – Sexism Identification in Memes:* Similar to Subtask 1.1, Subtask 2.1 is a binary classification task where participants must determine when a meme contains or describes sexist expressions or behaviors (i.e., it is sexist itself, describes a sexist situation, or criticizes a sexist behavior). *Subtask 2.2 – Source Intention in Tweets:* This subtask aims to categorize sexist messages according to the intention of the author in one of the following categories: (i) direct sexist message, (ii) judgmental message. *Subtask 2.3 – Sexism Categorization in Memes:* Finally, this subtask addresses the problem of categorizing a sexist meme according to the type of sexism that it encloses: (i) ideological and inequality, (ii) stereotyping and dominance, (iii) objectification, (iv) sexual violence, and (v) misogyny and non-sexual violence. *Task 3 - Sexism Identification and Characterization in TikTok Videos Subtask 3.1 – Sexism Identification in Videos:* Similar to Subtasks 1.1 and 2.1, this subtask is a binary classification task where participants must determine when a meme contains or describes sexist expressions or behaviors (i.e., it is sexist itself, describes a sexist situation, or criticizes a sexist behavior). *Subtask 3.2 – Source Intention in Videos:* This subtask aims to categorize sexist messages according to the intention of the author in one of the following categories: (i) direct sexist message, (ii) judgmental message. *Subtask 3.3 – Sexism Categorization in Videos:* Finally, this subtask addresses the problem of categorizing a sexist meme according to the type of sexism that it encloses: (i) ideological and inequality, (ii) stereotyping and dominance, (iii) objectification, (iv) sexual violence, and (v) misogyny and non-sexual violence.

ImageCLEF: Multimodal Challenge in CLEF[9] focused on evaluating technologies for annotating, indexing, classifying, retrieving, and generating multimodal data, providing access to large datasets across a variety of scenarios, including medical, social media, and internet-based applications. Building on the success of recent editions, it

---

[9] https://www.imageclef.org/2025.

encouraged interdisciplinary methods by engaging participants in diverse domains, providing large amounts of challenging multimodal data and providing an evaluation platform for a large number of use cases. This year's edition of ImageCLEF involved the following tasks: *Task 1 – ImageCLEFmedical:* In its 21st edition, the task continues all the medical sub-tasks from the last 2 years, namely: (i) the Caption task with medical concept detection and caption prediction, (ii) the GAN task focused on synthetic medical images, (iii) MEDVQA regarding Visual Question Answering for gastrointestinal data, and (iv) MEDIQA-MAGIC, introducing a new use-case on multimodal dermatology response generation. *Task 2 – Image Retrieval/Generation for Arguments:* As a joint task between Touché and ImageCLEF since 2022, the task aims to show the impact of images in arguments, making them more compelling. In this year's task, participants shall find suitable images that convey a given argument. Two submission styles are possible, either as a retrieval task or as prompt generation for an image generator. *Task 3 – ImageCLEFtoPicto:* The aim of this task is to convert either speech or text into a meaningful sequence of pictograms, aiding communication for people with language impairments, enhancing user understanding or helping with translation. Therefore, two sub-tasks are derived from this: (i) Text-to-Picto, involving generating pictograms starting from a French text and (ii) Speech-to-Picto, which focuses on translating speech to pictograms directly. *Task 4 – MultimodalReason:* This is a new task, focusing on Multilingual Visual Question Answering. Participants are given multiple-choice questions and corresponding images and are asked to identify the correct answer, in multiple languages, disciplines, and difficulty levels. The task aims to assess the reasoning abilities of modern LLMs across a wide range of real-world situations.

JOKER: Automatic Humour Analysis[10] aimed to foster interdisciplinary approaches to the (semi-)automatic analysis and processing of humor and wordplay. *Task 1 – Humor-Aware Information Retrieval:* For Task 1, the aim is to retrieve short humorous texts from a document collection based on a given query. The languages are English and Portuguese. *Task 2 – Wordplay Translation:* For Task 2, the goal is to translate English punning jokes into French. *Task 3 – Onomastic Wordplay:* For Task 3, the goal is to classify proper names according to whether they are humorous, and to translate them from English into French. *Task 4 – Controlled Creativity:* For Task 4, the goal is to identify the introduction of distorted or spurious content ("hallucinations") in generated creative texts.

LifeCLEF: Challenges on Species Presence Prediction and Identification, and Individual Animal Identification[11] focused on advancing AI-driven solutions for biodiversity monitoring through challenges on species and individuals recognition and prediction. *Task 1 – AnimalCLEF:* Multi-species individual animal identification. *Task 2 – BirdCLEF:* Bird species identification in soundscape recordings. *Task 3 – FungiCLEF:* Few-shot classification with rare fungi species. *Task 4 – GeoLifeCLEF:* Multimodal species prediction using remote sensing and large-scale biodiversity data. *Task 5 – PlantCLEF:* Multi-species plant identification in vegetation plot images.

---

[10] http://joker-project.com/.

[11] https://www.imageclef.org/LifeCLEF2025.

LongEval: Longitudinal Evaluation of Model Performance[12] aimed to ignite the development of Information Retrieval systems that can handle temporal data evolution. The retrieval systems evaluated in this task are expected to be persistent in their retrieval efficiency over time, as Web documents and Web queries evolve. To evaluate such features of systems, we rely on collections of documents and queries, corresponding to real data acquired from actual Web search engines. LongEval 2025 included two tasks: *Task 1 – WebRetrieval:* This task uses evolving Web data to evaluate IR system longitudinally, namely, it assesses whether the IR system performance is persistent over time. *Task 2 – SciRetrieval:* Similarly to Task 1, this task aims to examine how IR systems' effectiveness changes over time, when the underlying document collection changes, where the documents are scientific publications.

PAN: Lab on Stylometry and Digital Text Forensics[13] aimed to advance the state of the art and provide for an objective evaluation on newly developed benchmark datasets in those areas. The tasks proposed by PAN Lab this year included: *Task 1 – Generated Content Analysis:* Given a document, decide whether it was written by a human, an AI, or both. *Task 2 – Multilingual Text Detoxification:* Given a toxic piece of text, re-write it in a non-toxic way while saving the main content as much as possible. *Task 3 – Multi-author Writing Style Analysis:* Given a document, determine at which positions the author changes. *Task 4 – Generated Plagiarism Detection:* Given a generated and a human-written source document, identify the passages of reused text between them.

QuantumCLEF: Quantum Computing at CLEF[14] The second edition of the QuantumCLEF lab was composed of three tasks and aimed at: Discovering and evaluating Quantum Annealing approaches compared to their traditional counterpart; Identifying new ways of formulating Information Retrieval and Recommender Systems algorithms and methods, so that they can be solved with Quantum Annealing; Establishing collaborations among researchers from different fields to harness their knowledge and skills to solve the considered challenges and promote the usage of Quantum Annealing. This lab allowed participants to use real quantum computers provided by CINECA, one of the most important computing centers worldwide. *Task 1 – Feature Selection:* The task focuses on formulating the well-known NP-Hard Feature Selection problem and solving it with quantum annealers. Feature Selection is a widespread problem for both Information Retrieval and Recommender systems which requires identification of a subset of the available features (e.g., the most informative, less noisy, etc.) to train a learning model. This problem is very impacting since many of these systems involve the optimization of learning models, and reducing the dimensionality and noise of the input data can improve their performance. *Task 2 – Instance Selection:* The task focuses on formulating the Instance Selection problem to solve it through Quantum Annealing. Currently, transformer-based architectures, including 1st- and 2nd-generation transformers (e.g., RoBERTa) as well as current large language models (e.g., Llama3), are used and considered state of the art in several fields. Given the LLMs' high-cost application, one of the big challenges is to fine-tune these models efficiently. Instance Selection focuses on selecting a representative subset of instances from a dataset to make the training of these

---

[12] https://clef-longeval.github.io/.

[13] http://pan.webis.de/.

[14] https://qclef.dei.unipd.it/.

models faster while maintaining a high level of effectiveness of the trained model. *Task 3 – Clustering:* The task focuses on the formulation of the clustering problem to solve it with a quantum annealer. Clustering is a relevant problem for Information Retrieval and Recommender systems which involves grouping items together according to their characteristics. Clustering can be helpful for organizing large collections, helping users to explore a collection and providing similar results to a query. It can also be used to divide users according to their interests or build user models with the cluster centroids boosting efficiency or effectiveness for users with limited data.

SimpleText: Simplify Scientific Text (and Nothing More)[15] aimed at improving accessibility to scientific information for everyone, developing corpora, evaluation measures, and new IR/NL models able to reduce scientific text complexity with strict faithfulness to the original text. *Task 1 – Text Simplification: simplify scientific text:* The task aims to simplify scientific text, using aligned biomedical abstracts and lay summaries for sentence-level, paragraph-level, and document-level text simplification. *Task 2 – Controlled Creativity: identify and avoid hallucination:* The task aims to identify and avoid hallucination, by either post-hoc detection on CLEF submissions with overgeneration, or by avoiding creative license of models by design. *Task 3 – SimpleText 2024 Revisited: selected tasks by popular request:* The task aims to rerun selected tasks by popular request, on scientific passage retrieval and complex terminology detection, and on tracking the state of the art p in scholarly papers.

TalentCLEF: Skill and Job Title Intelligence for Human Capital Management[16] aimed to drive technological advancement in Human Capital Management by establishing a public benchmark for NLP models that facilitates their application in real-world Human Resources (HR) scenarios, incorporating evaluation criteria including multilingualism, fairness, and cross-industry adaptability. The lab also sought to build a community for researchers and practitioners to generate, evaluate, and discuss ideas on the use of AI in Human Resources, pushing the state-of-the-art of NLP applications for Human Resources. *Task 1 – Multilingual Job Title Matching:* The task involves the development of systems that can identify and rank job titles most similar to a given one. For each job title in a provided test set, participants must generate a ranked list of similar job titles from a specified knowledge base. The task includes multilingual and cross-lingual tracks, requiring participants to develop systems adapted to English, Spanish, German, and optionally Chinese. *Task 2 – Job Title-Based Skill Prediction:* The task involves developing systems capable of retrieving relevant skills associated with a given job title. Participants must train models that can retrieve a list of relevant skills from a provided knowledge base, ranking them according to their relevance to the job title. This task is in English.

Touché: Argumentation Systems[17] focused on computational argumentation and causality. Touché 2025 included 4 tasks. *Task 1 – Retrieval-Augmented Debating:* The task serves to develop generative retrieval systems that argue against their users to support users in forming or confirming opinions or to train their debating skills. *Task 2 – Ideology and Power Identification in Parliamentary Debates:* The task is concerned

---

[15] http://simpletext-project.com/.

[16] https://talentclef.github.io/talentclef/.

[17] https://touche.webis.de/.

with predicting ideology and power in parliamentary debates on a multi-lingual, multi-country dataset. *Task 3 – Image Retrieval/Generation for Arguments (Joint task with ImageCLEF):* The task is to find images that support a particular point of view. *Task 4 – Advertisement in Retrieval-Augmented Generation:* The task is to analyse possibilities and counter-measures for advertisements in retrieval-augmented search results.

The success of CLEF 2025 would not have been possible without the huge effort of several people and organizations, including the CLEF Association[18], the Program Committee, the Lab Organizing Committee, the reviewers, and the many students and volunteers who contributed.

The Friends of SIGIR program covered the registration fees for a number of students. UNED contributed to the event by providing funds to cover coffee breaks, along with institutional support and access to the venues at the Faculties of Education and Psychology. The event was also generously sponsored by the HiTZ Chair of Artificial Intelligence and Language Technology at the University of the Basque Country. We sincerely thank all our sponsors for their invaluable support in making this event possible.

July 2025

Jorge Carrillo-de-Albornoz
Alba García Seco de Herrera
Julio Gonzalo
Laura Plaza
Josiane Mothe
Florina Piroi
Paolo Rosso
Damiano Spina
Guglielmo Faggioli
Nicola Ferro

---

[18] https://www.clef-initiative.eu/#association.

Organization

General Chairs

Jorge Carrillo-de-Albornoz	Universidad Nacional de Educación a Distancia, Spain
Alba García Seco de Herrera	Universidad Nacional de Educación a Distancia, Spain
Julio Gonzalo	Universidad Nacional de Educación a Distancia, Spain
Laura Plaza	Universidad Nacional de Educación a Distancia, Spain

Program Chairs

Josiane Mothe	Université de Toulouse, France
Florina Piroi	Technische Universität Wien, Austria

Lab Chairs

Paolo Rosso	Universitat Politècnica de València, Spain
Damiano Spina	RMIT University, Australia

Lab Mentorship Chairs

Liana Ermakova	Université de Bretagne Occidentale, France
Florina Piroi	TU Wien, Austria

Proceedings Chairs

Guglielmo Faggioli	University of Padua, Italy
Nicola Ferro	University of Padua, Italy

Local Organization Committee

Víctor Fresno   Universidad Nacional de Educación a Distancia,
                                   Spain
Enrique Amigó   Universidad Nacional de Educación a Distancia,
                                   Spain

Supporters and Sponsors

HiTZ Chair
of Artificial Intelligence
and Language Technology
University
of the Basque
Country

CLEF Steering Committee

Steering Committee Chair

Nicola Ferro   University of Padua, Italy

Steering Committee Co-Chairs

Alba García Seco de Herrera   Universidad Nacional de Educación a Distancia,
                                    Spain
Alberto Barrón-Cedeño   University of Bologna, Italy

Deputy Steering Committee Chair for the Conference

Paolo Rosso   Universitat Politècnica de València, Spain

Deputy Steering Committee Chair for the Evaluation Labs

Martin Braschler   Zurich University of Applied Sciences,
                                    Switzerland

Members

Avi Arampatzis   Democritus University of Thrace, Greece
Khalid Choukri   Evaluations and Language resources Distribution
                                    Agency, France
Fabio Crestani   Università della Svizzera italiana, Switzerland
Carsten Eickhoff   University of Tübingen, Germany
Norbert Fuhr   University of Duisburg-Essen, Germany
Petra Galuščáková   University of Stavanger, Norway
Anastasia Giachanou   Utrecht University, The Netherlands
Lorraine Goeuriot   Université Grenoble Alpes, France
Julio Gonzalo   National Distance Education University, Spain
Donna Harman   National Institute for Standards and Technology,
                                    USA
Bogdan Ionescu   University "Politehnica" of Bucharest, Romania
Evangelos Kanoulas   University of Amsterdam, The Netherlands

Birger Larsen	University of Aalborg, Denmark
Maria Maistro	University of Copenhagen, Denmark
Josiane Mothe	IRIT, Université de Toulouse, France
Henning Müller	University of Applied Sciences Western Switzerland (HES-SO), Switzerland
Jian-Yun Nie	Université de Montréal, Canada
Gabriella Pasi	University of Milano-Bicocca, Italy
Eric SanJuan	University of Avignon, France
Laure Soulier	Sorbonne Université, France
Theodora Tsikrika	Information Technologies Institute, Centre for Research and Technology Hellas, Greece

Past Members

Paul Clough	University of Sheffield, UK
Djoerd Hiemstra	Radboud University, The Netherlands
Jaana Kekäläinen	University of Tampere, Finland
Séamus Lawless	Trinity College Dublin, Ireland
David E. Losada	Universidade de Santiago de Compostela, Spain
Mihai Lupu	Vienna University of Technology, Austria
Carol Peters (Steering Committee Chair 2000–2009)	ISTI, National Council of Research (CNR), Italy
Emanuele Pianta	Centre for the Evaluation of Language and Communication Technologies, Italy
Maarten de Rijke	University of Amsterdam, The Netherlands
Giuseppe Santucci	Sapienza University of Rome, Italy
Jacques Savoy	University of Neuchâtel, Switzerland
Alan Smeaton	Dublin City University, Ireland
Christa Womser-Hacker	University of Hildesheim, Germany

Contents

Condensed Labs Overviews

Conference Papers

From Uniform to Unique: Adaptive K–12 Assessment Using Large Language Models

Lokesh Goenka[1]([✉])[ID], Ajay Mukund S[2][ID], and P. Sunil Kumar[1][ID]

[1] Department of Mathematics and Computer Science, Sri Sathya Sai Institute
of Higher Learning, Andhra Pradesh, India
goenkalokesh@gmail.com, psunilkumar@sssihl.edu.in
[2] Department of Computer Science and Engineering, College of Engineering Guindy,
Anna University, Chennai, India

Abstract. This paper presents a novel approach to personalized K–12 education through the design and deployment of a Personalised Learning Assistant (PLA) powered by a fine-tuned Large Language Model (LLaMA 3.1). The PLA adaptively evaluates student understanding via a multi-level diagnostic test aligned with Bloom's Taxonomy. Using probabilistic inference and difficulty-aware weighted scoring, the system generates a learner-specific conceptual profile while minimizing the influence of random guessing. We introduce a fixed-attempt policy to ensure fairness in progression and score normalization across students with varying test trajectories. The PLA was deployed through a functional web prototype, with MCQs generated using Gemini 1.0 and a fine-tuned LLM hosted on Kaggle and a Vercel-based frontend for seamless interaction. Our evaluation demonstrates the framework's robustness in differentiating learner capabilities and its potential to inform targeted remediation. This work bridges generative AI, adaptive assessment, and cognitive pedagogy to enable scalable and equitable learning in classroom and self-paced environments.

Keywords: Adaptive Testing · Large Language Models · Personalized Learning · Bloom's Taxonomy · Probabilistic Scoring · Educational AI

1 Introduction

The integration of Artificial Intelligence (AI) into K–12 education is reshaping how instruction and assessment are delivered. Traditional one-size-fits-all approaches overlook the diversity in learners' cognitive abilities, prior knowledge, and learning styles. This motivates the need for adaptive educational systems that personalize learning pathways.

Recent advances in Large Language Models (LLMs) have made it feasible to build scalable, personalized learning environments. In this paper, we propose a *Personalised Learning Assistant (PLA)* that adapts to student performance through diagnostic testing and generative question delivery. The system profiles learner knowledge, identifies conceptual gaps, and provides tailored remediation.

© The Author(s), under exclusive license to Springer Nature Switzerland AG 2026
J. Carrillo-de-Albornoz et al. (Eds.): CLEF 2025, LNCS 16089, pp. 3–16, 2026.
https://doi.org/10.1007/978-3-032-04354-2_1

The PLA framework is composed of three key components:

1. **Learner Profiling:** Captures interaction data to model understanding.
2. **Adaptive Content Generation:** Uses fine-tuned LLMs to generate MCQs aligned with Bloom's levels (Remember, Apply, Evaluate).
3. **Multi-Level Diagnostic Testing:** Estimates conceptual mastery using weighted scoring, Bayesian inference, and guessing suppression.

This work is motivated by recent AI-in-education efforts such as Harvard's AI teaching assistant [10] and Taiwan's national adaptive learning platform [11]. These highlight the value of AI in scaling personalized instruction and academic support. Our contributions include: (1) an LLM-based question generation pipeline using NCERT content; (2) a probabilistic learner evaluation model; and (3) a fully functional web-based prototype integrating LLaMA 3.1 8B Instruct via Kaggle and a Vercel frontend. We demonstrate that combining generative AI with adaptive diagnostics can yield scalable, data-driven learning assistants with real-world impact.

The Personalised Learning Assistant follows a simple two-stage process:

Stage I: MCQ Generation. Text passages were extracted from NCERT textbooks, cleaned to remove irrelevant content, and embedded with structured prompts. These were processed using Gemini 1.0 Pro to generate initial MCQs, which were then used to fine-tune Google's T5 and Meta's LLaMA 3.1 models. After human validation, LLaMA 3.1 emerged as the more effective model for producing accurate and context-aware MCQs.

Stage II: Test and Analysis. The validated MCQs were used to create subject-specific tests. When a student selects a subject, a test is generated using the fine-tuned LLaMA 3.1 model. Upon completion, responses are analyzed using a predefined algorithm to assess performance and identify learning gaps. A detailed report is then generated, outlining the student's strengths and weaknesses.[1]

2 Related Work

Recent advances in Artificial Intelligence (AI) have significantly transformed educational technologies, particularly through the emergence of Personalised Learning Assistants (PLAs), generative Large Language Models (LLMs), and adaptive testing methodologies. PLAs are designed to tailor instruction based on individual learner profiles, cognitive styles, and prior performance. Systems proposed by Sajja et al. [13], Xia [18], and Bonde [1] leverage machine learning and NLP for real-time personalization. These systems dynamically adapt content and feedback, building on learner modeling strategies that analyze interaction history to recommend targeted interventions [5,12,15]. The educational benefits of such systems, including enhanced engagement and academic performance,

---

[1] Dataset available at Hugging Face; Code available at GitHub.

have been documented in several studies [9, 14]. Early works like that of Dolog et al. [3] also contributed foundational insights into scalable personalization in e-learning environments.

The deployment of LLMs in adaptive learning has become increasingly prominent due to their generative capabilities and semantic awareness. Lee et al. [8] introduced a difficulty-aware contrastive framework using LLM-derived question complexity to improve knowledge tracing. Wang et al. [17] provided a comprehensive survey on LLM applications in education, while also highlighting key challenges such as model bias and lack of explainability. Dynamic neural architectures for learner response prediction, such as those by Delianidi et al. [2], have further improved personalized feedback accuracy in educational settings.

In the domain of generative AI, Faruqui et al. [6] developed SAMCares, a Retrieval-Augmented Generation (RAG) platform for adaptive tutoring that integrates retrieval pipelines with context-aware generation. Complementary to this, LearnLM-Tutor [7] emphasizes pedagogically-aligned AI tutors evaluated against custom educational benchmarks to ensure quality and instructional relevance. Adaptive testing continues to evolve as a robust tool for personalized assessment. El-Sabagh [4] presented an e-learning platform that aligns content with individual learning styles, while Tutor CoPilot [16] demonstrates how human-AI collaboration, augmented by Item Response Theory (IRT), can scale real-time instructional support.

In terms of real-world deployment, Liu et al. [10] reported the successful use of AI teaching assistants in Harvard's CS50 course, and Liu [11] analyzed Taiwan's national implementation of adaptive learning systems in elementary schools. These examples underscore the growing practical relevance of AI-driven personalization in education. Our proposed PLA builds upon these foundations by integrating Bloom-aligned content generation via fine-tuned LLMs, probabilistic understanding estimation through Bayesian reasoning, and difficulty-weighted scoring. This approach enables scalable and equitable personalized learning for K–12 environments.

3 Methodology

The proposed Personalised Learning Assistant (PLA) system is designed to adaptively assess student knowledge across multiple difficulty levels using a probabilistic framework. The methodology consists of three major components (Fig. 1):

1. **Question Generation and Bloom-Level Classification**
2. **Probabilistic Knowledge Diagnosis via Multi-level Testing**
3. **Score Aggregation with Weighted Confidence Estimation**

3.1 Question Generation and Bloom-Level Classification

Given a subject domain $\mathcal{D}$, we extract structured textual content from authoritative K12 (Kindergarten to Class/Grade 12) sources (e.g., NCERT). Using

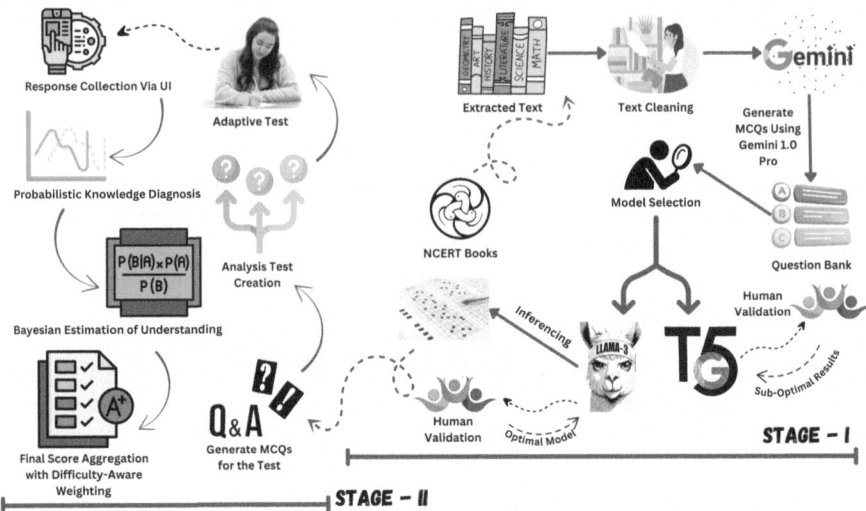

Fig. 1. End-to-End Architecture of the Personalised Learning Assistant (PLA). Stage I handles NCERT-based question generation and LLM fine-tuning; Stage II performs adaptive diagnostics using Bayesian inference and difficulty-weighted scoring to generate personalized feedback.

Google's Gemini 1.0 generative LLM, which was used via API for zero-shot MCQ generation aligned with Bloom's Taxonomy, we generate multiple-choice questions (MCQs) from passage-level input. Each question $q_i \in \mathcal{Q}$ is then classified into one of the following Bloom's Taxonomy levels:

- $\mathcal{B}_1$: Remember/Understand (Easy)
- $\mathcal{B}_2$: Apply/Analyze (Medium)
- $\mathcal{B}_3$: Evaluate/Create (Hard)

Let $\mathcal{Q}_\ell \subset \mathcal{Q}$ denote the set of MCQs at level $\mathcal{B}_\ell$, where $\ell \in \{1, 2, 3\}$.

3.2   Probabilistic Knowledge Diagnosis via Multi-Level Testing

We design a three-level diagnostic test $T = \{L_1, L_2, L_3\}$, where each level L_ℓ contains a primary set P_ℓ of 3 questions and a secondary set S_ℓ of 2 fallback questions.

Let the random variable $Z \in \{0, 1\}$ represent whether the student actually understands the concept ($Z = 1$) or not ($Z = 0$). Each question q_i is answered correctly ($R_i = 1$) or incorrectly ($R_i = 0$). Assuming 4-option MCQs, the probability of guessing correctly without understanding is $p_g = 0.25$.

Likelihood of Correct Answer Under Different States:

$$
\begin{aligned}
\mathbb{P}(R_i = 1 \mid Z = 1) &= \alpha \\
\mathbb{P}(R_i = 1 \mid Z = 0) &= p_g = 0.25
\end{aligned}
\tag{1}
$$

Here, α denotes the probability of answering correctly when the student has true knowledge of the concept, typically in the range $[0.85, 0.95]$. The guessing probability p_g is assumed to be 0.25 for four-option MCQs.

Assuming conditional independence between responses, the likelihood of a response vector $\mathbf{R}_\ell = [R_1, R_2, R_3]$ at level ℓ is:

$$\mathcal{L}(\mathbf{R}_\ell | Z) = \prod_{i=1}^{3} \mathbb{P}(R_i | Z)$$

Bayesian Estimation of Understanding at Level ℓ: Assuming a prior $\mathbb{P}(Z = 1) = \pi$, the posterior probability that a student understands the topic at level ℓ given responses $\mathbf{R}_\ell$ is:

$$\mathbb{P}(Z = 1 | \mathbf{R}_\ell) = \frac{\mathcal{L}(\mathbf{R}_\ell | Z = 1) \cdot \pi}{\mathcal{L}(\mathbf{R}_\ell | Z = 1) \cdot \pi + \mathcal{L}(\mathbf{R}_\ell | Z = 0) \cdot (1 - \pi)}$$

If this probability exceeds a threshold θ_ℓ (e.g., 0.8), the system advances to the next level. Otherwise, it triggers secondary questions or stops the test, based on the decision rules (Let us define them propery . . .).

3.3 Score Aggregation with Difficulty-Aware Weighting

To compute a student's overall confidence score in a topic, we extend the basic weighted score formula to incorporate the difficulty level of questions. This ensures that performance on harder questions contributes more significantly to the final score, reflecting the depth of understanding.

Let:

- $\ell \in \{1, 2, 3\}$: Difficulty level, where $1 =$ Easy, $2 =$ Medium, $3 =$ Hard.
- D_ℓ: Weight assigned to each difficulty level (e.g., $D_1 = 1.0, D_2 = 1.5, D_3 = 2.0$).
- CP_ℓ, CS_ℓ: Correct Primary and Secondary responses at level ℓ.
- AP_ℓ, AS_ℓ: Attempted Primary (fixed at 3) and Secondary questions at level ℓ.
- WP, WS: Weights for Primary and Secondary questions, respectively (e.g., $WP = 1.0, WS = 0.5$).

Then, the final difficulty-aware confidence score is given by:

$$\text{Score} = \frac{\sum_{\ell=1}^{3} D_\ell \cdot (CP_\ell \cdot WP + CS_\ell \cdot WS)}{\sum_{\ell=1}^{3} D_\ell \cdot (AP_\ell \cdot WP + AS_\ell \cdot WS)} \quad (2)$$

This formulation adjusts both correct responses and total attempts by their respective difficulty weights, ensuring that:

- Correct answers at higher levels (e.g., Hard) contribute more to the score.
- Students relying only on easy-level questions receive proportionally lower scores.
- The score remains normalized between 0 and 1.

3.4 Algorithmic Flow of the Adaptive Diagnostic Test

To implement adaptive testing effectively, we formalize the diagnostic flow as an algorithm that handles sequential responses, assesses knowledge probabilistically, and guides learners through Bloom's levels. Combining Bayesian estimation with rule-based thresholds, the system minimizes false positives and ensures meaningful progression.

Algorithm 1 outlines the full decision logic, including question sequencing, level transitions, and score computation using difficulty-weighted aggregation—enabling PLA to adapt in real time and generate accurate learner profiles.

3.5 Summary of Decision Logic

The adaptive diagnostic flow in PLA is structured to progressively assess conceptual understanding while minimizing false positives due to random guessing. The decision logic proceeds as follows:

1. **Initialization:** The student starts at level $\ell = 1$, aligned with Bloom's *Remember/Understand* (Easy).
2. **Primary Assessment:** At each level, the student attempts 3 primary questions P_ℓ, with CP_ℓ correct responses:
   - $CP_\ell = 3$: advance to $\ell + 1$
   - $CP_\ell = 0$: terminate test; issue remedial learning plan
   - $CP_\ell = 1$ or 2: administer 2 fallback questions S_ℓ
3. **Secondary Evaluation:** Let CS_ℓ be the number of correct fallback responses. If $CP_\ell + CS_\ell \geq 3$, progress to the next level; else, terminate and suggest support.
4. **Termination and Output:** On reaching level $\ell = 3$ or upon failure, the system ends and provides:
   - A difficulty-weighted confidence score (see Section IV-C)
   - A level-wise performance summary
   - A personalized learning path

This structured progression ensures that advancement is contingent upon demonstrable comprehension. By requiring 3 out of 5 correct answers per level and incorporating a secondary diagnostic path, the system reduces the likelihood of false advancement. The probabilistic design integrates robustness against guessing (see Section IV-D), while maintaining student motivation by building from foundational to advanced conceptual checks.

Algorithm 1. Adaptive Diagnostic Test with Bayesian and Rule-Based Decisions

---

Require: MCQ bank $\mathcal{Q}$ categorized by Bloom level $\ell \in \{1, 2, 3\}$, primary question count $= 3$, secondary $= 2$

1: Initialize: $\ell \leftarrow 1$, max_level $\leftarrow 3$, final_score $\leftarrow 0$
2: Set constants: $WP \leftarrow 1.0$, $WS \leftarrow 0.5$,
         $D_1 \leftarrow 1.0, D_2 \leftarrow 1.5, D_3 \leftarrow 2.0$
3: Initialize tracking variables: $CP_\ell, CS_\ell, AP_\ell, AS_\ell \leftarrow 0$ for all ℓ
4: **while** $\ell \leq$ max_level **do**
5:     Present 3 primary questions P_ℓ from $\mathcal{Q}_\ell$
6:     Record student responses $\mathbf{R}_\ell = [R_1, R_2, R_3]$
7:     $CP_\ell \leftarrow \sum R_i$, $AP_\ell \leftarrow 3$
8:     **if** $CP_\ell = 3$ **then**
9:        $\ell \leftarrow \ell + 1$ // Advance to next level
10:    **else if** $CP_\ell = 0$ **then**
11:       **Break:** terminate test and generate learning plan
12:    **else**
13:       Present 2 secondary questions S_ℓ from $\mathcal{Q}_\ell$
14:       Record responses $\mathbf{R}'_\ell = [R_4, R_5]$
15:       $CS_\ell \leftarrow \sum R'_i$, $AS_\ell \leftarrow 2$
16:       **if** $CP_\ell + CS_\ell \geq 3$ **then**
17:         $\ell \leftarrow \ell + 1$ // Advance to next level
18:       **else**
19:         **Break:** terminate test and generate learning plan
20:       **end if**
21:    **end if**
22: **end while**
23: Compute confidence score   // Equation 2

$$\text{final\_score} = \frac{\sum_{\ell=1}^{3} D_\ell \cdot (CP_\ell \cdot WP + CS_\ell \cdot WS)}{\sum_{\ell=1}^{3} D_\ell \cdot (AP_\ell \cdot WP + AS_\ell \cdot WS)}$$

24: Return final_score and recommended personalized learning path

---

4   Results and Discussion

To assess the practical viability and effectiveness of the proposed Personalised Learning Assistant (PLA), we developed and deployed a fully functional web prototype. The system integrates a fine-tuned LLaMA 3.1 8B Instruct model for domain-specific question generation and adaptive testing, with an interactive user interface hosted via Vercel.

The PLA system uses a lightweight, scalable setup. The frontend is a Vercel-hosted responsive web app enabling adaptive Q/A, real-time feedback, and score visualization. The backend runs a fine-tuned LLaMA 3.1 8B Instruct model on a GPU-enabled Kaggle notebook to generate Bloom-aligned MCQs and difficulty levels. Frontend-backend communication is handled via FastAPI-based HTTP APIs within a persistent container, ensuring efficient and reliable inference.

4.1   Robustness Against Random Guessing

A key objective of the adaptive framework is to reduce progression through random guessing. For 4-option MCQs, the guessing probability is $p_g = 0.25$. To counter this, the system uses a two-tier check per level: 3 primary questions and 2 secondary ones if needed. Learners must answer at least 3 out of 5 correctly to advance.

The chance of passing a level by guessing is the cumulative probability of getting 3 or more correct out of 5 trials, modeled by a binomial distribution with $n = 5$, $p = 0.25$ (see Fig. 2).

$$\mathbb{P}(\text{Pass}_{\text{guess}}) = \sum_{k=3}^{5} \binom{5}{k} \cdot p_g^k \cdot (1 - p_g)^{5-k} \tag{3}$$

Evaluating this yields:

$$\mathbb{P}(\text{Pass}_{\text{guess}}) \approx 0.103516$$

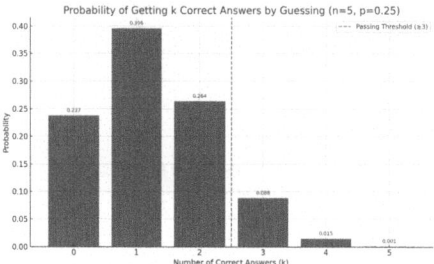

Fig. 2. Binomial distribution of correct guesses out of 5 questions with $p = 0.25$. The shaded region ($k \geq 3$) denotes the threshold required to pass a level, highlighting the low likelihood of advancement through guessing.

While this is already low, the novelty of our design lies in requiring the learner to sequentially clear **all three levels**—Easy, Medium, and Hard—each independently requiring at least 3 correct answers out of 5. Assuming independence of levels and guessing at each stage, the cumulative probability of passing all three levels purely via random guessing becomes:

$$\mathbb{P}_{\text{topic}} = (\mathbb{P}(\text{Pass}_{\text{guess}}))^3 \approx (0.103516)^3 \approx 0.0011$$

This translates to a mere 0.11% chance of a student successfully "clearing" a topic without understanding, thereby establishing strong resistance to noise induced by lucky guesses.

Comparison with TALP: In contrast, the Taiwan Adaptive Learning Platform (TALP) [11] adopts a top-down strategy wherein a student is first presented with the most difficult question. If answered correctly, the system concludes mastery of the entire topic and skips subordinate questions. However, it does not account for the probabilistic possibility of a correct response due to guessing. This omission may result in overestimating learner proficiency, especially in low-performing students who guess correctly once.

Our Contribution: The PLA framework overcomes this limitation by explicitly quantifying the probability of false-positive advancement and structuring the test such that cumulative guessing-based progression is statistically negligible. This design choice ensures that the adaptive system favors true understanding and penalizes random guessing, thereby yielding more reliable learner profiles.

4.2 Fairness in Progression: Differentiating Parallel Learners via Fixed Denominator

Table 1 presents a comparative case study of two learners—Student A and Student B—who achieved the same number of correct responses at early levels but diverged in adaptive progression. Both students answered five questions correctly; however, Student A demonstrated additional depth by answering one secondary question at Level 2 and advancing to Level 3, while Student B did not trigger any secondary questions and was halted after Level 2.

Table 1. Comparison of Adaptive Progression Between Parallel Learners With Equal Primary Accuracy

Student	L1 (P+S)	L2 (P+S)	L3 (P+S)	Correct	Score
A	3+0	2+1	0+0	6.75/15	0.45
B	3+0	2+0	–	6.00/15	0.40

Without a fixed denominator policy, Student B would have had a misleadingly higher score:

$$\text{Score}_{\text{B (unfixed)}} = \frac{6.00}{9.0} = 0.66$$

This inflation would wrongly suggest that Student B outperformed Student A, despite not handling more cognitively demanding items or reaching the Hard level.

Fixed Attempt Policy: To prevent such misrepresentations, our scoring mechanism enforces a fixed primary attempt count ($AP_\ell = 3$) at all levels, even if the learner does not reach them. This design ensures:

– Fair comparison across students regardless of progression depth.
– Penalization for not reaching higher Bloom levels.
– Emphasis on depth of understanding over shallow performance.

This policy ensures that learners like Student A—who engage with more challenging material—are appropriately rewarded, while avoiding artificially inflated scores for learners with limited progression. It maintains the scoring integrity of the adaptive framework and aligns with the educational goal of promoting higher-order thinking.

4.3   Empirical Evaluation with School Students

To assess the real-world applicability and learner-level interpretability of the PLA system, we deployed the platform using a Vercel-based frontend and invited almost 30 students from Grades 8 to 12 to participate in a self-paced diagnostic test. Each student selected three topics of their choice, and for each topic, they self-reported their prior understanding as a percentage. The system then administered the adaptive test and computed the difficulty-aware confidence score (also expressed as a percentage) using the methodology discussed earlier. The deviation between the self-assessed prior and the system-generated score provides a measure of either underestimation or overestimation in self-perception, thus offering insight into the diagnostic value of the PLA framework.

The visual trends in Fig. 3a depict the distribution of PLA scores in comparison to students' prior beliefs, highlighting several cases of over- and underestimation. Meanwhile, the heatmap in Fig. 3b offers a topic-wise deviation matrix

Table 2. Empirical Evaluation: Prior vs. Score Across Topics

Student	Grade	Topic	Prior (%)	Score (%)	Deviation
Aditi Verma	12	Geography	60	75	−15.00
Aditiya	10	History	70	90	−20.00
Anchal	12	History	50	66	−16.00
Arun	12	History	60	66	−6.00
Aryan	11	Economics	50	66	−16.00
Ashish	12	History	80	66	14.00
Ayush	10	History	80	100	−20.00
Bhavya Kapoor	6	History	70	100	−30.00
Dedeepya	9	Geography	60	33	27.00
Dhanush	4	Economics	80	66	14.00
Divyansh	12	History	70	70	0.00
Harshit	10	History	70	66	4.00
K.Sai Prashant	9	Geography	70	75	−5.00
..	..	..	..	..	..
Average Deviation Across All Topics					**−9.84**

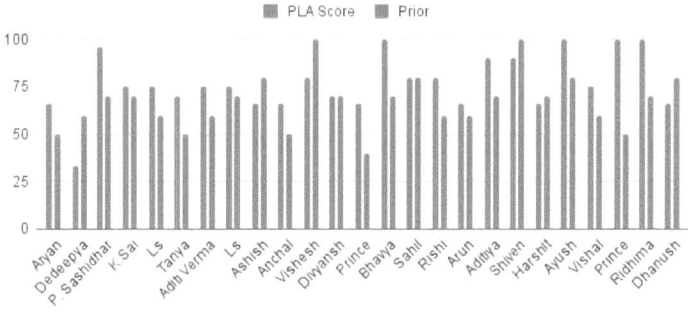

(a) Prior vs. PLA Score across student-topic pairs with data labels.

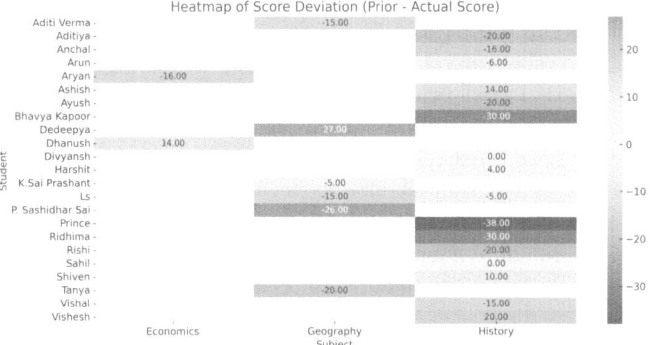

(b) Deviation heatmap (PLA Score Prior). Negative values indicate overestimation.

Fig. 3. Visual analysis of diagnostic results from the PLA system: alignment and mismatch between student self-assessment and difficulty-aware scoring.

that helps isolate consistent patterns in self-assessment inaccuracies, especially in subjects like Agriculture and Economics. Together, these visualizations provide an interpretable and granular evaluation of how the system reveals latent strengths and corrects student misconceptions.

The results in Table 2 demonstrate several key insights:

- Students with high self-assessed prior knowledge (e.g., Drishti and Adarsh in Climate and Economics) sometimes overestimated their understanding, as reflected by negative deviations.
- Substantial underestimations (positive deviations) occurred where the system revealed deeper understanding than expected, suggesting confidence gaps or lack of self-awareness.
- The average deviation of −**5.69%** indicates a slight trend toward underestimation, validating the role of PLA in revealing latent conceptual strengths.

Educational Implication: This empirical evaluation demonstrates that PLA not only diagnoses gaps but also corrects student misconceptions about their own preparedness, enabling more precise remediation and goal-setting.

5   Conclusion and Future Work

This paper introduced the Personalised Learning Assistant (PLA)—an adaptive diagnostic framework for K–12 education that leverages Large Language Models (LLMs) and pedagogical structure from Bloom's Taxonomy. The system generates questions aligned with cognitive levels and uses a probabilistic, difficulty-aware scoring model to create learner profiles that reflect both correctness and depth of understanding.

PLA addresses key limitations in systems like TALP, which rely on single-question assessments vulnerable to random guessing. By incorporating multi-tiered testing, Bayesian inference, and secondary validation, PLA ensures robust diagnosis and merit-based progression. Design elements such as fixed attempt limits and tiered fallback questions further enhance reliability. Simulation results show that PLA distinguishes learners with deeper conceptual understanding, beyond surface-level accuracy. A prototype, built using a fine-tuned LLaMA 3.1 model on Kaggle and deployed with a Vercel frontend, validates the feasibility of real-time, scalable deployment.

Future Work. We plan to expand PLA's capabilities to handle long-form and subjective responses, integrate Retrieval-Augmented Generation (RAG) for personalized and context-aware question generation, and explore multimodal affective feedback using speech and vision data. Further work will involve benchmarking PLA against standardized assessments and deploying it in real-world classrooms for longitudinal evaluation. Ultimately, we envision PLA evolving into a Human–AI collaborative assistant that empowers educators to design more inclusive, adaptive, and learner-centered experiences—advancing the goal of equitable, high-quality education through AI.

Acknowledgments. The authors gratefully acknowledge the divine inspiration of Bhagawan Sri Sathya Sai Baba, whose values and vision guided this work. We also thank our families, peers, and the Sri Sathya Sai Institute of Higher Learning for their constant support and encouragement throughout this project.

Disclosure of Interests. The authors declare that they have no competing interests relevant to the content of this article.

References

1. Bonde, L.: A generative artificial intelligence based tutor for personalized learning. In: 2024 IEEE SmartBlock4Africa, pp. 1–10 (2024). https://doi.org/10.1109/SmartBlock4Africa61928.2024.10779525

2. Delianidi, M., Diamantaras, K., Chrysogonidis, G., Nikiforidis, V.: Student performance prediction using dynamic neural models. arXiv preprint arXiv:2106.00524 (2021)

3. Dolog, P., Henze, N., Nejdl, W., Sintek, M.: Personalization in distributed e-learning environments. In: Proceedings of the 13th International World Wide Web Conference (WWW 2004), pp. 170–179. ACM (2004). https://doi.org/10.1145/1013367.1013395

4. El-Sabagh, H.A.: Adaptive e-learning environment based on learning styles and its impact on development students' engagement. Int. J. Educ. Technol. High. Educ. **18**(1), 53 (2021)

5. Fariani, R.I., Junus, K., Santoso, H.: A systematic literature review on personalised learning in the higher education context. Technol. Knowl. Learn. **28**, 449–476 (2022). https://doi.org/10.1007/s10758-022-09628-4

6. Faruqui, S.H.A., Tasnim, N., Basith, I.I., Obeidat, S., Yildiz, F.: Integrating AI in higher education: Protocol for a pilot study with'samcares: An adaptive learning hub'. arXiv preprint arXiv:2405.00330 (2024)

7. Jurenka, I., et al.: Towards responsible development of generative AI for education: An evaluation-driven approach. arXiv preprint arXiv:2407.12687 (2024)

8. Lee, U., et al.: Difficulty-focused contrastive learning for knowledge tracing with a large language model-based difficulty prediction. arXiv preprint arXiv:2312.11890 (2023)

9. Li, K., Wong, B.: Features and trends of personalised learning: a review of journal publications from 2001 to 2018. Interact. Learn. Environ. **29**, 182–195 (2020). https://doi.org/10.1080/10494820.2020.1811735

10. Liu, R., Zenke, C., Liu, C., Holmes, A., Thornton, P., Malan, D.J.: Teaching cs50 with AI: leveraging generative artificial intelligence in computer science education. In: Proceedings of the 55th ACM Technical Symposium on Computer Science Education, Vol. 1, pp. 750–756 (2024)

11. Liu, T.C.: A case study of the adaptive learning platform in a Taiwanese elementary school: precision education from teachers' perspectives. Educ. Inf. Technol. **27**(5), 6295–6316 (2022)

12. Prihar, E., Haim, A., Sales, A.C., Heffernan, N.: Automatic interpretable personalized learning. In: Proceedings of the Ninth ACM Conference on Learning @ Scale (2022). https://doi.org/10.1145/3491140.3528267

13. Sajja, R., Sermet, Y., Cikmaz, M., Cwiertny, D., Demir, I.: Artificial intelligence-enabled intelligent assistant for personalized and adaptive learning in higher education. ArXiv abs/2309.10892 (2023). https://doi.org/10.48550/arXiv.2309.10892

14. Sambrani, Y., Lamani, M., Kumar, P., Borah, B., Pragati: Personalised learning assistance system for slow learners. International Journal of Innovative Science and Research Technology (IJISRT) (2024). https://doi.org/10.38124/ijisrt/ijisrt24apr1485

15. Shemshack, A., Kinshuk, Spector, J.: A comprehensive analysis of personalized learning components. J. Comput. Educ. **8**, 485–503 (2021). https://doi.org/10.1007/s40692-021-00188-7

16. Wang, R.E., Ribeiro, A.T., Robinson, C.D., Loeb, S., Demszky, D.: Tutor copilot: A human-AI approach for scaling real-time expertise. arXiv preprint arXiv:2410.03017 (2024)
17. Wang, S., et al.: Large language models for education: A survey and outlook. arXiv preprint arXiv:2403.18105 (2024)
18. Xia, P.: Design of personalized intelligent learning assistant system under artificial intelligence background. In: Proceedings of the 2020 International Conference on Machine Learning and Big Data Analytics for IoT Security and Privacy (SPIoT 2020), Volume 1, pp. 194–200. Springer (2021)

Selective Search as a First-Stage Retriever

Gijs Hendriksen$^{(\boxtimes)}$ (ID), Djoerd Hiemstra (ID), and Arjen P. de Vries (ID)

Radboud University, Nijmegen, The Netherlands
{gijs.hendriksen,djoerd.hiemstra,arjen.devries}@ru.nl

Abstract. Selective search assumes a document collection can be partitioned into topical index shards in such a way that individual search requests would be satisfied with a few shards only. Previous work has considered primarily the retrieval effectiveness of selective search architectures in an early precision setting. In this work, we instead consider selective search as the first stage in a multi-stage pipeline, and therefore focus on obtaining high recall. We reproduce the most important algorithms from the selective search literature, and show that they can match the recall level of exhaustive search while reducing the required resources by 50%. We compare the different types of resource selection algorithms, and conclude that the more straightforward strategies that can select shards at a low cost actually outperform the more involved algorithms, in terms of reliably obtaining high recall with fewer shards.

Keywords: Selective Search · Federated Search · Resource Selection

1 Introduction

As re-rankers become more effective (but also more expensive), search engine architectures increasingly comprise multi-stage retrieval pipelines. The first stage of such pipelines is designed to produce a high-quality set of candidate documents efficiently. Subsequent stages re-rank this candidate set using more sophisticated models, to ensure that relevant documents are pushed to the top of the ranking. While these subsequent stages focus heavily on early precision, the first-stage retriever should mostly worry about obtaining high recall: an effective re-ranker will place relevant documents at the top of the ranking, but to do so, those documents need to be present in the candidate set of documents it receives.

Selective search is a technique developed to make large-scale search more efficient, by dividing a document collection into topical index shards and only searching the most relevant shards at query time. Its ability to efficiently handle massive document collections makes it a suitable option for a fast first-stage retrieval system. However, most previous work has evaluated selective search architectures in early precision settings, using metrics such as P@10, nDCG or MAP [1,2,29–31,41]. Some papers [16,17,37] acknowledge that selective search may hurt recall and evaluate their systems using deeper metrics. Still, they use nDCG and MAP for their evaluation, while metrics such as R@1000 or RBR [36] would align better with the recall-focused setting of a first-stage retriever.

© The Author(s), under exclusive license to Springer Nature Switzerland AG 2026
J. Carrillo-de-Albornoz et al. (Eds.): CLEF 2025, LNCS 16089, pp. 17–33, 2026.
https://doi.org/10.1007/978-3-032-04354-2_2

In this paper, we reproduce the key algorithms from the selective search literature. We compare their ability to match the recall level of exhaustive search at a reduced cost, finding that existing algorithms can be used to achieve high recall using 50% of the resources. When considering the cost and effectiveness of resource selection, we argue that vocabulary-based methods should be preferred over sample-based algorithms, especially when high recall is desired.

To strengthen the claims made in this paper and foster reproducibility, we publicly release our implementations of the used algorithms.[1][2]

2 Selective Search

To enable efficient retrieval over large document collections in a *distributed search* setup, indexes are split into *shards* and distributed over several machines. A *broker* receives user queries and distributes them among each of the machines, which run the query on their dedicated part of the index. After receiving a response from all workers, the broker merges the results to obtain a final ranking.

A specific instance of distributed search is *federated search*, in which no single party controls all machines. Instead, the broker maintains a list of so-called *resources* (specific third-party search engines). Upon receiving a query, the broker determines which resources are most likely to return relevant results (so-called *resource selection*), and forwards the query to those third parties.

Selective search combines ideas from standard distributed search and federated search. It assumes a fully cooperative setup, in which all machines are controlled and maintained by a single party. Collections are divided into topical index shards, and queries are answered using a few shards only. The resource selection algorithms used for federated search have proven to be effective in a selective search setup as well. The fully cooperative setup has access to more complete information (e.g. index statistics or shard sizes) from each of the shards, obviating the need to sample documents from third-party servers. Additionally, the ranking algorithm (and used statistics) may span all of the shards, ensuring the returned scores are similar and the results merging step becomes trivial.

Below, we discuss the two main design criteria of selective search: allocation of documents to shards (Sect. 2.1 and 2.2), and the choice of resource selection algorithm (Sect. 2.3).

2.1 Shard Map Creation

The distribution of documents across index shards (a so-called *shard map*) is a crucial component in selective search systems. The goal is to assign documents with similar topics to the same shards, such that relevant documents for a given query may cluster together in a low number of shards.

A simple strategy would be to assign documents to shards based on their *source*, assuming that different documents from the same source (e.g. website)

---

[1] https://gitlab.science.ru.nl/informagus/document-clustering/
[2] https://gitlab.science.ru.nl/informagus/duckdb-selective-search/

are likely to discuss similar topics. However, Kulkarni and Callan [29] found that source-based shard maps perform poorly compared to topic-based shard maps.

Xu and Croft [45] and Kulkarni and Callan [29] represent the documents in a collection using their smoothed unigram language models and use the K-means algorithm to perform unsupervised clustering. As a similarity measure, they compute the Kullback-Leibler divergence between the unigram language models of the document and that of the centroid of each cluster.

Dai et al. [17] improve on this approach by recognizing that some terms are more likely to be used by users, and thus should be weighed more heavily in clustering. Their QKLD distance measure scales the KL-divergence score for a term with a query-biased weight estimated by the term frequency in a supplied query log. They also introduce QInit, a centroid initialization strategy that performs agglomerative clustering on the (Word2Vec) word embeddings of terms in the query log and uses the resulting representations as initial K-means centroids.

Recently, large neural language models have also been increasingly used in topic modelling and clustering (e.g. BERTopic [23]). However, the computational costs associated with running these models make them as of yet unsuitable for use on large-scale collections that require selective search.

2.2   Shard Map Evaluation

Early work on selective search [29] evaluated document-shard allocation policies by setting up end-to-end selective search systems and computing standard information retrieval metrics. However, with this approach, end-to-end performance is influenced by more than just the quality of the shard map (e.g. the choice of resource selection or retrieval model), making it hard to determine shard map quality on its own. Dai et al. [17] additionally reported the coverage of relevant documents in the top $t\%$ of shards for each query, using an oracle resource selection method. While this allows for isolated estimation of shard map quality, it still requires access to (expensive) relevance assessments.

Kim and Callan [27] solved this issue with the AUReC measure by recognizing that selective search systems are intended to match the performance of exhaustive retrieval systems. They use the top k $(=1000)$ results D_k from a strong ranker as pseudo-relevant documents, rank the shards based on the number of documents from D_k they contain, and compute the area under the corresponding recall curve. Hendriksen et al. [25] incorporated shard sizes in the measure to reduce bias towards larger shards when shard sizes are skewed.

2.3   Resource Selection

The task of a resource selection algorithm in selective search is to determine, given an input query, which shards are most likely to contain relevant documents for this query. Most resource selection algorithms in the literature can be classified into *sample-based* and *vocabulary-based* methods. Sample-based resource selection algorithms [31, 41–43] create a central sample index (CSI) from a representative sample of documents from each shard (usually less than 5%). When a query comes in, it is issued to the CSI, and the documents at the top of the resulting ranking are used to score the shards they were assigned to. Vocabulary-based algorithms [1,4] create term-based shard representations and rank the shards based on the query terms. Other resource selection algorithms include those based on decision theory [22] or learning-to-rank [16,20].

3   Experimental Setup

In our experiments, we evaluate the ability of selective search architectures to meet the recall level of exhaustive search. To do so, we implemented the most important algorithms from the selective search literature. We first validate their correctness on standard test collections, after which we extend our evaluation to more datasets and the recall-oriented setting.

3.1   Datasets

To independently evaluate (our implementations of) the resource selection algorithms under review, we use the TREC Federated Web Search (FedWeb) Track [18,19] (specifically, the Resource Selection (RS) task).

In the past, selective search architectures have primarily been evaluated on the ClueWeb09-B[3] and GOV2[4] collections, using the TREC Web Track 2009–2012 [7–10] and TREC Terabyte Track 2004–2006 [3,6,11] topics, respectively. We also use these collections, as well as the ClueWeb12-B collection[5] and the associated TREC Web Track 2013–2014 [12,13] topics. The ClueWeb12-B collection has to date not yet been used for the evaluation of selective search architectures. A full overview of the test collections used in this paper can be found in Table 1.

3.2   Query Log

Several components in our setup require an external query log. For these use cases, we derived several query sets from the ORCAS query log [14]. In our 'lenient' split, we created a heldout query set by sampling 1000 queries from ORCAS, ensuring coverage of both popular and unpopular query terms by sorting the queries by their term frequencies and sampling uniformly from that list. In our 'strict' split, we removed all queries that had at least one query term in common with any of the TREC queries (roughly 40% of all 10M queries) before we sampled the heldout query set. For both splits, we repeat the above procedure three times to create three folds, to make our experimental findings more robust. The heldout sets were used for computing AUReC scores and training a learning-to-rank resource selection model from exhaustive search results (Sect. 3.5). The remainder of the queries are used for clustering (Sect. 3.3).

---

[3] https://lemurproject.org/clueweb09/, last accessed May 20th, 2025.

[4] http://ir.dcs.gla.ac.uk/test_collections/access_to_data.html, last accessed May 20th, 2025.

[5] https://lemurproject.org/clueweb12/, last accessed May 20th, 2025.

Table 1. Overview of the datasets used in this paper.

Federated search (resource selection only)

Name	TREC collections	#docs	#topics	#shards	avg. shard size
FedWeb 13	FedWeb '13	–	50	157	–
FedWeb 14	FedWeb '14	–	50	148	–

Selective search

Name	TREC collections	#docs	#topics	#shards	avg. shard size
ClueWeb09-B	Web '09-'12	50M	200	107–112	461k ± 99k
ClueWeb12-B	Web '13-'14	52M	100	103–106	503k ± 101k
GOV2	Terabyte '04-'06	25M	150	127–128	197k ± 49k

Note that our 'strict' split is artificially strict and pessimistic: in practice, query logs used for training are likely to overlap with queries seen in the future. By preventing leakage altogether, though, we address the concerns pointed out in [21] while still providing (conservative) effectiveness estimates of our models.

3.3 Clustering

To strengthen the claims in this reproduction study, and to be able to extend our evaluations to the ClueWeb12-B corpus, we do not use the shard maps published by Dai et al. [17] for our main experiments. Instead, we implemented our own versions of the KLD-Rand, QKLD-Rand and QKLD-QInit algorithms [24].

We use (part of) the ORCAS query set [14] for initializing the query-biased weights (see Sect. 3.2), and GloVe word embeddings [39] to initialize the centroids with QInit. Following Kulkarni and Callan [29], we sample a 1% sample of each collection to determine the cluster centroids, after which we map the remaining documents to the nearest centroids. Following Dai et al. [17], we use the hyperparameters $\lambda = 0.1$, $\mu = 0.1$, and $b = 1/16$. Like Kulkarni and Callan [29], we choose the number of clusters K for the ClueWeb collections such that each shard contains 500k documents on average. For GOV2, we aim for an average shard size of 170k documents, following Dai et al. [17]. The average number of shards produced by QKLD-QInit for each collection can be found in Table 1.

To validate the correctness of our implementation, we compare the resulting shard maps to those published by Dai et al. [17]. We do so by computing the (weighted) AUReC [25, 27], as well as measuring the end-to-end performance of selective search system built on each of the shard maps.

3.4 Retrieval

We implemented our retrieval systems in DuckDB [40], an analytical database engine. We followed the approach by Mühleisen et al. [38] to represent the inverted file and implement BM25 as main retrieval model. We use BM25 parameters $k_1 = 0.9$ and $b = 0.4$ (the defaults of Anserini [46]) and removed the stopwords using the stopword list in DuckDB's full-text search extension. For selective search, we use global term/document statistics, so the scores for documents in each shard are comparable and result merging becomes trivial.

By using a relational database to represent the inverted file, we could easily experiment with different shard maps, representations, and resource selection algorithms. Shards can be assigned by adding a 'shard' column to the documents table; statistics for vocabulary-based resource selection can be computed through simple aggregate queries; a CSI can be created by sampling from the documents table; and so forth. The existing resource selection algorithms can also be expressed naturally as SQL queries.

We implemented PyTerrier [35] transformer interfaces for all our models. That way, we could use PyTerrier and its integration with `ir_datasets` [34] and `ir_measures` [32] as a front-end for running our experiments.

3.5   Resource Selection

To evaluate selective search architectures in a deep, recall-oriented setting, we implemented the following popular resource selection algorithms:

CORI [4], a vocabulary-based method that represents (the dictionary of) each shard as a single large document. Shards are ranked by performing retrieval over the resulting representations using the INQUERY ranking formula.

ReDDE [42], a sample-based method that counts the number of times each shard appears in the top k (in our case, 1000) results from the CSI, after which they scale the shard scores based on their size. We use the ReDDE.top variant, which sums the relevance scores of the returned documents instead of counting them, and does not apply size-based shard weights.

Taily [1], a vocabulary-based method that estimates the distribution of unigram language model scores for the query terms by fitting gamma distributions on term-based score statistics. Using these, Taily estimates how likely each shard is to contain at least v documents out of the top n_c documents for a given query. We use the default parameters: $n_c = 400$, $v = 50$, and $\mu = 2500$.

Rank-S [31], a sample-based method that lets each document retrieved from the CSI vote for the shard they were found in. The voting strength is based on the relevance score and decays exponentially with the document's rank. The parameter B controls the decay rate; we use the default $B = 50$.

L2R [16], a method based on learning-to-rank. We use the following features from Dai et al. [16]: shard popularity, champion list scores, and the (binned) CORI, ReDDE, Rank-S and Taily scores. Following Arguello et al. [2] and Dai et al. [16], the exhaustive retrieval results provide training data for the learning-to-rank model. In our case, we use BM25 to retrieve the top 1000 results for all queries in one of the three heldout subsets of the ORCAS query log per split type (Sect. 3.2). The L2R model is trained to rank the shards based on the number of pseudo-relevant documents they contain. We split the heldout query set into a training set and a validation set containing 750 and 250 queries, respectively. For the implementation of the learning-to-rank model, we used the XGBoost [5] implementation of LambdaMART [44].

We compare these algorithms to the **Oracle** baseline, which ranks shards by the number of relevant documents they contain.[6] For sample-based methods, we repeat each experiment with three different CSIs (using a 1% document sample).

3.6   Evaluation

To confirm the correctness of our implementation of each resource selection algorithm, we evaluate them on the FedWeb collections using the official FedWeb metrics: nDCG@20,[7] as well the normalized graded precision (nP@k) at ranks 1 and 5 [18,19]. The nP@k metric is computed as the graded precision of the top k selected resources divided by the graded precision of the best k resources.

For the evaluation of selective search as first-stage retriever, we focus on high recall. Moffat et al. [36] have argued that standard recall (e.g. R@1000) is unsuitable for evaluating early-stage retrieval, as it does not distinguish between degrees of relevance; missing the most important document should be penalized more heavily than missing a slightly relevant document. They introduced rank-biased recall (RBR), which places more importance on high ranks in the reference ranking, and suggest appropriate values of the parameter ϕ to be between 0.8 and 0.98. We use RBR@1000 with $\phi = 0.9$ as primary evaluation metric.

To measure the efficiency of a selective search system, we use the Cost-in-Postings (CiP) measure introduced by Kulkarni [28]. CiP measures the number of postings accessed during *retrieval* (CiP_R), plus the number of postings used for *shard selection* (CiP_{SS}). For sample-based resource selection algorithms (ReDDE and Rank-S), CiP_{SS} is equal to the number of postings used from the CSI. For CORI, it is equal to the number of postings used from its index of large document representations. For Taily, it equals the number of term statistics accessed for all shard representations. For L2R, we sum the CiP_{SS} of all input features.

To test whether our results are statistically significant, we follow Kim [26, Chapter 2], who observed that standard paired t-tests are not suitable for testing whether a selective search system matches the effectiveness of exhaustive search. Failing to reject the null hypothesis $\mathcal{H}_0 : \mu_A = \mu_B$ does not necessarily imply that the two systems perform equivalently. Kim [26, Chapter 5] instead applies a non-inferiority test (commonly used in the medical field) to determine whether the effectiveness of the selective search system is at most δ worse than exhaustive search. We set δ to 5% of the mean of the exhaustive search effectiveness and use a significance threshold of $\alpha = 0.01$ (i.e. a confidence level of 99%). We use the highest p-value across folds in our significance tests (a conservative choice) and apply Bonferroni correction to correct for multiple hypothesis testing.

---

[6] This oracle is usually referred to as relevance-based ranking (RBR) in the literature. We refer to it as 'oracle' to prevent confusion with the rank-biased recall metric.

[7] We use the `trec_eval` nDCG implementation, which was also used in FedWeb 2014.

Table 2. Comparison between shard maps produced by our implementation and the ones published by Dai et al. [17].

(a) ClueWeb09-B

		AUReC	wAUReC	Taily			Rank-S		
				P@10	nDCG	MAP	P@10	nDCG	MAP
Dai et al. [17]	Exhaustive	-	-	0.26	0.28	0.19	0.26	0.28	0.19
	KLD-Rand	0.94	0.92	0.23	0.21	0.13	0.25	0.19	0.11
	QKLD-Rand	**0.95**	0.94	0.25	**0.22**	**0.14**	0.24	0.20	0.12
	QKLD-QInit	0.94	0.93	0.24	0.22	0.14	**0.25**	**0.21**	**0.13**
	KLD-Rand	0.93	0.92	0.24	0.20	0.12	0.23	0.18	0.11
Ours (lenient)	QKLD-Rand	0.94	**0.94**	0.25	0.21	0.14	0.24	0.20	0.12
	QKLD-QInit	0.94	0.94	0.25	0.22	0.14	0.24	0.20	0.13
	KLD-Rand	0.92	0.92	0.23	0.19	0.12	0.23	0.18	0.10
Ours (strict)	QKLD-Rand	0.94	0.94	**0.25**	0.21	0.13	0.24	0.20	0.12
	QKLD-QInit	0.94	0.93	0.24	0.21	0.13	0.24	0.20	0.13

(b) GOV2

		AUReC	wAUReC	Taily			Rank-S		
				P@10	nDCG	MAP	P@10	nDCG	MAP
Dai et al. [17]	Exhaustive	-	-	0.58	0.45	0.30	0.58	0.45	0.30
	KLD-Rand	0.96	0.96	0.44	0.31	0.18	0.47	0.32	0.17
	QKLD-Rand	**0.97**	0.96	0.46	0.33	0.19	**0.50**	**0.34**	0.19
	QKLD-QInit	0.97	**0.96**	**0.49**	**0.36**	**0.21**	0.49	0.33	**0.19**
	KLD-Rand	0.94	0.94	0.44	0.30	0.16	0.48	0.30	0.15
Ours (lenient)	QKLD-Rand	0.95	0.95	0.46	0.32	0.18	0.49	0.32	0.17
	QKLD-QInit	0.95	0.95	0.45	0.31	0.17	0.49	0.32	0.17
	KLD-Rand	0.94	0.94	0.43	0.29	0.16	0.48	0.30	0.15
Ours (strict)	QKLD-Rand	0.95	0.95	0.46	0.33	0.19	0.49	0.32	0.17
	QKLD-QInit	0.94	0.95	0.46	0.32	0.19	0.49	0.32	0.17

4   Results

In this section, we discuss our experimental findings. First, we evaluate the correctness of our clustering and resource selection algorithms. Then, we compare the different selective search architectures in a recall-oriented setting.

Table 2 compares the shard maps constructed using our implementation of the clustering algorithms to the ones published by Dai et al. [17] for the ClueWeb09-B and GOV2 collections[8]. We compare the shard maps on (w)AUReC scores and selective search performance with Taily or Rank-S

---

[8] Downloaded from https://boston.lti.cs.cmu.edu/appendices/CIKM2016-Dai/, last accessed May 20th, 2025.

resource selection. We report the same metrics as Dai et al.: P@10, nDCG@100, MAP@1000.

If we compare our Rank-S results on the Dai et al. shard maps, we see that our scores are slightly lower than the scores they reported [17, Tables 7 and 8]. Our exhaustive scores are similar, though, so it is unlikely that the difference can be explained by the difference in retrieval model (SDM instead of BM25).

The shard maps produced by our implementation obtain similar scores to the ones published by Dai et al. [17]. Both our shard maps and the ones published by Dai et al. obtain very high (w)AUReC scores on both collections. On average, our shard maps perform slightly worse in the end-to-end selective search setting. This could (partially) be explained by the fact that we used different embeddings: GLoVe instead of Word2Vec. Overall, we conclude that our implementation is consistent with the results published by Dai et al. [17].

The scores obtained using our 'strict' query splits are very similar to those obtained with the 'lenient' splits. This implies that removing overlapping queries from the TREC and heldout query sets does not negatively impact the clustering algorithm; it is robust enough to still form meaningful clusters. In the rest of the paper, we use shard maps made with QKLD-QInit and the 'strict' query splits.

4.1 Resource Selection

Table 3 shows the effectiveness of (our implementation of) the resource selection algorithms introduced in Sect. 3.5 on the Resource Selection (RS) task of the Fed-Web collections [18,19]. All our models perform in the expected range of resource selection algorithms that only use the sample documents provided by the Fed-Web track [18,19]. For both collections, Rank-S outperforms the other methods on all measures. If we compare our Taily implementation to the FedWeb 13 utTailyM400 run, the scores are close: we obtain an nP@1 of 0.22 (vs. 0.20) and nP@5 of 0.20 (vs. 0.23). The same holds for the FedWeb 14 run UTTailyG2000: we achieve an nDCG@20 of 0.34 (vs. 0.32), nP@1 of 0.12 (vs. 0.14) and nP@5 of 0.19 (vs. 0.22). We can draw similar conclusions if we compare our ReDDE implementation to the drexelRS2 run on FedWeb 14. From these results, we conclude that our implementations provide results consistent with previous work.

Table 3. Performance of the different resource selection algorithms on the FedWeb "Resource Selection" task. Best results per metric are marked in bold.

	FedWeb 13			FedWeb 14		
	nDCG@20	nP@1	nP@5	nDCG@20	nP@1	nP@5
CORI	0.30	0.12	0.17	0.27	0.09	0.13
ReDDE	0.37	0.15	0.23	0.33	0.15	0.16
Taily	0.38	0.22	0.20	0.34	0.12	0.19
Rank-S	**0.42**	**0.26**	**0.29**	**0.39**	**0.23**	**0.30**
L2R	0.37	0.15	0.23	0.33	0.15	0.16

Interestingly, the L2R model performs worse than some individual resource selection algorithms (especially Rank-S), even though it uses all of them as inputs. This implies that the procedure of generating training data from exhaustive retrieval results is insufficiently reliable for the federated search setting.

Finally, CORI performs notably poorer than the other algorithms. Since we only have the sample documents available, the large document representations generated by CORI may not be fully representative of the corresponding resources. If CORI has access to the full collection (e.g. in a selective search setting), it can create a better representation (see Sect. 4.2). Taily is not impacted by this issue, even though it also builds term-based representations.

4.2   Selective Search

To determine whether selective search can be used as a high-recall first-stage retriever in a multi-stage pipeline, we compare the performance of selective search architectures with the different resource selection algorithms to that of exhaustive search. Unlike the common practice in prior work, we do not report the effectiveness of each system for a set of chosen parameters. Instead, we focus on the idea that selective search should match the recall of exhaustive search, but at higher efficiency. We thus plot the effectiveness (RBR) against the efficiency (CiP) for different configurations (i.e. different parameter settings) of each resource selection algorithm. Figure 1 summarizes the results.

For all collections, the L2R model performs best; it can reach effectiveness within 5% of exhaustive search using only 30–40% of the resources. CORI and ReDDE follow, with CORI being substantially more effective for GOV2. Interestingly, both Taily and Rank-S fall short and are unable to meet the recall level of exhaustive search, likely due to their dynamic shard cutoff. Even with very conservative parameters (e.g. low v for Taily and low B for Rank-S), they stop selecting new shards before reaching the desired recall level. Taily does show a promising efficiency-effectiveness curve, so using Taily for shard ranking without applying a dynamic cutoff might be a suitable setup.[9] The fact that Taily and CORI perform better on GOV2 (relative to the sample-based methods ReDDE and Rank-S) implies that vocabulary-based methods perform better on average when a shard map contains more (and thus, on average smaller) shards.

The oracle run shows that there is still much room for improvement. With the optimal shard ranking, we can obtain the desired recall level with less than 10% of the required resources – much less than the other methods. In fact, the oracle's choice of shards can even outperform the exhaustive baseline due to a smaller number of false positives in the result list. These results highlight that there is indeed high potential for selective search as a high-recall first-stage retriever, but that existing resource selection algorithms struggle to rank the shards optimally.

It is noteworthy that the sample-based method ReDDE never meets the recall level of exhaustive search, even when it is allowed to return all shards in the shard

---

[9] Initial experiments confirm that Taily with a static cutoff outperforms all other methods on GOV2, obtaining high recall with less than 25% of the resources.

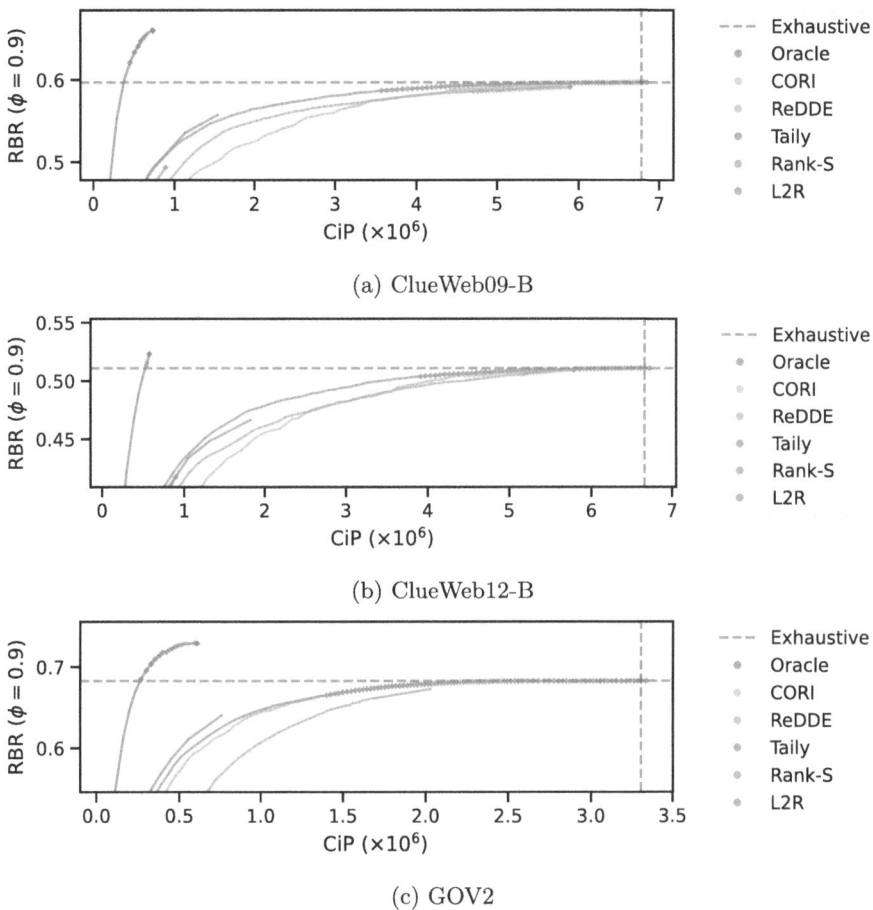

Fig. 1. Efficiency-effectiveness tradeoff curves of selective search architectures using different resource selection algorithms. The gray area indicates effectiveness within a 5% margin of exhaustive search. Results that are significantly non-inferior to exhaustive retrieval (with $p < 0.01$ and $\delta < 5\%$) are marked with $\diamond$. (Color figure online)

map. This can happen if the CSI documents sampled from a particular shard do not contain any of the query terms, even though some other documents in the shard do. As a result, the shard never contributes any documents to the top CSI results, and the shard is never selected. This seems to be a larger problem in our recall-oriented scenario than in a precision-focused setting [15].

4.3 Discussion

While we have shown that selective search architectures can reliably achieve recall within a 5% margin of the recall level achieved by exhaustive search, it appears difficult for resource selection algorithms to rank shards optimally for our high-recall requirement. The oracle method shows that it is possible to out-perform exhaustive search in terms of recall with only a fraction of the resources, but the existing resource selection algorithms struggle to even meet exhaustive search recall level – let alone surpass it. This indicates that the representations used to rank the shards are suboptimal for consistently selecting all shards with relevant documents before selecting other shards (even though they have been proven to be effective for settings that focus on early precision).

As we have seen in the previous section, a CSI may be constructed in such a way that some shards cannot be found by sample-based methods – even if they do contain relevant documents. In our experiments, this was the case for 11 queries for ClueWeb09-B, and 17 queries for GOV2. To prevent this from happening, a different sampling strategy or larger sample size may be necessary to ensure all vocabulary terms are represented in the CSI. A simpler (and likely more effective) strategy would be to use vocabulary-based methods (like CORI and Taily) only. These have the added benefit that they are orders of magnitude more efficient to compute, although they seem to perform slightly worse than the L2R model when using fewer, larger shards (e.g. for the ClueWeb collections).

Due to the filtering nature of selective search, we will likely miss relevant documents for a query sometimes (e.g. because it was assigned to an unrepresentative shard). It would be interesting to see how recall-improving methods, like adaptive re-ranking [33] or query expansion, would affect selective search.

5   Conclusion

We have re-implemented several popular algorithms from the selective search literature and verified their correctness on standard collections. Then, we compared them in a new light: a recall-oriented setting, in which the selective search system is assumed to be deployed as a first-stage retriever. We compared the different systems on their ability to match the recall level of an exhaustive search system. The main findings of our paper can be summarized as follows:

– Selective search is a viable strategy in the high-recall setting of first-stage retrieval. Existing resource selection algorithms need to select more shards than in the early-precision setting usually considered by previous work, but are still able to reduce the resources required for completing a search request by over 50% while still reaching recall non-inferior to exhaustive search.
– Vocabulary-based resource selection algorithms seem more suitable for recall-oriented selective search. They are orders of magnitude more efficient than sample-based methods and do not introduce indeterminism due to sampling.
– The L2R model, trained on pseudo-relevance feedback from exhaustive retrieval, performs well for selective search but fails for federated search.

Acknowledgments. This work has received funding from the European Union's Horizon Europe research and innovation programme under grant agreement No 101070014 (OpenWebSearch.EU, https://doi.org/10.3030/101070014.

References

1. Aly, R., Hiemstra, D., Demeester, T.: Taily: shard selection using the tail of score distributions. In: Proceedings of the 36th International ACM SIGIR Conference on Research and Development in Information Retrieval, pp. 673–682. SIGIR '13, Association for Computing Machinery, New York, NY, USA (2013). https://doi.org/10.1145/2484028.2484033

2. Arguello, J., Callan, J., Diaz, F.: Classification-based resource selection. In: Proceedings of the 18th ACM Conference on Information and Knowledge Management, pp. 1277–1286. CIKM '09, Association for Computing Machinery, New York, NY, USA (2009). https://doi.org/10.1145/1645953.1646115

3. Büttcher, S., Clarke, C.L.A., Soboroff, I.: The TREC 2006 terabyte track. In: Voorhees, E.M., Buckland, L.P. (eds.) Proceedings of the Fifteenth Text REtrieval Conference, TREC 2006, Gaithersburg, Maryland, USA, November 14-17, 2006. NIST Special Publication, vol. 500–272. National Institute of Standards and Technology (NIST) (2006)

4. Callan, J.P., Lu, Z., Croft, W.B.: Searching distributed collections with inference networks. In: Proceedings of the 18th Annual International ACM SIGIR Conference on Research and Development in Information Retrieval, pp. 21–28. SIGIR '95, Association for Computing Machinery, New York, NY, USA (1995). https://doi.org/10.1145/215206.215328

5. Chen, T., Guestrin, C.: XGBoost: a scalable tree boosting system. In: Proceedings of the 22nd ACM SIGKDD International Conference on Knowledge Discovery and Data Mining, pp. 785–794. KDD '16, Association for Computing Machinery, New York, NY, USA (2016). https://doi.org/10.1145/2939672.2939785

6. Clarke, C.L.A., Craswell, N., Soboroff, I.: Overview of the TREC 2004 terabyte track. In: Voorhees, E.M., Buckland, L.P. (eds.) Proceedings of the Thirteenth Text REtrieval Conference, TREC 2004, Gaithersburg, Maryland, USA, November 16-19, 2004. NIST Special Publication, vol. 500–261. National Institute of Standards and Technology (NIST) (2004)

7. Clarke, C.L.A., Craswell, N., Soboroff, I.: Overview of the TREC 2009 web track. In: Voorhees, E.M., Buckland, L.P. (eds.) Proceedings of the Eighteenth Text REtrieval Conference, TREC 2009, Gaithersburg, Maryland, USA, November 17-20, 2009. NIST Special Publication, vol. 500–278. National Institute of Standards and Technology (NIST) (2009)

8. Clarke, C.L.A., Craswell, N., Soboroff, I., Cormack, G.V.: Overview of the TREC 2010 web track. In: Voorhees, E.M., Buckland, L.P. (eds.) Proceedings of the Nineteenth Text REtrieval Conference, TREC 2010, Gaithersburg, Maryland, USA, November 16-19, 2010. NIST Special Publication, vol. 500–294. National Institute of Standards and Technology (NIST) (2010)

9. Clarke, C.L.A., Craswell, N., Soboroff, I., Voorhees, E.M.: Overview of the TREC 2011 web track. In: Voorhees, E.M., Buckland, L.P. (eds.) Proceedings of the Twentieth Text REtrieval Conference, TREC 2011, Gaithersburg, Maryland, USA, November 15-18, 2011. NIST Special Publication, vol. 500–296. National Institute of Standards and Technology (NIST) (2011)

10. Clarke, C.L.A., Craswell, N., Voorhees, E.M.: Overview of the TREC 2012 web track. In: Voorhees, E.M., Buckland, L.P. (eds.) Proceedings of the Twenty-First Text REtrieval Conference, TREC 2012, Gaithersburg, Maryland, USA, November 6-9, 2012. NIST Special Publication, vol. 500–298. National Institute of Standards and Technology (NIST) (2012)

11. Clarke, C.L.A., Scholer, F., Soboroff, I.: The TREC 2005 terabyte track. In: Voorhees, E.M., Buckland, L.P. (eds.) Proceedings of the Fourteenth Text REtrieval Conference, TREC 2005, Gaithersburg, Maryland, USA, November 15-18, 2005. NIST Special Publication, vol. 500–266. National Institute of Standards and Technology (NIST) (2005)

12. Collins-Thompson, K., Bennett, P.N., Diaz, F., Clarke, C.L.A., Voorhees, E.M.: TREC 2013 web track overview. In: Voorhees, E.M. (ed.) Proceedings of the Twenty-Second Text Retrieval Conference, TREC 2013, Gaithersburg, Maryland, USA, November 19-22, 2013. NIST Special Publication, vol. 500–302. National Institute of Standards and Technology (NIST) (2013)

13. Collins-Thompson, K., Macdonald, C., Bennett, P.N., Diaz, F., Voorhees, E.M.: TREC 2014 web track overview. In: Voorhees, E.M., Ellis, A. (eds.) Proceedings of the Twenty-Third Text Retrieval Conference, TREC 2014, Gaithersburg, Maryland, USA, November 19-21, 2014. NIST Special Publication, vol. 500–308. National Institute of Standards and Technology (NIST) (2014)

14. Craswell, N., Campos, D., Mitra, B., Yilmaz, E., Billerbeck, B.: ORCAS: 18 million clicked query-document pairs for analyzing search. In: Proceedings of the 29th ACM International Conference on Information & Knowledge Management, pp. 2983–2989. CIKM '20, Association for Computing Machinery, New York, NY, USA (2020). https://doi.org/10.1145/3340531.3412779

15. Dai, Z., Kim, Y., Callan, J.: How random decisions affect selective distributed search. In: Proceedings of the 38th International ACM SIGIR Conference on Research and Development in Information Retrieval, pp. 771–774. SIGIR '15, Association for Computing Machinery, New York, NY, USA (2015). https://doi.org/10.1145/2766462.2767796

16. Dai, Z., Kim, Y., Callan, J.: Learning to rank resources. In: Proceedings of the 40th International ACM SIGIR Conference on Research and Development in Information Retrieval, pp. 837–840. SIGIR '17, Association for Computing Machinery, New York, NY, USA (2017). https://doi.org/10.1145/3077136.3080657

17. Dai, Z., Xiong, C., Callan, J.: Query-biased partitioning for selective search. In: Proceedings of the 25th ACM International on Conference on Information and Knowledge Management, pp. 1119–1128. CIKM '16, Association for Computing Machinery, New York, NY, USA (2016). https://doi.org/10.1145/2983323.2983706

18. Demeester, T., Trieschnigg, D., Nguyen, D., Hiemstra, D.: Overview of the TREC 2013 federated web search track. In: Voorhees, E.M. (ed.) Proceedings of the Twenty-Second Text Retrieval Conference, TREC 2013, Gaithersburg, Maryland, USA, November 19-22, 2013. NIST Special Publication, vol. 500–302. National Institute of Standards and Technology (NIST) (2013)

19. Demeester, T., Trieschnigg, D., Nguyen, D., Hiemstra, D., Zhou, K.: Overview of the TREC 2014 federated web search track. In: Voorhees, E.M., Ellis, A. (eds.) Proceedings of the Twenty-Third Text Retrieval Conference, TREC 2014, Gaithersburg, Maryland, USA, November 19-21, 2014. NIST Special Publication, vol. 500–308. National Institute of Standards and Technology (NIST) (2014)

20. Ergashev, U., Dragut, E., Meng, W.: Learning to rank resources with GNN. In: Proceedings of the ACM Web Conference 2023, pp. 3247–3256. WWW '23, Association for Computing Machinery, New York, NY, USA (2023). https://doi.org/10.1145/3543507.3583360

21. Fröbe, M., Akiki, C., Potthast, M., Hagen, M.: How train–test leakage affects zero-shot retrieval. In: Arroyuelo, D., Poblete, B. (eds.) String Processing and Information Retrieval, pp. 147–161. Springer International Publishing, Cham (2022). https://doi.org/10.1007/978-3-031-20643-6_11
22. Fuhr, N.: A decision-theoretic approach to database selection in networked IR. ACM Trans. Inf. Syst. **17**(3), 229–249 (1999). https://doi.org/10.1145/314516.314517
23. Grootendorst, M.: BERTopic: Neural topic modeling with a class-based TF-IDF procedure (2022). https://doi.org/10.48550/arXiv.2203.05794
24. Hendriksen, G., Hiemstra, D., de Vries, A.P.: An open source implementation of web clustering algorithms for selective search. In: Proceedings 6th International Open Search Symposium #ossym2024, p. 79 (2024)
25. Hendriksen, G., Hiemstra, D., de Vries, A.P.: Weighted AUReC: handling skew in shard map quality estimation for selective search. In: Advances in Information Retrieval: 46th European Conference on Information Retrieval, ECIR 2024, Glasgow, UK, March 24–28, 2024, Proceedings, Part IV, pp. 87–96. Springer-Verlag, Berlin, Heidelberg (2024). https://doi.org/10.1007/978-3-031-56066-8_10
26. Kim, Y.: Robust Selective Search. Ph.D. thesis, Carnegie Mellon University (2019)
27. Kim, Y., Callan, J.: Measuring the effectiveness of selective search index partitions without supervision. In: Proceedings of the 2018 ACM SIGIR International Conference on Theory of Information Retrieval, pp. 91–98. ICTIR '18, Association for Computing Machinery, New York, NY, USA (2018). https://doi.org/10.1145/3234944.3234952
28. Kulkarni, A.: Efficient and Effective Large-scale Search. Ph.D. thesis, Carnegie Mellon University (2013)
29. Kulkarni, A., Callan, J.: Document allocation policies for selective searching of distributed indexes. In: Proceedings of the 19th ACM International Conference on Information and Knowledge Management, pp. 449–458. CIKM '10, Association for Computing Machinery, New York, NY, USA (2010). https://doi.org/10.1145/1871437.1871497
30. Kulkarni, A., Callan, J.: Selective search: efficient and effective search of large textual collections. ACM Trans. Inf. Syst. **33**(4), 17:1–17:33 (2015). https://doi.org/10.1145/2738035
31. Kulkarni, A., Tigelaar, A.S., Hiemstra, D., Callan, J.: Shard ranking and cutoff estimation for topically partitioned collections. In: Proceedings of the 21st ACM International Conference on Information and Knowledge Management, pp. 555–564. ACM, Maui Hawaii USA (2012). https://doi.org/10.1145/2396761.2396833
32. MacAvaney, S., Macdonald, C., Ounis, I.: Streamlining evaluation with ir-measures. In: Hagen, M., Verberne, S., Macdonald, C., Seifert, C., Balog, K., Nørvåg, K., Setty, V. (eds.) Advances in Information Retrieval, pp. 305–310. Springer International Publishing, Cham (2022). https://doi.org/10.1007/978-3-030-99739-7_38
33. MacAvaney, S., Tonellotto, N., Macdonald, C.: Adaptive re-ranking with a corpus graph. In: Hasan, M.A., Xiong, L. (eds.) Proceedings of the 31st ACM International Conference on Information & Knowledge Management, Atlanta, GA, USA, October 17-21, 2022, pp. 1491–1500. ACM (2022). https://doi.org/10.1145/3511808.3557231
34. MacAvaney, S., Yates, A., Feldman, S., Downey, D., Cohan, A., Goharian, N.: Simplified data wrangling with ir_datasets. In: Proceedings of the 44th International ACM SIGIR Conference on Research and Development in Information Retrieval, pp. 2429–2436. SIGIR '21, Association for Computing Machinery, New York, NY, USA (2021). https://doi.org/10.1145/3404835.3463254

35. Macdonald, C., Tonellotto, N.: Declarative experimentation in information retrieval using PyTerrier. In: Proceedings of the 2020 ACM SIGIR on International Conference on Theory of Information Retrieval, pp. 161–168 (2020). https://doi.org/10.1145/3409256.3409829

36. Moffat, A., Mackenzie, J., Mallia, A., Petri, M.: Rank-biased quality measurement for sets and rankings. In: Proceedings of the 2024 Annual International ACM SIGIR Conference on Research and Development in Information Retrieval in the Asia Pacific Region, pp. 135–144. SIGIR-AP 2024, Association for Computing Machinery, New York, NY, USA (2024). https://doi.org/10.1145/3673791.3698405

37. Mohammad, H.R., Xu, K., Callan, J., Culpepper, J.S.: Dynamic shard cutoff prediction for selective search. In: The 41st International ACM SIGIR Conference on Research & Development in Information Retrieval, pp. 85–94. SIGIR '18, Association for Computing Machinery, New York, NY, USA (2018). https://doi.org/10.1145/3209978.3210005

38. Mühleisen, H., Samar, T., Lin, J., de Vries, A.: Old dogs are great at new tricks: column stores for IR prototyping. In: Proceedings of the 37th International ACM SIGIR Conference on Research & Development in Information Retrieval, pp. 863–866. SIGIR '14, Association for Computing Machinery, New York, NY, USA (2014). https://doi.org/10.1145/2600428.2609460

39. Pennington, J., Socher, R., Manning, C.: GloVe: global vectors for word representation. In: Moschitti, A., Pang, B., Daelemans, W. (eds.) Proceedings of the 2014 Conference on Empirical Methods in Natural Language Processing (EMNLP), pp. 1532–1543. Association for Computational Linguistics, Doha, Qatar (2014). https://doi.org/10.3115/v1/D14-1162

40. Raasveldt, M., Mühleisen, H.: DuckDB: an embeddable analytical database. In: Proceedings of the 2019 International Conference on Management of Data, pp. 1981–1984. SIGMOD '19, Association for Computing Machinery, New York, NY, USA (2019). https://doi.org/10.1145/3299869.3320212

41. Shokouhi, M.: Central-rank-based collection selection in uncooperative distributed information retrieval. In: Amati, G., Carpineto, C., Romano, G. (eds.) Advances in Information Retrieval, pp. 160–172. Springer, Berlin, Heidelberg (2007). https://doi.org/10.1007/978-3-540-71496-5_17

42. Si, L., Callan, J.: Relevant document distribution estimation method for resource selection. In: Proceedings of the 26th Annual International ACM SIGIR Conference on Research and Development in Information Retrieval, pp. 298–305. SIGIR '03, Association for Computing Machinery, New York, NY, USA (2003). https://doi.org/10.1145/860435.860490

43. Thomas, P., Shokouhi, M.: SUSHI: Scoring scaled samples for server selection. In: Proceedings of the 32nd International ACM SIGIR Conference on Research and Development in Information Retrieval, pp. 419–426. SIGIR '09, Association for Computing Machinery, New York, NY, USA (2009). https://doi.org/10.1145/1571941.1572014

44. Wu, Q., Burges, C.J.C., Svore, K.M., Gao, J.: Adapting boosting for information retrieval measures. Inf. Retrieval 13(3), 254–270 (2010). https://doi.org/10.1007/s10791-009-9112-1

45. Xu, J., Croft, W.B.: Cluster-based language models for distributed retrieval. In: Proceedings of the 22nd Annual International ACM SIGIR Conference on Research and Development in Information Retrieval, pp. 254–261. SIGIR '99, Association for Computing Machinery, New York, NY, USA (1999). https://doi.org/10.1145/312624.312687

46. Yang, P., Fang, H., Lin, J.: Anserini: enabling the use of lucene for information retrieval research. In: Proceedings of the 40th International ACM SIGIR Conference on Research and Development in Information Retrieval, pp. 1253–1256. SIGIR '17, Association for Computing Machinery, New York, NY, USA (2017). https://doi.org/10.1145/3077136.3080721

Spatially Grounded Explanations in VisionLanguage Models for Document Visual Question Answering

Maximiliano Hormazábal[1,2]([✉]) [ID], Héctor Cerezo-Costas[2] [ID],
and Dimosthenis Karatzas[1] [ID]

[1] Computer Vision Center, Universitat Autònoma de Barcelona, Bellaterra,
Cataluña, Spain
{mhormazabal,dimos}@cvc.uab.es
[2] Gradiant, Vigo, Galicia, Spain
hcerezo@gradiant.org

Abstract. We introduce EaGERS, a fully training-free and model-agnostic pipeline that (1) generates natural language rationales via a vision language model, (2) grounds these rationales to spatial sub-regions by computing multimodal embedding similarities over a configurable grid with majority voting, and (3) restricts the generation of responses only from the relevant regions selected in the masked image. Experiments on the DocVQA dataset demonstrate that our best configuration not only outperforms the base model on exact match accuracy and Average Normalized Levenshtein Similarity metrics but also enhances transparency and reproducibility in DocVQA without additional model fine-tuning. Code available at: https://github.com/maxhormazabal/EaGERS-DVQA

Keywords: Document Intelligence · Visual Question Answering · Multimodal Reasoning · Explainability

1 Introduction

Document Visual Question Answering (DocVQA) has advanced rapidly with Transformer-based methods that integrate OCR, layout modeling, and domain adaptation [1,4,7,14]. Concurrently, general-purpose Vision Language Models (VLMs) [8,11,19] achieve strong document understanding without explicit DocVQA training.

Deploying off-the-shelf vision–language models in enterprise pipelines often involves costly fine-tuning, unstable prompt engineering and a lack of clear grounding between answers and source regions [21]. To address these challenges, we introduce Explanation-Guided Region Selection (EaGERS), a fully model-agnostic, training-free DocVQA pipeline that (i) generates natural language explanations, (ii) selects the top sub-regions over a configurable grid via multimodal embedding similarities and majority voting, and (iii) re-queries the model

© The Author(s), under exclusive license to Springer Nature Switzerland AG 2026
J. Carrillo-de-Albornoz et al. (Eds.): CLEF 2025, LNCS 16089, pp. 34–41, 2026.
https://doi.org/10.1007/978-3-032-04354-2_3

on a masked image so that answers derive solely from those validated regions ensuring transparency and reproducibility without additional model training.

The problem we address is: how to enforce that the answer can be reconstructed solely from document regions that are explicitly grounded and verbalised, without any additional training of the VLM.

The main contributions of this work are:

1. EaGERS: A fully model-agnostic and training-free DocVQA pipeline capable of generating answers on masked document images using general-purpose multimodal models.
2. Integration of text explanations and visual masking to contribute to the traceability and explainability of inferences.

2   Related Work

2.1   DocVQA

In recent years, DocVQA systems have achieved solid baselines on the DocVQA dataset [12], some approaches combine OCR and QA modules such LayoutLM [20] and TILT, yet still lag behind human performance. Other transformer-based models, such as DocFormer and Donut, adopt end-to-end, OCR-free architectures, and supervised-attention methods like M4C [5] integrate textual, positional, and visual cues to boost retrieval accuracy; however, their interpretability remains stuck to attention-weight analysis. Today, state-of-the-art OCR-free approaches reach near-human accuracy but offer no natural language rationales. Meanwhile, multimodal compression frameworks like mPLUG-DocOwl 2.0 improve scalability but still lack explanations grounded in specific document regions.

2.2   Explainability in DocVQA

In the area of spatial explainability, methods such as DocXplain [17] apply ablation techniques on document segments to measure their impact on predictions, demonstrating higher fidelity than Grad-CAM [18] at the cost of multiple inferences. Hybrid models like DLaVA [13] combine textual answers with bounding boxes, achieving Intersection over Union (IoU) [16] above 0.5 on DocVQA, while MRVQA [10] introduces textual rationales alongside visual highlights and proposes specific visual-text coherence metrics.

2.3   Modal Alignment and Multimodal Embeddings

Modal alignment projects text and image representations into a shared latent space, allowing direct comparison via geometric metrics such as cosine similarity. Pretrained models like CLIP [15], ALIGN [6], and BLIP [9] employ a multimodal contrastive objective to bring semantically related pairs closer together. In our pipeline, we use embeddings from BLIP, CLIP, and ALIGN to vectorize both natural language explanations and document sub-regions, enabling the multimodal similarity measurements that drive masking and focused re-querying.

3   Methodology

EaGERS-DocVQA leverages the knowledge of general-purpose multimodal models in document understanding without requiring dedicated training. Figure 1 shows the overall architecture of the proposed system, which consists of three main stages: A) Explanation generation, B) Region selection, and C) Answer generation. In our experiments, we use the Qwen2.5VL-3B model as the core component; however, the proposed system is essentially model-agnostic.

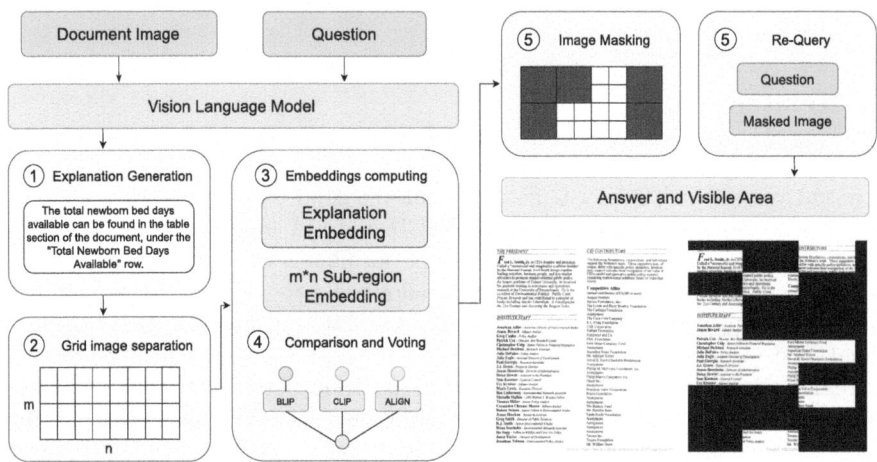

Fig. 1. EaGERS Document VQA pipeline: (1) the multimodal model generates a spatial natural language explanation from the image and the question; (2) the image is segmented into an $m \times n$ grid; (3) embeddings of the explanation and each sub-region are obtained using BLIP, CLIP, and ALIGN; (4) majority voting selects the most relevant regions; (5) the image is masked to retain only those regions, and the model is re-queried with the question to generate the final answer.

3.1   Explanation Generation

The document image and question are passed to a vision–language model, which uses them to generate a natural language explanation of how to obtain the requested answer in visual terms. This explanation is not the final answer but serves as a guide to locate the relevant information in the image. We employ this inference in subsequent steps as a semantic tool for comparing the image sub-regions with the generated explanation. Although these spatial explanations typically exhibit good alignment with relevant regions, there are occasional cases where the generated explanations may inaccurately refer to irrelevant areas, potentially impacting subsequent region selection.

3.2 Region Selection via Similarity

The document image is divided into an $m \times n$ grid, yielding $m \cdot n$ sub-regions, each of which may or may not be relevant for obtaining the answer.

Each sub-region is converted into a vector representation in order to compare it with the model's spatial explanation by an ensemble of three multimodal models: BLIP, CLIP, and ALIGN to generate embeddings in a complementary way in order to mitigate specific biases and improve robustness across heterogeneous document layouts.

Once we obtain a similarity score based on cosine similarity between reasoning and sub-region embeddings for each embedder, we select the top k for the final answer, where k is 30% (rounded up) of sub-regions based on preliminary experiments, as this provided a good compromise between spatial granularity for evidence localization. Although an adaptive grid might better handle irregular layouts, the fixed grid simplifies reproducibility and reduces complexity for this initial study.

The final ranking is determined by majority voting across the rankings from each embedder. In the event of ties during majority voting, we resolve these by prioritizing regions based on their average cosine similarity scores across all embedders, thus favoring sub-regions with more consistent overall relevance. The resulting list of selected sub-regions $\mathcal{R}$ defines the visible area used in the subsequent answer generation stage.

3.3 Masking and Re-query

We create a masked version of the original image in which all grid cells outside $\mathcal{R}$ are filled with black. The question and the masked image are then reintroduced to the same multimodal model, which must generate the answer using only the information within the justified regions, **without access** to the previously generated explanation except implicitly through the defined region mask.

In the following section we evaluate how effectively this approach recovers ground-truth answers under different grid and margin configurations.

4 Experimentation

4.1 Datasets and Evaluation Protocols

In our experiments, we use the validation split of the DocVQA Single Page dataset. We applied resizing preprocessing (preserving the aspect ratio) to compress images and optimize inferences. For spatial partitioning, we divide each image into a uniform grid of 5 columns and 5 rows (25 cells) in an initial series of tests, and an alternative configuration of 5 columns and 10 rows (50 cells) to explore the impact of granularity on relevant-region selection. We also have tested a margin expansion of the 15% of the unmasked sub-region for both amount of cells.

We use Exact Match (EM) and Average Normalized Levenshtein Similarity (ANLS) [2] which mitigates the impact of misrecognition errors by thresholding normalized edit distances.

4.2   Main Results

The Table 1 presents the performance results (ANLS and EM) of the different pipeline configurations. For comparison we have also run the Qwen2.5-VL-3B model directly in the DocVQA dataset to compare its performance with the pipeline. In addition to this, the unit inference time (model-only and EaGERS pipeline) have been measured to calculate the average inference time for each approach and their coefficients of variation (CV) have been calculated to gain insights of the similarity of times inference between document[1].

Table 1. Overall performance (EM and ANLS) on validation for different grid and margin settings.

Configuration				Performance			
Model	Cols	Rows	Margin	EM (%)	ANLS	Avg Time (s)	CV (%)
$EaGERS_{25\|0}$	5	5	0%	64.20	75.25	16.98	37.08
$EaGERS_{50\|0}$	5	10	0%	66.67	77.94	17.49	20.78
$EaGERS_{25\|15}$	5	5	15%	72.72	82.52	16.72	24.04
$EaGERS_{50\|15}$	5	10	15%	**74.50**	**83.31**	17.48	20.72
Qwen2.5-VL-3B*	*	*		71.17	82.90	**7.21**	30.08

As shown, adding a 15% masking margin consistently improves both EM and ANLS across grid sizes. In particular, the 50 cells grid with a 15% margin yields the best performance, suggesting that finer spatial granularity combined with slight overlap enhances the localization and understanding of regions relevant to the answer. This suggests that using a fixed distribution grid across the entire image can introduce complications when measuring cosine distance between embeddings. Specifically, relevant zones may fall on the borders between sub-regions, which could explain discrepancies (such as a margin of 0 versus 15) when the grid slightly shifts and either includes or excludes the answer location. This indicates clear future steps in improving the system towards a more flexible grid.

It is also relevant to see that the performance of the model "as-is" has gained explainability without experiencing loss in its performance level, but in the best version $EaGERS_{50\|15}$ presents a modest improvement pointing out that restricting the viewing space of the model so that it has more clarity of the answer is an interesting research direction.

Although this first study does not aspire to a head-to-head comparison with the state-of-the-art methods yet. It is also relevant to comment that EaGERS is above the solutions initially proposed in DocVQA [12] which have shown ANLS results in VQA models such as LoRRA=0.110 or M4C=0.385, BERT QA

---

[1] The GPU-hardware used for the experiments were 48GB NVIDIA RTX 6000 Ada Generation.

systems around 0.655 and multimodal architectures such as LayoutLMv2-BASE with an ANLS of 0.7421. However, next steps will be to evaluate this model against state-of-the-art solutions, which achieve even higher ANLS values.

5   Limitations and Future Work

Among the limitations of this study that suggest promising directions for future work: relying on fixed grid configurations may not generalize well to documents with irregular layouts or variable aspect ratios; object detectors may be useful for subdividing relevant regions [3]. Another important limitation is the dependence on spatial explanations generated by VLMs, which may occasionally produce inaccurate rationales, leading to incorrect region selections. Our pipeline assumes spatial accuracy of VLM explanations; however, when these do not match ground truth, the final fidelity may degrade which an aspect we will address in future work.

Future work should include systematic evaluations of the frequency and impact of such inaccuracies on the final results, we will evaluate more robust fusion strategies, such as Reciprocal Rank Fusion (RRF), and quantify agreement using Krippendorff alpha (α), in order to analyze in depth the internal consistency of spatial selections.

It would also be valuable to extend experiments to datasets that explicitly involve answer-localization annotations, enabling the use of quantitative fidelity metrics, such as (IoU) or visual–text coherence scores. Moreover, efficiency improvements are needed considering the increase in inference time reported in Table 1. Finally, we plan to explore methods to assess properly the level of explainability in comparison with alternatives.

6   Conclusions

We have proposed a pipeline that unifies natural language explanations with quantitative region selection and masked re-querying to ensure answers derive only from validated document regions. Our approach yields significant improvements in Exact Match and ANLS over standard baselines, demonstrating enhanced transparency and reproducibility. In future work, we will investigate adaptive grid partitioning to better handle structural variability, conduct comprehensive ablation studies to optimize embedder selection, and directly compare performance with more advanced state-of-the-art models. We also plan to integrate quantitative explainability metrics and carry out user studies to assess the clarity and reliability of the generated explanations.

References

1. Appalaraju, S., Jasani, B., Kota, B.U., Xie, Y., Manmatha, R.: DocFormer: end-to-end transformer for document understanding. In: Proceedings of the IEEE/CVF International Conference on Computer Vision, pp. 993–1003 (2021)
2. Biten, A., et al.: Scene text visual question answering, pp. 4290–4300. Proceedings of the IEEE International Conference on Computer Vision, Institute of Electrical and Electronics Engineers Inc., United States (2019). https://doi.org/10.1109/ICCV.2019.00439, funding Information: This work has been supported by projects TIN2017-89779-P, Marie-Curie (712949 TECNIOspring PLUS), aBSINTHE (Fundacion BBVA 2017), the CERCA Programme / Generalitat de Catalunya, a European Social Fund grant (CCI: 2014ES05SFOP007), NVIDIA Corporation and PhD scholarships from AGAUR (2019-FIB01233) and the UAB. Publisher Copyright: 2019 IEEE
3. Gómez, L., et al.: Multimodal grid features and cell pointers for scene text visual question answering. Pattern Recogn. Lett. **150**, 242–249 (2021). https://doi.org/10.1016/j.patrec.2021.06.026, https://www.sciencedirect.com/science/article/pii/S0167865521002336
4. Hu, A., et al.: mPLUG-DocOwl2: High-resolution compressing for OCR-free multipage document understanding (2024). https://arxiv.org/abs/2409.03420
5. Hu, R., Singh, A., Darrell, T., Rohrbach, M.: Iterative answer prediction with pointer-augmented multimodal transformers for TextVQA . In: 2020 IEEE/CVF Conference on Computer Vision and Pattern Recognition (CVPR), pp. 9989–9999. IEEE Computer Society, Los Alamitos, CA, USA (2020). https://doi.org/10.1109/CVPR42600.2020.01001
6. Jia, C., et al.: Scaling up visual and vision-language representation learning with noisy text supervision. In: Meila, M., Zhang, T. (eds.) Proceedings of the 38th International Conference on Machine Learning. Proceedings of Machine Learning Research, vol. 139, pp. 4904–4916. PMLR (2021). https://proceedings.mlr.press/v139/jia21b.html
7. Kim, G., et al.: OCR-free document understanding transformer. In: European Conference on Computer Vision, pp. 498–517. Springer (2022)
8. Li, J., Li, D., Savarese, S., Hoi, S.: BLIP-2: bootstrapping language-image pretraining with frozen image encoders and large language models. In: Krause, A., Brunskill, E., Cho, K., Engelhardt, B., Sabato, S., Scarlett, J. (eds.) Proceedings of the 40th International Conference on Machine Learning. Proceedings of Machine Learning Research, vol. 202, pp. 19730–19742. PMLR (2023). https://proceedings.mlr.press/v202/li23q.html
9. Li, J., Li, D., Xiong, C., Hoi, S.: BLIP: bootstrapping language-image pre-training for unified vision-language understanding and generation. In: Chaudhuri, K., Jegelka, S., Song, L., Szepesvari, C., Niu, G., Sabato, S. (eds.) Proceedings of the 39th International Conference on Machine Learning. Proceedings of Machine Learning Research, vol. 162, pp. 12888–12900. PMLR (2022). https://proceedings.mlr.press/v162/li22n.html
10. Li, K., Vosselman, G., Yang, M.Y.: Convincing rationales for visual question answering reasoning (2025). https://arxiv.org/abs/2402.03896
11. Liu, H., Li, C., Wu, Q., Lee, Y.J.: Visual instruction tuning. In: Oh, A., Naumann, T., Globerson, A., Saenko, K., Hardt, M., Levine, S. (eds.) Advances in Neural Information Processing Systems. vol. 36, pp. 34892–34916. Curran Associates, Inc. (2023). https://proceedings.neurips.cc/paper_files/paper/2023/file/6dcf277ea32ce3288914faf369fe6de0-Paper-Conference.pdf

12. Mathew, M., Karatzas, D., Jawahar, C.: DocVQA: a dataset for VQA on document images. In: Proceedings of the IEEE/CVF Winter Conference on Applications of Computer Vision, pp. 2200–2209 (2021)
13. Mohammadshirazi, A., Neogi, P.P.G., Lim, S.N., Ramnath, R.: DLaVA: Document language and vision assistant for answer localization with enhanced interpretability and trustworthiness (2024). https://arxiv.org/abs/2412.00151
14. Powalski, R., Borchmann, Ł, Jurkiewicz, D., Dwojak, T., Pietruszka, M., Pałka, G.: Going Full-TILT boogie on document understanding with text-image-layout transformer. In: Lladós, J., Lopresti, D., Uchida, S. (eds.) ICDAR 2021. LNCS, vol. 12822, pp. 732–747. Springer, Cham (2021). https://doi.org/10.1007/978-3-030-86331-9_47
15. Radford, A., et al.: Learning transferable visual models from natural language supervision. In: Meila, M., Zhang, T. (eds.) Proceedings of the 38th International Conference on Machine Learning. Proceedings of Machine Learning Research, vol. 139, pp. 8748–8763. PMLR (2021). https://proceedings.mlr.press/v139/radford21a.html
16. Rezatofighi, H., Tsoi, N., Gwak, J., Sadeghian, A., Reid, I., Savarese, S.: Generalized intersection over union: a metric and a loss for bounding box regression . In: 2019 IEEE/CVF Conference on Computer Vision and Pattern Recognition (CVPR), pp. 658–666. IEEE Computer Society, Los Alamitos, CA, USA (2019). https://doi.org/10.1109/CVPR.2019.00075
17. Saifullah, S., Agne, S., Dengel, A., Ahmed, S.: DocXplain: a novel model-agnostic explainability method for document image classification. In: International Conference on Document Analysis and Recognition, pp. 103–123. Springer (2024)
18. Selvaraju, R.R., Cogswell, M., Das, A., Vedantam, R., Parikh, D., Batra, D.: Grad-CAM: visual explanations from deep networks via gradient-based localization. In: 2017 IEEE International Conference on Computer Vision (ICCV), pp. 618–626 (2017). https://doi.org/10.1109/ICCV.2017.74
19. Singh, A., et al.: FLAVA: a foundational language and vision alignment model. In: 2022 IEEE/CVF Conference on Computer Vision and Pattern Recognition (CVPR), pp. 15617–15629 (2022). https://doi.org/10.1109/CVPR52688.2022.01519
20. Xu, Y., Li, M., Cui, L., Huang, S., Wei, F., Zhou, M.: LayoutLM: pre-training of text and layout for document image understanding. In: Proceedings of the 26th ACM SIGKDD International Conference on Knowledge Discovery & Data Mining, pp. 1192–1200. KDD '20, ACM (2020). https://doi.org/10.1145/3394486.3403172
21. Zhou, Y., et al.: Large language models are human-level prompt engineers. In: The Eleventh International Conference on Learning Representations (2022)

Better Call Claude: Can LLMs Detect Changes of Writing Style?

Johannes Römisch[1], Svetlana Gorovaia[3], Mariia Halchynska[1],
Gleb Schmidt[2(✉)], and Ivan P. Yamshchikov[1]

[1] Center for Artificial Intelligence, Technical University of Applied Sciences
Würzburg-Schweinfurt, Münzstraße 12, 97070 Würzburg, Germany
[2] Humanities Lab, Faculaty of Arts, Radboud University, Houtlaan 4,
6525 Nijmegen, XZ, The Netherlands
gleb.schmidt@ru.nl
[3] LEYA Lab, School of Computer Science, Physics and Technology, HSE University,
6, 25th Liniya, Vasilievsky Ostrov, 199004 St Petersburg, Russia

Abstract. This article explores the zero-shot performance of state-of-the-art large language models (LLMs) on one of the most challenging tasks in authorship analysis: sentence-level style change detection. Benchmarking four LLMs on the official PAN 2024 and 2025 "Multi-Author Writing Style Analysis" datasets, we present several observations. First, state-of-the-art generative models are sensitive to variations in writing style—even at the granular level of individual sentences. Second, their accuracy establishes a challenging baseline for the task, outperforming suggested baselines of the PAN competition. Finally, we explore the influence of semantics on model predictions and present evidence suggesting that the latest generation of LLMs may be more sensitive to content-independent and purely stylistic signals than previously reported.

Keywords: AI-assisted authorship analysis · style change detection · large language models · authorship analysis · semantic similarity

1 Introduction

Style change detection is one of the most challenging problems within the broader field of authorship analysis, with numerous academic and industrial applications ranging from philological and historical research to anti-plagiarism, copyright protection, forensics, cybersecurity, and governance. It is hardly a coincidence, therefore, that specific instances of this problem—such as author diarization or multi-author style analysis—have been featured among the PAN shared tasks since 2016, longer than any other task.

Approaches to this task primarily relied on feature engineering and classical machine learning and, more recently, deep learning methods, especially pre-trained model based solutions. Despite the ever-increasing quality and performance of these models, developing and using such systems has become prohibitively expensive, requiring large amounts of rare labeled data and advanced

J. Carrillo-de-Albornoz et al. (Eds.): CLEF 2025, LNCS 16089, pp. 42–56, 2026.
https://doi.org/10.1007/978-3-032-04354-2_4

technical expertise. At the same time, these systems are often difficult to interpret—an important limitation for many use cases, particularly in academic research within the humanities and social sciences, as well as in forensic applications.

The release of a pleiade of LLMs since 2022 is believed to have shifted the paradigm. Trained on extensive corpora, these models can detect even subtle regularities in texts, resulting in impressive performance in text generation and a variety of other downstream tasks.

In the field of authorship analysis too, LLMs gave birth to entirely new fields of analysis, and the wave of enthusiasm among scholars suggests that LLMs are expected to soon bridge the efficiency of deep learning systems and the need for interpretability.

Aiming to contribute to a better understanding of LLMs as tools for authorship analysis and to encourage their adoption, we present an analysis of their performance on sentence-level style change detection—a subfield of authorship analysis in which the capabilities of LLMs have so far remained underexplored.

1.1 Task Definition

The core task of style change detection is to determine whether a given text was written by multiple authors and, if so, to identify the positions at which authorship changes occur[1].

1.2 Goals

The first goal of this article is to survey the zero-shot performance of state-of-the-art generative AI models on the task. Our second objective is to explore various factors that influence the models' predictions. We use the Hamming distance as a measure of response correctness and examine its correlation with parameters of the problems (e.g., number of authors, number of changes, length of the problem) and their semantic features.

2 Related Work

2.1 Style Change Detection

In early editions of PAN, the task was approached through manual and corpus-specific engineering of stylistic features subsequently used for unsupervised clustering of predefined text segments [7,13,22,27]. Since the late 2010s, methodologies shifted towards neural networks [2], and by 2023, transformer-based architectures dominated approaches presented at PAN.

Leveraging rich linguistic knowledge from pre-training on massive corpora, transformer backbones consistently achieve exceptional performance, regularly

---

[1] The task can be formulated at different levels of granularity (sentence, paragraph, etc.), but can also be presented as a clustering problem.

surpassing 80% accuracy even on challenging datasets with uniform topic signal. Apart from traditional classification [8,18], particularly popular strategies include contrastive learning [31,32,34], ensembling of models fine-tuned on various aspects of style.

Less conventional yet insightful methodologies also were presented by employing manually-constructed prompts with masked language modeling [37]; Gao et al. [4] openly "hacked" the task using extrinsic web clues.

Additionally, closely related tasks such as detecting AI-generated or Human-AI co-authored texts share these methodologies as well as boundary detection and segment-level attribution. For instance, [35] introduced a modular framework combining segment detection with sentence-level classification using various Transformer architectures [9,19,36,38], while GigaCheck [28] combined sentence-level binary classification with DETR-style character-level span prediction for detailed LLM attribution. Other recent approaches have exploited structural and attributional signals beyond superficial content, such as TopFormer's integration of topological features via Topological Data Analysis (TDA) [29], persistent homology derived from attention maps [15], and token-level log-probability dynamics aggregated through CNN and self-attention mechanisms [30]. These methods, especially those focused on stylometric shifts, segmentation, or local attribution, are readily adaptable to style change detection.

2.2 AI-Assisted Authorship Analysis

The surge of GenAI gave birth to a new field within authorship analysis—AI-assisted authorship analysis. New LLM-based methodologies emerged using LLMs for feature extraction, data augmentation, and direct stylistic analysis [11].

Relying on the robust abilities of LLMs to describe writing styles, [21] introduced a system producing interpretable style embeddings. Similarly, [23] used GPT-4-Turbo to generate structured stylistic descriptions as training data for a smaller Llama-3-8B model, enhancing interpretability.

Directly prompting LLMs with authorship-related queries, [10,26] report promising reliability in authorship verification, although the latter highlights potential bias due to semantic similarity in historical languages. Explicitly guiding models with linguistically-informed prompts (LIP) notably boosts performance and analytical quality, particularly in English texts [10]. The PromptAV framework similarly directs LLMs through stylometric reasoning for authorship tasks [12].

Our experiment thus aligns with broader explorations of LLM capabilities in authorship analysis, contributing specifically to the niche but increasingly relevant task of sentence-level style change detection.

3 Methodology

3.1 Zero-Short Prompting

Given the proven effectiveness of zero-shot prompting strategies in various tasks [14,24], including those involving complex reasoning and style-based predictions, we also adopted a zero-shot approach in our experiments, prompting the models with a "problem"—a sequence of sentences—and a task: predict style changes occurring in pairs of adjacent sentences (0–1, 1–2, etc.). Following the insights from the literature emphasizing that the carefully engineered prompts are essential for high-quality output [25], we tailored ours in two ways.

First, after initial experiments with minimal prompt, we tried to employ what can be described as "strategic guessing". We began by establishing random baselines[2] and carefully exploring the data to identify the strategies that could improve the models' odds independently of actual textual content. It turned out that hypothesizing 3 or 4 authors (that is, 2 or 3 style switches per problem) would cover most cases in the data (see Fig. 1 in Appendix B). Therefore, our prompt (see Appendix A) included an explicit recommendation to assume that there are "approximately 3" authors in the problem. We then trained an XGBoost classifier to predict the number of authors (see Appendix B) and injected its output into the prompt, but it overwhelmingly predicted three authors. As a result, there was no significant performance difference between the static prompt and the dynamic, classifier-driven prompt.

Table 1. Random and constant baselines.

Baseline	F1 (macro)
all changes	0.1570
no changes	0.4426
3 random changes	0.4946
4 random changes	0.4982

The second axis of our prompting strategy involved providing explicit instructions about the stylistic features expected to be useful for distinguishing between authors[3]. For contemporary and well-resourced languages, at least, such a strat-

---

[2] Random baselines consist of three trivial strategies: never predicting an author change, always predicting a change at every boundary, and randomly selecting a fixed number of change points (e.g. 1, 2, or 3). F1 macro was computed for each strategy, see Table 1.

[3] Our prompt contained the following addition: "Analyze the writing styles of the input texts, disregarding the differences in topic and content. Base your decision on linguistic features such as: phrasal verbs; modal verbs punctuation; rare words; affixes; quantities; humor; sarcasm; typographical errors; misspellings.".

egy has been reported to be beneficial [10][4], which builds on the conceptual framework outlined in [5].

3.2 Measure of "Correctness": Hamming Distance

To assess the factors influencing model predictions within a single problem, one needs a measure to quantify how closely the predicted sequence of author change labels matches the ground truth. Hamming distance is a metric used to measure the difference between two strings of equal length. It is defined as the number of positions at which the corresponding symbols differ. This makes it particularly suitable for evaluating the overall correctness of a model's output in sentence-level style change detection, where each sentence is assigned an author label. For easier interpretation and cross-problem comparison, we use the normalized Hamming distance, obtained by dividing the raw distance by the total number of sentences. A lower value indicates higher prediction accuracy, while a value of zero denotes a perfect match with the ground truth.

3.3 Measure of Semantic Similarity

With a clear indicator of overall prediction accuracy for individual problems, we further examine its correlation with various measures of semantic similarity within each problem. We used `sentence-transformers/all-MiniLM-L6-v2` and `styleDistance/styledistance` to vectorize the sentences and compute the semantic similarity and correlations. Specifically, we consider several related metrics:

1. Average cosine similarity of sentences within a problem;
2. average cosine similarity of pairs with an author switch;
3. average cosine similarity of adjacent pairs;
4. mean pairwise cosine distance within a problem.

3.4 Data

To ensure the comparability of our results with existing approaches, we rely on the widely recognized evaluation framework of PAN and used their official "Multi-Author Writing Style Analysis" datasets from 2024 and 2025. The data consists of lists of sentences (such lists are called "problems") and arrays of binary labels for each pair of adjacent sentences in the problem. The texts represent continuous discussions in Reddit threads. 2025 data is divided into sentences, while 2024 into paragraphs. Both datasets are served at three levels of difficulty: `easy`, `medium`, and `hard`, each `aligned` into `train` (70%), `validation` (15%), and `test` (15%) sets. In this study, all exploratory work is done on `train` set, while `validation` is only used for prompting LLMs.

---

[4] However, it remains an open question whether this approach retains its effectiveness when applied to historical languages [26]. With reservations, own research suggests the opposite.

3.5   Models and Experiments

Our experiment comprises three stages. In the first stage, we randomly sample 250 problems from the `validation` splits of three datasets and submit them as prompts to state-of-the-art models: `GPT-4o`, `Claude 3.7 Sonnet`, `DeepSeek-R1`, and `Meta-Llama-3.1-405B-Instruct`. In the second stage, we prompt the entire `validation` splits of each dataset (900 problems) to the best-performing model identified in the first stage. In the third stage, having obtained predictions from that model, we compute the Hamming distance, semantic similarity, and correlations as described above.

4   Results

The results of the first evaluation stage are summarized in Table 2. Among the evaluated models, `Claude-3.7-Sonnet` consistently outperformed its competitors—`GPT-4o`, `Deepseek-R1`, and `Llama-3.1-405B-Instruct`—across all difficulty levels and despite the distinct characteristics of the three datasets (see Appendix B). Even in the zero-shot setting, the model's reasoning allowed it to achieve competitive accuracy. Table 3 presents detailed performance metrics for the best-performing model, `Claude-3.7-sonnet`, evaluated on the full validation splits of all three datasets. To contextualize Claude's performance, we fine-tuned a `styleDistance/styledistance` transformer combined with an LSTM and MLP (127 m) for adjacent sentence classificationusing combined PAN 2024 and PAN 2025 training data[5]. Remarkably, Claude's zero-shot prompting performance nearly surpasses that of the fine-tuned transformer on the `medium` dataset and remains close behind on the other two. These results align with additional observations made on the PAN 2024 paragraph-level style-change dataset, we simply adapted the prompt by replacing the term "sentence" with "paragraph". There, Claude achieved a score of 0.618 on the `hard` dataset, outperforming not only baseline models but also four official submissions [1]. Unsurprisingly, Claude's results on `easy` and `medium` datasets were even more impressive, reaching 0.83.

Table 2. Performance (F1 macro) on 250 randomly sampled problems (validation split).

Model	Easy	Medium	Hard
`Claude-3.7-Sonnet`	0.8638	0.8412	0.6580
`DeepSeek-R1`	0.6332	0.6082	0.5263
`LLaMA-3.1-40B-Instruct`	0.6817	0.6609	0.5614
`GPT-4o`	0.5540	0.5557	0.5105

---

[5] For training we used the paragraph-level annotations to generate labeled pairs by splitting paragraphs into adjacent units, treating them as sentence-like segments. This allowed us to augment the data and obtain style-change labels between pairs.

Table 3. Performance of `claude-3.7-sonnet` on full validation data of the three tasks.

Model	Easy	Medium	Hard
`Claude-3.7-Sonnet`	0.8559	0.8182	0.6612
`StyleDistance + LSTM + MLP`	0.9231	0.8276	0.7240

5   Discussion

5.1   When Does Claude Fail?

Measuring Hamming distance allowed us to examine how prediction accuracy is influenced by various problem-level parameters. Table 4 presents correlations between Hamming distance and four parameters: the number of authors per problem, the number of actual author changes, the number of predicted changes, and the total number of sentences in the problem.

The number of predicted changes shows a strong positive correlation with Hamming distance across all three datasets. This indicates that the more style changes Claude predicts—whether correctly or not—the more likely it is to deviate from the ground truth. This suggests this tendency of the model to over-segment is a key source of error.

The number of actual style changes is only weakly correlated with Hamming distance, and significantly so only in the easy and medium datasets. This indicates that the intrinsic complexity of the text (in terms of true author switches) only slightly contributes to model confusion, suggesting that the local stylistic context may play a bigger role than the dynamics in the problem in general.

The medium dataset stands out as particularly challenging. The correlation between predicted changes and Hamming distance is strongest here, and the correlation with the number of authors is also positive and significant, unlike in the easy and hard sets. This suggests that the medium dataset, while not explicitly labeled as most difficult, may contain greater internal variance—particularly in the number and distribution of authors—leading to less reliable prediction.

In sum, these correlations seem to bring to light that prediction errors are less a function of the data complexity and more a result of the model's internal heuristics—especially its tendency to overpredict change points. These findings suggest that there probably remains a room for improvement of the prompt and the entire pipeline, in which the LLM will be a crucial but not only participant.

5.2   Correlation Between Style Change and Semantics

A comparison between correlations of semantic similarity with predicted and true authorship changes reveals a striking pattern. Importantly, note that a predicted value of 1 indicates a change of author, which is expected to correspond to low cosine similarity between adjacent sentences; conversely, a value of 0 (no change) should align with high similarity. Therefore, a negative correlation implies that model predictions are consistent with this expectation—the lower the similarity, the more likely the model predicts a change.

Table 4. Correlation between Hamming distance of predicted vs. actual style switches and problem-level parameters.

Dataset	Feature	Spearman	Pearson	Kendall
easy	num_authors	−0.056	−0.078	−0.045
	num_changes	−0.092	−0.104	−0.073
	num_changes_pred	**0.400**	**0.387**	**0.312**
	num_sentences	−0.098	−0.122	−0.099
hard	num_authors	0.048	0.036	0.037
	num_changes	0.056	0.051	0.043
	num_changes_pred	**0.337**	**0.303**	**0.257**
	num_sentences	−0.260	−0.277	−0.197
medium	num_authors	**0.168**	**0.141**	**0.134**
	num_changes	**0.178**	**0.176**	**0.134**
	num_changes_pred	**0.455**	**0.400**	**0.346**
	num_sentences	**0.140**	**0.132**	**0.083**

Across all models, we observe a moderate to strong correlation on the easy split, suggesting that stylistic shifts are often accompanied by semantic divergence in simpler cases. However, this correlation becomes much weaker or disappears entirely on the hard split, indicating that in more complex texts, stylistic changes do not always show as measurable differences in sentence embeddings, see Table 5.

Moreover, in the ground truth, the interesting trend is observed in harder cases: true authorship changes often occur between semantically similar sentences, possibly due to topic continuity or deliberate stylistic imitation.

Table 5. Spearman, Pearson, and Kendall correlation between sentence similarity and a switch prediction.

Dataset	Spearman	Pearson	Kendall
easy	**−0.239**	**−0.205**	**−0.195**
medium	**−0.160**	**−0.148**	**−0.131**
hard	**−0.117**	**−0.102**	**−0.096**

5.3 Generalizability of Suggested Approach

In our prompting strategy, we explicitly instructed the model to assume the presence of approximately three authors per problem. This decision was informed by a statistical analysis of the PAN datasets, which revealed that the majority of documents contain between two and four authors (see Appendix B, Fig. 1). However, we acknowledge that this heuristic is highly dependent on the specific distribution of the PAN data. When applying the same approach to corpora from other domains—such as literary texts, social media outside Reddit, or historical documents—it would be necessary to conduct a similar exploratory analysis to determine an appropriate prior. While effective in our case, such assumptions should be treated as dataset-specific and not be blindly generalized across tasks.

5.4   LLMs: A Baseline for PAN?

An important consideration in interpreting our results is the source of the data. The documents used in this task originate from Reddit, a platform whose content has probably been extensively represented in the training data of the LLMs. This raises the possibility that the models, despite being used as black boxes, may benefit from prior exposure to the domain and its stylistic conventions. However, rather than undermining our findings, this observation reinforces a central claim of our study: LLMs perform surprisingly well on the multi-author style change detection task in a zero-shot setting, relying solely on prompt-based guidance and without any task-specific fine-tuning. On this basis, we argue that LLMs should be regarded as a strong and competitive baseline for this task.

This observation, in turn, raises broader questions about task design. If the goal is to rigorously assess the generalization capabilities of LLMs, particularly in unfamiliar stylistic or genre contexts, future benchmarks may need to draw on curated or out-of-domain sources. Historical corpora, literary texts, or non-mainstream digital genres could provide more robust tests of the models' true stylistic sensitivity, independent of memorization or domain familiarity.

5.5   Hallucinations

LLMs are increasingly optimized to minimize hallucinations. Prominent models such as GPT and Claude incorporate internal safety mechanisms aiming at ethical alignment and factual reliability. While these safeguards are not always thoroughly documented in technical papers, they can be broadly understood as a set of training procedures and architectural decisions geared towards reducing harmful, irrelevant, or incorrect outputs.

A central component of these efforts is reinforcement learning from human feedback (RLHF) [16,20], which plays a critical role in aligning model behavior with human preferences, including truthfulness and ethical reasoning. This alignment process has been shown to reduce hallucinations and improve the model's ability to reference factual content with greater accuracy.

Recent research has begun to explore the extent to which LLMs can attribute their outputs to external sources—a capability closely related to hallucination minimization. For example, Gao et al. [3] introduce ALCE, a benchmark for evaluating citation quality in LLM-generated answers. ALCE combines metrics for fluency, factuality, and attribution across QA datasets. Similarly, AttrScore

[33] offers a framework for evaluating whether model-generated claims are supported by cited references, using classification labels such as attributable, extrapolatory, and contradictory. This approach integrates both prompting and fine-tuning strategies, and is validated on both synthetic and real-world Bing search outputs.

Further, a comprehensive survey by Li et al.[17] reviews existing techniques for tracing LLM outputs back to source materials, presenting an overview of available datasets, evaluation metrics, error typologies, and outstanding challenges in grounding generated content in verifiable evidence. Complementary to this, Guo et al.[6] propose a unified framework for automated fact-checking, encompassing claim detection, evidence retrieval, verdict prediction, and justification generation, along with a systematic review of associated resources.

It is reasonable to hypothesize that a model's attribution capability may correlate with its hallucination rate, with stronger grounding mechanisms contributing to reduced factual errors. However, validating this hypothesis is difficult due to several factors: the proprietary nature of leading LLMs, limited transparency in model architecture and training data, and the lack of standardized, robust benchmarks for hallucination detection in state-of-the-art systems. This remains a critical limitation for future research, especially in high-stakes applications requiring verifiability and trustworthiness.

A   Prompt Template Used for LLM Inference

Listing 1.1. LLM Prompt Template

```
PROMPT_TEMPLATE = """You are an expert in authorship
    attribution. You notice changes in writing style, topic
    and especially in tone and sentiment and use this
    information to complete your task.
Your task is to analyze a sequence of sentences and determine
    a binary sequence indicating where the author changes at
    sentence boundaries (1 = change, 0 = no change).
The input is a json formatted list of sentences, which can
    include quotes, which are escaped with backslashes.
There is always at least one change present.
Return your response in the following JSON format:
{ "changes": [<0s and 1s indicating sentence-boundary changes
    >] }

Example 1:
Input: ["Sentence one.", "Sentence two.", "Sentence three.",
    "Sentence four."]
Output: { "changes": [0, 1, 0] }

Example 2:
Input: ["This is written by author A.", "Another sentence by
    author A.", "Alright... now, a new author starts.", "Yet
    another author change here."]
```

```
Output: { "changes": [0, 1, 1] }

Now analyze the following text:"""

DYNAMIC_PROMPT_PART = (
    "There should be exactly {} sentences resulting in a
    changes array of lenght {}.

    Keep in mind, that the sentences originate from reddit,
        so consecutive sentences, that agree with each other
        do not always have the same author, if the tone or
        style changes. Still, consider that they may follow
        each other but always assume around 3 changes.
        Analyze the writing styles of the input texts,
        disregarding the differences in topic and content.
        Base your decision on linguistic features such as:
        phrasal verbs; modal verbs punctuation; rare words;
        affixes; quantities; humor; sarcasm; typographical
        errors; misspellings."
)
```

B   Data Exploration

We analyzed the distribution of authors per document, revealing a balanced composition primarily dominated by documents featuring two to four authors, thus typically containing one to three style changes. This finding informed our prompt design, as we explicitly indicated the expected number of style switches to guide model predictions.

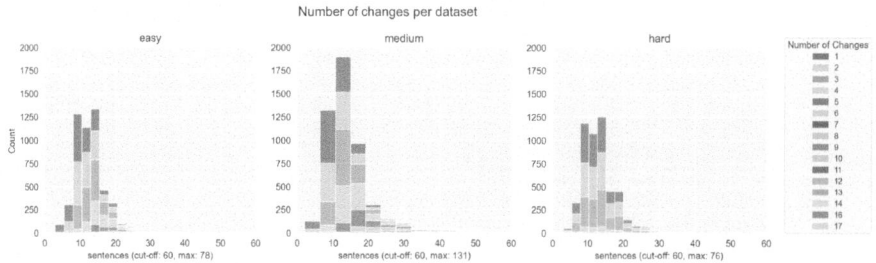

Fig. 1. Number of changes per dataset.

Notably, categorical labels representing the presence of either "1–2" or "3–4" authors per document could be predicted with reasonable accuracy using only superficial features. In our experiment, a general description of cosine similarity distributions between adjacent sentences within a problem together with a fixed-size binned representation of this distribution and basic stylistic features extracted via the `textstat` package allowed us to obtain reasonable results on `easy` and `medium` datasets.

Table 6 indicates significant potential for developing a system that provides models with highly reliable and precise information regarding the expected number of authors.

Table 6. Performance of `XGBoostClassifier` on the number-of-changes prediction tasks (as categorical variable). Mean over 100 stratified shuffled splits (20% of the data).

Dataset	Metric	1–2	3–4	5+
Easy	Precision	0.832	0.680	0.543
	Recall	0.818	0.716	0.437
	F1-score	0.825	0.698	0.484
	Support	845	598	87
Medium	Precision	0.738	0.581	0.677
	Recall	0.679	0.682	0.526
	F1-score	0.707	0.628	0.592
	Support	660	623	247
Hard	Precision	0.742	0.624	0.143
	Recall	0.861	0.446	0.071
	F1-score	0.797	0.520	0.095
	Support	985	531	14

Additionally, we computed pairwise cosine similarities between adjacent sentences, stratified by dataset difficulty levels. Mean cosine similarity generally decreased as the number of style changes increased, although this trend was less pronounced in the hardest dataset.

References

1. Ayele, A.A., et al.: Overview of PAN 2024: multi-author writing style analysis, multilingual text detoxification, oppositional thinking analysis, and generative AI authorship verification. In: Goeuriot, L., et al. (eds.) Experimental IR Meets Multilinguality, Multimodality, and Interaction. 15th International Conference of the CLEF Association (CLEF 2024). Lecture Notes in Computer Science, Springer, Berlin Heidelberg New York (2024)
2. Bagnall, D.: Authorship clustering using multi-headed recurrent neural networks (2016). https://doi.org/10.48550/ARXIV.1608.04485, https://arxiv.org/abs/1608.04485, publisher: arXiv Version Number: 1

3. Gao, T., Yen, H., Yu, J., Chen, D.: Enabling large language models to generate text with citations. In: Bouamor, H., Pino, J., Bali, K. (eds.) Proceedings of the 2023 Conference on Empirical Methods in Natural Language Processing, pp. 6465–6488. Association for Computational Linguistics, Singapore (2023). https://doi.org/10.18653/v1/2023.emnlp-main.398

4. Graner, L., Ranly, P.: An unorthodox approach for style change detection. In: CLEF (Working Notes), pp. 2455–2466 (2022)

5. Grant, T.: The Idea of Progress in Forensic Authorship Analysis. Elements in Forensic Linguistics, Cambridge University Press (2022)

6. Guo, Z., Schlichtkrull, M., Vlachos, A.: A survey on automated fact-checking. Trans. Assoc. Comput. Linguist. **10**, 178–206 (2022). https://doi.org/10.1162/tacl_a_00454, https://aclanthology.org/2022.tacl-1.11/

7. Gómez-Adorno, H., Aleman, Y., Ayala, D.V., Sanchez-Perez, M.A., Pinto, D., Sidorov, G.: Author Clustering using Hierarchical Clustering Analysis. In: CLEF (Working notes) (2017)

8. Hashemi, A., Shi, W.: Enhancing writing style change detection using transformer-based models and data augmentation. In: CLEF (Working Notes), pp. 2613–2621 (2023)

9. He, P., Liu, X., Gao, J., Chen, W.: DeBERTa: decoding-enhanced BERT with disentangled attention. In: International Conference on Learning Representations (2021). https://openreview.net/forum?id=XPZIaotutsD

10. Huang, B., Chen, C., Shu, K.: Can Large Language Models Identify Authorship? (2024). https://doi.org/10.48550/arXiv.2403.08213, arXiv:2403.08213 [cs]

11. Huang, B., Chen, C., Shu, K.: Authorship Attribution in the Era of LLMs: Problems, Methodologies, and Challenges (2025). https://arxiv.org/abs/2408.08946, _eprint: 2408.08946

12. Hung, C.Y., Hu, Z., Hu, Y., Lee, R.: Who wrote it and why? Prompting large-language models for authorship verification. In: Bouamor, H., Pino, J., Bali, K. (eds.) Findings of the Association for Computational Linguistics: EMNLP 2023, pp. 14078–14084. Association for Computational Linguistics (2023). https://doi.org/10.18653/v1/2023.findings-emnlp.937

13. Kocher, M., Savoy, J.: Author clustering with an adaptive threshold. In: Jones, G.J.F., et al. (eds.) CLEF 2017. LNCS, vol. 10456, pp. 186–198. Springer, Cham (2017). https://doi.org/10.1007/978-3-319-65813-1_19

14. Kojima, T., Gu, S.S., Reid, M., Matsuo, Y., Iwasawa, Y.: Large language models are zero-shot reasoners. In: Proceedings of the 36th International Conference on Neural Information Processing Systems. NIPS '22, Curran Associates Inc., Red Hook, NY, USA (2022)

15. Kushnareva, L., et al.: Artificial text detection via examining the topology of attention maps. In: Moens, M.F., Huang, X., Specia, L., Yih, S.W.t. (eds.) Proceedings of the 2021 Conference on Empirical Methods in Natural Language Processing, pp. 635–649. Association for Computational Linguistics, Online and Punta Cana, Dominican Republic (2021). https://doi.org/10.18653/v1/2021.emnlp-main.50

16. Lambert, N., Castricato, L., von Werra, L., Havrilla, A.: Illustrating reinforcement learning from human feedback (RLHF). Hugging Face Blog (2022). https://huggingface.co/blog/rlhf

17. Li, D., et al.: A Survey of Large Language Models Attribution (2023). https://arxiv.org/abs/2311.03731, _eprint: 2311.03731

18. Lin, T., Wu, Y., Lee, L.: Team NYCU-NLP at PAN 2024: integrating transformers with similarity adjustments for multi-author writing style analysis. Working Notes of CLEF (2024)

19. Lo, K., Jin, Y., Tan, W., Liu, M., Du, L., Buntine, W.: Transformer over pre-trained transformer for neural text segmentation with enhanced topic coherence. In: Moens, M.F., Huang, X., Specia, L., Yih, S.W.t. (eds.) Findings of the Association for Computational Linguistics: EMNLP 2021, pp. 3334–3340. Association for Computational Linguistics (2021). https://doi.org/10.18653/v1/2021.findings-emnlp.283

20. Ouyang, L., et al.: Training language models to follow instructions with human feedback. arXiv (2022), https://arxiv.org/abs/2203.02155

21. Patel, A., Rao, D., Kothary, A., McKeown, K., Callison-Burch, C.: Learning Interpretable Style Embeddings via Prompting LLMs (2023). http://arxiv.org/abs/2305.12696, arXiv:2305.12696 [cs]

22. Potthast, M., Rangel, F., Tschuggnall, M., Stamatatos, E., Rosso, P., Stein, B.: Overview of PAN 2017: author identification, author profiling, and author obfuscation. In: Jones, G.J.F., Lawless, S., Gonzalo, J., Kelly, L., Goeuriot, L., Mandl, T., Cappellato, L., Ferro, N. (eds.) Experimental IR Meets Multilinguality, Multimodality, and Interaction. 8th International Conference of the CLEF Initiative (CLEF 2017). Lecture Notes in Computer Science, vol. 10456, pp. 275–290. Springer, Berlin Heidelberg New York (2017)

23. Ramnath, S., Pandey, K., Boschee, E., Ren, X.: CAVE: Controllable Authorship Verification Explanations. arXiv preprint arXiv:2406.16672 (2024)

24. Reynolds, L., McDonell, K.: Prompt programming for large language models: beyond the few-shot paradigm. In: Extended Abstracts of the 2021 CHI Conference on Human Factors in Computing Systems. CHI EA '21, Association for Computing Machinery, New York, NY, USA (2021). https://doi.org/10.1145/3411763.3451760

25. Sahoo, P., Singh, A.K., Saha, S., Jain, V., Mondal, S., Chadha, A.: A systematic survey of prompt engineering in large language models: Techniques and applications (2025). https://arxiv.org/abs/2402.07927

26. Schmidt, G., Gorovaia, S., Yamshchikov, I.: Sui Generis: Large Language Models for Authorship Attribution and Verification in Latin. FL (Nov, Miami (2024)

27. Stamatatos, E., et al.: Clustering by authorship within and across documents. In: Working Notes Papers of the CLEF 2016 Evaluation Labs. CEUR Workshop Proceedings/Balog, Krisztian [edit.]; et al., pp. 691–715 (2016)

28. Tolstykh, I., Tsybina, A., Yakubson, S., Gordeev, A., Dokholyan, V., Kuprashevich, M.: GigaCheck: Detecting LLM-generated content (2024). https://arxiv.org/abs/2410.23728

29. Uchendu, A., Le, T., Lee, D.: TopFormer: Topology-Aware Authorship Attribution of Deepfake Texts with Diverse Writing Styles (2024). https://doi.org/10.3233/FAIA240647

30. Wang, P., Li, L., Ren, K., Jiang, B., Zhang, D., Qiu, X.: SeqXGPT: sentence-level AI-generated text detection. In: Bouamor, H., Pino, J., Bali, K. (eds.) Proceedings of the 2023 Conference on Empirical Methods in Natural Language Processing, pp. 1144–1156. Association for Computational Linguistics, Singapore (2023). https://aclanthology.org/2023.emnlp-main.73/

31. Wu, Q., Kong, L., Ye, Z.: Team bingezzzleep at PAN: a writing style change analysis model based on RoBERTa encoding and contrastive learning for multi-author writing style analysis. In: Faggioli, G., Ferro, N., Galuščáková, P., Herrera, A.G.S. (eds.) Working Notes Papers of the CLEF 2024 Evaluation Labs, pp. 2963–2968. CEUR-WS.org (2024). http://ceur-ws.org/Vol-3740/paper-288.pdf

32. Ye, Z., Zhong, C., Qi, H., Han, Y.: Supervised contrastive learning for multi-author writing style analysis. In: CLEF (Working Notes), pp. 2817–2822 (2023)

33. Yue, X., Wang, B., Chen, Z., Zhang, K., Su, Y., Sun, H.: Automatic evaluation of attribution by large language models. In: Bouamor, H., Pino, J., Bali, K. (eds.) Findings of the Association for Computational Linguistics: EMNLP 2023, pp. 4615–4635. Association for Computational Linguistics (2023). https://doi.org/10.18653/v1/2023.findings-emnlp.307
34. Zangerle, E., Mayerl, M., Potthast, M., Stein, B.: Overview of the multi-author writing style analysis task at PAN 2023. In: Aliannejadi, M., Faggioli, G., Ferro, N., Vlachos, M. (eds.) Working Notes of the Conference and Labs of the Evaluation Forum (CLEF 2023). CEUR Workshop Proceedings, vol. 3497, pp. 2513–2522 (2023). https://ceur-ws.org/Vol-3497/paper-201.pdf
35. Zeng, Z., et al.: Detecting AI-generated sentences in human-AI collaborative hybrid texts: challenges, strategies, and insights. In: Proceedings of the Thirty-Third International Joint Conference on Artificial Intelligence. IJCAI '24 (2024). https://doi.org/10.24963/ijcai.2024/835
36. Zeng, Z., et al.: Detecting AI-generated sentences in human-AI collaborative hybrid texts: Challenges, strategies, and insights (2024). https://arxiv.org/abs/2403.03506
37. Zhang, Z., Han, Z., Kong, L.: Style change detection based on prompt. In: CLEF (Working Notes), pp. 2753–2756 (2022)
38. Zhuang, L., Wayne, L., Ya, S., Jun, Z.: A robustly optimized BERT pre-training approach with post-training. In: Li, S., Sun, M., Liu, Y., Wu, H., Liu, K., Che, W., He, S., Rao, G. (eds.) Proceedings of the 20th Chinese National Conference on Computational Linguistics, pp. 1218–1227. Chinese Information Processing Society of China (2021). https://aclanthology.org/2021.ccl-1.108/

MedAID-ML: A Multilingual Dataset of Biomedical Texts for Detecting AI-Generated Content

Patrick Styll[1]([✉])[ID], Leonardo Campillos-Llanos[2][ID], Jorge Fernández-García[3][ID], and Isabel Segura-Bédmar[3][ID]

[1] TU Wien Informatics, Vienna, Austria
`patrick.styll@tuwien.ac.at`
[2] ILLA, CSIC, Madrid, Spain
`leonardo.campillos@csic.es`
[3] Universidad Carlos III de Madrid, Madrid, Spain
`isegura@uc3m.es`

Abstract. Improper use of synthetic medical texts created with generative artificial intelligence (AI) poses the risk of spread of fake content and disinformation. Detection of AI-generated text is a need for protecting non-specialist readers from manipulation. We introduce MedAID-ML, a multilingual dataset of biomedical texts for the detection of AI-generated text and authorship attribution. We gathered human-written texts in English, French, German and Spanish from authorized medical sources, and we generated artificial counterparts using three large language models: GPT-4o, MISTRAL and LLaMA-3.1. The current version includes 50% AI-generated and 50% human-written texts, and amounts to 13292 documents (3795449 tokens). We endeavoured to prevent data leakage by gathering a dedicated test set composed exclusively of texts published in 2025 and generated texts using model checkpoints from 2024. We present results from human evaluations and baseline experiments using a statistical classifier and state-of-the-art multilingual language models. By comparing performance metrics on test sets with and without potential data leakage, we provide evidence that prior work in this area might have reported inflated metric scores. In addition, we applied Integrated Gradients and SHAP to analyze model behavior to see whether classification decisions rely on linguistic markers specific to LLMs (https://github.com/Padraig20/MedAID-ML).

Keywords: AI-Generated Text · Biomedical Natural Language Processing · Generative AI · Explainable AI

1 Introduction

The advancement of Large Language Models (LLMs) has boosted the productivity of text-related tasks, with the counterpart of enabling any *novice* user to easily generate realistic synthetic texts—even for improper use. In a highstakes domain such as medicine, texts created with LLMs pose risks to non-specialists readers. These include

J. Carrillo-de-Albornoz et al. (Eds.): CLEF 2025, LNCS 16089, pp. 57–73, 2026.
https://doi.org/10.1007/978-3-032-04354-2_5

the spread of fake medical content and malicious influence campaigns, identity theft of content authorship, publication of AI-written articles in scientific journals, health disinformation and manipulation of public opinion [6,7,31,41].

The academic and societal consequences of synthetic text have increased the interest in research on automatic approaches to distinguish human-written and AI-generated text. Still, most research has focused on web content and English data from the general domain [36,42]. Although multilingual corpora are available [34], they lack parallel translations or rely on machine translations. Moreover, previous data collection efforts have not adequately paid attention to the risk that evaluation data may overlap with training data [28]. LLM providers do not transparently disclose their training datasets, making it likely that the available web contents have already been included in model training. Research has highlighted *data leakage* as a problem in LLM evaluation [2].

In this context, we present MedAID-ML (Medical Artificial Intelligence text Detection in Multilingual settings), a dataset of human-written and AI-generated medical texts in English, German, French and Spanish. To our knowledge, no prior research explored the automatic detection and authorship attribution of AI-generated texts in the biomedical domain using parallel and comparable texts across these languages.

Our contributions include:

- A new multilingual dataset of biomedical texts sourced from publicly available authorized sources, encompassing 3202149 tokens within 10792 documents for training, and 593300 tokens within 2500 documents for testing.
- An evaluation that prevents *data leakage* from the LLMs by verifying that texts in the test data (i) were not automatically generated, and (ii) were published after the release of the LLMs we used for generating artificial texts.
- Baseline experiments for the detection of AI-generated texts using both simple statistical classifiers and state-of-the-art multilingual language models. In addition, we compare model performance with human evaluation.
- An explainable AI (XAI) analysis of the models' reliance on linguistic markers of AI-generated texts.

2   Related Work

Several endeavors have curated datasets specifically for training and evaluating models for AI-generated text detection [5] and authorship attribution [18]. Domains encompass academic texts [4,24], news [38,42], web texts [36] or creative writing and student essays [17,20,40]; and some datasets gather heterogeneous types of text [15,22]. There have also been efforts to identify AI-generated content in technical research articles across scientific domains [19,26,31]. Comprehensive analyses of language have shed light on the features that help recognize AI-generated texts [8]. Tools such as GLTR [13] aim at detecting artifacts found in synthetic texts.

In terms of multilingualism, few resources are available for other languages beyond English, although there are exceptions in Chinese [15], Dutch [9] and Russian [35]. The AuTexTification 2023 and 2024 datasets [33,34] gather data in English, Spanish, Catalan, Galician, Basque and Portuguese across five domains (legal documents, news articles, reviews, tweets and how-to articles). In addition, the MULTITuDE dataset covers a total of 11 languages from articles available in the MassiveSumm dataset [39], providing more information on how LLMs behave in different languages [27].

Still, no dataset exists with (bio)medical texts with parallel data in the four languages considered. MedAID-ML fills a gap on research on AI-generated text detection of multilingual medical contents. In addition, our collection method tried to avoid *data leakage*, which enables to examine whether this affected former studies in future work.

3   Methods

Figure 1 summarizes our approach. We collected human-authored contents from reliable medical sources (§3.1), then we generated artificial counterparts by means of three open-source and commercial LLMs (§3.2), but only versions from 2024. We split the data a into a train and a test set, and restricted the data in the held-out set to texts published in 2025 to avoid *data leakage*. For baseline experiments, we tested a total of seven approaches to detect whether texts were human-authored or AI-generated (§3.3). Lastly, we conducted an explainable AI (XAI) analysis to investigate how classifiers distinguish between AI-generated and human-written texts (§ 3.4), and we further evaluate human performance in identifying AI-generated content (§ 3.5).

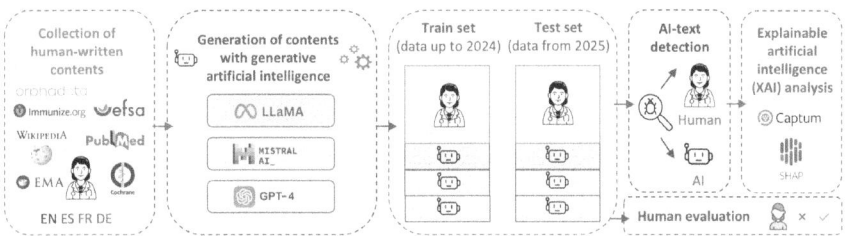

Fig. 1. Steps for dataset creation and model evaluation.(source of icons: The Noun Project)

3.1   Data Collection

We scraped human-authored corpora from public authorized health websites where parallel or comparable texts were available in the four languages:

- **Cochrane**: a database of meta-analyses and systematic reviews of updated results of clinical studies. We used abstracts of systematic reviews in all four languages.
- **European Clinical Trials (EUCT)**: we used protocols issued in 2025, which makes it unlikely to have been used to train the LLMs for our experiments.
- **European Food Safety Authority (EFSA)**: it gathers information about food- and chemical-related diseases; we used human disease topics in all target languages.
- **European Medicines Agency (EMA)**: we downloaded reports from new medicinal products released in the four languages from January 2025.

- **European Vaccination Information Portal (EVIP)**: this website provides information on vaccines, and disease factsheets in all target languages.
- **German Ministry of Health (BfG)**: the *Migration und Gesundheit* portal provides health information for migrants and refugees in our four working languages.
- **Immunize.org**: this organization provides vaccine educational information; we used the Vaccine Information Statements (VIS) for all relevant languages.
- **Orphadata (INSERM)**: Orphanet curates a multilingual knowledge base of rare diseases and orphan drugs; we gathered definitions about 4200 rare diseases.
- **PubMed (National Library of Medicine)**: we scraped abstracts from medical journals that were available in all target languages.
- **Wikipedia**: we collected biomedical articles, ensuring they were published before the release of ChatGPT 3.5 (November 2022) to avoid AI-generated content.

After initially scraping the data, manual and automated cleaning of the data was necessary, which involved several steps to ensure clean text (e.g., removing leading and trailing whitespace characters, extra spaces or newline characters). We followed recommendations by previous research [14] and the observation that very short texts are difficult to classify in this task [40]. Thus, we removed all texts shorter than 100 words. In order to not bias classifiers to long/short texts, very long texts (i.e., ≥ 1350 words, which is 50% of the originally largest text in the dataset) have been either removed or split into two distinct texts if the individual parts are still semantically coherent. When splitting or removing documents, we made sure that documents still remained parallel among languages.

For Wikipedia articles, each section was treated as a separate text. To provide context, the title of the entire article, along with the section title, was placed at the beginning of the text. Since the number of sections varies across languages, only sections with roughly the same contents were preserved to ensure parallel topics.[1] As for Orphadata, each text contains a medical term (rare disease) with one or two sentences as a definition with the same format (`<term>: <definition>`). To match the length of other texts in the dataset, we grouped five randomly selected term-definition pairs into each document. For detailed information on the data see Appendix A.2, Table 5.

3.2 Data Generation

Following the approach by [33], we generated texts by means of LLMs and using as a basis the human-written content previously collected. We first partitioned each human-authored text $\mathcal{H}$ into two parts with an equal number of sentences: a prefix ($\mathcal{H}_p$) and a continuation ($\mathcal{H}_c$), segmenting sentences with spaCy.[2] We chose to split documents at the sentence level, rather than at the token level as in [33], to avoid unnatural texts. The prefix $\mathcal{H}_p$ was then used as a prompt for an LLM (Mistral, GPT-4o and LLAMA-3.1) to generate a synthetic continuation $\mathcal{G}_c$, resulting in a AI-generated text represented as $\mathcal{G} = \mathcal{H}_p \oplus \mathcal{G}_c$. To ensure the generated text closely mimicked human writing, we used nucleus sampling with p uniformly sampled between 0.9 and 1.0, as [42] demonstrated that AI-generated text detection is most difficult within this parameter range.

---

[1] Thus, Wikipedia texts are not always parallel in our data.

[2] https://spacy.io/.

Similarly to [27,33], we observed that LLMs often hallucinated, providing outputs that were not in the target language. We thus applied the langdetect[3] language identifier on sentence level for all $\mathcal{G}$—if one sentence was not in the target language, the specific document was re-generated until it passed all filters. Unfinished sentences at the end of the text were removed, and we left the previous sentences.

The final dataset consists of 50% human-generated and 50% AI-generated text. For each human prefix, one LLM has been applied to obtain an AI-generated counterpart. AI-generated texts are equally distributed among MISTRAL (`mistral-7b-instruct`), LLAMA-3.1 (`llama3.1-70b`), and GPT-4o (`gpt-4o-2024-08-06`), each contributing one-third of the total 50% of artificial texts.

We realized that, despite prompting the LLMs to keep the length of $\mathcal{G}_c$ similar to $\mathcal{H}_p$ (see Appendix A.1), both $\mathcal{H}$ and $\mathcal{G}$ were often quite dissimilar in length. In fact, a classifier taking only the word-count into consideration would achieve an accuracy of 67.3% for detecting AI-generated documents on Llama-generated training data. Therefore, we ensured that both $\mathcal{H}_c$ and $\mathcal{G}_c$ were of comparable length by iteratively removing sentences from the longer segment of $\mathcal{H}_c$ or $\mathcal{G}_c$ until the length difference was reduced to 10%. As a result, the performance of the mentioned word count-based classifier was reduced to 53.9%.

The dataset includes 13292 texts for four languages—English (3346 texts), French (3234), German (3418) and Spanish (3294)—with 50% being human-written and 50% AI-generated. Table 5 in the Appendix reports the descriptive statistics of the dataset.

3.3 Baseline Experiments for AI-Text Detection

We report baseline experiments by using statistical methods as well as pre-trained language models that were highly performant in former works [11,34].

First, we used as a baseline a traditional statistical classifier that was trained with features such as: average sentence length, average document length, average word length, lexical diversity (type-token ratio, TTR), stopword ratio, punctuation density and the average inverse document frequency (IDF). We experimented with several traditional machine learning models, and finally chose a random forest classifier, since it obtained the best results.

We also used encoder transformers: XLM-RoBERTa (`xlm-roberta-base`) [23], Multilingual BERT (`bert-base-multilingual-cased`) [32], and multilingual DeBERTa (`mdeberta-v3-base`) [16].

We also used a decoder transformer, GPT-2 [30] (124M parameters). In addition, we experimented with Medical mT5 (`Medical-mT5-large`) [12], an encoder-decoder transformer trained on medical texts for English, Spanish, French and Italian.

Furthermore, we benchmarked Fast-DetectGPT [3], a zero-shot classifier that improves upon DetectGPT [29] in both speed and performance. We used `falcon-7b` [1] as the backend, which is reported in the original article to yield the best performance for AI-generated text identification. Lastly, we tested an ensemble

---
[3] https://tinyurl.com/langdetect.

that combines the output of all transformer models and aggregates them with a majority voting approach.

We run the experiments on a Tesla T4 GPU from Google Colab by applying five different initialization seeds with the same hyperparameters (see Table 4 in Appendix A.2). The dataset was split into three parts: 70% for training, 15% for basic hyperparameter tuning, and 15% for validation. In addition, we evaluated performance on a separate test set, in which we sought to avoid *data leakage*; this test set only gathers texts published in 2025, after the release of the LLMs we used to generate artificial texts. This allows for comparison between models exposed to potential *data leakage* and those tested on entirely unseen data. We report the average macro precision, recall, F1 score and the standard deviation of the five experimental rounds.

3.4 Explainability

We attempted to shed more light on how a model makes its classification decisions to discriminate between AI-generated and human-written content. We hypothesize that certain phrases (*AI markers*) may lead the model to classify a text as AI-generated. To analyze this, we computed token attributions using Integrated Gradients [37] (implemented in the Captum library [21]) on an n-gram level ($n \in \{1, 5, 10\}$) to capture local contextual contexts. We manually examined the top 200 extracted excerpts per language from the test set. Subsequently, we determined and recorded the frequency of *AI marker phrases* in AI-generated versus human-written texts. Finally, we used SHAP [25] to generate local explanations for correctly classified AI-generated texts to assess whether the classifiers indeed rely on the previously identified *AI marker phrases* in practice. For all attribution and explanation analyses, we used a multilingual BERT model as described in § 3.3, as it outperformed other BERT-based models in our evaluations.

3.5 Human Evaluation

We conducted an assessment on the capabilities of human evaluators to discriminate human-written and AI-generated texts. Eight evaluators with a background in biology or medicine (two per language) were provided with 60 texts (six from each source, 30 written by humans and 30 by LLMs, with an equal part for each LLM). Evaluators were tasked to label each text as human or machine, and did not know how texts were created. These results are a baseline to compare human skills with automatic approaches.

4 Results

Table 1. Results on the test set (mean and standard deviation of 5 experimental rounds).

Model	Precision	Recall	F1	Accuracy
Random Forest	$0.62_{\pm 0.01}$	$0.62_{\pm 0.01}$	$0.62_{\pm 0.01}$	$0.62_{\pm 0.01}$
Multilingual-BERT	$0.77_{\pm 0.01}$	$0.75_{\pm 0.02}$	$0.75_{\pm 0.02}$	$0.75_{\pm 0.02}$
XLM-RoBERTa	$0.65_{\pm 0.23}$	$0.62_{\pm 0.17}$	$0.62_{\pm 0.17}$	$0.67_{\pm 0.10}$
mdeberta-v3-base	$0.74_{\pm 0.01}$	$0.65_{\pm 0.02}$	$0.65_{\pm 0.02}$	$0.67_{\pm 0.01}$
GPT-2	$0.69_{\pm 0.01}$	$0.68_{\pm 0.01}$	$0.68_{\pm 0.01}$	$0.68_{\pm 0.01}$
Medical mT5	$0.60_{\pm 0.13}$	$0.55_{\pm 0.15}$	$0.55_{\pm 0.15}$	$0.59_{\pm 0.12}$
Fast-DetectGPT	$0.77_{\pm 0.00}$	$0.77_{\pm 0.00}$	$0.77_{\pm 0.00}$	$0.77_{\pm 0.00}$
Ensemble Transformer	$\mathbf{0.80_{\pm 0.01}}$	$\mathbf{0.78_{\pm 0.02}}$	$\mathbf{0.78_{\pm 0.02}}$	$\mathbf{0.78_{\pm 0.02}}$

Overall, the ensemble transformer achieved the highest performance, closely followed by Fast-DetectGPT and the multilingual variant of BERT (Table 1). Despite being pretrained on medical texts, the Medical mT5 model obtained worse results (even below the Random Forest classifier), contrary to what we expected. A Kruskall-Wallis test showed a statistical difference in the F1 score of the models on the test set without data leakage ($p < 0.0001$). However, a Dunn's multiple comparison test showed that only the differences between certain models were significant: mDeBERTa-v3 vs. FastDetect-GPT and vs. Ensemble; and medical MT5 vs. FastDetect-GPT and vs. Ensemble.

Regarding the differences across languages, English generally yielded the best results (Fig. 4). In contrast, variance with respect to the LLM used to generate the synthetic texts is negligible. We observe a considerable drop in accuracy for human-written documents, which suggests that these are frequently misclassified as AI-generated.

As for the human evaluation (Table 2), reviewers' performance was rather poor (close to the random guess), which shows that automatic methods were more effective.

We also identified a set of linguistic cues that are indicative of AI-generated text (see Table 3). These phrases were assigned high attribution scores by Integrated Gradients and were found to occur more frequently in AI-generated texts compared to human-written ones. As seen in Fig. 2, local SHAP explanations for correctly classified LLM-generated documents reveal that classifiers might tend to rely on these markers.

Table 2. Results of the human evaluation for 8 different reviewers, 2 for each language.

Language	Reviewer 1			Reviewer 2			Overall
	Correct	Incorrect	Accuracy	Correct	Incorrect	Accuracy	
English	30	30	50.0%	32	28	53.3%	51.6%
German	34	26	56.7%	35	25	58.3%	57.5%
French	41	19	68.3%	34	26	56.6%	62.5%
Spanish	31	29	51.7%	40	20	66.7%	59.2%
Mean Overall Accuracy							**57.7%**

Since 2010, genomic selection in Swiss cattle breeding has led to significant progress in the two local Braunvieh populations, Brown Swiss and Original Braunvieh. However, it has also contributed to further inbreeding, leading to an increased risk of outbreaks of monogenic recessive defects. Some long-known inherited diseases such as arachnomelia, spinal muscular atrophy, spinal dysmyelination, Weaver syndrome and renal dysplasia are now of little clinical importance, while new haplotypes and gene variants associated with reproductive disorders have recently been described. These include so-called fertility haplotypes and genetic diseases that rarely or never occur homozygous because the affected animals die shortly after birth or early in pregnancy. This work provides an overview of the known recessive genetic defects in Swiss Braunvieh cattle. The identification of these defects is crucial for managing genetic diversity and ensuring healthy cattle populations. Genomic tools have allowed breeders to screen for these haplotypes and reduce their frequency within the populations. Breeding programs are increasingly incorporating genomic information to make informed decisions that balance genetic gain with the control of inbreeding. As technology advances, there is a potential to further refine these strategies and develop novel solutions for maintaining genetic health. Continued research and collaboration among breeders, geneticists, and veterinarians are essential to address these challenges effectively.

Welche Risiken sind mit Siiltibcy verbunden? Siiltibcy kann eine Reihe von Nebenwirkungen verursachen, darunter Übelkeit, Kopfschmerzen und Schwindel. In einigen Fällen können auch schwerwiegendere Risiken wie Leberschäden oder Herzrhythmusstörungen auftreten. Allergische Reaktionen sind selten, aber möglich, und erfordern sofortige ärztliche Hilfe. Es ist wichtig, vor der Einnahme von Siiltibcy mit einem Arzt über alle potenziellen Risiken und Wechselwirkungen zu sprechen.

la indicación de vertebrectomía total en metástasis vertebral única está cada vez más cuestionada, especialmente en metástasis con tumor primario desconocido. en este artículo describimos un caso de metástasis vertebral de primario desconocido tratado mediante vertebrectomía total y se presenta una revisión de la literatura. A pesar de los avances en técnicas quirúrgicas, los riesgos asociados a la cirugía compleja siguen siendo significativos. Es crucial evaluar cuidadosamente los beneficios frente a los riesgos en pacientes con metástasis vertebral única.

La prise en charge des pathologies des voies respiratoires supérieures est caractérisée par le partage des voies aériennes supérieures entre les équipes anesthésiologique et chirurgicale. De nombreuses stratégies ont été développées afin de garantir une oxygénation et/ou une ventilation efficace mais doivent constamment être adaptées à l'évolution des impératifs chirurgicaux. Les techniques incluent l'utilisation de dispositifs spécialisés pour assurer la perméabilité des voies aériennes. Une communication étroite entre les équipes est essentielle pour prévenir les complications et assurer la sécurité du patient.

Fig. 2. Excerpts from local SHAP explanations of correctly classified LLM-generated text show that classifiers rely some linguistic markers (see Table 3).

5 Discussion

In this study, we evaluated the performance of several classifiers for distinguishing AI-generated from human-written texts in the bio-medical domain, with a focus on multilingual settings. In our experiments, the ensemble outperformed the other models; however, Fast-DetectGPT was slightly below its scores, and it would be a reasonable choice for this task in terms of performance and computational costs. Multilingual-BERT performed slightly below Fast-DetectGPT. However, Medical mT-5 did not yield performant results and even lied below the Random Forest classifier (see Fig. 1), despite supporting most target languages (with the exception being German) and being specific to the (bio)medical domain. The reason for this is most likely the sub-optimal architecture; text-to-text transformers that are directly used for classification often yield worse results than models with a dedicated classification head (like BERT or RoBERTa) [10].

As shown in Fig. 4, English documents were classified correctly more frequently than those in other languages, with the exception being the statistical baseline classifier. This is a direct consequence of the fact that most language models are pre-trained

Table 3. Language markers and attribution scores of the n-grams in which they appear. The frequency ('Freq') is the ratio of a phrase to the total number of texts containing it.

English			German		
Phrase	Score	AI % Freq.	Phrase	Score	AI % Freq.
is crucial	0.269	92.2% $95/103$	*ist wichtig*	0.199	94.5% $121/128$
moreover	0.090	84.0% $37/44$	*ist entscheidend*	0.287	94.1% $48/51$
is essential	0.264	80.0% $92/115$	*zudem*	0.199	68.3% $112/164$
challenge	0.271	78.3% $108/138$	*problem*	0.269	65.2% $1269/1946$
ensur[e]	0.279	77.6% $267/344$	*erfordert*	0.363	64.1% $233/363$

French			Spanish		
Phrase	Score	AI % Freq.	Phrase	Score	AI % Freq.
est essentiel	0.411	91.7% $133/145$	*es crucial*	0.272	91.1% $92/101$
ouv[re | rir] la voie [à]	0.241	81.4% $22/27$	*instrucciones*	0.234	72.7% $24/33$
comprendre	0.294	80.7% $130/161$	*importancia*	0.284	67.7% $86/127$
necessite	0.169	75.4% $156/206$	*importante*	0.290	66.4% $681/875$
mieux	0.251	71.8% $74/103$	*especialmente*	0.222	62.0% $237/382$

on large-scale English corpora, while many other languages remain under-resourced in NLP. As a result, performance in non-English languages often lags behind due to limited representation in the training data. Furthermore, we observed only minimal performance differences across source LLMs used to generate synthetic texts. The exception is Fast-DetectGPT, which achieved its best results on documents generated by MISTRAL, while showing reduced accuracy on texts produced by larger models such as LLaMA 3.1 and GPT-4o. Given that Fast-DetectGPT is the only zero-shot classifier in our experiments, obtaining these results might suggest that detecting outputs from more powerful LLMs is a greater challenge when not provided with task-specific training.

As described in § 3.3, we evaluated models both on a validation set as well as a held-out test set, where we attempted to avoid the *data leakage* problem. In Fig. 3, we can see a drop in performance for the test set in relation to the validation set of about 5–10%, indicating that models might have indeed suffered from *data leakage*. Therefore, we show some evidence that prior experiments might have reported inflated results. Trivially, this does not apply to the statistical baseline classifier, which suffers only a negligible loss of performance of roughly 2%. Interestingly, Fast-DetectGPT did not suffer from this loss in performance and, in fact, exhibited an increase of roughly 4%, possibly due to its zero-shot classification setup. Nonetheless, the consistent decline in performance observed in the other models seems to indicate that *data leakage* remains a critical issue in this task.

Our preliminary XAI analysis of AI-generated texts showed that sentences from the middle to the end (i.e. $\mathcal{G}_c$) generally had stronger SHAP scores. Interestingly, the analysis revealed a set of linguistic cues that seem characteristic of AI-generated texts

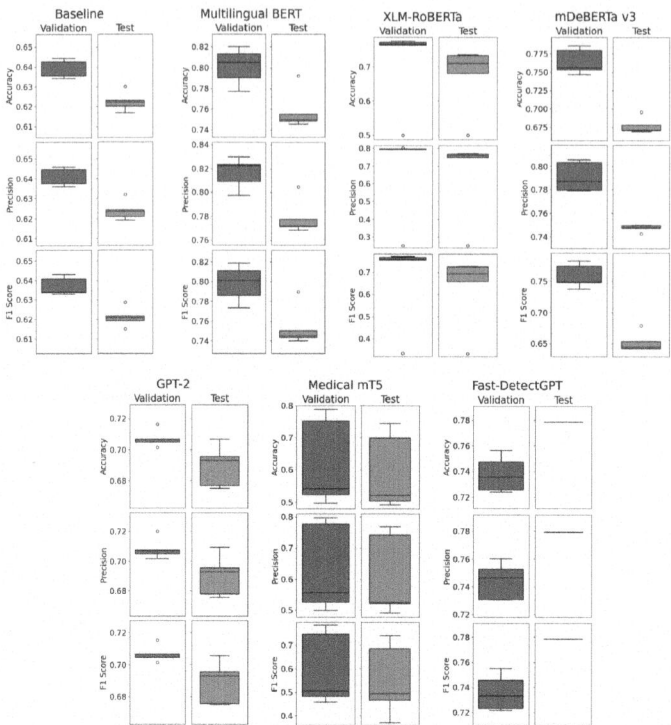

Fig. 3. Average model performances across 5 different seeds between the validation set (blue), and the separate test set (orange), designed to avoid data leakage.

(Table 3). Notably, we found that some of these phrases were the same across all four target languages (e.g., *is crucial*, *ist wichtig*, *est essentiel* and *es crucial*). By analyzing local SHAP explanations of correctly classified LLM-generated documents (Fig. 2), we found supporting evidence for our hypothesis that classifiers indeed rely on a certain number of marker phrases when identifying AI-generated content.

5.1 Limitations

We intentionally confined to use only LLMs released in 2024 and constructed the test set using texts published in 2025. This was done to minimize the risk that the LLMs had seen the test content during training and therefore leading to *data leakage*. Furthermore, we used OpenAI's GPT-4o API to generate artificial texts in order to avoid the model from being fed with the contents of the prompts. Even so, given that the sources of the training data are not completely disclosed for all LLMs we used, we cannot totally guarantee that no *data leakage* had occurred in a surreptitious way. Nonetheless, the restricted criteria we used are fine contributions towards a more solid, grounded evaluation of this task. To the best of our knowledge, it has not been applied in similar works. Moreover, this dataset paves the way to test current approaches to LLM authorship verification, which was also out of the scope of the present work.

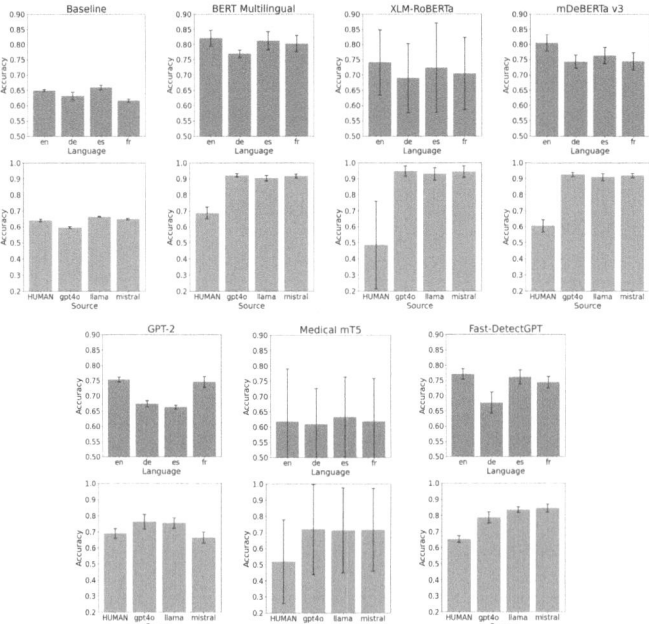

Fig. 4. Average model performances with standard deviation across 5 different seeds, grouped by language (blue) and the LLMs used to generate texts (red).

5.2 Ethical Considerations

Given the sensitive nature of our data, we are not releasing the complete AI-generated texts openly. The correctness of the contents were not checked by health professionals, and texts may contain misleading information or hallucinations. We thus avoid the uncontrolled dissemination of misinformation and potential misuse. Researchers who are interested in our dataset can request access via our institutional repositories.[4]

6 Conclusion

Automatic detection of AI-generated texts can help protect readers from fake content in synthetic text. For this task in the (bio)medical domain, we presented MedAID-ML, a dataset in English, French, German, and Spanish with human-written and AI-generated texts. To our knowledge, no similar resource exists with parallel contents and with a collection protocol to avoid the *data leakage* problem. We reported baseline experiments in which we achieved up to $F1 = 0.78$. In addition, an XAI analysis provided evidence that classifiers rely on certain linguistic cues which are indicative of AI-generated texts.

---

[4] https://digital.csic.es/handle/10261/389309
https://researchdata.tuwien.ac.at/records/xr34x-tsp12.

Acknowledgments. Isabel Segura's work was conducted within the HUMAN_AI project (PID2023-148577OB-C21), funded by MICIU/AEI/10.13039/501100011033 and by FEDER/UE. We greatly thank the eight annotators who conducted the human evaluation.

Disclosure of Interests. The authors have no competing interests to declare.

A   Appendix

A.1   LLM Prompts

Text Continuation

You are a helpful assistant generating text in the necessary language. Make sure to only return your own generated text.
Continue this text with **{num_sents}** sentences in **{language}**. Return only raw text, without numbering, lists, or newline characters: **{text}**

Generating Definitions from Terms

You are a helpful assistant generating text definitions in the necessary language. Make sure to only return your own generated text. You are given **{num_defs}** disease definitions in **{language}**.
{text}
Give me **{num_new_defs}** definitions for the following diseases in the same format, return only raw text, without enumerations etc: **{definitions}**

Generating Answers for Questions

You are a doctor answering questions about a specific drug. Make sure to only return your own generated text.
Answer this question about the drug '**{drug_name}**' with **{num_sents}** sentences in **{language}**.
Return only raw text, without numbering, lists, or newline characters: **{question}**

A.2   Experimental Details and Data Statistics

Table 4. Hyperparameters to fine-tune all models.

Batch size	32	Optimizer	Adam
Learning rate	2e-5	Patience	5
Maximum epochs	3	Seeds	1, 2, 3, 4, 5

Table 5. Descriptive statistics (*avg*: 'average'; *SD*: 'standard deviation')

	Source	Human Tokens	Human Avg (SD)	GPT4 Tokens	GPT4 Avg (SD)	LLAMA Tokens	LLAMA Avg (SD)	MISTRAL Tokens	MISTRAL Avg (SD)	Files
EN (train)	Abstracts	9758	$238.00_{\pm112.66}$	3354	$239.57_{\pm125.44}$	2863	$260.27_{\pm121.19}$	3624	$226.50_{\pm86.31}$	82
	BfG	8558	$237.72_{\pm68.65}$	3685	$245.67_{\pm46.02}$	2469	$274.33_{\pm89.18}$	2556	$213.00_{\pm63.61}$	72
	EFSA	28913	$473.98_{\pm211.61}$	10764	$398.67_{\pm165.65}$	6508	$433.87_{\pm178.71}$	11013	$579.63_{\pm238.26}$	122
	EVIP	8101	$405.05_{\pm120.13}$	1649	$329.80_{\pm50.37}$	4658	$423.45_{\pm120.87}$	1768	$442.00_{\pm150.08}$	40
	Immunize	16786	$799.33_{\pm126.25}$	5694	$813.43_{\pm82.07}$	4830	$805.00_{\pm161.74}$	6600	$825.00_{\pm121.09}$	42
	Orphadata	216411	$257.63_{\pm69.72}$	75309	$270.90_{\pm73.17}$	70768	$245.72_{\pm69.33}$	69645	$254.18_{\pm66.75}$	1680
	Wikipedia	103835	$314.65_{\pm172.48}$	28181	$279.02_{\pm143.40}$	38074	$317.28_{\pm167.22}$	37806	$346.84_{\pm171.34}$	660
	Total train	**392362**		**128636**		**130170**		**133012**		**2698**
EN (test)	EUCTrials	16992	$679.68_{\pm194.73}$	5201	$743.00_{\pm211.25}$	6652	$665.20_{\pm101.05}$	5457	$682.12_{\pm263.69}$	50
	Cochrane	7498	$624.83_{\pm94.09}$	1793	$597.67_{\pm138.23}$	3462	$577.00_{\pm49.23}$	2037	$679.00_{\pm159.35}$	24
	EMA	14582	$131.37_{\pm52.94}$	5569	$123.76_{\pm43.67}$	4614	$148.84_{\pm48.21}$	5087	$145.34_{\pm51.27}$	222
	EFSA	2473	$618.25_{\pm306.40}$	880	$880.00_{\pm0.00}$	927	$463.50_{\pm239.71}$	639	$639.00_{\pm0.00}$	8
	Abstracts	34199	$198.83_{\pm71.09}$	11475	$197.84_{\pm66.30}$	10792	$192.71_{\pm63.49}$	12389	$213.60_{\pm78.96}$	344
	Total test	**75744**		**24918**		**26447**		**25609**		**648**
ES (train)	Abstracts	10979	$267.78_{\pm129.45}$	2571	$233.73_{\pm105.69}$	2758	$229.83_{\pm81.30}$	6230	$346.11_{\pm159.79}$	82
	BfG	8827	$245.19_{\pm77.34}$	3800	$253.33_{\pm75.39}$	3585	$239.00_{\pm71.70}$	1393	$232.17_{\pm45.41}$	72
	EFSA	34627	$567.66_{\pm264.07}$	10761	$489.14_{\pm231.24}$	8825	$630.36_{\pm238.48}$	13892	$555.68_{\pm245.10}$	122
	EVIP	9407	$470.35_{\pm168.98}$	3855	$385.50_{\pm111.71}$	1713	$428.25_{\pm63.12}$	3647	$607.83_{\pm240.20}$	40
	Immunize	20229	$963.29_{\pm158.22}$	4147	$829.40_{\pm149.12}$	12237	$941.31_{\pm130.56}$	3417	$1139.00_{\pm222.69}$	42
	Orphadata	255523	$304.19_{\pm80.22}$	92799	$322.22_{\pm82.07}$	81840	$295.45_{\pm83.96}$	79884	$290.49_{\pm76.91}$	1680
	Wikipedia	91731	$277.97_{\pm156.01}$	28281	$257.10_{\pm152.83}$	32049	$281.13_{\pm132.76}$	32419	$305.84_{\pm163.42}$	660
	Total train	**431323**		**146214**		**143007**		**140882**		**2698**
ES (test)	EUCTrials	18433	$768.04_{\pm225.37}$	8039	$669.92_{\pm178.72}$	2860	$715.00_{\pm237.14}$	7141	$892.62_{\pm234.73}$	48
	Cochrane	9428	$785.67_{\pm183.24}$	3058	$764.50_{\pm170.55}$	1493	$746.50_{\pm44.55}$	5539	$923.17_{\pm281.37}$	24
	EMA	14499	$130.62_{\pm58.62}$	4385	$125.29_{\pm47.49}$	5473	$136.82_{\pm47.71}$	4882	$135.61_{\pm75.31}$	222
	EFSA	2759	$689.75_{\pm264.31}$	889	$889.00_{\pm0.00}$	1565	$782.50_{\pm40.31}$	291	$291.00_{\pm0.00}$	8
	Abstracts	27059	$184.07_{\pm63.99}$	8685	$180.94_{\pm70.30}$	9719	$186.90_{\pm53.37}$	9217	$196.11_{\pm55.10}$	294
	Total test	**72178**		**25056**		**21110**		**27070**		**596**
FR (train)	Abstracts	11694	$285.22_{\pm137.60}$	2777	$308.56_{\pm96.39}$	3906	$355.09_{\pm178.56}$	5369	$255.67_{\pm112.65}$	82
	BfG	9446	$262.39_{\pm79.49}$	2043	$291.86_{\pm38.86}$	2502	$227.45_{\pm92.68}$	5146	$285.89_{\pm83.25}$	72
	EFSA	35589	$583.43_{\pm248.48}$	11205	$509.32_{\pm212.38}$	10287	$605.12_{\pm268.23}$	13542	$615.55_{\pm246.70}$	122
	EVIP	9323	$466.15_{\pm158.19}$	2992	$427.43_{\pm130.64}$	4626	$420.55_{\pm95.90}$	1403	$701.50_{\pm221.32}$	40
	Immunize	22040	$1049.52_{\pm159.80}$	9033	$1129.12_{\pm167.84}$	1075	$1075.00_{\pm0.00}$	12489	$1040.75_{\pm169.54}$	42
	Orphadata	252548	$300.65_{\pm82.41}$	85892	$311.20_{\pm81.61}$	85252	$298.08_{\pm87.25}$	82236	$295.81_{\pm74.36}$	1680
	Wikipedia	84506	$256.08_{\pm143.82}$	25981	$249.82_{\pm140.19}$	28036	$243.79_{\pm124.76}$	32605	$293.74_{\pm176.05}$	660
	Total train	**425146**		**139923**		**135684**		**152790**		**2698**
FR (test)	EUCTrials	23008	$766.93_{\pm211.74}$	9443	$786.92_{\pm170.03}$	7461	$678.27_{\pm219.11}$	5793	$827.57_{\pm257.01}$	60
	Cochrane	10972	$914.33_{\pm297.74}$	4721	$944.20_{\pm306.26}$	4706	$784.33_{\pm244.02}$	1117	$1117.00_{\pm0.00}$	24
	EMA	16123	$145.25_{\pm72.54}$	3956	$131.87_{\pm37.91}$	5178	$156.91_{\pm58.80}$	7067	$147.23_{\pm75.21}$	222
	EFSA	2551	$637.75_{\pm241.78}$	1588	$794.00_{\pm32.53}$	327	$327.00_{\pm0.00}$	791	$791.00_{\pm0.00}$	8
	Abstracts	14753	$132.91_{\pm79.02}$	4259	$125.26_{\pm69.53}$	4815	$150.47_{\pm81.00}$	6187	$137.49_{\pm83.90}$	222
	Total test	**67407**		**23967**		**22487**		**20955**		**536**
DE (train)	Abstracts	10103	$246.41_{\pm111.78}$	3864	$241.50_{\pm101.34}$	3744	$267.43_{\pm114.50}$	2163	$196.64_{\pm90.98}$	82
	BfG	7947	$220.75_{\pm63.68}$	2633	$202.54_{\pm54.93}$	2738	$228.17_{\pm76.33}$	2424	$220.36_{\pm47.83}$	72
	EFSA	28222	$462.66_{\pm211.83}$	9379	$446.62_{\pm167.07}$	8139	$406.95_{\pm194.60}$	10957	$547.85_{\pm197.75}$	122
	EVIP	7611	$380.51_{\pm104.86}$	2857	$357.12_{\pm122.99}$	646	$323.00_{\pm94.75}$	4129	$412.90_{\pm110.48}$	40
	Immunize	18591	$885.29_{\pm159.56}$	7565	$840.56_{\pm192.47}$	5918	$845.43_{\pm165.77}$	4920	$984.00_{\pm78.87}$	42
	Orphadata	201307	$239.65_{\pm66.76}$	68907	$251.49_{\pm67.47}$	68825	$238.98_{\pm66.13}$	63614	$228.83_{\pm63.09}$	1680
	Wikipedia	77683	$235.40_{\pm128.52}$	26456	$224.20_{\pm139.77}$	25091	$248.43_{\pm103.68}$	26567	$239.34_{\pm134.18}$	660
	Total train	**351464**		**121661**		**115101**		**114774**		**2698**
DE (test)	EUCTrials	16144	$645.76_{\pm183.12}$	4518	$645.43_{\pm240.93}$	6395	$639.50_{\pm193.00}$	5337	$667.12_{\pm160.31}$	50
	Cochrane	8406	$700.50_{\pm174.93}$	3998	$799.60_{\pm157.28}$	1904	$634.67_{\pm153.00}$	2775	$693.75_{\pm209.68}$	24
	EMA	13275	$119.59_{\pm49.51}$	3247	$124.88_{\pm58.00}$	4193	$116.47_{\pm31.31}$	5923	$120.88_{\pm48.57}$	222
	EFSA	2446	$611.50_{\pm260.76}$	–	–	1674	$558.00_{\pm259.83}$	706	$706.00_{\pm0.00}$	8
	Abstracts	39507	$189.94_{\pm43.84}$	12567	$182.13_{\pm35.65}$	13769	$193.93_{\pm46.00}$	13568	$199.53_{\pm41.14}$	416
	Total test	**79778**		**24330**		**27935**		**28309**		**720**

References

1. Almazrouei, E., et al.: Falcon-40B: an open large language model with state-of-the-art performance (2023). https://arxiv.org/pdf/2311.16867
2. Balloccu, S., Schmidtová, P., Lango, M., Dusek, O.: Leak, cheat, repeat: data contamination and evaluation malpractices in closed-source LLMs. In: Graham, Y., Purver, M. (eds.) Proceedings of the 18th Conference of the European Chapter of the Association for Computational Linguistics (Volume 1: Long Papers), pp. 67–93. Association for Computational Linguistics, St. Julian's, Malta (2024). https://aclanthology.org/2024.eacl-long.5/
3. Bao, G., Zhao, Y., Teng, Z., Yang, L., Zhang, Y.: Fast-DetectGPT: Efficient Zero-Shot Detection of Machine-Generated Text via Conditional Probability Curvature (2024). https://arxiv.org/abs/2310.05130
4. Bartoli, A., De Lorenzo, A., Medvet, E., Tarlao, F.: Your paper has been accepted, rejected, or whatever: automatic generation of scientific paper reviews. In: Availability, Reliability, and Security in Information Systems: IFIP WG 8.4, 8.9, TC 5 International Cross-Domain Conference, CD-ARES 2016, and Workshop on Privacy Aware Machine Learning for Health Data Science, PAML 2016, Salzburg, Austria, August 31-September 2, 2016, Proceedings, pp. 19–28. Springer (2016). https://inria.hal.science/hal-01635011/file/430962_1_En_2_Chapter.pdf
5. Bevendorff, J., et al.: Overview of PAN 2025: generative AI authorship verification, multi-author writing style analysis, multilingual text detoxification, and generative plagiarism detection. In: Experimental IR Meets Multilinguality, Multimodality, and Interaction. Proceedings of the Fourteenth International Conference of the CLEF Association (CLEF 2025). Lecture Notes in Computer Science, Springer, Berlin Heidelberg New York (2025)
6. Cabanac, G., Labbé, C.: Prevalence of nonsensical algorithmically generated papers in the scientific literature. J. Assoc. Inf. Sci. Technol. 72(12), 1461–1476 (2021). https://asistdl.onlinelibrary.wiley.com/doi/pdf/10.1002/asi.24495
7. Crothers, E.N., Japkowicz, N., Viktor, H.L.: Machine-generated text: a comprehensive survey of threat models and detection methods. IEEE Access 11, 70977–71002 (2023). https://ieeexplore.ieee.org/iel7/6287639/6514899/10177704.pdf
8. Dou, Y., Forbes, M., Koncel-Kedziorski, R., Smith, N.A., Choi, Y.: Is GPT-3 text indistinguishable from human text? Scarecrow: A framework for scrutinizing machine text. In: Proc. of the 60th Annual Meeting of the Association for Computational Linguistics, pp. 7250–7274. Association for Computational Linguistics (2022). https://aclanthology.org/2022.acl-long.501.pdf
9. Fivez, P., et al.: The CLIN33 shared task on the detection of text generated by large language models. Comput. Linguist. Netherlands J. 13, 233–259 (2024). https://clinjournal.org/clinj/article/download/182/198
10. Galke, L., et al.: Are we really making much progress in text classification? a comparative review (2025). https://arxiv.org/abs/2204.03954
11. García, J.F., Segura-Bedmar, I.: Human after all: using transformer based models to identify automatically generated text. In: Proceedings of the IberLEF (2024). https://ceur-ws.org/Vol-3756/IberAuTexTification2024_paper4.pdf
12. García-Ferrero, I., et al.: MedMT5: an open-source multilingual text-to-text LLM for the medical domain. In: Calzolari, N., Kan, M.Y., Hoste, V., Lenci, A., Sakti, S., Xue, N. (eds.) Proceedings of the 2024 Joint International Conference on Computational Linguistics, Language Resources and Evaluation (LREC-COLING 2024), pp. 11165–11177. ELRA and ICCL, Torino, Italia (2024). https://aclanthology.org/2024.lrec-main.974/
13. Gehrmann, S., Strobelt, H., Rush, A.: GLTR: statistical detection and visualization of generated text. In: Costa-jussà, M.R., Alfonseca, E. (eds.) Proceedings of the 57th Annual Meeting of the Association for Computational Linguistics: System Demonstrations, pp. 111–116.

Association for Computational Linguistics, Florence, Italy (2019). https://doi.org/10.18653/v1/P19-3019, https://aclanthology.org/P19-3019/

14. Gritsai, G., Voznyuk, A., Grabovoy, A., Chekhovich, Y.: Are AI detectors good enough? a survey on quality of datasets with machine-generated texts (2025). https://arxiv.org/abs/2410.14677

15. Guo, B., et al.: How close is ChatGPT to human experts? Comparison corpus, evaluation, and detection (2023). https://arxiv.org/abs/2301.07597

16. He, P., Gao, J., Chen, W.: DeBERTaV3: improving DeBERTa using ELECTRA-style pre-training with gradient-disentangled embedding sharing. In: Proceedings of ICLR (2023). https://arxiv.org/pdf/2111.09543

17. Herbold, S., Hautli-Janisz, A., Heuer, U., Kikteva, Z., Trautsch, A.: A large-scale comparison of human-written versus ChatGPT-generated essays. Sci. Rep. 13(1), 18617 (2023). https://www.nature.com/articles/s41598-023-45644-9.pdf

18. Jawahar, G., Abdul-Mageed, M., Lakshmanan, V.S., L.: Automatic detection of machine generated text: a critical survey. In: Scott, D., Bel, N., Zong, C. (eds.) Proceedings of the 28th International Conference on Computational Linguistics, pp. 2296–2309. International Committee on Computational Linguistics, Barcelona, Spain (Online) (2020). https://doi.org/10.18653/v1/2020.coling-main.208

19. Kashnitsky, Y., Herrmannova, D., de Waard, A., Tsatsaronis, G., Fennell, C., Labbé, C.: Overview of the DAGPap22 shared task on detecting automatically generated scientific papers. In: Third Workshop on Scholarly Document Processing (2022). https://hal.science/hal-03828597v1/file/SDP_Workshop___DAGPap22___Overview_2022.pdf

20. Koike, R., Kaneko, M., Okazaki, N.: Outfox: LLM-generated essay detection through in-context learning with adversarially generated examples. In: Proceedings of the AAAI Conference on Artificial Intelligence. vol. 38, pp. 21258–21266 (2024). https://ojs.aaai.org/index.php/AAAI/article/view/30120/31980

21. Kokhlikyan, N., et al.: Captum: A unified and generic model interpretability library for PyTorch (2020). https://arxiv.org/abs/2009.07896

22. Li, Y., et al.: MAGE: machine-generated text detection in the wild. In: Ku, L.W., Martins, A., Srikumar, V. (eds.) Proceedings of the 62nd Annual Meeting of the Association for Computational Linguistics (Volume 1: Long Papers), pp. 36–53. Association for Computational Linguistics, Bangkok, Thailand (2024). https://doi.org/10.18653/v1/2024.acl-long.3

23. Liu, Y., et al.: RoBERTa: A robustly optimized BERT pretraining approach (2019). https://arxiv.org/abs/1907.11692

24. Liyanage, V., Buscaldi, D., Nazarenko, A.: A benchmark corpus for the detection of automatically generated text in academic publications. In: Calzolari, N., et al. (eds.) Proceedings of the Thirteenth Language Resources and Evaluation Conference, pp. 4692–4700. European Language Resources Association, Marseille, France (2022). https://aclanthology.org/2022.lrec-1.501/

25. Lundberg, S., Lee, S.I.: A unified approach to interpreting model predictions (2017). https://arxiv.org/abs/1705.07874

26. Ma, Y., et al.: AI vs. Human–differentiation analysis of scientific content generation. arXiv preprint arXiv:2301.10416 (2023). https://arxiv.org/pdf/2301.10416

27. Macko, D., et al.: MULTITuDE: large-scale multilingual machine-generated text detection benchmark. In: Proceedings of the 2023 Conference on Empirical Methods in Natural Language Processing, pp. 9960–9987. Association for Computational Linguistics (2023). https://doi.org/10.18653/v1/2023.emnlp-main.616

28. Magar, I., Schwartz, R.: Data contamination: from memorization to exploitation. In: Muresan, S., Nakov, P., Villavicencio, A. (eds.) Proceedings of the 60th Annual Meeting of the

Association for Computational Linguistics (Volume 2: Short Papers), pp. 157–165. Association for Computational Linguistics, Dublin, Ireland (2022). https://doi.org/10.18653/v1/2022.acl-short.18

29. Mitchell, E., Lee, Y., Khazatsky, A., Manning, C.D., Finn, C.: DetectGPT: Zero-Shot Machine-Generated Text Detection using Probability Curvature (2023). https://arxiv.org/abs/2301.11305

30. Radford, A., Wu, J., Child, R., Luan, D., Amodei, D., Sutskever, I.: Language models are unsupervised multitask learners (2019). https://api.semanticscholar.org/CorpusID:160025533

31. Rodriguez, J.D., Hay, T., Gros, D., Shamsi, Z., Srinivasan, R.: Cross-domain detection of GPT-2-generated technical text. In: Carpuat, M., de Marneffe, M.C., Meza Ruiz, I.V. (eds.) Proceedings of the 2022 Conference of the North American Chapter of the Association for Computational Linguistics: Human Language Technologies, pp. 1213–1233. Association for Computational Linguistics, Seattle, United States (2022). https://doi.org/10.18653/v1/2022.naacl-main.88

32. Sanh, V., Debut, L., Chaumond, J., Wolf, T.: DistilBERT, a distilled version of BERT: smaller, faster, cheaper and lighter (2020). https://arxiv.org/abs/1910.01108

33. Sarvazyan, A.M., González, J.Á., Franco-Salvador, M., Rangel, F., Chulvi, B., Rosso, P.: Overview of AuTexTification at IberLEF 2023: Detection and attribution of machine-generated text in multiple domains. Procesamiento de lenguaje natural **71**, 275–288 (2023). http://journal.sepln.org/sepln/ojs/ojs/index.php/pln/article/download/6559/3959

34. Sarvazyan, A.M., González, J.Á., Rangel, F., Rosso, P., Franco-Salvador, M.: Overview of IberAuTexTification at IberLEF 2024: detection and attribution of machine-generated text on languages of the Iberian Peninsula. Procesamiento del Lenguaje Natural (73), 421–434 (2024). http://journal.sepln.org/sepln/ojs/ojs/index.php/pln/article/view/6628/4020

35. Shamardina, T., et al.: CoAT: corpus of artificial texts. Nat. Lang. Process. **31**(1), 150–175 (2025). https://www.cambridge.org/core/services/aop-cambridge-core/content/view/7E2CA97E21663CC031FB6BAFE56E0046/S2977042424000384a.pdf/coat-corpus-of-artificial-texts.pdf

36. Solaiman, I., et al.: Release strategies and the social impacts of language models (2019). https://arxiv.org/abs/1908.09203

37. Sundararajan, M., Taly, A., Yan, Q.: Axiomatic attribution for deep networks. In: Proceedings of International Conference on Machine Learning, pp. 3319–3328. PMLR (2017)

38. Uchendu, A., Le, T., Shu, K., Lee, D.: Authorship attribution for neural text generation. In: Proceedings of the 2020 Conference on Empirical Methods in Natural Language Processing (EMNLP), pp. 8384–8395 (2020). https://aclanthology.org/2020.emnlp-main.673.pdf

39. Varab, D., Schluter, N.: MassiveSumm: a very large-scale, very multilingual, news summarisation dataset. In: Moens, M.F., Huang, X., Specia, L., Yih, S.W.t. (eds.) Proceedings of the 2021 Conference on Empirical Methods in Natural Language Processing, pp. 10150–10161. Association for Computational Linguistics, Online and Punta Cana, Dominican Republic (2021). https://doi.org/10.18653/v1/2021.emnlp-main.797

40. Verma, V., Fleisig, E., Tomlin, N., Klein, D.: Ghostbuster: detecting text ghostwritten by large language models. In: Duh, K., Gomez, H., Bethard, S. (eds.) Proceedings of the 2024 Conference of the North American Chapter of the Association for Computational Linguistics: Human Language Technologies (Volume 1: Long Papers), pp. 1702–1717. Association for Computational Linguistics, Mexico City, Mexico (2024). https://doi.org/10.18653/v1/2024.naacl-long.95

41. Wadden, D., et al.: Fact or fiction: verifying scientific claims. In: Webber, B., Cohn, T., He, Y., Liu, Y. (eds.) Proceedings of the 2020 Conference on Empirical Methods in Natural Language Processing (EMNLP), pp. 7534–7550. Association for Computational Linguistics, Online (2020). https://doi.org/10.18653/v1/2020.emnlp-main.609
42. Zellers, R., et al.: Defending against neural fake news. Adv. Neural Inf. Process. Syst. **32** (2019). https://proceedings.neurips.cc/paper/2019/file/3e9f0fc9b2f89e043bc623 3994dfcf76-Paper.pdf

Taxonomy Generation for Scientific Concepts Using Large Language Models

Yue Zhang⬤, Zi Long Zhu⬤, Artemis Capari⬤, Hosein Azarbonyad$^{(\boxtimes)}$⬤, Zubair Afzal⬤, and George Tsatsaronis⬤

Elsevier, Amsterdam, The Netherlands
{z.zhu,a.capari,h.azarbonyad,zubair.afzal,
g.tsatsaronis}@elsevier.com

Abstract. Traditional data-driven automatic taxonomy generation methods struggle with complex, large, and domain-specific datasets. To address these issues, this study leverages Large Language Models (LLMs) to automate key stages of taxonomy generation, focusing on scientific concepts. Our approach employs LLMs at several stages of the taxonomy generation process, including extracting candidate concepts and organizing keywords into taxonomies centered around chosen scientific concepts. By incorporating LLMs, we aim to enhance depth, accuracy, and coherence of generated taxonomies. Comparative analyses show that the proposed LLM-based taxonomy generation method outperforms state-of-the-art taxonomy generation methods on several metrics, such as concept coherence and coverage. Using a hybrid evaluation framework that combines automatic and human assessments, we demonstrate that our LLM-based solution is scalable, adaptable, and capable of generating high-quality taxonomies tailored to specific scientific concepts.

Keywords: Automatic Taxonomy Construction · Scientific Document Processing · LLMs for Taxonomy Construction

1 Introduction

The exponential growth of scientific literature and digital content has intensified the need for effective organization of unstructured textual data. One widely adopted solution is the construction of taxonomies–hierarchical structures that categorize concepts into parent–child relationships, thereby structuring knowledge from general to specific [21]. Taxonomies support a variety of applications, such as semantic search, document classification, and recommendation systems [4]. In academic and scientific domains, they provide structured overviews of complex fields, helping researchers explore topics, identify emerging areas, and understand relationships among concepts.

Despite their utility, traditional taxonomy construction methods are labor-intensive and struggle to scale across diverse scientific domains. They typically follow a three-stage process: (1) term extraction, where domain-relevant terms or phrases are identified from a corpus; (2) relationship discovery, where terms are grouped or connected

---

* Note: This paper is accepted as a poster.

© The Author(s), under exclusive license to Springer Nature Switzerland AG 2026
J. Carrillo-de-Albornoz et al. (Eds.): CLEF 2025, LNCS 16089, pp. 74–86, 2026.
https://doi.org/10.1007/978-3-032-04354-2_6

based on semantic similarity or co-occurrence; and (3) hierarchy construction, where the discovered relationships are structured into a taxonomy, often with manual intervention or rule-based heuristics [14, 18]. These methods face persistent challenges, such as determining the appropriate level of granularity, maintaining coherence in the hierarchy, and ensuring domain relevance of the extracted terms. To address these challenges, researchers have begun exploring Large Language Models (LLMs) as tools for automatic taxonomy generation. LLMs exhibit strong capabilities in semantic understanding, abstraction, and reasoning, making them promising candidates for automating various stages of the taxonomy construction pipeline. Prior work has shown that LLMs can organize predefined concept sets into hierarchical structures using prompting or fine-tuning strategies [10]. However, these approaches often assume that the input concepts are already available and domain-relevant. In contrast, this paper tackles a more challenging and practical problem: constructing taxonomies from scratch, given only a corpus of text centered around a scientific concept (e.g., "Machine Learning" or "Quantum Computing"). This setting removes the assumption of a predefined concept list and requires the system to both identify relevant concepts and organize them hierarchically, reflecting real-world scenarios where such structured resources are unavailable.

We investigate how LLMs can be used to construct accurate, coherent, and useful scientific taxonomies directly from unstructured text corpora. Our goal is twofold: (1) to design methods that enhance the taxonomy construction process using LLMs at different stages, and (2) to systematically evaluate the effectiveness of these methods in generating high-quality taxonomies. To this end, we propose and compare three LLM-based taxonomy construction pipelines, each differing in how it interacts with the traditional taxonomy generation process: LLM-centric taxonomy generation, hybrid approach, and prompt engineering (corpus-free).

The LLM-centric taxonomy generation method replaces all stages of the traditional process. This approach first clusters the given corpus into clusters of semantically similar documents. Then, the LLM is prompted to generate a taxonomy directly per cluster, extracting both concepts and their relationships. Finally, these clusters are merged to form the final taxonomy. This approach explores the LLM's capability for joint inference over content understanding and structure generation. The hybrid approach combines traditional and LLM-based methods to leverage their respective strengths. Since traditional methods perform well at term extraction, while LLMs excel at structural reasoning, we separate the process into two stages. First, we employ conventional techniques (e.g., TF-IDF) to extract key terms from the corpus. In the second stage, the LLM is prompted to organize these terms into a hierarchical taxonomy. This modular pipeline offers a balance between the precision and scalability of traditional term extraction methods with the semantic reasoning and abstraction capabilities of LLMs. In the prompt engineering approach, the LLM is provided only with the name of a scientific concept (e.g., "Graph Theory") and asked to generate a taxonomy based purely on its internal knowledge. This method bypasses corpus processing entirely and assesses the extent to which LLMs can function as standalone taxonomists.

Although these pipelines vary in what they extract—terms vs. relationships—they all share the same end goal: to produce coherent, thematically accurate, and practically useful taxonomies. These design variations allow us to explore (1) the impact of different initial representations, and (2) the comparative performance of LLMs when used independently versus in combination with traditional methods. We evaluate the resulting taxonomies using both automatic metrics (e.g., structural integrity, coverage, coherence) and human assessments (e.g., interpretability and usefulness). Our experiments demonstrate that LLM-based approaches–particularly the hybrid approach–outperform traditional methods across most metrics. Notably, the prompt engineering-based method also achieves strong results in human evaluations, highlighting its potential as a rapid and accessible solution for taxonomy generation. These findings indicate that LLMs not only match, but often surpass traditional methods and provide taxonomy structures that are not only hierarchically coherent but also adaptive and extensible.

In summary, our main contributions are:

- We propose and systematically compare three LLM-based pipelines for automatic taxonomy generation from scientific corpora.
- We introduce a hybrid evaluation framework combining automatic and human assessments.
- We demonstrate, through extensive experiments, that LLM-based methods outperform traditional approaches in taxonomy quality and scalability.

2   Related Work

Taxonomy generation progressed from manual, expert-driven approaches [18] to rule-based algorithms [14], and later to machine learning based techniques such as KMeans clustering and Latent Dirichlet Allocation [5,7,16]. Recent advancements emphasize scalability and precision, utilizing Hidden Markov Models for cross-domain accuracy [15] and platforms like Apache Spark for large-scale processing [1]. These advances laid the groundwork for LLM-based taxonomy construction by enabling scalable pipelines.

The introduction of LLMs, particularly GPT [25], has significantly improved taxonomy generation, offering greater accuracy and context awareness. These models excel in complex domains such as healthcare and legal studies [2,29], with ongoing research in domain-specific fine-tuning and multilingual capabilities [10,26]. Our work focuses on using LLMs for automatic taxonomy construction and comparing such methods with traditional taxonomy construction methods such as TaxoCom [19] and NetTaxo [27].

Despite progress, challenges persist in evaluation methodologies, which often rely on subjective human judgments [22,28]. There is a need for robust frameworks that combine qualitative and quantitative metrics [27,30]. Furthermore, while powerful, LLMs are resource-intensive and require substantial fine-tuning for domain-specific applications [11,31]. In response, this paper introduces a robust evaluation framework to assess the performance of LLMs across different stages of taxonomy creation.

3   Taxonomy Generation Pipelines

This work investigates how LLMs can be integrated into the taxonomy construction process around scientific concepts, starting from a collection of relevant text snippets.

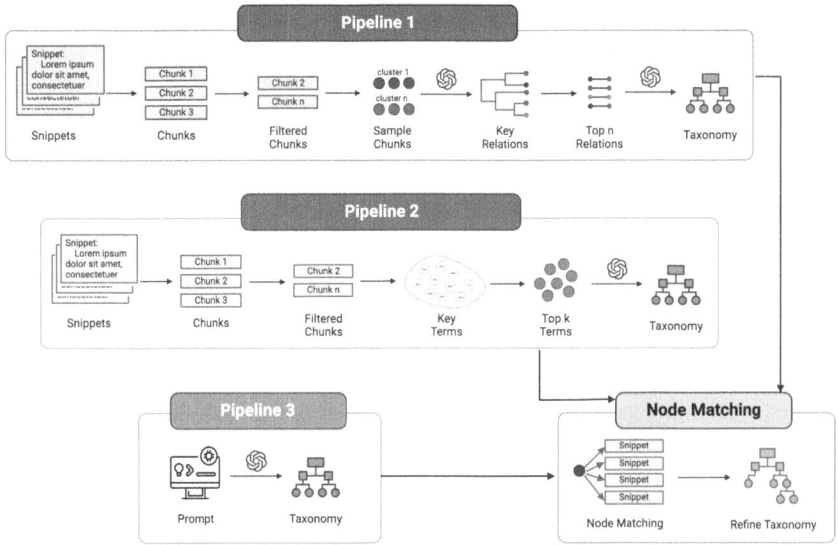

Fig. 1. Three pipeline designs for taxonomy generation using LLMs. Pipeline 1 focuses on LLM-centric taxonomy generation, Pipeline 2 uses a hybrid approach, and Pipeline 3 explores prompt engineering for taxonomy generation. The final step in each pipeline involves node matching and taxonomy refinement.

We design and evaluate three taxonomy generation pipelines, each placing the LLM at a different stage of the process, to assess its effectiveness in relation to traditional techniques.

The input to each pipeline is a set of text snippets relevant to a specific scientific concept (e.g., "Machine Learning"). The goal in all cases is to generate a hierarchical taxonomy that meaningfully organizes the concepts and themes present in the snippets.

Pipeline 1 (LLM-Centric Taxonomy Generation): In this pipeline, LLMs replace traditional methods at every stage of taxonomy generation, as shown in the upper part of Fig. 1. The aim is to test the LLM's ability to autonomously perform all major stages of taxonomy generation—term identification, relationship discovery, and hierarchy structuring. The process commences with the segmentation of raw text snippets into smaller chunks, followed by a filtering process to retain only the most relevant content. Subsequently, the filtered chunks are grouped using a clustering algorithm. From each cluster, the representative chunks are selected and analyzed to extract relational patterns between terms, which are then refined into the top n relationships. These relationships form the foundation of the final taxonomy structure, which is created by prompting an LLM to build the taxonomy given the relationships.

Pipeline 2 (Hybrid Approach): The second pipeline combines traditional keyword extraction techniques with LLM-driven taxonomy generation. As depicted in the middle section of Fig. 1, the process begins similarly with the chunking and filtering of snippets. However, instead of immediate clustering, this pipeline introduces a step where key terms are extracted from the filtered chunks using traditional methods such as TF-IDF or LDA. These key terms are then processed by LLMs, which organize them into the top k most significant terms. The LLM is then prompted to organize the selected terms into a coherent taxonomy. This approach evaluates the LLM's ability to build a taxonomy when provided with pre-processed, high-quality data inputs.

Pipeline 3 (LLM Prompt Engineering): The third pipeline explores an approach where LLMs generate taxonomies based solely on prompts, without predefined corpus input, as illustrated in the lower section of Fig. 1. In this scenario, the LLMs receive a prompt that instructs the model to rely only on its pre-trained knowledge when constructing a taxonomy. This pipeline tests an LLM's adaptability and creative capabilities, assessing their ability to generate meaningful taxonomies with minimal input data.

Common Final Stage: Node Matching and Refinement. Across all three pipelines, a common final step involves node matching and refinement, shown in the rightmost part of Fig. 1. Here, the initial taxonomy is validated by matching text snippets with the generated nodes. The outcome of this matching process guides a final refinement phase, during which underdeveloped branches are expanded and irrelevant nodes are pruned. This process helps to ensure that the final taxonomy is both representative of the underlying scientific source material and practical for real-world applications.

This detailed methodological approach examines the effectiveness of LLMs in various stages of taxonomy generation and highlights the potential for integrating deep learning techniques with traditional data processing methods.

4   Experimental Setup and Implementation Details

This section outlines our research questions and describes the experimental framework for constructing scientific taxonomies using large language models (LLMs).

Research Questions: This study addresses the following research questions:

RQ1: How does the Use of LLMs in Taxonomy Generation Compare to Traditional Methods? This question evaluates the role of LLMs in enhancing or replacing traditional taxonomy generation methods, focusing on differences in accuracy, efficiency, scalability, and semantic quality. We compare the performance of the proposed pipelines to the performance TaxoCom and NetTaxo methods.

RQ2: What is the Effect of Applying LLMs at Different Stages of the Taxonomy Generation Process? This question examines how LLMs influence the quality of the taxonomy when applied at different stages of the taxonomy generation process, such as feature extraction, clustering, and taxonomy construction.

Dataset: The dataset for this work is drawn from Elsevier's Topic Pages database[1] containing snippets extracted from academic articles (published in over 2,700 journals) and 43,000 books centered around scientific concepts [3,8,9]. We use "Machine Learning" as the primary domain concept with a dataset of 98,223 text snippets. These are snippets (sections) of text that mention the concept or one of its synonyms. The selected snippets are segmented into 512-token chunks to standardize input length.

Data Filtering: To filter for relevance, we first encode each snippet using a Sentence Transformer model[2]. We then compute the cosine similarity between each snippet embedding and the embbeding of the query "Machine Learning". The snippets are ranked by similarity, and the top-5% of most relevant chunks are retained, yielding a set of 4,912 text chunks for further analysis.

Term Extraction and Refinement: The **LLM-centric approach** (Pipeline 1) involves clustering the filtered chunks and using LLMs for taxonomy generation. We apply UMAP [24] for dimensionality reduction, followed by KMeans clustering ($k = 100$) [7]. Finally, we prompt an LLM to create a structured taxonomy based on the identified cluster representations. The representations per cluster are selected by fitting a KMeans model ($k = 40\%$ of the snippets in the cluster) to the embeddings of snippets within the cluster and selecting the centroids. The **hybrid approach** (Pipeline 2) bypasses clustering by employing KeyBERT [13] to extract key terms directly from the filtered text chunks. These terms are ranked by frequency, with the top-200 terms used as input for LLM-based taxonomy generation.

Taxonomy Generation Prompt: For each pipeline, we craft prompts to guide the LLM in generating taxonomies. These prompts reflect the extracted relationships and terms, and produce a structured JSON output specifying categories, subcategories, and their interconnections. An example for the LLM-centric approach is shown in Fig. 2. As LLM, we use GPT-3.5-turbo with the temperature set to zero in all pipelines.

Node Matching and Taxonomy Refinement: After generating taxonomies, we match each node to relevant text snippets based on the cosine similarity between node and snippet embeddings (using the *all-MiniLM-L6-v2* model). A similarity threshold of 0.7 is empirically chosen to identify matches. Each snippet with a similarity score above this threshold is assigned to the corresponding node. Nodes are further evaluated based on their snippet support (i.e. the number of matched snippets), with irrelevant nodes (support < 5 snippets) pruned, while highly-supported nodes (support > 50 snippets) are expanded. This process ensures that the taxonomy is both accurate and reflective of the conceptual structure of the domain.

5   Results

This section presents a comprehensive evaluation of our proposed LLM-based taxonomy generation pipelines. We compare their performance against established baseline

---

[1] https://www.sciencedirect.com/topics.

[2] https://huggingface.co/sentence-transformers/all-MiniLM-L6-v2.

```
You are an expert diagram creator. Your task is to provide an overview of a topic area based
on a concept and relevant snippets.

CONCEPT: <concept>
SNIPPETS: <snippets>
INSTRUCTIONS:
1. Identify the most relevant keywords relevant to the concept, then group them
into categories and give each category a label.
2. Think of subcategories for each category and give each subcategory a label.
3. Use the syntax provided in the EXAMPLE for the diagram output.

EXAMPLE OUTPUT:
[
  {"category": "null", "subcategory": "Architectural Features", "snippet_id": },
  {"category": "Architectural Features", "subcategory": "Structural Design", "snippet_id": },
  {"category": "Architectural Features", "subcategory": "Facade Design", "snippet_id": },
  {"category": "Architectural Features", "subcategory": "Sustainability", "snippet_id": },
  {"category": "Architectural Features", "subcategory": "Seismic Retrofitting",
  "snippet_id": },
  {"category": "Architectural Features", "subcategory": "Trends", "snippet_id": },
  {"category": "Structural Design", "subcategory": "Structural Material", "snippet_id": },
  {"category": "Structural Design", "subcategory": "Variety Of Loads", "snippet_id": },
  {"category": "Structural Design", "subcategory": "Damping Systems", "snippet_id": },
  {"category": "Facade Design", "subcategory": "Enhancement Trend", "snippet_id": },
  {"category": "Facade Design", "subcategory": "Historical Building", "snippet_id": },
  {"category": "Sustainability", "subcategory": "Green Building", "snippet_id": },
  {"category": "Sustainability", "subcategory": "Energy Efficiency", "snippet_id": }
]

REMEMBER
- Generate a diagram with up to 12 keywords inspired by the provided keywords that are directly
relevant to the concep.
- Avoid using the conjunction word or symbol, and use noun phrases.
- The format of "snippet_id" must be the same as the format provided.
- It is not necessary to use all snippet to build the taxonomy, the most important thing is to
make sure that the final taxonomy is clear and structured.
```

Fig. 2. Prompt used for generating the taxonomy using the LLM-centric approach. The taxonomy is created based on the sampled snippets per snippet cluster.

methods, specifically TaxoCom [19] and NetTaxo [27], across multiple metrics assessing their effectiveness in organizing and structuring knowledge domains.

Detailed explanations of the evaluation metrics are provided in Table 1. Our evaluation framework integrates both automated metrics and human-centered assessments, categorized into four key dimensions:

Structural Integrity: Measures hierarchical relationships and structure through parent-child similarity [27], node quantity, and relation accuracy in the constructed taxonomy.

Coverage and Comprehensiveness: Assesses how thoroughly the taxonomy covers domain concepts, evaluated by concept coverage, key concept quantity, and soft recall [12].

Coherence and Consistency: Evaluates semantic relatedness among sibling concepts via Concept Coherence [19].

Usability and Accessibility: Focuses on how easily users can navigate and retrieve information within the taxonomy. This is assessed by human evaluators through metrics such as Understandability and Usefulness.

The human evaluation involved **4** Subject Matter Experts (SMEs) in **Computer Science**, who independently assessed the taxonomy based on Coverage, Understandability, and Usefulness. Reported scores represent the average across SMEs per metric.

Finally, we compare our LLM-based pipelines (1, 2 and 3) against baseline methods (TaxoCom and NetTaxo) across several domains, starting with "Machine Learning" as a primary concept.

5.1 Automatic Evaluation

Table 2 presents the results of the automated evaluation. Our LLM-based methods consistently outperform traditional baselines in parent-child similarity, relation accuracy, and concept coverage, demonstrating their ability to generate more coherent and comprehensive taxonomies. TaxoCom and NetTaxo, despite their sophistication, struggle to capture the breadth and structure of the domain, as evidenced by their low snippet coverage (107 and 713, respectively) and poor relation accuracy.

Notably, the hybrid pipeline (2) achieves the highest scores across most metrics, with a parent-child similarity of 0.579 and a relation accuracy of 0.525, highlighting its strength in accurately capturing hierarchical relationships. The LLM-centric pipeline, which combines clustering with prompting, performs well in structure-related metrics (e.g., node quantity), but slightly lags behind in semantic coverage and coherence compared to the hybrid pipeline. This suggests that while clustering can help group similar content, it may introduce noise or fragmentation when used without careful integration.

All LLM-based pipelines show marked improvement in aligning sibling nodes under meaningful parent categories, a traditional weakness of automated methods. For instance, while NetTaxo achieves decent coherence (0.419), it falls short in overall coverage and granularity. TaxoCom's extremely low node quantity (42) and concept coverage (0.049) suggest a narrow view of the domain, likely due to reliance on statistical signals rather than semantic understanding. NetTaxo performs moderately better but still falls short in taxonomy breadth and relational correctness.

These results confirm the efficacy of our methods in generating well-structured and contextually accurate taxonomies that surpass traditional approaches. The findings collectively demonstrate that LLMs are not only capable of understanding domain-relevant concepts but also of organizing them into hierarchies that are both broad and semantically valid—a challenging task for traditional models.

5.2 Human-Based Evaluation

SME evaluation results, shown in Table 3, reveal that our methods significantly outperform the baselines regarding coverage, understandability, and usefulness. These results

Table 1. Detailed Explanations of Evaluation Metrics

Metric	Explanation
Parent-Child Similarity	Measures semantic alignment between parent and child nodes. Unlike NetTaxo [27], which uses human evaluators, we compute cosine similarity between fastText [6] embeddings of parent and child terms to evaluate semantic alignment.
Node Quantity	Counts the total number of nodes in the taxonomy, reflecting its depth and breadth. While higher counts indicate more coverage, quantity does not necessarily imply quality. This metric helps analyze the balance between domain representation and structural complexity.
Relation Accuracy	Inspired by TopicExpan's [20] 'precision for novel topic discovery', which measures the precision of newly identified topic relationships. We verify parent-child entities in the taxonomy have a "subclass of" relationship in Wikidata, validated via cosine similarity (>0.7) between their semantic embeddings. The average accuracy score is reported. Results are then aggregated to derive an average Relation Accuracy score.
Concept Coverage	Measures the proportion of concepts from a ground-truth taxonomy are covered by the generated taxonomy. We extract a specific range of concepts from the OmniScience tree [23], which serve as our ground truth for comparison. We embed each concept in this reference list and all terms in our taxonomy using *all-MiniLM-L12-v2 model*. The concepts are then matched by the cosine similarity of their embeddings (>0.8). The proportion of these matched to the ground truth taxonomy serve as the *Concept Coverage* metric.
Key Concept Quantity	Counts terms in the taxonomy that exactly match terms in the Wikipedia taxonomy, reflecting the taxonomy's ability to precisely capture essential concepts.
Soft Recall	An adaptation of traditional recall metric, considering partial matches or semantically related concepts, following [12].
Concept Coherence	Inspired by the human-assessed coherence metrics used in TaxoCom [19], it evaluates semantic consistency of sibling nodes under each parent. Calculated by averaging cosine similarities among fastText [6] embeddings of child nodes, then averaging across all parent groups.
Understandability	Human-rated on a 1–10 scale by SMEs, where 1 indicates very difficult to understand and 10 indicates very easy to understand.
Usefulness	Human-rated on a 1–10 scale by SMEs, assessing how effectively the taxonomy supports information retrieval and navigation.

suggest that human evaluators found the taxonomy produced by the prompt engineering method (Pipeline 3) to be the most comprehensive and user-friendly, despite it not having the highest automatic scores in snippet or concept coverage. This highlights an

Table 2. Performance of different taxonomy generation methods.

Metric	TaxoCom	NetTaxo	Pipeline 1	Pipeline 2	Pipeline 3
Snippets Coverage	107	713	832	**6305**	3125
Parent-Child Similarity	0.162	0.426	0.551	**0.579**	0.398
Node Quantity	42	25	**98**	76	63
Relation Accuracy	0.000	0.125	0.375	**0.525**	0.375
Concept Coherence	0.176	0.419	0.494	**0.502**	0.491
Concept Coverage	0.049	0.063	0.148	**0.204**	0.120
Key Concept Quantity	6	10	18	**19**	16
Soft Recall	0.027	0.312	0.768	**0.881**	0.667

important point: human perception of quality does not always align with raw coverage or node count, and instead prioritizes clarity and organization. However, this pipeline covers much less snippets than the hybrid pipeline, showing that the generated taxonomy might not fit the data very well.

The hybrid pipeline is a close second, particularly with respect to its wide domain coverage (8.25) and high clarity (8.50). This pipeline's strong alignment with both semantic structure and domain relevance may explain its high scores in usefulness (8.75), indicating that it is particularly effective for knowledge retrieval or educational tools. The LLM-centric pipeline, though solid in structure, ranks slightly lower in human evaluation. SMEs noted some inconsistencies in the grouping of terms, possibly introduced by the clustering stage. These artifacts may reduce perceived coherence, despite decent automatic metrics. TaxoCom and NetTaxo received the lowest ratings across all human-centered metrics. Qualitative comments from domain experts pointed to vague categorization, lack of depth, and outdated or inconsistent terminology as key reasons for their lower scores.

The inter-rater agreement, measured by Krippendorff's Alpha (Interval) [17], was 0.3984, indicating a moderate level of agreement among evaluators. Qualitative feedback from domain experts highlighted the strengths and weaknesses of each approach. While TaxoCom and NetTaxo were noted for occasionally including irrelevant information, our LLM-based methods, particularly the LLM-centric and hybrid methods, were found to provide higher clarity and comprehensive coverage.

The differences between the three different pipelines reveal that the stage at which LLMs are introduced significantly affects performance. Prompting over clean keyphrases (Pipeline 2) leads to better semantic coverage, while multi-stage prompting with some pruning (Pipeline 3) results in better clarity and usability. While Pipeline 2 achieves the best coverage and semantic metrics, Pipeline 3 is favored by human experts for interpretability and navigation. This suggests that for end-user applications, a balance between completeness and clarity is crucial.

Table 3. SME scores on the taxonomies resulted from different taxonomy generation methods.

Metric	TaxoCom	NetTaxo	Pipeline 1	Pipeline 2	Pipeline 3
Coverage	4.00	7.50	7.75	8.25	**9.00**
Understandability	4.75	8.00	8.00	8.50	**9.25**
Usefulness	4.50	8.00	8.25	8.75	**9.25**

Table 4. Performance of the hybrid pipeline (Pipeline 2) across different concepts.

Metric	ML	DBMS	IoT	DM
Snippets Coverage	0.064	0.052	0.047	0.067
Parent-Child Similarity	0.579	0.636	0.569	0.547
Node Quantity	76	45	70	47
Relation Accuracy	0.525	0.200	0.321	0.240
Concept Coherence	0.502	0.479	0.485	0.538
Concept Coverage	0.204	0.412	0.222	0.333
Key Concept Quantity	18	14	17	11
Soft Recall	0.768	0.841	0.886	0.795

5.3   Evaluation Across Different Concepts

To validate the generalizability of our approach, we extended the evaluation to additional concepts: Internet of Things (IoT), DataBase Management Systems (DBMS), and Diabetes Mellitus (DM). The results in Table 4 demonstrate that the hybrid pipeline maintains strong performance across different concepts, confirming its adaptability and effectiveness in different fields. We only evaluate the hybrid pipeline for this experiment as based on the automatic and human-based evaluation this pipeline outperformed the other ones.

The pipeline preserved strong conceptual organization, as reflected by high soft recall scores across all domains, indicating that it successfully captured a large portion of the relevant concepts even when precise hierarchical relations were less reliable. Interestingly, the concept coherence metric remained stable, suggesting that the generated taxonomies still formed semantically coherent groupings, even in less technical domains. The high performance in DM, particularly in concept coherence and snippet coverage, highlights the pipeline's potential to extend beyond Computer Science. The consistent high performance across diverse domains demonstrates the versatility and robustness of our hybrid taxonomy generation approach, supporting its broad applicability.

6   Conclusion

This paper introduced novel pipelines for scientific taxonomy generation by leveraging the capabilities of Large Language Models (LLMs). Through a comprehensive evaluation, we demonstrated that LLM-based approaches substantially outperform traditional methods across a wide range of structural and semantic metrics. In particular, the hybrid method (Pipeline 2) delivered significant improvements in both structural integrity and conceptual coverage, with metrics such as parent-child similarity and concept coverage far exceeding those of the strongest baseline (TaxoCom). Our comparative analysis of different integration strategies revealed that each LLM-driven pipeline offers unique strengths. The LLM-Centric taxonomy generation (Pipeline 1) and the hybrid method (Pipeline 2) showed superior performance on automated metrics, while the prompt-driven approach (Pipeline 3) was favored in human evaluations for producing more understandable and practically useful taxonomies. The robustness of the hybrid method across diverse concepts demonstrates its strong generalizability and potential for broad adoption. Its consistent performance across technical and biomedical domains under-lines the promise of this method in building high-quality taxonomies. Future work could explore the integration of external domain-specific ontologies and knowledge graphs to further enhance structural accuracy and interpretability. Additionally, expanding the evaluation to interdisciplinary and low-resource domains may uncover new opportunities and challenges for LLM-based taxonomy generation.

References

1. Aalijah, K., Irfan, R.: Scalable taxonomy generation and evolution on apache spark. In: International Conference on Dependable, Autonomic and Secure Computing, International Conference on Pervasive Intelligence and Computing, International Conference on Cloud and Big Data Computing, International Conference on Cyber Science and Technology Congress (DASC/PiCom/CBDCom/CyberSciTech), pp. 634–639 (2020)
2. Alowais, S.A., et al.: Revolutionizing healthcare: the role of artificial intelligence in clinical practice. BMC Med. Educ. **23**(1), 689 (2023)
3. Azarbonyad, H., Afzal, Z., Tsatsaronis, G.: Generating topic pages for scientific concepts using scientific publications. In: ECIR, pp. 341–349 (2023)
4. Bagherifard, K., Rahmani, M., Rafe, V., Nilashi, M.: A recommendation method based on semantic similarity and complementarity using weighted taxonomy: a case on construction materials dataset. J. Inf. Knowl. Manag. **17**, 1850010 (2018)
5. Blei, D.M., Ng, A.Y., Jordan, M.I.: Latent dirichlet allocation. J. Mach. Learn. Res. **3**(Jan), 993–1022 (2003)
6. Bojanowski, P., Grave, E., Joulin, A., Mikolov, T.: Enriching word vectors with subword information. Trans. Assoc. Comput. Linguist. **5**, 135–146 (2017)
7. Bottou, L., Bengio, Y.: Convergence properties of the k-means algorithms. Adv. Neural Inf. Process. Syst. **7** (1994)
8. Capari, A., Azarbonyad, H., Tsatsaronis, G., Afzal, Z., Dunham, J.: Sciencedirect topic pages: a knowledge base of scientific concepts across various science domains. In: SIGIR, pp. 2976–2980 (2024)
9. Capari, A., et al.: Knowledge acquisition passage retrieval: corpus, ranking models, and evaluation resources. In: CLEF, pp. 74–87 (2024)
10. Chen, B., Yi, F., Varró, D.: Prompting or fine-tuning? a comparative study of large language models for taxonomy construction. In: MODELS-C, pp. 588–596 (2023)
11. Ding, B., et al.: Data augmentation using LLMs: data perspectives, learning paradigms and challenges. In: ACL, pp. 1679–1705 (2024)
12. Fränti, P., Mariescu-Istodor, R.: Soft precision and recall. Pattern Recogn. Lett. **167**, 115–121 (2023)

13. Grootendorst, M.: KeyBERT: Minimal keyword extraction with BERT (2020). https://doi.org/10.5281/zenodo.4461265
14. Hsu, S.H., Hsia, T.C., Wu, M.C.: A flexible classification method for evaluating the utility of automated workpiece classification system. Int. J. Adv. Manufact. Technol. **13**(9), 637–648 (1997)
15. Iloga, S., Romain, O., Tchuenté, M.: An efficient generic approach for automatic taxonomy generation using HMMs. Pattern Anal. Appl. **24**(1), 243–262 (2021)
16. Ji, S., Ye, J.: Linear dimensionality reduction for multi-label classification. In: IJCAI, pp. 1077–1082. Morgan Kaufmann Publishers Inc. (2009)
17. Krippendorff, K.: Computing krippendorff's alpha-reliability (2011)
18. Lamparter, S., Ehrig, M., Tempich, C.: Knowledge extraction from classification schemas. In: CoopIS, DOA, and ODBASE, pp. 618–636 (2004)
19. Lee, D., Shen, J., Kang, S., Yoon, S., Han, J., Yu, H.: TaxoCom: topic taxonomy completion with hierarchical discovery of novel topic clusters. In: WWW, pp. 2819–2829 (2022)
20. Lee, D., Shen, J., Lee, S., Yoon, S., Yu, H., Han, J.: Topic taxonomy expansion via hierarchy-aware topic phrase generation. In: EMNLP, pp. 1687–1700 (2022)
21. Li, T., Anand, S.S.: Exploiting domain knowledge by automated taxonomy generation in recommender systems. In: EC-WEB, pp. 120–131 (2009)
22. Li, X., et al.: Inducing taxonomy from tags: an agglomerative hierarchical clustering framework. In: ADMA, pp. 64–77 (2012)
23. Malaisé, V., Otten, A., Coupet, P.: Omniscience and extensions–lessons learned from designing a multi-domain, multi-use case knowledge representation system. In: EKAW, pp. 228–242 (2018)
24. McInnes, L., Healy, J., Saul, N., Großberger, L.: UMAP: uniform manifold approximation and projection. J. Open Source Softw. **3**(29), 861 (2018)
25. Radford, A., Wu, J., Child, R., Luan, D., Amodei, D., Sutskever, I., et al.: Language models are unsupervised multitask learners. OpenAI Blog **1**(8), 9 (2019)
26. Shah, C., et al.: Using large language models to generate, validate, and apply user intent taxonomies. ACM Trans. Web (2025). just Accepted
27. Shang, J., Zhang, X., Liu, L., Li, S., Han, J.: NetTaxo: automated topic taxonomy construction from text-rich network. In: WWW, pp. 1908–1919 (2020)
28. Treeratpituk, P., Khabsa, M., Giles, C.L.: Graph-based approach to automatic taxonomy generation (GraBTax). arXiv preprint arXiv:1307.1718 (2013)
29. Wang, Y., Qian, W., Zhou, H., Chen, J., Tan, K.: Exploring new frontiers of deep learning in legal practice: a case study of large language models. Int. J. Comput. Sci. Inf. Technol. **1**(1), 131–138 (2023)
30. Weng, M.H., Wu, S., Dyer, M.: Identification and visualization of key topics in scientific publications with transformer-based language models and document clustering methods. Appl. Sci. **12**(21) (2022)
31. Zhang, Y., Yang, R., Xu, X., Li, R., Xiao, J., Shen, J., Han, J.: TELEClass: taxonomy enrichment and LLM-enhanced hierarchical text classification with minimal supervision. In: The Web Conference, pp. 2032–2042 (2025)

Best of CLEF 2024 Labs

Robustness of Misinformation Classification Systems to Adversarial Examples Through BeamAttack

Arnisa Fazla[1]([✉]), Lucas Krauter[2], David Guzman Piedrahita[2], and Andrianos Michail[3]

[1] Department of Medical Informatics, Amsterdam UMC, Amsterdam, The Netherlands
`a.fazla@amsterdamumc.nl`
[2] Department of Informatics, University of Zurich, Zürich, Switzerland
`{lucassteffen.krauter,david.guzmanpiedrahita}@uzh.ch`
[3] Department of Computational Linguistics, University of Zurich, Zürich, Switzerland
`andrianos.michail@cl.uzh.ch`

Abstract. We extend BeamAttack, an adversarial attack algorithm designed to evaluate the robustness of text classification systems through word-level modifications guided by beam search. Our extensions include support for word deletions and the option to skip substitutions, enabling the discovery of minimal modifications that alter model predictions. We also integrate LIME to better prioritize word replacements. Evaluated across multiple datasets and victim models (BiLSTM, BERT, and adversarially trained RoBERTa) within the BODEGA framework, our approach achieves over a 99% attack success rate while preserving the semantic and lexical similarity of the original texts. Through both quantitative and qualitative analysis, we highlight BeamAttack's effectiveness and its limitations. Our implementation is available at https://github.com/LucK1Y/BeamAttack.

Keywords: Model Robustness · Adversarial Attack · Beam search · Masked Language Models

1 Introduction

Social media platforms increasingly rely on machine learning algorithms for content filtering to identify misleading, harmful, or illegal content. Despite advancements, these models remain vulnerable to adversarial attacks, wherein the input text is manipulated to deceive the model. An adversarial sample in the text classification domain is generated from the original sample by applying the smallest possible change that leads to a misclassification by the target model, known as the victim classifier. The goal is to alter the sample just enough to trick the classifier into making an incorrect prediction, while preserving the original meaning

A. Fazla, L, Krauter and D. G. Piedrahita—These authors contributed equally.

for human interpretation [22]. Adversarial training (continued training of the models with adversarial samples to improve the resilience against such attacks) can make the victim models more robust, but novel attack methods can succeed even when tested on adversarially trained models. As adversaries continue to devise novel attack strategies that exploit vulnerabilities models were not trained to handle, we need to develop new attack techniques for assessing the robustness of text classification systems and finding potential weaknesses. These weaknesses can then be mitigated through adversarial training with samples from these novel techniques.

Previous adversarial attacks on text classification systems have explored character-level modifications such as DeepWordBug [5], word-level modifications such as TextBugger [11], BERT-ATTACK [12] and BeamAttack [23], and sentence level paraphrasing of the input text such as SEAs [20]. While these approaches have shown promise, there remains a need for comprehensive evaluation frameworks that assess attack effectiveness across diverse datasets and victim models, particularly those that have undergone adversarial training.

BeamAttack [23] runs a beam search with the heuristic of maximizing the victim model's output probability for the target class while minimizing the number of word modifications. It represents an effective grey-box approach that balances attack success with semantic preservation. Experiments show that BeamAttack achieves trade-off results compared with previous works, with adversarial exam-

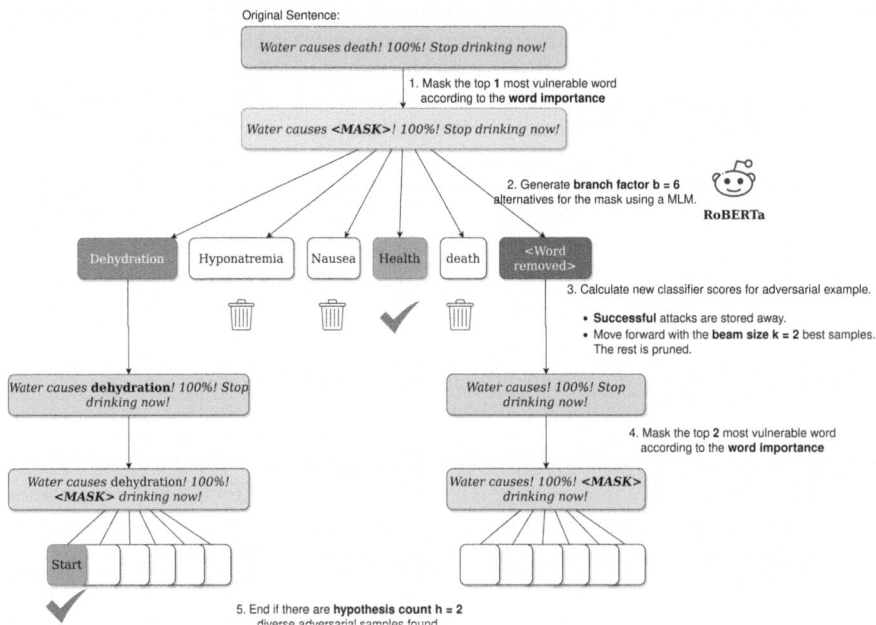

Fig. 1. Diagram depicting the series of actions in our BeamAttack algorithm, where beam size $k = 2$, branching factor $b = 6$, hypothesis count $h = 2$.

ples crafted by BeamAttack demonstrating high semantic similarity, low perturbation, and good transferability.

However, BeamAttack relies on logit-based word importance ranking to determine which words to target for substitution. LIME (Local Interpretable Model-agnostic Explanations) [19] offers a more precise alternative by training local linear models to estimate how each token influences a specific prediction. Unlike global explanation methods, LIME's local approach is ideal for adversarial attacks since we need to understand token importance for individual samples rather than model behavior overall. By prioritizing words with the highest LIME-derived importance scores, we hypothesize that attacks can achieve misclassification with fewer modifications, focusing perturbations on tokens that most directly influence the model's decision for that specific input.

We present an extended version of BeamAttack, evaluated within the BODEGA framework [17], which provides standardized metrics and datasets for benchmarking adversarial attacks on text classifiers. Our implementation was originally developed as part of our submission [15] to Task 6 of the CheckThat! Lab at CLEF 2024 [16]. Like the original BeamAttack, our approach uses beam search to identify optimal word-level modifications and generate multiple adversarial hypotheses capable of flipping the model's prediction. We rely on a Masked Language Model (MLM), specifically RoBERTa [13], to produce contextually appropriate substitution candidates.

Although our core algorithmic approach is similar to BeamAttack, our work makes several unique contributions. We have developed extensive evaluation frameworks and novel analysis methodologies that provide insights into the effectiveness of beam search-based adversarial attacks in various contexts. Our comprehensive analysis aims to deepen the understanding of adversarial attack mechanisms and identify opportunities to improve BeamAttack. Our key contributions are the following:

- **LIME for Word Importance Ranking:** Beyond the standard logit-based calculation of word importance, which is common in word-level adversarial attacks [11,12,23], we introduce and evaluate LIME-based [19] word importance ranking as an alternative strategy for identifying vulnerable words in an input text. We demonstrate that, for certain specific tasks (*HN* and *FC*) and transformer based encoders, using LIME to determine which words to replace achieves better results than the logit-based importance method, albeit with a significantly higher number of queries.
- **Token Removal:** We extend the original BeamAttack by incorporating token removal as an additional modification strategy, examining its impact on attack success rates and human assessment of semantic preservation of the original text.
- **Cross-Model Evaluation:** We provide a comprehensive analysis of BeamAttack across diverse victim models (BiLSTM, BERT, and adversarially trained RoBERTa) and datasets. This analysis offers insights into attack transferability and model-specific vulnerabilities.

 – **Linguistic and Quality Analysis:** We conduct both automatic and manual evaluations of successful adversarial samples, focusing on word substitution rates (WSR) and part-of-speech (POS) tag changes. By analyzing POS transition patterns across tasks and victim models, we gain insights into how linguistic structures are manipulated during attacks, how these affect perceived quality, and what they reveal about task difficulty and model robustness—informing future defense strategies.

2   Related Work

Adversarial text attacks can be categorized by their granularity of modification. Character-level methods change individual characters, such as DeepWordBug [5], which replaces characters to make words unrecognizable while remaining human-readable. Word-level methods modify text by substituting, adding, or removing words, exemplified by approaches that replace words with synonyms or out-of-vocabulary terms [8,12]. Sentence-level methods paraphrase entire sentences, such as SCPN [7], which generates adversarial examples through whole-text paraphrasing. Most recent work follows a search-based paradigm that ranks words by importance and searches for effective replacements using strategies like synonym substitution, masked language models [6,12], or evolutionary algorithms [2].

Attacks are further categorized by objective and access level. Targeted attacks aim to mislead models into specific incorrect classes, while untargeted attacks seek any misclassification. Regarding access, white-box attacks leverage full model access for gradient-based optimization [4], black-box attacks use only predictions with surrogate models [14], and grey-box attacks access both predictions and probability distributions [12,17].

Our method follows the established search-based research paradigm, operating as a grey-box, word-level, untargeted attack that leverages probability distributions to guide adversarial example generation.

Our method follows the established search-based paradigm as a grey-box, word-level, untargeted attack. We evaluate using the BODEGA score [17], which is a composite metric incorporating three components: the confusion score, which indicates whether the victim classifier's decision has changed and the attack was successful; the semantic score, which measures the similarity between the original and adversarial examples using BLEURT [21]; and the character score, which calculates the Levenshtein distance [9] as a character-level similarity measure. All three metrics range from 0 to 1, with higher values indicating greater success or similarity. Evaluation is conducted within the InCrediblAE framework from CheckThat! lab at CLEF 2024 [3,16].

3   Methodology

We build upon BeamAttack [23], extending it with additional word importance methods and replacement strategies. Our approach reframes word-level attacks as search problems, where the search space is determined by the number of words to be replaced and valid replacements for each word.[1]

Most existing techniques employ greedy search, replacing words sequentially by importance order until the model's prediction changes [8,12,14,18], which can lead to sub-optimal adversarial samples requiring unnecessary word changes.

We adapt beam search for adversarial text generation, where it explores multiple solutions concurrently in a tree structure. Each node represents a partial adversarial text and branches signify word substitutions or removals. Beam search maintains k most promising partial solutions at each step, balancing exploration with computational efficiency.

Our implementation operates by iteratively expanding promising nodes until reaching final adversarial examples. The root node represents the original input text, and each subsequent level adds one word modification. At each depth, we evaluate $k \times b$ nodes (k retained nodes from the previous level, each expanded with b word replacement candidates) and prune back to the k most promising solutions. Node selection prioritizes effectiveness in reducing the original class probability while increasing the target class probability.

The search continues until generating h successful adversarial samples that flip the model's classification. The final adversarial example is selected as the hypothesis with highest semantic similarity to the original text, measured by BLEURT scores [21]. This approach generates adversarial samples with fewer modifications while maintaining semantic coherence.

Our extended BeamAttack algorithm consists of two core components, both integrated with the beam search procedure: a method for determining the order in which words are modified (word importance ranking) and a strategy for how those words are replaced (word replacement strategy). An overview of the full algorithm is provided in Fig. 1. We enhance the original BeamAttack by incorporating LIME to compute word importance scores and by introducing word deletion as an additional replacement option.

3.1 Word Replacement Order

We evaluate two techniques for identifying the most important words: LIME [19] and logit-based importance scores [12].

LIME provides local explanations by perturbing input data and observing output changes. For each word, it generates perturbed sentence versions with word replacements, then fits an interpretable model to estimate word importance based on output variations.

Logit-based importance measures each word's influence on the model's output logits by removing the word and observing the resulting change in the logits. For input sentence $S = [w_0, \ldots, w_i, \ldots]$ and logit output $o_y(S)$ for the correct label y. The importance score I_{w_i} for each word w_i is defined as the difference in logits when the word is masked:

---

[1] The pseudocode for the algorithm can be found in Appendix B.

$$I_{w_i} = o_y(S) - o_y(S \setminus w_i),$$

where $S \setminus w_i$ is the sentence with w_i replaced by a [MASK] token. This score quantifies how much the presence of w_i contributes to the model's prediction.

3.2   Word Replacement Strategy

For each individual replacement, we enable the option to either keep the word intact or remove it altogether at any depth of the beam search. This added flexibility allows us to preserve the original word if replacements have no impact on changing the probability, thus reducing similarity and edit distances. Conversely, removing the word entirely may be the best strategy to confuse the model in more challenging cases, even though it might introduce grammar errors and inconsistencies into the adversarial sample. Following BeamAttack [23], we use a Masked Language Model (MLM) to leverage the model's pretraining to ensure the suggestions fit the context fluently.

For each depth of beam search, we evaluate the top b (branching factor) highest likelihood word replacements as suggested by the masked language model. This strategy ensures that the potential replacements not only fit well within the context of the sentence but also maintain the overall semantic coherence of the text.

3.3   Hyperparameters

For each dataset and victim model combination, we determined distinct hyperparameters for our BeamAttack approach through an initial exploration on the development sets (Appendix A). While this approach may not yield globally optimal combinations, it balanced effectiveness with computational constraints. Complete hyperparameter configurations and resource usage details are provided in Appendices C and D, respectively.

To reduce query overhead, we also lowered LIME's sample count from 5000 to 500. Initial observations showed this had minimal impact on attack quality for shorter texts, though it may affect longer samples. We leave further tuning of LIME's kernel parameters for future work.

4   Results

We report the automatic evaluation results of BeamAttack in Table 1, compared to the two strongest methods from previous work, BERT-ATTACK [12] and DeepWordBug [5], as well as OpenFact [10], the top-performing system submitted to the CheckThat! 2024 Lab Task 6. The complete results, including all baseline methods and all submissions to the shared task, can be found in the shared task overview paper [16].

Table 1. Comparison of adversarial attack results on BiLSTM, BERT, and RoBERTa classifiers across five misinformation detection tasks. Evaluation measures include BODEGA score (B.), confusion score (con), semantic score (sem), character score (char) and number of queries to the attacked model (Q.). The best scores for each classifier are in boldface. **Key observations**: (1) DeepWordBug consistently achieves the highest character preservation scores; (2) BeamAttack typically attains the highest confusion scores (attack success rates); (3) For the critical BODEGA metric, OpenFact outperforms other methods against more sophisticated models like BERT and RoBERTa due to superior semantic preservation while maintaining high character scores.

Task	Method	BiLSTM				BERT				RoBERTa			
		B.	con	sem	char.	B.	con	sem	char	B.	con	sem	char
HN	BERT-ATTACK	0.64	0.98	0.66	0.99	0.60	0.96	0.64	0.97	0.38	0.67	0.60	0.95
	DeepWordBug	0.41	0.53	0.77	**1.00**	0.22	0.29	0.78	**1.00**	0.16	0.21	0.76	**1.00**
	BeamAttack	**0.90**	**1.00**	0.91	0.99	0.84	**1.00**	0.86	0.97	0.67	**1.00**	0.72	0.91
	OpenFact	0.89	0.97	**0.93**	0.99	**0.91**	**1.00**	**0.92**	0.99	**0.83**	0.99	**0.86**	0.97
PR	BERT-ATTACK	0.53	0.80	0.72	**0.91**	0.43	0.70	0.68	0.90	0.20	0.32	0.69	0.91
	DeepWordBug	0.29	0.38	0.79	0.96	0.28	0.36	**0.79**	0.96	0.13	0.17	**0.81**	**0.96**
	BeamAttack	**0.70**	**0.97**	**0.80**	0.90	**0.69**	**0.98**	0.77	0.89	0.45	**0.97**	0.55	0.79
	OpenFact	0.65	0.94	0.77	0.89	0.68	0.97	0.77	0.89	**0.62**	0.93	0.75	0.87
FC	BERT-ATTACK	0.60	0.86	0.73	0.95	0.53	0.77	0.73	0.95	0.56	0.79	0.73	0.96
	DeepWordBug	0.48	0.58	**0.85**	**0.98**	0.44	0.53	**0.84**	**0.98**	0.37	0.46	0.83	**0.98**
	BeamAttack	0.76	**1.00**	0.81	0.94	0.79	**1.00**	0.83	0.96	**0.82**	**1.00**	**0.84**	0.97
	OpenFact	**0.80**	0.98	0.84	0.97	**0.80**	**1.00**	0.83	0.97	0.80	**1.00**	0.82	0.97
RD	BERT-ATTACK	0.29	0.79	0.41	0.89	0.18	0.44	0.43	0.96	0.17	0.41	0.42	0.95
	DeepWordBug	0.16	0.24	0.68	**0.99**	0.16	0.23	0.70	**0.99**	0.12	0.18	0.69	**0.99**
	BeamAttack	0.83	**1.00**	0.87	0.96	0.59	**0.80**	0.80	0.91	0.54	**0.87**	0.69	0.83
	OpenFact	**0.84**	0.95	**0.91**	0.98	**0.65**	0.78	**0.86**	0.95	**0.55**	0.71	**0.82**	0.93
C19	BERT-ATTACK	0.50	0.84	0.62	0.95	0.42	0.74	0.60	0.95	0.37	0.68	0.58	0.93
	DeepWordBug	0.33	0.48	0.70	**0.99**	0.27	0.39	0.71	**0.99**	0.20	0.28	0.72	**0.98**
	BeamAttack	**0.72**	**0.99**	0.78	0.92	0.71	**0.98**	0.78	0.92	0.66	**1.00**	0.72	0.91
	OpenFact	**0.72**	0.91	**0.83**	0.96	**0.72**	0.91	**0.82**	0.96	**0.72**	0.99	**0.78**	0.93

On automatic evaluation, our algorithm outperforms all other algorithms on BERT and BiLSTM victims, demonstrating the superior effectiveness of BeamAttack. While other algorithms may generate adversarial samples with better character or automatically measured semantic preservation score in specific tasks, they fail to achieve this consistently across the entire dataset. Moreover, our confusion scores (success rates) are always superior to those of other techniques.

Nevertheless, our algorithm needs many victim queries to achieve this goal. For example, in the task *RD* on the BERT classifier, our BeamAttack takes roughly double the number of queries compared to the Genetic algorithm (The details on number of queries to the victim model are reported in Appendix

Table 5). This highlights a trade-off between the quality of the generated adversarial samples and the computational resources required to achieve them.

4.1 Manual Evaluation Results

In contrast to the automatic evaluation results, the human evaluation of semantic preservation revealed a different trend. Our team, TextTrojaners, received the lowest scores in this round, with only 7% of our adversarial samples categorized as `Preserving the Original Meaning`, while 63% were labeled as `Changing the Original Meaning`, and 30% as `Nonsensical` [16]. Notably, the manually evaluated subset consisted solely of samples from the *FC* dataset, with 96% targeting the adversarially trained RoBERTa model. This outcome indicates that our attack strategy, although effective against automated systems, often altered the original semantics.

One possible explanation for the 63% of samples labeled as semantically changed is the use of masked language models (MLMs) for word replacement. While these replacements are contextually and grammatically plausible, they do not explicitly aim to preserve the meaning of the original token, often introducing substitutions such as antonyms or semantically divergent alternatives. For the 30% of samples deemed nonsensical, we suspect that our word removal strategy significantly contributes to syntactic breakdowns. To better understand these failure cases, we conduct an in-depth manual analysis of these ungrammatical cases in Sect. 5.1, identifying specific patterns and causes of structural degradation.

5 Analysis of Output Adversarial Samples and Discussion

We analyze structural differences between original and adversarial samples through POS tag sequences and word substitution ratios. Our analysis covers 6,503 successful adversarial examples (97.16% success rate) to examine the magnitude and nature of perturbations.

Figure 2 presents the distribution of word-level edit ratios required to fool each victim model, serving as a proxy for task difficulty. Longer texts (*HN*: 1,972 words) show lower relative substitution rates (4.8%) than shorter texts, indicating that adversarial perturbations must be more concentrated in longer documents to achieve success.

We label original and adversarial texts using the `flair/pos-english` model[2], selected for its contextual understanding and fine-grained tagging capabilities [1]. Our analysis focuses on 2,130 single-word substitutions that preserve POS sequence length (90.48% of single substitutions), as deviations likely indicate tagging noise.[3] Among these samples, 933 (43.8%) show no POS tag changes,

---

[2] Accessible at: https://huggingface.co/flair/pos-english.

[3] Examples of outliers where a single-word substitution led to a five-token POS tag change in one case, and an eight-token change in another, can be found in Appendix G.

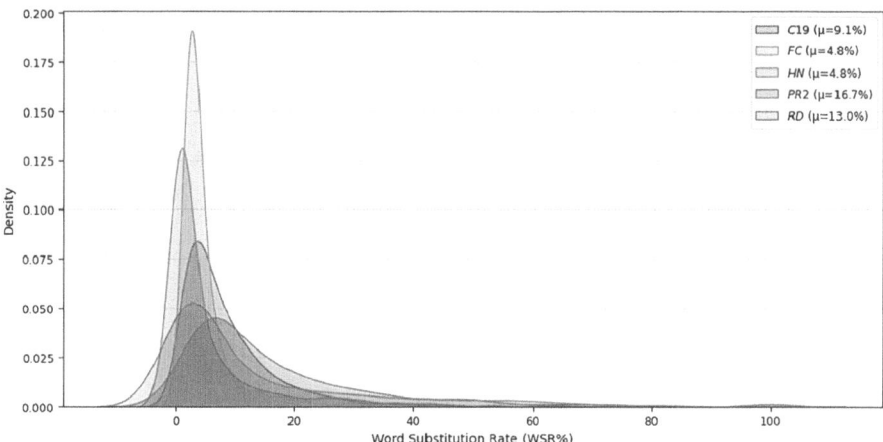

Fig. 2. Kernel Density Estimation (KDE) plot showing the density distribution of Word Substitution Rate (WSR (%)) [23] per adversarial sample, grouped by task and aggregated over all victim models. Mean WSR for each task is indicated in the legend. Simpler tasks such as *FC* and *HN* exhibit both lower mean WSR and smaller standard deviations compared to the other three tasks.

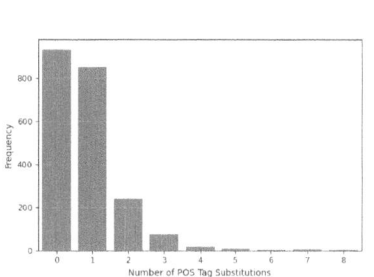
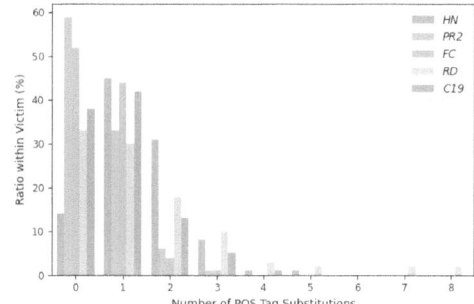

(a) Distribution of number of POS tag substitutions for successful samples with a single word-substitution, where the number of POS tags remained unchanged, aggregated across all tasks and victim models.

(b) Normalized distribution of number of POS tag substitutions for successful samples with a single word substitution, where the number of POS tags remained unchanged. Frequencies are normalized within each task and aggregated over all victim models.

Fig. 3. Histograms of POS tag changes across and within tasks.

while 851 (39.95%) contain exactly one change, following a long-tailed distribution where most edits are syntactically minimal (Fig. 3a). This indicates that lexical replacements often preserve syntactic structure.

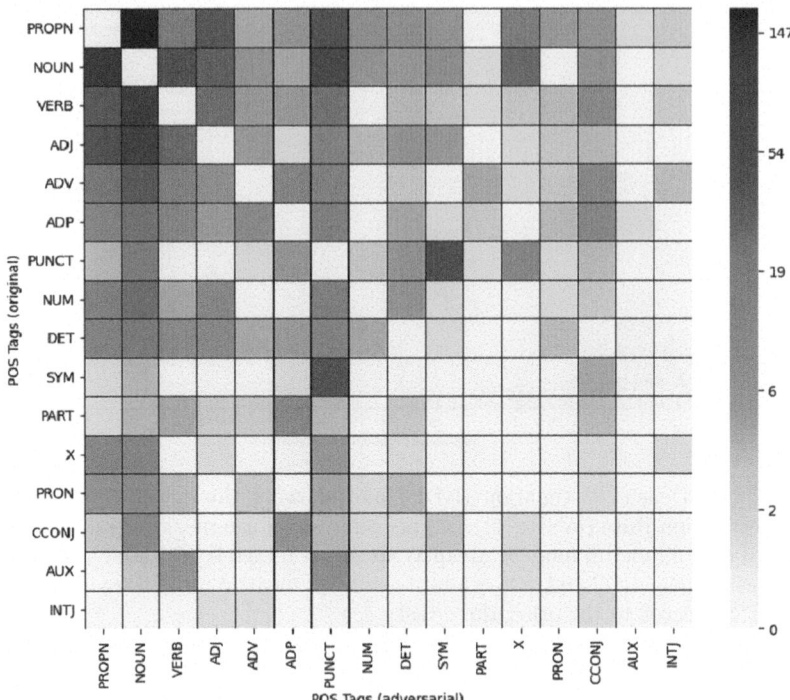

Fig. 4. POS transition matrix for the successful samples where the length of POS tags remained the same, and only one POS tag was substituted to another, aggregated over all tasks and victim models.

Task-specific analysis (Fig. 3b) reveals notable differences across datasets. *PR2* samples exhibit the fewest changes, with 58% remaining unchanged, suggesting that adversarial outputs often retain grammatical consistency. In contrast, *RD* shows a broader distribution, with higher frequencies of 2- and 3-token POS tag substitutions, and a relatively higher occurrence of samples with 7 or 8 changes. This may be attributed to differences in dataset sources: *PR2* (Propaganda Detection) texts are drawn from structured news articles, while *RD* (Rumor Detection) texts originate from informal Twitter threads, which are often unstructured and include links, user mentions, and varied formatting. These characteristics likely make it more challenging for the MLM to substitute tokens while preserving grammatical structure.

Figure 4 displays the POS transition matrix from samples with single POS tag substitutions. The matrix appears roughly symmetrical, indicating that similar POS tags can be switched interchangeably. Notable patterns include frequent transitions between PROPN (Proper Nouns) and NOUN, as well as between SYM (Symbols) and PUNC (Punctuation), suggesting these categories are particularly susceptible to adversarial substitution.

5.1   Manual Evaluation Analysis

We analyze adversarial samples labeled as `Nonsensical` to identify failure modes. Our analysis reveals two primary substitution categories: (1) grammatically-preserving noun replacements that maintain structure while introducing semantic absurdity, and (2) structure-breaking replacements that alter grammatical integrity.

`Original:`   "Fanaticism   is   a   monster   that   pretends   to   be   the   child of religion" Voltaire #JeSuisCharlie

`Adversarial:`   "". is   a   monster   page   pretends to   Advertements   the advertisement advertisement Posted" a #JeSuisCharlie

`Original:`   Rage Against the Machine stopped being together .

`Adversarial:`   Rage Against the Machine stopped being pants .

Fig. 5. The first example demonstrates the substitution of multiple words with "advertisements", while the second illustrates a nonsensical noun substitution that preserves grammatical structure but introduces semantic absurdity.

Several consistent substitution patterns emerge from our analysis of adversarial examples. For instance, the word "advertisement" was introduced in 488 cases, likely to signal commercial intent and shift model predictions. Conversely, "facts" appeared in 559 cases, perhaps to suggest factual content and steer the model in the opposite direction. Politically charged terms like "nazis" (137 instances) and "trump" (70) were also frequently inserted, likely probing for bias. Other recurring substitutions include instruction-like words such as "caps" (301 instances) and semantically absurd but grammatically valid words like "pants", which distort meaning while preserving syntax (see Fig. 5).

These patterns reflect the behavior of the BeamAttack algorithm. For each masked token, RoBERTa generated candidate words, and beam search selected the one that caused the largest change in the target model's output logits. This strategy favored substitutions with the highest impact on predictions, regardless of whether they made the sentence more coherent or more absurd.

Overall, the attack effectively identifies sensitive target words but often selects replacements with low semantic quality. As a result, the adversarial samples remain syntactically plausible while altering meaning or introducing cues that manipulate model behavior.

6   Conclusion

We presented BeamAttack, a beam search-based adversarial text generation method that we extend and evaluate using the BODEGA framework. While the core algorithm builds on the original BeamAttack [23], our work explores several improvements and new directions, including alternative word importance ranking via LIME, task and model-specific hyperparameter optimization, and an in-depth analysis of attack quality through both automatic and manual evaluation.

Our results show that BeamAttack achieves very high success rates across diverse models and datasets, but often at the cost of lower semantic and character similarity, and very high query counts—particularly when attacking robust models like adversarially-trained RoBERTa. These tradeoffs stem from the method's brute-force nature and the large beam width, which increases search coverage but also introduces less coherent substitutions. We found that the optimal hyperparameter configuration, including the choice between LIME and logit-based word importance methods, depends on both dataset characteristics (input text length, writing style ranging from formal to informal social media content) and target model capabilities. More robust models like adversarially-trained RoBERTa require stronger hyperparameter settings that increase success rates but reduce adversarial sample quality.

This creates a fundamental trade-off that can be controlled through hyperparameter tuning: practitioners can prioritize either higher success rates with lower quality samples, or focus on generating only high-quality adversarial examples with potentially lower success rates. The choice also affects computational efficiency and query budget constraints when targeting black-box models.

Manual evaluation highlighted issues with incoherent substitutions, likely caused by overgeneration in RoBERTa due to large beam width. We suggest that constraining substitutions—e.g., by limiting POS tag group changes—could improve fluency, though at a potential cost to success rate. Nevertheless, both manual quality and attack effectiveness could be improved through targeted modifications informed by analysis, such as restricting specific transition patterns or optimizing branching behavior for more efficient and plausible adversarial generation.

7   Limitations

Our approach has several limitations that point to directions for future improvement. First, removing words can lead to syntactically incorrect sentences. In addition, RoBERTa's context-based alternatives may produce semantically opposite replacements since they ignore the original word's meaning. Future work could address both issues by using large language models (LLMs) for substitution, or applying similarity-based filtering to preserve semantic intent.

Second, scoring beams at each search step—rather than only at the final hypothesis—could improve semantic coherence. Targeted attacks, which aim to push the model toward a specific class, may also benefit from these enhanced candidate generation methods.

Finally, while we showed that LIME provides more effective word importance estimates than a logit-based method for specific datasets and models, computational constraints and a large hyperparameter space prevented full ablation studies. Although we developed intuitions for hyperparameter tuning based on factors such as input length and model robustness (such as adversarially trained RoBERTa requiring a higher branching factor b), comprehensive studies are needed to generalize these findings across architectures and tasks.

Acknowledgments. The initial version of this work was prepared during the Machine Learning for NLP2 course at the University of Zurich. We thank Simon Clematide, who teaches the course, and the Department of Computational Linguistics for their support.

Disclosure of Interests. The authors have no competing interests to declare that are relevant to the content of this article.

A   Dataset Statistics

The basic statistics of the provided datasets are summarized in Table 2. Each dataset is divided into three subsets: training, development (dev), and attack. The training and dev subsets are used to train the classifier, while the attack subset serves as the evaluation dataset, where the effectiveness of our attack is measured. Each dataset is designed for a binary classification task, and therefore, the positive rate corresponds to the portion of samples belonging to class 1, and is therefore also a measure of class imbalance. Notably, we observed significant variations in the length of samples across the datasets, measured in terms of the number of words and characters, which are also reported in the table.

Next, we briefly outline the different text domains described in the Check-That! lab at CLEF 2024 [3]:

- **Style-based news bias assessment (*HN*):** Categorizing news articles as either credible or non-credible based on stylistic cues.
- **Propaganda detection (*PR2*):** Text passages from news articles that employ propaganda techniques to influence readers.
- **Fact checking (*FC*):** Evaluating the accuracy of news articles by considering contextual information from given related Wikipedia snippets.
- **Rumor detection (*RD*):** Identifying Twitter threads that disseminate information without a reliable source.
- **COVID-19 misinformation detection (*C19*):** Comprises social media messages that convey either factual information or misinformation about the COVID-19 pandemic. The classifier must rely on subtle cues, such as writing styles reminiscent of those found in high-quality news sources (*HN*), to make its assessments.

Table 2. Datasets used in BODEGA, described by the task, number of instances in training, attack and development subsets, the overall percentage of positive (non-credible) class, and the average sample length measured in both words and characters.

Task	Training	Attack	Dev	Positive	Avg. Length (Words)	Avg. Length (Characters)
RD	8,683	415	2,070	34.24%	146.16	1111.63
FC	172,763	405	19,010	51.27%	47.23	257.16
HN	60,234	400	3,600	50.00%	313.98	1972.45
PR2	11,546	407	3,186	71.03%	21.82	128.01
C19	1,130	595	–	34.96%	43.06	131.96

B   BeamAttack Algorithm

Input: $S, Y, k, b, h, model$
Output: Adversarial example S_{adv}
1 Initialize beam as priority queue with $(S, model.predict\_proba(S)[Y])$
2 Compute word importance scores using LIME or logit-based method
3 Sort words by descending importance

4 **while** *beam not empty* **do**
5   Initialize empty list for new candidates
6   **foreach** $(S_{cur}, score)$ *in beam* **do**
7   **foreach** *important word* w_j **do**
8   Obtain top b replacements for w_j using masked language model
9   **foreach** *replacement* **do**
10   Create new sentence S_{new} by replacing w_j
11   Compute $new\_score = model.predict\_proba(S_{new})[Y]$
12   Add (S_{new}, new_score) to current candidates

13   Sort current candidates by score and keep top k in beam
14   **foreach** $(S_{cand}, score)$ *in beam* **do**
15   **if** $model.predict(S_{cand}) \neq Y$ **then**
16   Add $(S_{cand}, score)$ to successful candidates

17   **if** *# successful candidates* $\geq h$ **then**
18   **break**

19 Select S_{adv} from successful candidates using BLEURT similarity with original S
20 **return** S_{adv} if found, else None

Algorithm 1: BeamAttack: A beam search-based adversarial attack. Inputs: tokenized sentence $S = [w_0, \ldots, w_n]$, gold label Y, beam size k, branching factor b, and required successful hypotheses h. The attack perturbs important words using replacements from a masked language model and scores them using model confidence.

C   Hyperparameter Search

For each dataset and victim model combination, we determined a different set of hyperparameters for our BeamAttack approach. To achieve this, we employed an informed random search on subsets of 10–50 samples for each task and victim model. Our search strategy was guided by intuition and initial observations, allowing us to effectively fine-tune the hyperparameters. We considered the following ranges for the hyperparameter values:

1. **Beam size** k: We tested values between 10 to 100, increasing in steps of 5 (i.e., 10, 15, 20, ...). We increased the beam size until the improvement in the BODEGA score became negligible (less than 1 point). Figure 6 illustrates how this approach led to convergence.
2. **Branching factor** b: We tested a range of values similar to the chosen k values, within a range of -20 to +20 around the selected k values.
3. **Hypothesis count** h: Based on initial experiments, we decided to test hypothesis counts of 5, 10, and 20. This decision was made considering the difficulty of the task at hand and the BODEGA scores obtained during the hyper-parameter search. We observed that further increasing h did not significantly improve the scores, as explained in Sect. E.2.

Table 3. Selected hyper-parameter combinations for each combination of task and victim model.

Task	Victim	Beam size k	Hypotheses h	Branches b	Importances
PR2	BiLSTM	25	30	50	logit-based
	BERT	60	10	80	logit-based
	RoBERTa	60	10	80	logit-based
FC	BiLSTM	10	10	30	logit-based
	BERT	10	10	30	LIME
	RoBERTa	60	10	80	logit-based
RD	BiLSTM	10	10	30	logit-based
	BERT	10	10	10	logit-based
	RoBERTa	20	20	40	logit-based
HN	BiLSTM	20	10	20	logit-based
	BERT	20	5	20	LIME
	RoBERTa	10	10	10	LIME
C19	BiLSTM	15	10	30	logit-based
	BERT	10	10	30	logit-based
	RoBERTa	40	10	60	logit-based

4. **Word importance scoring method**: We compared the results of both word importance scoring methods, LIME and logit-based, in the initial experiments of each dataset and victim model combination.

We report the selected parameters for each scenario in Table 3.

For the Mask Language Model we used RoBERTa-large[4] to replace the masks, with one exception: for the Covid-19 (*C19*) task on the BERT classifier, we utilized the `vinai/bertweet-large`[5] model instead. This decision was motivated by the intuition that, given the Twitter domain of the COVID-19 dataset, this model might perform better.

D Computational Resources

This chapter outlines the computational resources utilized and the hyper-parameter tuning strategies employed to optimize our BeamAttack approach, which demanded substantial computational power. To overcome the limitations imposed by computational and time constraints, we resorted to a random search strategy to determine the optimal parameter set for each scenario, recognizing that this approach may not necessarily result in the identification of the globally optimal combination.

---

[4] https://huggingface.co/FacebookAI/roberta-large.
[5] https://huggingface.co/vinai/bertweet-large.

Table 4. Running time statistics (in seconds) for each combination of task and victim model, measured on the entire dataset.

Task	Victim	Time per Example (s)	Total Attack Time (s)	Total Evaluation Time (s)
PR2	BiLSTM	1.513	629.279	15.843
	BERT	12.586	5235.587	15.986
	RoBERTa	72.192	30031.747	20.433
FC	BiLSTM	2.111	854.859	54.206
	BERT	7.988	3235.195	51.033
	RoBERTa	3.467	1403.967	44.047
RD	BiLSTM	13.248	5497.992	560.019
	BERT	70.347	29193.831	488.945
	RoBERTa	72.192	30031.747	20.433
HN	BiLSTM	9.578	3831.309	615.956
	BERT	86.049	34419.532	660.498
	RoBERTa	144.188	57675.011	544.444
C19	BiLSTM	2.260	1344.562	95.838
	BERT	12.453	7409.627	103.194
	RoBERTa	54.731	32564.714	117.173

Table 5. Comparison of number queries on BiLSTM, BERT, and RoBERTa classifiers across five misinformation detection tasks. OpenFact did not publish their number of queries.

Task	Method	BiLSTM	BERT	RoBERTa
HN	BERT-ATTACK	487.9	648.4	1782.0
	DeepWordBug	396.2	395.9	384.3
	BeamAttack	937.0	4327.7	4596.6
PR	BERT-ATTACK	61.4	80.2	117.6
	DeepWordBug	27.5	27.4	26.9
	BeamAttack	593.4	1373.4	10286.8
FC	BERT-ATTACK	132.8	146.7	164.1
	DeepWordBug	54.4	54.3	53.4
	BeamAttack	1549.1	1390.8	498.9
RD	BERT-ATTACK	985.5	774.3	951.9
	DeepWordBug	232.8	232.7	229.6
	BeamAttack	3831.7	10618.9	15458.1
C19	BERT-ATTACK	127.2	161.7	198.3
	DeepWordBug	61.2	61.0	61.0
	BeamAttack	837.4	2628.9	6491.4

We primarily used Kaggle's free GPU infrastructure, which offers NVIDIA Tesla P100s and T4x.[6] We report the running time statistics for the final runs of each model and dataset combination in Table 4, and the number of victim queries in Table 5.

E   Ablation Experiments

We conduct a small-scale ablation study to investigate the impact of various hyperparameters on the performance of the BeamAttack algorithm, shedding some light on the relationship between these parameters and the algorithm's effectiveness. Specifically, we investigate the impact of logit-based versus LIME, beam size k, branching factor b and the hypothesis count h on the algorithm's performance, exploring each of these parameters in the context of specific scenarios of task and victim model.

E.1   Word Importance

Besides the logit-based approach from BERT-Attack [12], we utilized LIME, a state-of-the-art interpretability framework, to identify crucial words in a sample that influence the victim's decision-making process. By modifying these words, we increased the victim's susceptibility to our adversarial attack.

We report some comparisons of LIME in Table 6. From this table, we can gather multiple insights. For instance, for the BERT victim the choice of importance method depends on the task dataset, with RD favoring the logit-based method and FC preferring LIME. For the BiLSTM, the logit-based method is superior on all datasets. For RoBERTa, which is architecturally similar to BERT, the choice of importance method also depends on the dataset, with FC favoring LIME.

Furthermore, we observed that for the RoBERTa victim and $C19$ task, the choice between LIME and logit-based depends on other hyperparameters such as beam size, branching factors, and hypothesis count. More specifically, our results suggest transformer-based classifiers, namely classifiers BERT and RoBERTa are generally more vulnerable to LIME over logit-based for the HN and FC datasets. Both these datasets classify whether a statement contains facts, with HN using subtle clues in writing styles and FC using external knowledge from Wikipedia. We leave the investigation behind this observation to future work.

Our results suggest that exploring explainable AI frameworks for adversarial attacks is a promising direction of research. However, our experiments show that LIME only improved upon the logit-based approach on a few datasets. Additionally, LIME requires more queries, and when the improvement was marginal, we opted for the logit-based approach due to resource constraints.

---

[6] For the submission phase, we used an NVIDIA Tesla T4 GPU.

Table 6. Evaluation metrics for different importance methods on different tasks and victims. We always bold the better BODEGA score with the same settings for either LIME or BERT. k refers to the selected beam size, h to hypothesis counts and b to the branching factor.

Task	Victim	Importance Method	k, h, b	Queries per example	Success	Semantic	Character	BODEGA
FC	BiLSTM	LIME	10, 10, 30	2597.3	0.99	0.8111	0.9332	0.7511
FC	BiLSTM	logit-based	10, 10, 30	1549.1	1.00	0.8059	0.9410	**0.7642**
HN	BiLSTM	LIME	10, 10, 20	1529.04	1.00	0.9043	0.9837	0.8905
HN	BiLSTM	logit-based	10, 10, 20	665.72	1.00	0.9112	0.9893	**0.9031**
RD	BERT	LIME	10, 10, 10	16342.9	0.70	0.8160	0.9174	0.5380
RD	BERT	logit-based	10, 10, 10	5947.0	0.90	0.8189	0.9130	**0.6874**
FC	BERT	LIME	10, 10, 10	5520.44	1.00	0.8392	0.9546	**0.8044**
FC	BERT	logit-based	10, 10, 10	732.66	1.00	0.8188	0.9466	0.7787
FC	RoBERTa	LIME	10, 10, 20	557.08	1.00	0.8398	0.9644	**0.8110**
FC	RoBERTa	logit-based	10, 10, 20	112.4	1.00	0.8392	0.9641	0.8105
C19	RoBERTa	LIME	40, 10, 40	9714.4	1.00	0.5625	0.8602	0.4860
C19	RoBERTa	logit-based	40, 10, 40	5505.0	1.00	0.6541	0.8939	**0.5853**
C19	RoBERTa	LIME	10, 10, 20	1858.0	1.00	0.6304	0.8669	**0.5500**
C19	RoBERTa	logit-based	10, 10, 20	972.7	1.00	0.6120	0.8677	0.5316

E.2   Hypothesis Count H

Table 7. Results of different hypothesis counts h with the **RoBERTa** victim and different tasks.

Task	hypothesis count h	beam size k	branching factor b	Subset size	Queries per example	Success	Semantic	Character	BODEGA
RD	20	20	40	20	37576.3	0.95	0.6645	0.7952	0.5501
RD	10	20	40	20	37764.7	0.95	0.6654	0.7951	0.5508
PR2	20	40	40	30	3744.8	1.00	0.5996	0.8235	0.5101
PR2	10	40	40	30	3656.9	1.00	0.5993	0.8280	0.5130
C19	20	40	40	10	5505.0	1.00	0.6541	0.8939	0.5853
C19	10	40	40	10	5505.0	1.00	0.6541	0.8939	0.5853

Our algorithm's flexibility in replacing, skipping, or removing words, combined with the use of a masked language model, ensures that the resulting sentence remains semantically correct. The beam search approach enables us to efficiently query for multiple adversarial samples and use the one that is most semantically close to our original sentence.

We conducted an ablation study on the hypothesis count h, which defines how many adversarial samples to use before selecting the closest. We explored different scenarios, incrementing the parameter from 10 to 20, but observed only minimal effects, resulting in a negligible improvement or deterioration of less than 1% in the BODEGA score. We provide a small-scale report of our findings for the **RoBERTa** victim and various tasks in Table 7. Notably, in some cases, it came at the cost of requiring more queries. Therefore, we opted to maintain a constant value of 10 for the hypothesis count in most cases.

E.3 Beam Size K

Table 8. Scores for Different Beam Sizes k for task **PR2** and victim **RoBERTa**. Here hypothesis counts h is always 10 and $b = k$ for each case.

beam size k branching factor b	Subset size	Queries per example	Success	Semantic	Character	BODEGA
10	30	**530.53**	0.93	0.58	0.78	0.44
20	30	1334.07	1.00	0.56	0.78	0.47
30	30	2451.77	1.00	0.58	0.79	0.49
35	30	3090.13	1.00	0.59	0.81	0.50
40	30	3656.93	1.00	0.60	0.83	0.51
50	30	5308.23	1.00	0.60	**0.84**	0.52
60	30	7151.87	1.00	0.61	**0.84**	0.53
70	30	9279.90	1.00	0.61	0.83	0.53
80	30	11374.77	1.00	0.62	**0.84**	0.54
90	30	14546.27	1.00	0.62	**0.84**	0.54
100	30	16499.97	1.00	**0.63**	**0.84**	**0.55**

We also conducted an experiment to investigate the effect of the beam size k on our algorithm's performance. Specifically, we experimented on a subset of 30 samples from the*PR2* dataset using the **RoBERTa** victim model. The results are presented in Fig. 6, with the exact scores provided in Table 8. Our analysis reveals a consistent improvement in all scores as the beam size increases.

However, we encountered a significant limitation. As shown in the table, a beam size of 100 requires 16,500 queries per sample, which is impractical for limited GPU setup. Therefore, we sought an optimal trade-off where the scores would reach a plateau. In the case of *PR2* and RoBERTa, we determined that a beam size of 60 strikes a reasonable balance between performance and computational feasibility.

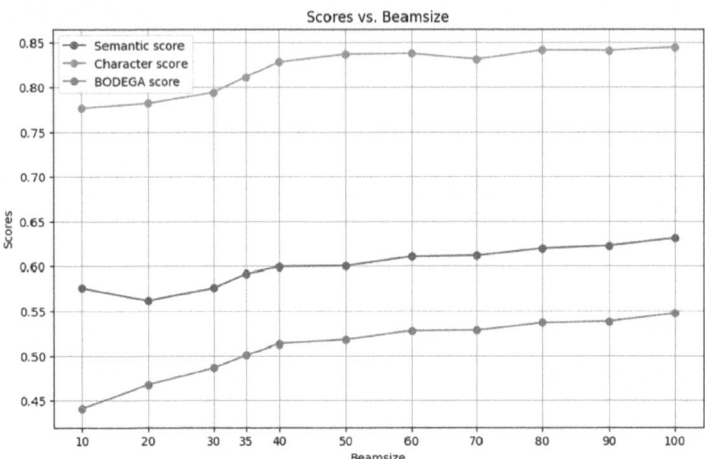

Fig. 6. Comparison of the semantic score, character score and confusion score across different settings for the beam size k.

Table 9. Scores for Different branching factors b with a fixed Beam size k for task **PR2** and victim **RoBERTa** with hypothesis counts $h = 10$.

branching factor b	Beam size k	Subset size	Queries per example	Success	Semantic	Character	BODEGA
30	40	30	**3032.733**	1.0	0.5835	0.7924	0.4904
40	40	30	3656.933	1.0	0.5993	0.8280	0.5130
50	40	30	4476.300	1.0	0.5957	**0.8331**	0.5115
60	40	30	4836.333	1.0	0.6058	0.8309	0.5210
70	40	30	5799.533	1.0	**0.6113**	0.8249	**0.5253**

E.4   Branching Factor B

We further explored the impact of the branching factor b on our algorithm's performance, using a fixed beam size of 40. We conducted an experiment on a subset of 30 samples from the *PR2* dataset, employing the **RoBERTa** victim model. The results are illustrated in Fig. 7, with the exact scores provided in Table 9. Our analysis reveals that increasing the branching factor b leads to a slight improvement in the BODEGA score.

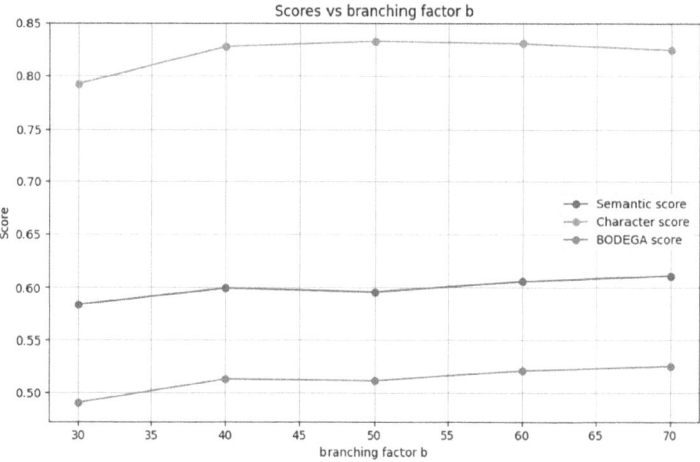

Fig. 7. Comparison of the semantic score, character score and confusion score across different settings for the branching factor *b*.

F   POS Tag Mapping

Table 10. Mapping from Penn Treebank POS tags (as used by Flair) to Universal POS tags.

Penn Tag	Description	Universal POS
ADD	Email address	X
AFX	Affix	X
CC	Coordinating conjunction	CCONJ
CD	Cardinal number	NUM
DT	Determiner	DET
EX	Existential there	PRON
FW	Foreign word	X
HYPH	Hyphen	PUNCT
IN	Preposition or subord. conjunction	ADP
JJ	Adjective	ADJ
JJR	Adjective, comparative	ADJ
JJS	Adjective, superlative	ADJ
LS	List item marker	X
MD	Modal	AUX
NFP	Superfluous punctuation	PUNCT
NN	Noun, singular or mass	NOUN
NNP	Proper noun, singular	PROPN

(*continued*)

Table 10. (*continued*)

Penn Tag	Description	Universal POS
NNPS	Proper noun, plural	PROPN
NNS	Noun, plural	NOUN
PDT	Predeterminer	DET
POS	Possessive ending	PART
PRP	Personal pronoun	PRON
PRP$	Possessive pronoun	PRON
RB	Adverb	ADV
RBR	Adverb, comparative	ADV
RBS	Adverb, superlative	ADV
RP	Particle	PART
SYM	Symbol	SYM
TO	to	PART
UH	Interjection	INTJ
VB	Verb, base form	VERB
VBD	Verb, past tense	VERB
VBG	Verb, gerund or present participle	VERB
VBN	Verb, past participle	VERB
VBP	Verb, non-3rd person present	VERB
VBZ	Verb, 3rd person singular present	VERB
WDT	Wh-determiner	DET
WP	Wh-pronoun	PRON
WP$	Possessive wh-pronoun	PRON
WRB	Wh-adverb	ADV
XX	Unknown	X
-LRB-	Left round bracket	PUNCT
-RRB-	Right round bracket	PUNCT
.	Sentence-final punctuation	PUNCT
,	Comma	PUNCT
:	Colon or ellipsis	PUNCT
``	Opening quotation mark	PUNCT
''	Closing quotation mark	PUNCT
''	Quotation mark	PUNCT
$	Dollar sign	SYM

The complete mapping from Penn Treebank POS tags to Universal POS categories is provided in Table 10. For the purpose of analysis, we grouped the fine-grained Penn Treebank POS tags produced by the Flair POS tagger[7] into the 17 Universal POS (UPOS) tags defined by the Universal Dependencies framework.[8] This abstraction allows for broader syntactic interpretation while preserving sufficient granularity for downstream analysis.

G   Inconsistencies with POS Tagging

G.1   Sample with Five POS Tag Changes as a Result of Single-Word Substitution

The following is an example of how the *length* of the POS tags can be changed by 5, with only a single word substitution (Figs. 8 and 9):

Original:
Breaking: Armed men attack offices of French satirical magazine Charlie Hebdo, killing at least 10, police say http://t.co/tYCeEMKwOo
RT @WSJ: Breaking: Armed men attack offices of French satirical magazine Charlie Hebdo killing at least 10 police say http://t.co/WGxWZgzDEN
@WSJ Paris has too much "skin in the game" to stay on sidelines. They need a spot at the tip of the bullet.

Adversarial:
Breaking: Armed men attack offices of French satirical magazine Charlie Hebdo, killing at least 10, police say .
RT @WSJ: Breaking: Armed men attack offices of French satirical magazine Charlie Hebdo killing at least 10 police say http://t.co/WGxWZgzDEN
@WSJ Paris has too much "skin in the game" to stay on sidelines. They need a spot at the tip of the bullet.

Fig. 8. Text comparison showing the replacement of a URL with a period

The five POS tag changes in this example resulted from replacing a URL (http://t.co/tYCeEMKwOo) with a simple period. The original URL was tagged as a sequence of five distinct POS tags:

- NN (Noun)
- NFP (Non-final punctuation)
- NNP (Proper noun)
- : (Colon)
- ADD (Email/URL)

When our adversarial attack replaced this URL with a single period, the POS tagger assigned it only a single tag: NFP (Non-final punctuation). This substitution resulted in a reduction from five POS tags to one, accounting for the observed length difference in the POS tag sequences.

---

[7] https://huggingface.co/flair/pos-english.
[8] https://universaldependencies.org/u/pos/.

Original POS Tags:
['VBG', ':', 'JJ', 'NNS', 'VBP', 'NNS', 'IN', 'JJ', 'JJ', 'NN',
'NNP', 'NNP', ',', 'VBG', 'IN', 'JJS', 'CD', ',', 'NNS', 'VBP',
'NN', 'NFP', 'NNP', ':', 'ADD', 'NFP', 'NN', 'IN', 'NNP', ':',
'NN', ':', 'JJ', 'NNS', 'VBP', 'NNS', 'IN', 'JJ', 'JJ', 'NN', 'NNP',
'NNP', 'VBG', 'IN', 'JJS', 'CD', 'NNS', 'VB', 'NN', 'NFP', 'NNP',
'SYM', 'NN', ':', 'ADD', 'IN', 'NNP', 'NNP', 'VBZ', 'RB', 'JJ', '"',
'NN', 'IN', 'DT', 'NN', '"', 'TO', 'VB', 'IN', 'PRP', 'VBP', 'DT',
'NN', 'IN', 'DT', 'NN', 'IN', 'DT', 'NN']

Adversarial POS Tags:
['VBG', ':', 'JJ', 'NNS', 'VBP', 'NNS', 'IN', 'JJ', 'JJ', 'NN',
'NNP', 'NNP', ',', 'VBG', 'IN', 'JJS', 'CD', ',', 'NNS', 'VBP',
'NFP', 'NNP', 'IN', 'NNP', ':', 'NN', ':', 'JJ', 'NNS', 'VBP',
'NNS', 'IN', 'JJ', 'JJ', 'NN', 'NNP', 'NNP', 'VBG', 'IN', 'JJS',
'CD', 'NNS', 'VB', 'NN', 'NFP', 'NNP', 'SYM', 'NNP', ':', 'ADD',
'IN', 'NNP', 'NNP', 'VBZ', 'RB', 'JJ', '"', 'NN', 'IN', 'DT', 'NN',
'"', 'TO', 'VB', 'IN', 'PRP', 'VBP', 'DT', 'NN', 'IN', 'DT', 'NN',
'IN', 'DT', 'NN']

Fig. 9. Comparison of POS tag sequences with differences highlighted

G.2   Sample with 8 POS Tag Changes as a Result of Single-Word Substitution

Original:
French media say Paris hostage-taker demanding freedom of #CharlieHebdo suspects #ParisAttacks http://t.co/ATmBpjwW3b http://t.co/MI7DYRuBDP @SkyNews http://t.co/HqPRuk7rhb
@SkyNews is that an M-1 Carbine? ...

Adversarial:
French media say Paris hostage-taker demanding freedom of #CharlieHebdo suspects #terrorism http://t.co/ATmBpjwW3b http://t.co/MI7DYRuBDP @SkyNews http://t.co/HqPRuk7rhb
@SkyNews is that an M-1 Carbine? ...

Fig. 10. Replacement of hashtag #ParisAttacks with #terrorism, which alters semantic framing

In this example, the adversarial substitution of #ParisAttacks with #terrorism led the POS tagger to incorrectly split the hashtag into three tokens, each tagged as NOUN. This artifact results in multiple apparent POS tag changes despite a single-word substitution, highlighting limitations of POS tagging in handling special tokens like hashtags. The rest of the POS tag changes happened later in the sentence, so they were cut off (Figs. 10 and 11).

```
Original POS Tags:
'ADJ', 'NOUN', 'VERB', 'PROPN', 'NOUN', 'VERB', 'NOUN', 'ADP',
'NOUN', 'PROPN', 'VERB', 'SYM' , 'PROPN' , 'PROPN' , 'SYM', 'PROPN',
'SYM', 'PROPN', ...


Adversarial POS Tags:
'ADJ', 'NOUN', 'VERB', 'PROPN', 'NOUN', 'VERB', 'NOUN', 'ADP',
'NOUN', 'PROPN', 'VERB', 'NOUN' , 'NOUN' , 'NOUN' , 'SYM', 'PROPN',
'SYM', 'PROPN', ...
```

Fig. 11. Comparison of POS tag sequences with differences highlighted

H   Victim-Level Analysis

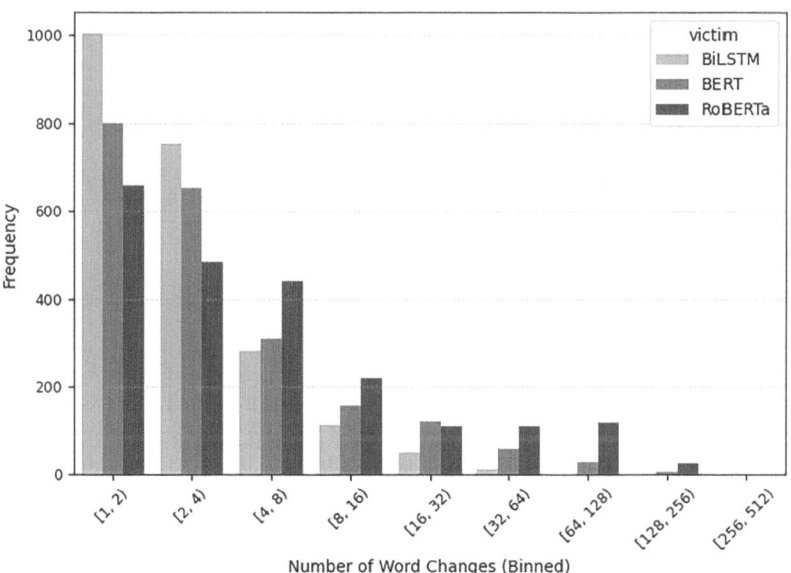

Fig. 12. Histogram showing the distribution of word changes per adversarial sample, grouped by victim model and aggregated over all tasks.

Figure 12 shows that the majority of successful examples require only a single word substitution, with two-word and three-word edits occurring less frequently, followed by a long-tailed distribution for higher edit counts. This trend holds consistently across victim models. For low-edit examples (1–3 changes), we observe the expected vulnerability order: BiLSTM > BERT > RoBERTa, confirming that simpler models are more susceptible to minimal perturbations. Interestingly, this ordering reverses for samples with four or more substitutions, suggesting that more robust models demand more extensive changes to be successfully misled (Fig. 13).

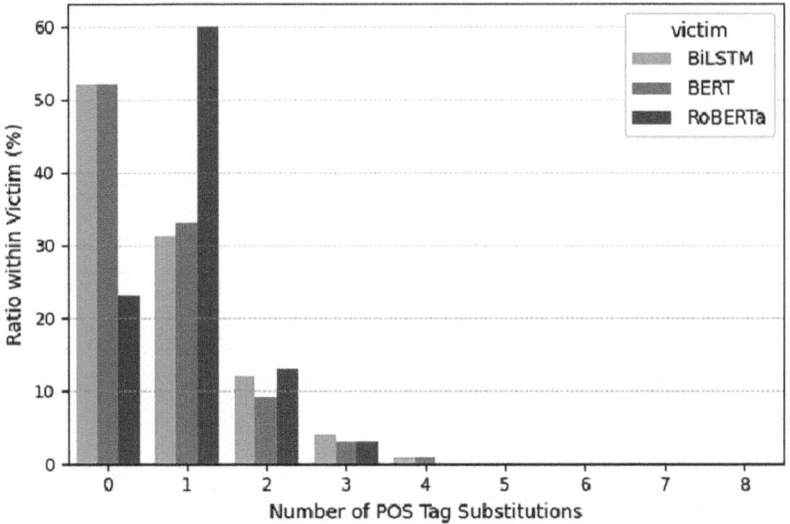

Fig. 13. Normalized distribution of number of POS tag substitutions for successful samples with a single word substitution, where the number of POS tags remained unchanged. Frequencies are normalized within each victim model and aggregated over the tasks.

References

1. Akbik, A., Bergmann, T., Blythe, D., Rasul, K., Schweter, S., Vollgraf, R.: FLAIR: an easy-to-use framework for state-of-the-art NLP. In: Ammar, W., Louis, A., Mostafazadeh, N. (eds.) Proceedings of the 2019 Conference of the North American Chapter of the Association for Computational Linguistics (Demonstrations), pp. 54–59. Association for Computational Linguistics, Minneapolis, Minnesota (2019). https://doi.org/10.18653/v1/N19-4010
2. Alzantot, M., Sharma, Y., Elgohary, A., Ho, B.J., Srivastava, M., Chang, K.W.: Generating natural language adversarial examples. In: Riloff, E., Chiang, D., Hockenmaier, J., Tsujii, J. (eds.) Proceedings of the 2018 Conference on Empirical

Methods in Natural Language Processing, pp. 2890–2896. Association for Computational Linguistics, Brussels, Belgium (2018). https://doi.org/10.18653/v1/D18-1316

3. Barrón-Cedeño, A., et al.: The CLEF-2024 CheckThat! Lab: check-worthiness, subjectivity, persuasion, roles, authorities, and adversarial robustness. In: Goharian, N., et al. (eds.) Advances in Information Retrieval, pp. 449–458. Springer Nature Switzerland, Cham (2024)

4. Ebrahimi, J., Rao, A., Lowd, D., Dou, D.: HotFlip: white-box adversarial examples for text classification. In: Gurevych, I., Miyao, Y. (eds.) Proceedings of the 56th Annual Meeting of the Association for Computational Linguistics (Volume 2: Short Papers), pp. 31–36. Association for Computational Linguistics, Melbourne, Australia (2018). https://doi.org/10.18653/v1/P18-2006

5. Gao, J., Lanchantin, J., Soffa, M.L., Qi, Y.: Black-box generation of adversarial text sequences to evade deep learning classifiers. In: 2018 IEEE Security and Privacy Workshops (SPW), pp. 50–56 (2018). https://doi.org/10.1109/SPW.2018.00016

6. Garg, S., Ramakrishnan, G.: BAE: BERT-based adversarial examples for text classification. In: Webber, B., Cohn, T., He, Y., Liu, Y. (eds.) Proceedings of the 2020 Conference on Empirical Methods in Natural Language Processing (EMNLP), pp. 6174–6181. Association for Computational Linguistics, Online (2020). https://doi.org/10.18653/v1/2020.emnlp-main.498

7. Iyyer, M., Wieting, J., Gimpel, K., Zettlemoyer, L.: Adversarial example generation with syntactically controlled paraphrase networks. In: Walker, M., Ji, H., Stent, A. (eds.) Proceedings of the 2018 Conference of the North American Chapter of the Association for Computational Linguistics: Human Language Technologies, Volume 1 (Long Papers), pp. 1875–1885. Association for Computational Linguistics, New Orleans, Louisiana (2018). https://doi.org/10.18653/v1/N18-1170

8. Jin, D., Jin, Z., Zhou, J.T., Szolovits, P.: Is BERT really robust? A strong baseline for natural language attack on text classification and entailment. Proc. AAAI Conf. Artif. Intell. **34**(05), 8018–8025 (2020). https://doi.org/10.1609/aaai.v34i05.6311

9. Levenshtein, V.I., et al.: Binary codes capable of correcting deletions, insertions, and reversals. In: Soviet physics doklady. vol. 10, pp. 707–710. Soviet Union (1966)

10. Lewoniewski, W., et al.: OpenFact at CheckThat! 2024: combining multiple attack methods for effective adversarial text generation. arXiv preprint arXiv:2409.02649 (2024)

11. Li, J., Ji, S., Du, T., Li, B., Wang, T.: TextBugger: generating adversarial text against real-world applications. In: Proceedings 2019 Network and Distributed System Security Symposium. NDSS 2019, Internet Society (2019). https://doi.org/10.14722/ndss.2019.23138

12. Li, L., Ma, R., Guo, Q., Xue, X., Qiu, X.: BERT-ATTACK: adversarial attack against BERT using BERT. In: Webber, B., Cohn, T., He, Y., Liu, Y. (eds.) Proceedings of the 2020 Conference on Empirical Methods in Natural Language Processing (EMNLP), pp. 6193–6202. Association for Computational Linguistics, Online (2020). https://doi.org/10.18653/v1/2020.emnlp-main.500

13. Liu, Y., et al.: RoBERTa: a robustly optimized BERT pretraining approach (2019). https://arxiv.org/abs/1907.11692

14. Maheshwary, R., Maheshwary, S., Pudi, V.: A strong baseline for query efficient attacks in a black box setting. In: Moens, M.F., Huang, X., Specia, L., Yih, S.W.t. (eds.) Proceedings of the 2021 Conference on Empirical Methods in Natural Language Processing, pp. 8396–8409. Association for Computational Linguistics, Online and Punta Cana, Dominican Republic (2021). https://doi.org/10.18653/v1/2021.emnlp-main.661

15. Piedrahita, D.G., Fazla, A., Krauter, L.: TextTrojaners at CheckThat! 2024: robustness of credibility assessment with adversarial examples through BeamAttack. In: Working Notes of CLEF 2024-Conference and Labs of the Evaluation Forum, CLEF (2024)

16. Przybyła, P., et al.: Overview of the CLEF-2024 CheckThat! Lab task 6 on robustness of credibility assessment with adversarial examples (incrediblae). In: Faggioli, G., Ferro, N., Galuščáková, P., García Seco de Herrera, A. (eds.) Working Notes of CLEF 2024 - Conference and Labs of the Evaluation Forum. CLEF 2024, Grenoble, France (2024)

17. Przybyła, P., Shvets, A., Saggion, H.: Verifying the robustness of automatic credibility assessment. Nat. Lang. Process. 1–29 (2024). https://doi.org/10.1017/nlp.2024.54

18. Ren, S., Deng, Y., He, K., Che, W.: Generating natural language adversarial examples through probability weighted word saliency. In: Korhonen, A., Traum, D., Màrquez, L. (eds.) Proceedings of the 57th Annual Meeting of the Association for Computational Linguistics, pp. 1085–1097. Association for Computational Linguistics, Florence, Italy (2019). https://doi.org/10.18653/v1/P19-1103

19. Ribeiro, M.T., Singh, S., Guestrin, C.: "why should I trust you?": Explaining the predictions of any classifier. In: Proceedings of the 22nd ACM SIGKDD International Conference on Knowledge Discovery and Data Mining, San Francisco, CA, USA, August 13-17, 2016, pp. 1135–1144 (2016)

20. Ribeiro, M.T., Singh, S., Guestrin, C.: Semantically equivalent adversarial rules for debugging NLP models. In: Gurevych, I., Miyao, Y. (eds.) Proceedings of the 56th Annual Meeting of the Association for Computational Linguistics (Volume 1: Long Papers), pp. 856–865. Association for Computational Linguistics, Melbourne, Australia (2018). https://doi.org/10.18653/v1/P18-1079

21. Sellam, T., Das, D., Parikh, A.: BLEURT: Learning robust metrics for text generation. In: Jurafsky, D., Chai, J., Schluter, N., Tetreault, J. (eds.) Proceedings of the 58th Annual Meeting of the Association for Computational Linguistics, pp. 7881–7892. Association for Computational Linguistics, Online (2020). https://doi.org/10.18653/v1/2020.acl-main.704

22. Zhang, W.E., Sheng, Q.Z., Alhazmi, A., Li, C.: Adversarial attacks on deep-learning models in natural language processing: a survey. ACM Trans. Intell. Syst. Technol. 11(3) (2020). https://doi.org/10.1145/3374217

23. Zhu, H., Zhao, Q., Wu, Y.: BeamAttack: generating high-quality textual adversarial examples through beam search and mixed semantic spaces. In: Kashima, H., Ide, T., Peng, W.C. (eds.) Advances in Knowledge Discovery and Data Mining, pp. 454–465. Springer Nature Switzerland, Cham (2023)

Simplified Longitudinal Retrieval Experiments: A Case Study on Query Expansion and Document Boosting

Jüri Keller[1]([✉]) [iD], Maik Fröbe[2] [iD], Gijs Hendriksen[3] [iD], Daria Alexander[3] [iD],
Martin Potthast[4] [iD], and Philipp Schaer[1] [iD]

[1] TH Köln - University of Applied Sciences, Cologne, Germany
`jueri.keller@th-koeln.de`
[2] Friedrich-Schiller-Universität Jena, Jena, Germany
[3] Radboud Universiteit Nijmegen, Nijmegen, The Netherlands
[4] University of Kassel, hessian.AI, ScaDS.AI, Kassel, Germany

Abstract. The longitudinal evaluation of retrieval systems aims to capture how information needs and documents evolve over time. However, classical Cranfield-style retrieval evaluations only consist of a static set of queries and documents and thereby miss time as an evaluation dimension. Therefore, longitudinal evaluations need to complement retrieval toolkits with custom logic. This custom logic increases the complexity of research software, which might reduce the reproducibility and extensibility of experiments. Based on our submissions to the 2024 edition of LongEval, we propose a custom extension of `ir_datasets` for longitudinal retrieval experiments. This extension allows for declaratively, instead of imperatively, describing important aspects of longitudinal retrieval experiments, e.g., which queries, documents, and/or relevance feedback are available at which point in time. We reimplement our submissions to LongEval 2024 against our new `ir_datasets` extension, and find that the declarative access can reduce the complexity of the code.

Keywords: Longitudinal Evaluation · Continuous Evaluation · Temporal Information Retrieval · ir_metadata · ir_datasets

 https://github.com/clef-longeval/ir-datasets-longeval

1 Introduction

Information Retrieval (IR) is centered on the task of finding relevant information to meet users' information needs. Therefore, systems must be able to cope with the ongoing flood of information. Current IR test collections abstract this complex task to enable a well-proven evaluation framework. This abstraction excludes certain aspects, such as temporal dynamics, that are inherent to real-world retrieval environments and tasks. The LongEval lab reintroduces some of these dynamics into offline evaluations with dynamic test collections [3,4,8]. That is, a changing document corpus, evolving information needs, and updated

J. Carrillo-de-Albornoz et al. (Eds.): CLEF 2025, LNCS 16089, pp. 117–127, 2026.
https://doi.org/10.1007/978-3-032-04354-2_8

relevance judgments are considered. This makes the evaluations more realistic in these regards and allows new directions for retrieval approaches. For example, if previous relevance judgments are available, based on user interactions, as in the LongEval test collections, they can be used as a relevance signal. Our previous works submitted to the lab have shown that such signals can strongly improve retrieval effectiveness at low cost [2,17,18].

While IR systems are exposed to changing data, users expect consistent, good effectiveness. To provide and maintain this, it is essential to regularly assess the system. Such longitudinal evaluations substantially increase the complexity compared to conventional Cranfield-style offline evaluations. With each new snapshot of a search setting, new versions of documents, queries, and qrels are introduced that need to be stored and maintained. Additionally, the retrieval approach and adequate baselines must take care to process only past data, while it would be helpful if they can easily process new and modified versions of the data. The number of necessary experiments increases with each variation of the parameters and snapshots. That means that the complexity, the demand for resources, and the propensity to errors of the experiments drastically increase.

These challenges of longitudinal evaluations make software submissions, as enabled by TIRA [12], difficult, as the submitted software becomes more complex and, as a result, the maintenance effort for organizers would increase (e.g., because every submission would come with different custom code to load the data). Still, in longitudinal settings, software submissions are especially interesting because they allow the application of the exact same approach to different snapshots.

To simplify longitudinal retrieval experiments and thereby facilitate software submissions in longitudinal scenarios, a standardized interface to dynamic test collections is necessary. We extend the `ir_datasets` framework [19] with methods tailored to dynamic test collections. We re-implement some of the proposed methods from last year's iteration of the LongEval lab and compare the results in terms of code complexity and retrieval effectiveness. Our `ir_datasets` extension provides a valuable benefit for software submissions, as the submitted software can focus on the main ideas of their approach and does not have to do dedicated data wrangling, thereby reducing the maintenance efforts of organizers. Even bugs in the `ir_datasets` extension can be handled without modifying the code of participants, e.g., by re-installing the extension in the submitted software. We hope that our `ir_datasets` extension also simplifies software submissions for longitudinal experiments in the future, as submitted software is less complex.

In summary, our core contributions are:

– Re-implementation of our original approaches submitted to LongEval 2024 [2,17,18], as PyTerrier transformers [20].
– An extension of `ir_datasets` for longitudinal evaluations.
– We added the LongEval Sci and LongEval Web datasets to the extension as the first dynamic test collections.
– A preliminary analysis of the re-implemented systems in terms of code complexity and retrieval effectiveness.

The `ir_datasets` extension[1] and our re-implemented approaches from LongEval 2024[2] are publicly available on GitHub.

2   Related Work

The observation that real-world search engines must operate in an evolving environment motivates longitudinal evaluations. Compared to traditional Cranfield-style evaluations, which use a static set of documents and relevance judgments, longitudinal retrieval evaluations must capture varying versions of relevance judgments and documents, thereby increasing the complexity of the experiments. Streamlining access to dynamic test beds should support researchers in conducting longitudinal evaluations. In this context, we discuss the related work.

Evolving Search Settings. While in traditional Cranfield-style evaluations most dynamics are abstracted, in real-world settings each component may change over time and thereby influence the retrieval results [16]. The document corpus or the content of the documents change [1,22]. Simultaneously, information needs evolve [9,22,28], ultimately affecting what users perceive as relevant [5].

Longitudinal Evaluations. Assessing how well retrieval systems work for users at different points in time requires repeated experiments. This significantly increases the complexity of IR experiments, as every point in time might have different restrictions on what information can be accessed. This evolving setting is investigated by the LongEval CLEF lab [3,4] that provides two dynamic test collections with up to 15 snapshots [8,13]. Beyond that, not many dynamic evolving test collections are available, so that studies often simulate dynamics or rely on other versioned datasets such as TREC-COVID, with five snapshots [29].

In many longitudinal experiments, each snapshot requires unique runs. This increases the complexity and computational resources of experiments. For instance, Keller et al. tested five different retrieval approaches on TREC-COVID, LongEval 2023, and a simulated dynamic test collection based on TripClick [16,23]. In total, 45 retrieval runs were created. Depending on the simulation strategy, many more snapshots can be created, resulting in even more retrieval runs [10].

Accessing Datasets. Thakur et al. propose the BEIR benchmark to investigate the out-of-distribution generalization of IR systems [27]. They enable straight forward evaluations over 18 datasets from various domains. MacAveney et al. introduce `ir_datasets`, a Python toolkit that provides a standardized interface to datasets typically used in IR research [19], which got also integrated into TIRA [11,12] to promote more reproducible shared tasks. The community can contribute to `ir_datasets` to extend the interface which is essentially what we did for longitudinal evaluations. Currently, the catalogue holds 55 datasets. Earlier works that provide interfaces to multi-dataset benchmarks are, for example, MultiReQA with eight datasets and a focus on question answering [15].

---

[1] https://github.com/clef-longeval/ir-datasets-longeval.

[2] https://github.com/clef-longeval/longeval-code/tree/main/clef25.

```
from ir_datasets_longeval import load

ir_datasets = load("longeval-sci/*")
for dataset in ir_datasets.get_datasets():
    run_experiment(dataset)
```

Listing 1. Processing all snapshots from the LongEval Sci collection.

3   Interfacing Dynamic Test Collections

Traditionally, test collections in IR consist of three components: a document corpus, a set of queries, and a set of relevance judgments [24]. In longitudinal evaluations, these three components evolve over time, yielding time as an additional dimension for evaluation. While this evolution is natural to all deployed IR systems, it is challenging to design offline experiments that account for it. Dynamic test collections capture this evolution in the form of versioned sub-collections [14,25]. The same test collection is captured repeatedly at different points in time, each with its own document corpus, queries, and qrels. We refer to such sub-collections as a snapshot, as they preserve one point in time of a search setting. Effective systems in these settings incorporate relevance informations from previous snapshots into their rankings, either by directly relying on previous relevance judgments or by training on previous queries and documents [3,4]. We extend the `ir_datasets` framework [19] to support dynamic test collections, and provide a unified access to different (subsets of) snapshots and their components.

The `ir_datasets` toolkit already models the access to datasets using hierarchical dataset IDs. We extend this by directly returning a meta dataset for IDs of higher order. The call for the id `longeval-sci/*` in Listing 1 provides a meta dataset containing all snapshots for the LongEval Sci test collection.

Additional IDs are available for accessing only the snapshots used as test data for a shared task. For example, the ID `longeval-sci/clef2025-test` provides a meta dataset capturing the snapshot `longeval-sci/2024-11` and the snapshot `longeval-sci/2025-01`. This allows for convenient testing of a retrieval approach on all snapshots.

Similar to the `get_datasets()` method of the meta datasets, we create a `get_prior_datasets()` method that returns all prior snapshots of a dataset (Listing 2). This allows for easy and recursive access to previous qrels and other components, so that no custom logic is needed to ensure approaches only use the allowed parts of the data. The prior snapshots are ordered by their timestamp, so that the most recent snapshot is first. Retrieval approaches can access the most recent snapshots independently of the elapsed time between or the aggregation level of the snapshots.

Listing 3 shows how we extend the dataset object with additional methods to provide metadata relevant to longitudinal evaluations. The `get_timestamp()` method returns the timestamp of the snapshot. This can be used as a reference

```
for prior_dataset in dataset.get_prior_datasets():
    prior_dataset.get_timestamp()
```

Listing 2. Accessing the prior datasets of a snapshots.

```
# At what time does/did a dataset take place?
dataset.get_timestamp()   # 2024-11

# What is the name of the dataset?
dataset.get_snapshot()   # "2024-11"

# When was a document published or updated?
store = dataset.docs_store()
for doc in store.docs_iter():
    doc.publishedDate   # 2016-02-10T00:00:00
    doc.updatedDate   # 2024-02-29T10:01:57
```

Listing 3. Accessing properties of datasets and documents.

point to dynamically calculate the recency of documents. The `get_snapshot()` method returns the snapshot name to locate a snapshot in the meta dataset. If adequate metadata is available, timestamps for documents can also be retrieved from the document store.

Since the search settings naturally evolve over time, for example when the corpus is updated, frequent changes can be expected. It is desirable to provide the same standardized interface to these newly evolved snapshots without the overhead of registering a new snapshots. Furthermore, retrieval experiments might run on modified versions of the original datasets. To facilitate this, custom, local datasets can be directly loaded from disk. These datasets need to follow the data structure of one of the official dynamic test collections and provide a metadata file with the information about timestamps and prior datasets. Listing 4 shows how a custom dataset can be loaded.

With these extensions, the `ir_datasets` framework provides a convenient interface to access dynamic test collections. It allows for easy access to previous snapshots and their components, and to use them in retrieval approaches.

4   Approaches and Re-implementation

In the LongEval scenario, we implemented various approaches that utilize prior snapshots of the same test collection to enhance retrieval effectiveness as part of our participation in LongEval 2024. They are based on the heuristic that prior snapshots with their rankings bear some kind of relevance information that can be used for the ranking of the current snapshots. Generally, we derive relevance signals for the ranking at the timestamp t_n from the snapshot at one or more earlier timestamps t_{n-1}. Compared to an earlier snapshot, a current

```
dataset = load("<PATH-TO-A-DIRECTORY-ON-YOUR-MACHINE>")
```

Listing 4. Loading custom datasets.

```
pipeline = bm25 >> QrelBoost(dataset, memory=1)
```

Listing 5. QREL Boost as PyTerrier Transformer applied to BM25.

ranking at t_n can contain new documents that are added after t_{n-1}, documents that were also present at t_{n-1}, and documents that were present at t_{n-1} but now have different content. This setting is streamlined with the `ir_datasets` integration as explained in the previous section. In this work, we focus especially on two approaches that were re-implemented using the `ir_datasets` extension and PyTerrier transformers [20], namely qrel boosting and relevance feedback.

4.1   Qrel Boost

The Qrel Boost approach directly uses the relevance label from a prior snapshot to boost documents that are known to be relevant. We boost documents d for a query q at t_n judged as $rel(q, d, t)$ at a previous snapshot t_{n-1} by:

$$\text{score}(q, d) \;=\; \text{score}_0 \;\times\; \prod_{t=t_1}^{t_k} \begin{cases} (1 - \lambda)^2, & \text{if } rel(q, d, t) = 0, \\ \lambda^2, & \text{if } rel(q, d, t) = 1, \\ \lambda^2 \mu, & \text{if } rel(q, d, t) = 2. \end{cases} \tag{1}$$

The parameters λ and μ are free and control the general weighting factor and additional weights for highly relevant documents. In the re-implementation, only the dataset needs to be passed to the PyTerrier transformer, examplified in Listing 5. The prior snapshots are accessed with `get_prior_datasets()`. How many prior snapshots should be used can additionally be controlled with the `memory` parameter.

4.2   Relevance Feedback

Our relevance feedback approach expands the queries with terms from previously relevant documents. Each query is expanded with the top k terms with the highest tf-idf scores. In the original implementation, the tf-idf scores were calculated using an additional database with all document texts. With the proposed `ir_datasets` extension, the document texts can be directly accessed through the `ir_datasets` document store. Still, the re-implementation relies directly on the tf-idf scores from the index of the previous snapshot since this is a theoretically better indicator for relevant terms. The directory of all prior indices is passed to the PyTerrier transformer (see Listing 6) and the correct index is identified by the `get_snapshot()` method.

```
indices = "path/to/indices"
pipeline = RF(dataset, indices, memory=1) >> bm25
```

Listing 6. Relevance Feedback as PyTerrier Transformer passed as query reformulation into BM25.

5   Experiments and Evaluation

As an experimental evaluation of our contributions, we reimplement our submissions to LongEval 2024 with our new `ir_datasets` extension and measure the code complexity and the reproduction of retrieval results.

5.1   Code Complexity

To analyze how the `ir_datasets` extension affects implementation and experimentation efforts of longitudinal retrieval experiments, we analyze the code complexity of the implementation of the retrieval approaches. Therefore, the initial version of the approach from [18] and its new re-implementation are compared using the code complexity measures from the Lizard toolkit.[3] Lizard provides metrics such as the total number of lines of code (NLOC), average NLOC per function (Avg.NLOC), the average cyclomatic complexity (AvgCCN), average number of tokens per function, and the number of functions (Fun Cnt). The Average NLOC is the total number of lines of code divided by the number of functions, and the Average CCN is the total cyclomatic complexity divided by the number of functions. The AvgCCN measures the average number of independent paths through the functions in the code. A lower value indicates simpler, more maintainable functions with fewer decision points [21].

A direct comparison of the code bases is difficult as the initial version is scattered across at least three files, with two of them shared. Additionally, a database was used to access document information that is no longer needed. Still, the results generally indicate a decreased complexity for the re-implementations. Table 1 provides an overview of the complexity measures.

For both approaches, the total NLOC and the Avg. NLOC is decreased. The AvgCCN is generally low, which indicates maintainable code in both cases (which is reasonable as the ideas are conceptually easy). However, it slightly increased for the qrel_boost approach, but decreased for relevance_feedback. Therefore, the new version of the relevance_feedback approach gained two more functions while the qrel_boost approach lost five functions.

These results indicate a generally decreased complexity. This is further supported by the fact that they now better adhere to PyTerrier and `ir_datasets`, two widely adopted frameworks in the community.

---

[3] https://github.com/terryyin/lizard.

Table 1. Code complexity metrics for the re-implemented approaches. The first row of each group shows the initial version of the approach from [18] and the second row shows the new re-implementation. The best values per group are highlighted in **bold**.

	Total NLOC	Avg. NLOC	AvgCCN	Avg. token	Fun Cnt
qrel_boost	250	17.8	**2.6**	142.2	12
qrel_boost new	**99**	**12.0**	2.9	**98.9**	7
relevance_feedback	231	23.8	3.0	170.6	8
relevance_feedback new	**197**	**17.3**	**2.5**	**124.8**	10

5.2 Replication of Retrieval Results

Finally, we want to ensure that the re-implementations of the approaches still achieve similar retrieval results to the original approaches. Currently, only the most recent LongEval datasets are added to the extension because they also contain all snapshots from 2023 and 2024. However, this year's version processed the data differently and also contained many more queries. For example, in the 2024 version of the dataset, new document IDs were assigned for each new version of a website. This is unified in the current dataset, which better supports temporal approaches that rely on prior snapshots. Given these changes, a direct comparison is difficult, and the focus needs to be on the general conclusions instead of absolute results.

We measured how well the re-implementations replicate the results of the original approaches according to the Delta Relative Improvement (ΔRI), Effect Ratio (ER) and the p-values of unpaired t-tests as proposed by Breuer et al. [6]. To compare the runs of different topic sets across implementations, with ΔRI and ER are the runs related to a reference system, BM25 in our case. Only the deltas are then compared across the original and the re-implemented approaches. The ΔRI describes the relative change in effectiveness of the re-implementation compared to the original approach. The further ΔRI diverges from zero, the weaker the replication. The ER measures how well the per-topic difference from the reference system is recovered in the re-implementation. The closer the ER is to one, the better the re-implementation recovers the differences from the reference system. The p-value of the unpaired t-test indicates whether the difference between the original and re-implementation is statistically significant. Since we aim to achieve similar results from the re-implementation, no significant differences with p-values over 0.05 are desirable. The experiments were carried out with `repro_eval` [7].

The replication results based on nDCG@10 are reported in Table 2. They indicate that the re-implementation most often achieves different results. This is most likely due to the differences in the collections. Given the low p-values, the results are significantly different. The ER always exceeds the ideal value of 1, meaning that the re-implementations are not replicating the same effect but achieve higher differences compared to the reference systems. This is further

Table 2. This table reports the replicability results based on ΔRI, ER, and p-values. Additionally, the average retrieval performance is reported for the re-implementation and also the original version in teal. All results are based on nDCG@10.

snapshot	system	ER	ΔRI	p-value	nDCG@10
2022-07	qrel_boost	2.307	-0.407	3.63e-13	0.245/0.343
2022-07	relevance_feedback	3.230	-0.535	8.91e-20	0.229/0.351
2022-09	qrel_boost	1.943	-0.458	2.25e-05	0.259/0.317
2022-09	relevance_feedback	2.224	-0.461	5.90e-06	0.243/0.303
2023-01	qrel_boost	4.925	-0.522	5.24e-36	0.234/0.445
2023-01	relevance_feedback	3.685	-0.312	1.30e-21	0.231/0.386
2023-06	qrel_boost	1.569	0.003	3.49e-13	0.268/0.421
2023-06	relevance_feedback	1.723	-0.028	1.65e-11	0.220/0.355
2023-08	qrel_boost	1.896	-0.071	1.75e-42	0.199/0.338
2023-08	relevance_feedback	2.018	-0.050	7.46e-34	0.171/0.287

supported by the almost always negative ΔRI. Additionally, the scores indicate that the relative improvement of the overall effect is relatively well replicated, especially on the later snapshots.

6   Conclusion

We describe how longitudinal retrieval experiments can be simplified with an extension for `ir_datasets` that improves the support for dynamic test collections. We demonstrated its utility by re-implementing previous approaches as PyTerrier transformers that make use of the added functionality. These changes simplify software submissions for the LongEval lab, piloting this year with the SciRetrieval task. We examined how the proposed changes affect longitudinal experiments in terms of code complexity and how well the re-implementations replicate the original retrieval effectiveness. The results indicate that the overall complexity decreased. Although the initial results could be only weakly replicated, the effectiveness of the approaches improved. This is most likely due to the updated test beds and the improvements made to the approaches.

Future works regard improvements and extensions of the proposed methods. The `ir_datasets` extension for dynamic test collections is a first step towards a standardized interface to longitudinal evaluations. While not many dynamic test collections are available yet, new datasets should be added as they become available. Furthermore, related datasets with versioned collections could be added, such as Soboroff's evolving version of the GOV2 dataset [25]. A limitation of the extension is that each snapshot requires a complete document corpus, even if it is only partially changed. This could be addressed in the future by implementing approaches that only store the deltas between snapshots, similar to [26]. Finally, it needs to be checked what features can be directly integrated into the `ir_datasets` framework and what should be kept as a separate extension.

With the proposed extension, we hope to streamline longitudinal experiments and help researchers to cope with the additional complexity that longitudinal experiments introduce.

Acknowledgments. We gratefully acknowledge the support of the German Research Foundation (DFG) through project grant No. 407518790.

Disclosure of Interests. The authors Jüri Keller, Maik Fröbe, and Philipp Schaer joined the LongEval organization team in 2025. All other authors have no competing interests to declare that are relevant to the content of this article.

References

1. Adar, E., Teevan, J., Dumais, S.T., Elsas, J.L.: The web changes everything: understanding the dynamics of web content. In: WSDM, pp. 282–291. ACM (2009)
2. Alexander, D., et al.: Team OpenWebSearch at CLEF 2024: LongEval. In: Working Notes Papers of the CLEF 2024 Evaluation Labs. CEUR Workshop Proceedings (2024)
3. Alkhalifa, R., et al.: Overview of the CLEF-2023 longeval lab on longitudinal evaluation of model performance. In: CLEF. Lecture Notes in Computer Science, vol. 14163, pp. 440–458. Springer, Cham (2023)
4. Alkhalifa, R., et al.: Overview of the CLEF 2024 longeval lab on longitudinal evaluation of model performance. In: CLEF (2). Lecture Notes in Computer Science, vol. 14959, pp. 208–230. Springer, Cham (2024)
5. Altingövde, I.S., Ozcan, R., Ulusoy, Ö.: Evolution of web search results within years. In: SIGIR, pp. 1237–1238. ACM (2011)
6. Breuer, T., et al.: How to measure the reproducibility of system-oriented IR experiments. In: SIGIR, pp. 349–358. ACM (2020)
7. Breuer, T., Ferro, N., Maistro, M., Schaer, P.: repro_eval: a python interface to reproducibility measures of system-oriented IR experiments. In: ECIR (2). Lecture Notes in Computer Science, vol. 12657, pp. 481–486. Springer, Cham (2021)
8. Cancellieri, M., et al.: Longeval at CLEF 2025: longitudinal evaluation of IR model performance. In: ECIR (5). Lecture Notes in Computer Science, vol. 15576, pp. 382–388. Springer, Cham (2025)
9. Dumais, S.T.: Putting searchers into search. In: SIGIR, pp. 1–2. ACM (2014)
10. El-Ebshihy, A., et al.: Predicting retrieval performance changes in evolving evaluation environments. In: CLEF. Lecture Notes in Computer Science, vol. 14163, pp. 21–33. Springer, Cham (2023)
11. Fröbe, M., et al.: The information retrieval experiment platform. In: Chen, H.H., Duh, W., Huang, H.H., Kato, M., Mothe, J., Poblete, B. (eds.) 46th International ACM SIGIR Conference on Research and Development in Information Retrieval (SIGIR 2023), pp. 2826–2836. ACM (2023). https://doi.org/10.1145/3539618.3591888

12. Fröbe, M., et al.: Continuous integration for reproducible shared tasks with tira.io. In: ECIR (3). Lecture Notes in Computer Science, vol. 13982, pp. 236–241. Springer, Cham (2023)
13. Galuscáková, P., et al.: Longeval-retrieval: French-English dynamic test collection for continuous web search evaluation. In: SIGIR, pp. 3086–3094. ACM (2023)
14. Gonzales Saéz, G.: Continuous evaluation framework for information retrieval systems. Ph.D. thesis, UNIVERSITÉ GRENOBLE ALPES, Grenoble (2023)
15. Guo, M., Yang, Y., Cer, D., Shen, Q., Constant, N.: Multireqa: a cross-domain evaluation for retrieval question answering models. CoRR abs/2005.02507 (2020)
16. Keller, J., Breuer, T., Schaer, P.: Evaluation of temporal change in IR test collections. In: ICTIR, pp. 3–13. ACM (2024)
17. Keller, J., Breuer, T., Schaer, P.: Leveraging prior relevance signals in web search. In: Faggioli, G., Ferro, N., Galuscáková, P., de Herrera, A.G.S. (eds.) Working Notes of the Conference and Labs of the Evaluation Forum (CLEF 2024), Grenoble, France, 9–12 September 2024. CEUR Workshop Proceedings, vol. 3740, pp. 2396–2406. CEUR-WS.org (2024). https://ceur-ws.org/Vol-3740/paper-220.pdf
18. Keller, J., et al.: Counterfactual query rewriting to use historical relevance feedback. In: Hauff, C., et al. (eds.) Advances in Information Retrieval - 47th European Conference on Information Retrieval, ECIR 2025, Lucca, Italy, 6–10 April 2025, Proceedings, Part III. Lecture Notes in Computer Science, vol. 15574, pp. 138–147. Springer, Heidelberg (2025). https://doi.org/10.1007/978-3-031-88714-7_11
19. MacAvaney, S., Yates, A., Feldman, S., Downey, D., Cohan, A., Goharian, N.: Simplified data wrangling with IR_datasets. In: SIGIR, pp. 2429–2436. ACM (2021)
20. Macdonald, C., Tonellotto, N.: Declarative experimentation in information retrieval using pyterrier. In: ICTIR, pp. 161–168. ACM (2020)
21. McCabe, T.J.: A complexity measure. IEEE Trans. Software Eng. **2**(4), 308–320 (1976)
22. Pass, G., Chowdhury, A., Torgeson, C.: A picture of search. In: Infoscale. ACM International Conference Proceeding Series, vol. 152, p. 1. ACM (2006)
23. Rekabsaz, N., Lesota, O., Schedl, M., Brassey, J., Eickhoff, C.: TripClick: the log files of a large health web search engine. In: Proceedings of the 44th International ACM SIGIR Conference on Research and Development in Information Retrieval, pp. 2507–2513. ACM, Virtual Event Canada (2021). https://doi.org/10.1145/3404835.3463242
24. Sanderson, M.: Test collection based evaluation of information retrieval systems. Found. Trends Inf. Retr. **4**(4), 247–375 (2010)
25. Soboroff, I.: Dynamic test collections: measuring search effectiveness on the live web. In: Efthimiadis, E.N., Dumais, S.T., Hawking, D., Järvelin, K. (eds.) SIGIR 2006: Proceedings of the 29th Annual International ACM SIGIR Conference on Research and Development in Information Retrieval, Seattle, Washington, USA, 6–11 August 2006, pp. 276–283. ACM (2006). https://doi.org/10.1145/1148170.1148220
26. Staudinger, M., Piroi, F., Rauber, A.: Reproducible hybrid time-travel retrieval in evolving corpora. In: SIGIR-AP, pp. 203–208. ACM (2024)
27. Thakur, N., Reimers, N., Rücklé, A., Srivastava, A., Gurevych, I.: BEIR: a heterogeneous benchmark for zero-shot evaluation of information retrieval models. In: NeurIPS Datasets and Benchmarks (2021)
28. Tyler, S.K., Teevan, J.: Large scale query log analysis of re-finding. In: WSDM, pp. 191–200. ACM (2010)
29. Voorhees, E.M., et al.: TREC-COVID: constructing a pandemic information retrieval test collection. SIGIR Forum **54**(1), 1:1–1:12 (2020)

SimpleText Best of Labs in CLEF-2024: Application of Large Language Models for Scientific Text Simplification

Nicholas Largey[ID], Reihaneh Maarefdoust[ID], Shea Durgin[ID], and Behrooz Mansouri[✉][ID]

University of Southern Maine, Portland, ME 04103, USA
{nicholas.largey,reihaneh.maarefdoust,shea.durgin,
behrooz.mansouri}@maine.edu

Abstract. Accessing and understanding scientific literature remains difficult due to its inherent complexity, characterized by specialized terminology and reliance on readers' prior knowledge. The CLEF 2024 SimpleText lab aims to improve accessibility to scientific information for a broader audience through advancements in information retrieval and natural language processing. This paper explores the application of large language models for three scientific text simplification tasks. The first task involves retrieving passages to include in a simplified summary. The proposed systems select candidates using TF-IDF with expanded queries via LLaMA3, followed by a bi-encoder, cross-encoder, and LLaMA3-based re-ranking. In the second task, the goal is to identify and explain difficult concepts. Three models are explored for this task using LLaMA3 and Mistral. Finally, for Task 3, which focuses on simplifying scientific text, similar to Task 2, LLaMA3 and Mistral are used with different prompting and fine-tuning approaches. The experimental results show that the proposed systems in Task 1 are the most effective, and for Tasks 2 and 3 are comparable with other systems proposed in the SimpleText lab.

Keywords: Scientific Text Simplification · Definition Extraction · Large Language Models

1 Introduction

Scientific publications are often dense and filled with specialized terminology, which can create barriers for readers outside a narrow expert community. Simplifying these texts helps bridge gaps between disciplines, accelerates literature reviews, and makes research accessible to students, practitioners, and the general public. By reducing linguistic complexity while preserving technical accuracy, text simplification supports more inclusive scientific communication and aids education, policy making, and interdisciplinary collaboration.

The CLEF 2024 SimpleText lab [2] is dedicated to enhancing accessibility to scientific information for all users, encompassing both information retrieval and natural language processing aspects. Unlike conventional text simplification approaches, where general texts are tailored for younger or lower-literacy

J. Carrillo-de-Albornoz et al. (Eds.): CLEF 2025, LNCS 16089, pp. 128–141, 2026.
https://doi.org/10.1007/978-3-032-04354-2_9

audiences, scientific text simplification targets the specialized vocabulary and complex sentence structures inherent in research articles. SimpleText'24 consider four tasks as follows:

1. **Content Selection:** Retrieving Passages to Include in a Simplified Summary.
2. **Complexity Spotting:** Identifying and Explaining Difficult Concepts.
3. **Text Simplification:** Simplify Scientific Text.
4. **SOTA?:** Tracking the State-of-the-Art in Scholarly Publications.

With advances in large language models (LLMs), many of the participating teams in this lab considered exploring LLMs' applications for the proposed tasks. This ranges from closed-LLMs like GPT-3.5 and GPT-4 Turbo [15, 16] to open-LLMs such as LLaMA and Mistral [17, 18]. The Artificial Intelligence and Information Retrieval (AIIR) lab participated in first three Tasks of the CLEF 2024 SimpleText lab. This paper explores the proposed approaches by our team, recognized as the best of lab team. Our proposed approaches mainly rely on two LLMs: LLaMA3 [19] and Mistral [1].

Task 1 [3] requires selecting relevant passages from a large collection of academic abstracts and bibliographic records that support understanding of a target article. For this task, we proposed ranges from a dual-encoder retrieval architectures to query expansion and passage re-ranking with advanced language systems. Task 2 [4], focuses on detecting technical terms within scientific abstracts that may obstruct comprehension and providing clear, concise explanations. Our team tackled both Subtask 2.1 (selecting up to five challenging terms per passage) and Subtask 2.2 (generating their explanations) using either LLaMA-3 or Mistral. Finally, Task 3 [5], involves producing simplified version of sentences extracted from scientific abstracts, given corresponding popular science articles and queries and matching abstracts of scientific papers. For this task, we employed fine-tuned LLaMA model alongside Mistral to generate simplified text that retain scientific fidelity.

The experimental results show that our proposed systems for all three Tasks are highly effective. For Task 1 and Subtask 3.2, our proposed models were the most effective ones in the lab, while for Task 2 and Subtask 3.1, they are comparable to the leading systems. In the next sections, we will describe our systems for each Task, followed by evaluation results and analysis.

2   Task 1: Retrieving Passages to Include in a Simplified Summary

This section first describes the data for Task 1 [3]. Then we describe our proposed systems, followed by the experimental results and analysis.

Table 1. System prompts used for query expansion and re-ranking with LLaMA3.

Task	Prompt
Query Expansion	Being a ranking model your first Task is to do query expansion. For an information need, you will add more context to it. Contextualize the query as best as you can in one or two short sentences, for a given information need and context.
Re-ranking	You are a ranking model for information retrieval. Given a query and two documents, you will say which one is more relevant. If Document 1 is more relevant say yes, otherwise say no.

2.1 Topic and Collection

Task 1 topics are selected from two sources: 1) the tech section of The Guardian[1] newspaper (topics G01 to G20), and 2) Tech Xplore[2] website (topics T01 to T20). Each topic represents a query selected from one of these resources. For instance, the query with Id 'T03.1', is "imitation learning", with its context being an article titled "RoboTurk: A crowdsourcing platform for imitation learning in robotics", from The Tech Xplore. Participants have access to the whole article, its title, and the query.

The main corpus consists of a large set of scientific abstracts and associated metadata in the field of computer science and engineering. The 12th version of the Citation Network Dataset [6], released in 2020, provides this data extracted from DBLP, ACM, MAG (Microsoft Academic Graph), and other sources. It contains 4,894,083 bibliographic references published before 2020, 4,232,520 English abstracts, 3,058,315 authors with affiliations, and 45,565,790 ACM citations.

2.2 Proposed Models

For Task 1, we evaluated five distinct retrieval models. Among these, two models leverage LLaMA-3 for query expansion, subsequently employing pre-trained bi-encoders or cross-encoders for document ranking. Two of our other models utilize LLaMA-3 as a pair-wise re-ranker. The specific prompts employed in conjunction with LLaMA-3 are shown in 1. Additionally, our evaluation includes a baseline run featuring a fine-tuned cross-encoder. The subsequent sections provide a detailed description of each model.

LLaMA BiEncoder/CrossEncoder: For Task 1, input queries are short keyword terms (e.g., "phototransistor", "NLP applications", "intelligent parking") selected from technical articles. To better contextualize these queries, we used LLaMA3-8B-Instruct model for query expansion. Following the approach proposed by Anand et al. [7], we provide the query and the article to the model, and use the system prompt (shown in Table 1) for query expansion. Then, we

---

[1] https://www.theguardian.com/uk/technology.
[2] https://techxplore.com/.

Table 2. Query rewriting/expansion using LLaMA3.

Initial Query	Expanded Query
phototransistor	Retina-inspired phototransistor sensors for adaptive visual perception in various illumination conditions
NLP applications	NLP applications in media companies and potential biases in algorithmic decision-making
intelligent parking	Intelligent parking systems using artificial neural networks inspired by biology

pass the query, the related article title, and context to LLaMA3 and expand the initial query. Table 2 shows examples of expanded queries. After this step, we use TF-IDF from PyTerrier [9] with default parameters to get the top-5000 results for each expanded query.

The candidates are then re-ranked with SentenceBERT [8], once using a bi-encoder architecture and once using a cross-encoder. For the bi-encoder, we use the 'all-mpnet-base-v2' model due to its demonstrated effectiveness in capturing semantic similarity between queries and documents across various information retrieval tasks. This model is used without further fine-tuning. Based on previous lab participation [10], we also consider another model with a fine-tuned cross-encoder, 'ms-marco-MiniLM-L-6-v2'. For fine-tuning, we use the data from previous years of the lab, splitting into 90% training and 10% validation sets. We fine-tune the model for 25 epochs, choosing the hyperparameters with the highest MRR@10 (Mean Reciprocal Rank) on the validation set.

For both models, we represent the input query as a combination of the initial query, related article title, and LLaMA-expanded query. Documents are represented as concatenation of the title and abstract of each passage. However, for the cross-encoder model, as we use fine-tuning, three special tokens are introduced to separate different text. Each query is represented as:

Initial Query + [TOP] + Article's Title + [CON] + Expanded Query

where the initial query is the query specified by the organizers, the Article's Title corresponds to the topic text, and the Expanded Query is the context generated by LLaMA3. Documents in the collection are also represented as 'title + [ABS] + abstract'. The **LLaMABiEncoder** model, re-rank the the results from TF-IDF model and returns the top-100 results for each query. The **LLaMACrossEncoder** model, re-rank the results from bi-encoder.

Re-ranking with LLaMA (LLaMA Re-Ranker): Our two other proposed models use LLaMA as pair-wise re-ranker. Following the approach proposed by Qin et al. [11], we used a system prompt for pair-wise re-ranking shown in Table 1. Two variations of this architecture were implemented, differing in the user message provided to LLaMA3. In one version (**LLaMAReranker**), the user message included the query, related article title, and context generated from the previous runs (i.e., the expanded query from the LLaMA3). The other version (**LLaMAReranker2**) omitted the context.

LLaMA3 was tasked with determining which of the two documents was more relevant to the query based on the provided information. We re-ranked the top-100 candidates retrieved by the bi-encoder model. We used a similar algorithm as Bubble Sort, to pass pairs of documents for a query to the LLM. However, as the LLaMA3's outputs in this context might not be suitable for direct confidence scores, we assigned a simple ranking based on enumeration. The highest-ranked document received a score of 100, with scores decreasing by 1 for lower ranks.

Fine-Tuned Cross-Encoder combined with ElasticSearch (CERRF): For our final run, we used the organizers' ElasticSearch setup to pull the top-100 documents for each topic by searching with both the query and the topic description. We then re-rank those 100 hits with a fine-tuned cross-encoder model (ms-marco-MiniLM-L-6-v2) with the data from previous CLEF labs. To prepare inputs, we combined the query and topic text as "<query> [QSP] <topic text>", while the articles were represented as "<title> [TSP] ". Here, [QSP] and [TSP] are special tokens separating the query text from the topic text and the paper title from its abstract, respectively. We picked topics G10 and G11 to tune hyperparameters and used the 2023 test set for final evaluation. Once the best settings were found, we trained the model on the full set of training topics.

In addition to the cross-encoder approach, we also perform a separate retrieval using Elasticsearch with only the query (without the topic text). In parallel, we ran a second ElasticSearch query using only the original query text. We then merged the two result lists with modified Reciprocal Rank Fusion (MRRF) [12], shown in Eq. 1. Here for each document d, s_m is its similarity score from method m, and r_m is its rank. By combining scores and ranks, MRRF gives higher priority to documents that appear near the top in both searches.

$$RRFscore(d \in D) = \sum_{m \in M} \frac{s_m(d)}{60 + r_m(d)} \qquad (1)$$

2.3   Experimental Results and Analysis

In SimpleText 2024, 30 queries were developed for testing for Task 1. Documents were assessed for topical relevance with a score of 0 to 2, based on their content alignment with the original articles. Table 3 shows the effectiveness of our proposed models. Except for P@20, the *LLaMABiEncoder* archives the highest effectiveness across all measures.

Looking at the *LLaMABiEncoder* results, for only 10% of topics, the MRR value is not 1. The lowest MRR is for the topic 'G02.C1', at 0.33 (P@10 of 0.7). For this topic, the query text by the organizers is defined as "concerns related to the handling of sensitive information by voice assistants". With LLaMA3, the expanded query is "voice assistants handling sensitive information concerns Apple Siri recordings", which does not seem to add any new useful terms to the original query. The top retrieved result for this topic is an article titled, "Poster: A First Look at the Privacy Risks of Voice Assistant Apps.", assessed as non-relevant. For topics like 'T11.1' the original query "character relationship"

Table 3. AIIRLab systems results for CLEF 2024 SimpleText Task 1 on the test Qrels (G01.C1-G10.C1 and T06-T11).

Model	MRR	P@10	P@20	NDCG@10	NDCG@20	Bpref	MAP
LLaMABiEncoder	**0.9444**	**0.8167**	0.5517	**0.6170**	**0.5166**	**0.3559**	**0.2304**
LLaMAReranker2	0.9300	0.7933	0.5417	0.5943	0.5004	0.3495	0.2177
LLaMAReranker	0.8944	0.7967	**0.5583**	0.5889	0.5011	0.3541	0.2200
LLaMACrossEncoder	0.7975	0.6933	0.5100	0.4745	0.4240	0.3404	0.1970
CERRF	0.7264	0.5033	0.4000	0.3584	0.3239	0.2204	0.1309

is expanded to "character relationship network map The Witcher", helping find more relevant results, leading to MRR and P@10 of 1.

Comparing our LLaMA3 re-ranking approach system, *LLaMAReranker2* against *LLaMABiEncoder*, there is no significant difference between the two systems, using Two-sided Paired Student's t-Test (p-value = 0.05). Interestingly, both models have the same topics for which they did not achieve MMR of 1. For topic 'G02.C1', the MMR drops to 0.2 with *LLaMAReranker2* (P@10 of 0.3). Investigating the results for this topic, LLaMA3 gave higher ranks to articles that have only titles (abstract missing) such as the article titled "Examining the Use of Voice Assistants: A Value-Focused Thinking Approach". With the article's abstract missing, these articles are assessed as non-relevant. Overall, using LLaMA3 for either re-ranking or query expansion showed similar effectiveness, while re-ranking with a bi-Encoder proved more efficient.

In addition to effectiveness metrics for topical relevance, Task 1 also evaluates the credibility and text complexity of the retrieved results. The metrics related to this aspect for our systems are shown in Table 4. For our runs with LLaMA, we provided two versions of the rankings: one based solely on topical relevance (denoted by ^{rel}), and another combining topical relevance with the Flesch-Kincaid Grade Level (FKGL) readability score (denoted by ^{comb}). The ^{comb} rank scores are computed as a weighted average of the topical relevance score (weight 0.75) and the FKGL score of the retrieved instance (weight 0.25). Most of the participating runs show similar FKGL scores around 15, which aligns with the complexity expected from university-level scientific content. However, the results from LLaMA bi/cross-encoders, when ranked using a combined scoring method, show noticeably higher FKGL scores. This increase in reading difficulty appears to be due to longer sentences in those texts, averaging 31 words per sentence compared to 23 words in texts ranked by relevance alone.

3   Task 2: Identifying and Explaining Difficult Concepts

This section describes the data for Task 2 [4], our proposed models, and evaluation results. We rely on LLaMA3 and Mistral [1] language models and propose two main systems for Subtasks 1 and 2.[3]

---

[3] We skip our third erroneous system for space.

Table 4. Evaluation of AIIRLab systems for complexity and credibility in Task 1 (over all 176 queries).

Model	Avg. #Refs	Avg. vocab size	Avg. FKGL
$LLaMABiEncoder^{rel}$	8.7	95.8	15.3
$LLaMABiEncoder^{comb}$	9.5	98.1	20.7
$LLaMACrossEncoder^{rel}$	10.7	104.3	15.5
$LLaMACrossEncoder^{comb}$	10.0	99.4	20.4
$LLaMAReranker^{rel}$	8.8	95.8	15.5
$LLaMAReranker^{comb}$	8.8	96.1	15.7
$LLaMAReranker2^{rel}$	8.6	94.0	15.3
$LLaMAReranker2^{comb}$	8.6	93.9	15.5
$CERRF^{comb}$	10.6	96.4	15.3

3.1 Training and Test Data:

For Task 2, 576 sentences from 115 documents are provided for training. For these sentences, 2590 annotated difficult terms are available. Subtask 2.2 leverages a dataset of 501 sentences across 55 documents, containing 2,006 explanations and 1,521 definitions. These documents are selected from high-ranked abstracts to the requests of Task 1. Participants are asked to detect difficult terms, along with the difficulty level for Subtask 2.1, and provide definitions and explanations of detected difficult terms for Subtask 2.2 [2].

3.2 Proposed Models

Our team participated in Task 2 with techniques based on LLaMA3 and Mistral. Here we describe our models:

LLaMA: Our first model uses LLaMA3-8B-Instruct, using a system prompt to instruct the model to act as a knowledgeable high school student (details in Table 5). This prompt achieved the best performance among those studied on the training data. We process each sentence from the test set using the following user message:

> For the sentence: SENTENCE, what are difficult terms (one to five consecutive terms)? What is the difficulty level? Your output is term or terms: difficulty level (e, m, or d). Do not provide explanation, just give the answer.

where SENTENCE represents the actual sentence. We specify the output format, as LLaMA can add unnecessary information. After identifying difficult terms, we again utilize LLaMA to generate definitions and explanations. As shown in Table 5, we instruct LLaMA to act as a technician with knowledge of technical terms and request definitions and explanations. The following user message is used for this step:

Table 5. System prompts used for detecting difficult terms and generating definitions and explanations with LLaMA3 and Mistral for Task 2.

Model	Prompt
Task 2.1 (Detecting Difficult Terms)	
LLaMA3	You are a high school student with good general knowledge. Given a sentence, you want to determine which terms are not clear. You choose the terms that should be defined in order to understand the sentence. This includes technical terms and abbreviations. You can choose one to five consecutive terms. You will also decide the difficulty level for each identified term, with labels easy (label e), medium (label m), hard (label d). Here is an example For sentence: CRISPR-Cas is a tool that is widely used for gene editing; you identified "CRISPR-Cas" with difficulty: d
Mistral	You are a helpful assistant. Given a sentence, you will just output the unclear technical term or terms (up to 5 terms). You choose the terms that should be defined in order to understand the sentence. Each sentence can have up to 5 phrases. You will decide the difficulty of unclear terms with scales easy (e), medium (m), hard (d). Note that easy does not include terms such as shown, pronouns, or numbers
Task 2.2 (Definition and Explanation)	
LLaMA3 & Mistral	You are a technician with knowledge of technical terms. Given a term, in a sentence provide definition of it. Then provide an explanation of that term. Your goal is to make sure other non technicians understand the sentence. Definition and explanation should be separate from each other

You have identified term "TERM" in the sentence: "SENTENCE" as an unclear term. Provide its definition and explain what it is. The output should be like:
Definition: {Give definition here}, Explanation: {Give explanation here}

where TERM represents the term identified earlier and SENTENCE is the sentence it originated from.

Mistral: Similar to our LLaMA3-based model, our approach with Mistral-7B leverages a system prompt (details in Table 5). This prompt instructs Mistral to identify difficult terms. We process training examples through a series of prompts and responses with Mistral to achieve this, as shown in Fig. 1. The examples used come from the training data, and as the final prompt, the test sentence is passed to the LLM. After detecting the difficult terms, we use a similar system prompt (shown in Table 5) as our first model to generate definitions and explanations of difficult terms with Mistral.

3.3   Experimental Results and Analysis

Our proposed systems results on the test set are summarized in Table 6. For each run, the organizers reported:

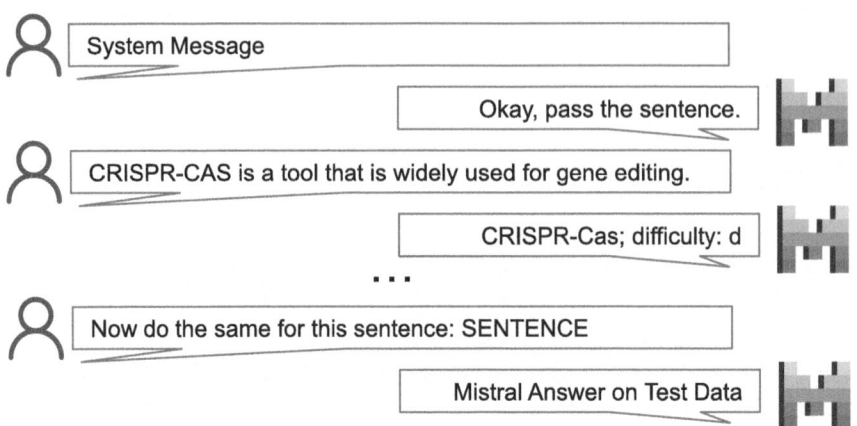

Fig. 1. Prompts used for Subtask 2.1 to extract difficult terms. The SENTENCE represents the test sentence passed to Mistral.

- Recall of all the terms, independently from the level of difficulty
- Precision of all the terms, independently from the level of difficulty
- Recall of the difficult terms
- Precision of the difficult terms
- BLEU score computed for bigrams

Our proposed Mistral approach provided better results compared to LLaMA3. Providing an example, for the sentence "Cryptocurrency was built initially as a possible implementation of digital currency, then various derivatives were created in a variety of fields such as financial transactions, capital management, and even nonmonetary applications." (sentence ID: G08.1_2972302621_1), Table 7 shows the ground-truth, and the results generated by Mistral and LLaMA, for Subtask 2.1. As can be seen, LLaMA tends to extract fewer terms for each sentence, leading to lower recall; however, the precision for correctly identifying difficulty level is more precise.

Another interesting aspect of Task 2 is duplicate sentences. The organizers have provided repeated sentences to study whether LLMs provide the same results. Our results show while Mistral mostly produces the same responses, LLaMA3 responses seem to differ each time. For a short sentence, "This is especially true for self-driving vehicles deployed in public transport services.", LLaMA3 once extracts the terms 'self-driving', 'vehicles', 'public transport' and the next time extracts 'self-driving', 'deployed'. Mistral extracted terms, however, remained the same.

4   Task 3: Simplify Scientific Text

This section describes the data, proposed models, and evaluation results for Task 3 [5]. LLaMA3-8B-Instruct was utilized for both Subtasks 3.1 and 3.2.

Table 6. AIIRLab systems results for CLEF 2024 SimpleText Task 2.

Model	Recall	Precision	Rec_Difficult	Prec_Difficult	Blue
Mistral	0.41	0.69	0.19	0.49	0.13
LLaMA	0.28	0.65	0.26	0.67	0.15

Table 7. Extracted difficult terms with their difficulty levels for sentence ID 'G08.1_2972302621_1' from SimpleText 2024. Letters 'd', 'm', and 'e' show difficult, medium, and easy terms, respectively.

Ground-truth		Mistral		LLaMA	
Term	Difficulty	Term	Difficulty	Term	Difficulty
cryptocurrency	m	cryptocurrency	d	cryptocurrency	d
digital currency	m	digital currency	m	digital currency	m
capital management	m	capital management	m	derivatives	m
nonmonetary applications	d	nonmonetary applications	m		
financial transactions	e	financial transactions	e		

4.1 Topic and Collection

The training data consists of a collection of parallel text passages (source and simplified versions). These simplified sentences are directly created from original scientific abstracts in the DBLP Citation Network Dataset for Computer Science, Google Scholar, and PubMed articles on Health and Medicine (all from 2023). The dataset includes 648 sentences for training and 245 sentences for testing. The simplification process involved either master's students in Technical Writing and Translation or a team of a computer scientist and a professional translator (native English speaker).

4.2 Proposed Models

AIIR Lab submitted a total of three main runs for both Subtasks 3.1 (sentence-level) and 3.2 (abstract-level) using prompt engineering with a fine-tuned LLaMA3-8B model. Our first three runs for this Task utilized LLaMA3-8B, which was fine-tuned with the provided training data for both the sentence and abstract levels. We used a split of 90:10 for training and validation. While fine-tuning LLaMA, we employed Quantized Low-Rank Adaptation (QLoRA), a method used while fine-tuning to reduce the amount of GPU virtual memory required and computational cost [13]. The model's weights are first converted from a 16-bit floating point number to a 4-bit "NormalFloat". These reduced-size weight matrices are then approximated to low-rank matrices by reducing the number of parameters in the matrices. This conversion results in speeding up computation time, and reducing the size of the data's footprint. These 4-bit embeddings then utilize NVIDIA's *unified memory* feature, which allows for automatic paging optimization before updating the weights. This paging optimization allows for the CPU RAM to be

Table 8. System prompts used for Task 3.

Model	Prompt
LLaMA3-8B Run 1	Simplify this text for English speaking science students in college. Maximize the use of simple words and short sentences, but include keywords from the original text. Optimize the output ROUGE, SARI, and BLEU scores
LLaMA3-8B Run 2	You are a skilled editor, known for your ability to simplify complex text while preserving its meaning. You have a strong understanding of readability principles and how to apply them to improve text comprehension.
LLaMA3-8B Run 3	Simplify the following scientific text for an average American citizen. Keep, but define, any keywords and subjects with less complex words and phrases.

accessed by the GPU directly for page-to-page transferring, preventing the possibility of running out of GPU memory space as long as sufficient system memory is available.

During the training, the data was first processed with QLoRA to resize the token embeddings. The hyperparameters were set as the following: an *alpha* of 32, a *dropout* of 0.1, a *Task type* of "CASUAL_LM", and an *R-value* of 8. The output data was then fed to LLaMA3-8B with the hyperparameters of a *learning rate* of e-4, a paged_adam_32 *optimization function*, a *batch size* of 8, and was trained for 20 *epochs*. We passed the training data for LLaMA3 fine-tuning as:

"Instruction:" + [P] + "Input: " + [S] + "Response: " + [T]

where prompt (P), for all training samples, was the prompt outlined for Run 1 (Table 8). The source (S) and target (T) values would be the output token embeddings from QLoRA. For prompt engineering, we focused on the average FKGL (Flesch-Kincaid Grade Level) score for the provided test sentences and abstracts. The data was passed into our fine-tuned model, and the training dataset's FKGL score was averaged at the end of each run.

4.3   Experimental Results and Analysis

Task 3 results are evaluated based on several metrics, with SARI [14] score against the human reference simplifications as the main measure. Table 9 shows our results for both Subtasks 3.1 (sentence-level) and 3.2 (abstract-level). While for Subtask 3.1, our team's runs are ranked second in terms of SARI score, we achieved the highest SARI score for Subtask 3.2 among the participating teams. Compared to the references, our models have high compression ratios and sentence splits, as LLaMA's outputs are lengthier. Table 10 shows an example where our simplified version of the original input text is compared against the ground-truth for Subtask 3.1.

For Subtask 3.1, all LLaMA3's Sari scores fell within a ±0.82 difference from one another. The Sari scores for Subtask 3.2 were similar to Subtask 3.1, in that

Table 9. AIIRLab systems results for CLEF 2024 SimpleText Task 3.

	Subtask 3.1			Subtask 3.2		
Model	FKGL	SARI	BLEU	FKGL	SARI	BLEU
LLaMA3-8B Run1	**8.39**	**40.58**	**7.53**	**9.07**	**43.44**	**11.73**
LLaMA3-8B Run3	9.47	40.36	6.26	10.17	43.21	11.03
LLaMA3-8B Run2	10.33	39.76	5.46	10.22	42.19	7.99

they varied by a relatively narrow margin of ±1.25. The original sentences have an FKGL of 13–14 corresponding to a university-level text, with the reference scores being 8.86 for Subtask 3.1 and 8.91 for Subtask 3.2. Our FKGL results for all runs in both Tasks fell within the 8.39 to 10.33 FKGL range, with our run 1 scores being 0.47 points below for Task 3.1 and 0.16 points above for Task 3.2 compared to the reference FKGL score.

Table 10. Our results for sentence-level simplification for sentence ID 'G01.1_1584012683_9' in Subtask 3.1.

Original text	We have applied VMDs to facilitate distributed analysis of electroencephalogram (EEG) data in Internet environment.
Simplification system	**Simplified Result**
Ground-truth	We used VMDs to help analyze data over the internet.
LLaMA3-8B Run 1	We used a special tool to help analyze brain wave data from the internet.
LLaMA3-8B Run 2	We used special software to help people analyze brain wave data from the internet.
LLaMA3-8B Run 3	We used a special tool called VMDs to help analyze EEG data on the internet.

5   Conclusion

This paper presented the AIIR Lab's participation in the SimpleText CLEF 2024 evaluation on Tasks 1 through 3, using LLaMA-3 and Mistral models. In Task 1, we explored applications of LLaMA for query expansion and re-ranking. Among all teams, our bi-encoder and LLaMA-based re-ranking approaches achieved the highest effectiveness. For Task 2, we compared two systemsone built on LLaMA and the other on Mistraland found that the Mistral-based system outperformed in recall and precision when identifying complex terms, while the LLaMA-based system more accurately assigned difficulty levels. In Task 3, we entered both subtasks with three runs that combined additional training and tailored prompts for LLaMA; these runs secured leading SARI scores in both simplification subtasks.

Looking ahead, we plan to extend our work by incorporating structured reasoning strategies, such as chain-of-thought techniques, to further explore and enhance the performance of these models on future SimpleText challenges.

References

1. Jiang, A., et al.: Mistral 7B. ArXiv Preprint ArXiv:2310.06825 (2023)
2. Ermakova, L., et al.: Overview of CLEF 2024 simpletext track on improving access to scientific texts. Experimental IR Meets Multilinguality, Multimodality, and Interaction. In: Proceedings of the Fifteenth International Conference of the CLEF Association (CLEF 2024) (2024)
3. SanJuan, E., et al.: Overview of the CLEF 2024 SimpleText task 1: retrieve passages to include in a simplified summary. In: Working Notes of the Conference and Labs of the Evaluation Forum (CLEF 2024) (2024)
4. Nunzio, G., et al.: Others Overview of the CLEF 2024 SimpleText task 2: identify and explain difficult concepts. In: Working Notes of the Conference and Labs of the Evaluation Forum (CLEF 2024) (2024)
5. Ermakova, L., et al.: Others overview of the CLEF 2024 SimpleText task 3: simplify scientific text. In: Working Notes of the Conference and Labs of the Evaluation Forum (CLEF 2024) (2024)
6. Tang, J., et al.: ArnetMiner: extraction and mining of academic social networks. In: Proceedings of the 14th ACM SIGKDD International Conference on Knowledge Discovery and Data Mining (2008)
7. Anand, A., et al.: Context aware query rewriting for text rankers using LLM. ArXiv Preprint ArXiv:2308.16753. (2023)
8. Reimers, N. Gurevych, I.: Sentence-BERT: sentence embeddings using Siamese BERT-networks. In: Proceedings of the 2019 Conference on Empirical Methods in Natural Language Processing and the 9th International Joint Conference On Natural Language Processing (EMNLP-IJCNLP) (2019)
9. Macdonald, C., Tonellotto, N.: Declarative experimentation in information retrieval using PyTerrier. In: Proceedings of the 2020 ACM SIGIR on International Conference on Theory of Information Retrieval (2020)
10. Mansouri, B., Durgin, S., Franklin, S., Fletcher, S., Campos, R.: AIIR and LIAAD labs systems for CLEF 2023 SimpleText. In: CLEF (Working Notes) (2023)
11. Qin, Z., et al.: Large language models are effective text rankers with pairwise ranking prompting. In: Findings of the Association For Computational Linguistics: NAACL 2024 (2024)
12. Mansouri, B., Oard, D. Zanibbi, R.: DPRL systems in the CLEF 2022 ARQMath Lab: introducing MathAMR for Math-Aware search. In: Proceedings of the CLEF 2022 (CEUR Working Notes) (2022)
13. Dettmers, T., Pagnoni, A., Holtzman, A., Zettlemoyer, L.: QLoRA: efficient fine-tuning of quantized LLMs. Adv. Neural. Inf. Process. Syst. **36**, 10088–10115 (2024)
14. Xu, W., Napoles, C., Pavlick, E., Chen, Q., Callison-Burch, C.: Optimizing statistical machine translation for text simplification. Trans. Assoc. Computat. Linguist. **4**, 401–415 (2016)
15. Ortiz-Zambrano, J., Espin-Riofrio, C., Montejo-Ráez, A.: SINAI participation in SimpleText task 2 at CLEF 2024: zero-shot prompting on GPT-4-turbo for lexical complexity prediction. In: Working Notes of the Conference and Labs of the Evaluation Forum (CLEF 2024), pp. 3288–3299 (2024)
16. Ali, S., et al.: Team sharingans at simpletext: fine-tuned LLM based approach to scientific text simplification. In: Working Notes of the Conference and Labs of the Evaluation Forum (CLEF 2024), pp. 3174–3181 (2024)
17. Mann, R., Mikulandric, T.: CLEF 2024 SimpleText tasks 1-3: use of LLaMA-2 for text simplification. In: Working Notes of the Conference and Labs of the Evaluation Forum (CLEF 2024), pp. 3274–3283 (2024)

18. Largey, N., Maarefdoust, R., Durgin, S., Mansouri, B.: AIIR lab systems for CLEF 2024 SimpleText: large language models for text simplification. In: Working Notes of the Conference and Labs of the Evaluation Forum (CLEF 2024), pp. 3261–3273 (2024)
19. Grattafiori, A., et al.: The Llama 3 herd of models. ArXiv Preprint ArXiv:2407.21783 (2024)

Language-Based Mixture of Transformers for Sexism Identification in Social Networks

Alexandru Petrescu[1,2], Ciprian-Octavian Truică[1,2(✉)],
and Elena-Simona Apostol[1,2]

[1] National University of Science and Technology Politehnica University Bucharest,
Splaiul Independenței 313, 060042 Bucharest, Romania
[2] Academy of Romanian Scientists, 3 Ilfov, Bucharest, Romania
ciprian.truica@upb.ro

Abstract. In this paper, we present an enhanced approach for sexism identification that employs a Mixture of Transformers (MoT) framework, which leverages the language performance of individual models. Our method incorporates straightforward yet effective preprocessing modules, which are integrated with state-of-the-art Transformer architectures. We systematically compare the effectiveness of general-purpose, task-specific, and data source-specific models, evaluating both English and multilingual variants for the EXIST 2024 shared tasks. This refined approach addresses key limitations observed in earlier methods, leading to improved results across all evaluated tasks. Notably, our model demonstrates superior performance in soft-label evaluations compared to hard-label assessments. We introduce three distinct types of model mixtures, each of which achieved optimal results on training data and exhibited strong generalization to unseen data. Furthermore, the proposed architecture is designed for easy upgrades, maintains reasonable resource requirements, and delivers robust overall performance in competitive settings. Our findings underline the value of combining multiple Transformer models tailored to specific language and task requirements, paving the way for more adaptable and effective natural language processing solutions.

Keywords: Sexism Identification · Mixture of Transformers · Learning with Disagreements · Text Classification · Social Media Analysis

1 Introduction

From sexist remarks and gender-based stereotypes to harassment and threats, social media platforms like X (formerly Twitter) have increasingly become hubs for content that perpetuates gender discrimination and hostility [30]. Sexist content is no longer viewed simply as inappropriate opinion but as a pervasive global issue that poses real risks to the mental, emotional, and physical well-being of

J. Carrillo-de-Albornoz et al. (Eds.): CLEF 2025, LNCS 16089, pp. 142–155, 2026.
https://doi.org/10.1007/978-3-032-04354-2_10

those targeted, particularly women and marginalized gender groups [11,33]. Identifying the posts and comments that spread gender-based toxicity is essential to reduce their influence and mitigate their harmful impact. By identifying sexist content, platforms can help protect users from emotional trauma and contribute to a safer and more inclusive online environment.

This article outlines the methodology and findings from Team Awakened's submission to the EXIST 2024 competition. Our approach, detailed further in accompanying publications [13,15,18,19], addresses the critical challenge of identifying sexism within social networks. EXIST, a prominent series of scientific events and shared tasks, is dedicated to advancing research in this field. Its main objective is to detect the full spectrum of sexist expressions, from explicit misogyny to the more subtle and implicit manifestations of sexist behavior often found in online interactions. Our participation aimed to contribute to this crucial effort by developing robust Transformers-based models capable of accurately identifying these nuanced forms of sexism.

For the EXIST 2024 competition, our team concentrated on the Natural Language Processing (NLP) challenges, deciding to address three of the six available tasks. We specifically skipped the corresponding Computer Vision tasks, which involved similar objectives but for visual content. Our chosen tasks included:

Task 1: Sexism Identification - binary classification: this task determines whether a given text contains sexism.
Task 2: Source Intention - multi-class (4) classification technique, leveraging the outcome of TASK 1, identifying the nature.
Task 3: Sexism Categorization - multi-label classification, showing the probability of each possible outcome.

Given the current emphasis on Generative AI within NLP, our proposed architecture offers a distinct approach. Instead of leveraging Large Language Models (LLMs) for text generation, our system employs a Mixture-of-Experts-like framework built upon Transformer-based Language Models (LMs). These LMs are specifically adapted to solve the three classification tasks at hand. Transformers have consistently demonstrated superior performance in diverse text-related operations, particularly in classification problems. We take advantage of these remarkable capabilities of transformers, making use of industry-trained models facilitated by the Hugging Face platform. To further optimize their performance for the specific nuances of sexism identification and categorization, these models are fine-tuned on our task-specific datasets.

Building upon our previous experience in this competition [13], our proposed methodology incorporates a diverse mixture of Transformer models. Specifically, our approach integrates: (i) general-purpose transformers, (ii) task-specific models: harmful speech detection, (iii) data-source specific models: trained on Tweets.

This article is structured as follows. Section 2 discusses the current research related to harmful speech detection. Section 4 presents the experimental setup and analyzes the results for the three tasks. Section 5 presents the conclusions and highlights the key outcomes. Section 6 outlines future directions.

2 Related Work

The tasks proposed for this lab aim at mitigating harmful speech, in detail the sexist one, from social networks and we plan on improving the proposed approach in the previous edition of our team [13], but leverage the idea of the latest AI trend, gen AI, namely MoE [22] (Mixture of Experts). MoE proposes training separately a multitude of models, reducing the required resources of training a model that combines everything.

The idea of leveraging multiple simple models is not new and has been previously used for this task successfully in [8], both for English and non-English tweets. Another approach that successfully uses multi-lingual transformers, namely [12] proposes some data augmentation in the pipeline also. An important hint that English-only embeddings might have good results in non-English tasks is provided in another working note from the previous edition, namely [20]. Other works focus on word embeddings [7], transformer embeddings [13,23], sentence transformers [28] or document embeddings [24] for detecting online harmful content. Other architectures for detecting harmful content focus on stacked deep neural networks [2] or integrating network information into their deep neural architectures [25,26]. The current literature also focuses on how harmful content is spread online [14,17,29] and how its effects can be mitigated on social platforms [1,16,27].

3 System Architecture

Our proposed system architecture, depicted in Fig. 1, is specifically engineered to effectively detect sexism in social media text by harnessing the strengths of multiple transformer-based models. The system integrates a total of nine distinct transformer models, including four that are capable of handling multiple languages.

A key feature of our approach is the dynamic selection of models based on the detected language of each input tweet. This language-aware routing mechanism ensures that each tweet is analyzed by the most suitable model or combination of models, thereby maximizing detection accuracy across diverse linguistic scenarios. Additionally, the system considers specific evaluation metrics, for each task (they will be presented in the next sections), when choosing models, further optimizing performance for different contexts and requirements.

By combining both English-only and multilingual transformers, our architecture is able to address the challenges posed by the varied linguistic landscape of social media, ensuring robust and adaptable sexism detection regardless of language.

The task Preprocessing-modules apply standard NLP transformations for text classification (lemmatization, discarding special-tokens, etc.) and specifically for training aggregates the labels provided into a the task-specific format, either binary classification, multi-class or multi-label.

Our selection of English-specific models for these tasks includes: 1) **twitter-roberta**: a model pre-trained on Twitter data, ideal for social media language;

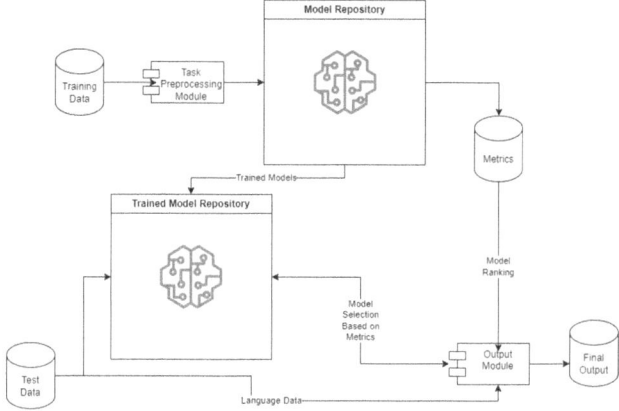

Fig. 1. The Proposed Architecture

2) **bert-toxic-comment-classification**: a BERT-based model fine-tuned for detecting toxic content; 3) textbfdistilbert-uncased-english: a smaller, faster, and more efficient version of BERT; 4) **MiniLM-L12-H384**: a highly compact but powerful language model; and 5) **roberta-hate-speech-dynabench-r4**: a RoBERTa variant specifically trained for hate speech detection.

For our system's multilingual functionality, we selected the following transformer models: 1) **twitter-xlm-roberta-base-sentiment**: an XLM-RoBERTa variant fine-tuned for sentiment analysis on Twitter data across multiple languages; 2) **twitter-xlm-roberta-base-sentiment-multilingual**: this model is similar to the above but fine-tuned on a broader, more diverse multilingual sentiment dataset; 3) **distilbert-base-multilingual-cased-sentiments**: a more compact and efficient version of BERT, pre-trained on a vast multilingual corpus and fine-tuned for sentiments; and 4) **xlm-roberta**: a multilingual model pre-trained on an enormous dataset covering over 100 languages.

Key to the architecture's adaptability is its output module, which strategically combines predictions from the best English and multi-lingual models using three distinct weight-based mixture strategies. The three implemented mixture strategies are: 1) **Half-Half**: This strategy assigns an equal 50% weight to both the English and multi-lingual models, regardless of the input language. 2) **Dominant-75%**: In this approach, the dominant model (English for English tweets, multi-lingual for others) contributes 75% of the weight to the final output, while the other model contributes the remaining 25%. This strategy gives a clear preference to the language-appropriate model while still incorporating some input from the secondary model. 3) **Dominant**: In this approach, the dominant model contributes 100% of the weight, effectively making the decision solely based on the predictions of the language-specific model.

This adaptive weighting mechanism allows our system to prioritize the most relevant model for a given input, enhancing overall accuracy and robustness.

4 Experiments

4.1 Exploratory Data Analysis

In order to better understand the task at hand, we propose a simple EDA, as we want to use a mixture of models, based on the language of the tweets. As shown in Table 1, the chosen train-test split of 79% and 21% preserves the same language distribution compared to the more conventional 80% and 20% split. Given the balanced representation of tweets across languages (53% and 47%), this approach supports meaningful comparisons between multi-lingual and English-only models. Consequently, we can expect the evaluation of these models to yield insightful results.

Table 1. Distribution of Tweets by Language for the Train/Test Split

DatasetSplit	Language	NumberOfItems	Percentage
Train	en	3 749	47%
Train	es	4 209	53%
Test	en	978	47%
Test	es	1 098	53%

4.2 Experimental Setup

For our experiments, we utilize a combination of English-only and multilingual transformer-based models, sourced from HuggingFace, as detailed in Table 2. This selection allows us to demonstrate our model architecture—which is illustrated in Fig. 1—and is tailored according to the language of the tweets.

Table 2. All the proposed models for our experiments

Run	Model Name	Hugging Face Model	IsMultiLingual
Exist2024	twitter-roberta [4,10]	cardiffnlp/twitter-roberta-base-sentiment-latest	No
Exist2024	twitter-xlm-roberta-base-sentiment-multilingual [4]	cardiffnlp/twitter-xlm-roberta-base-sentiment-multilingual	**Yes**
Exist2024	twitter-xlm-roberta-base-sentiment [3]	cardiffnlp/twitter-xlm-roberta-base-sentiment	**Yes**
Exist2024	bert-toxic-comment-classification [6]	JungleLee/bert-toxic-comment-classification	No
Exist2024	distilbert-uncased-english [21]	distilbert/distilbert-base-uncased-finetuned-sst-2-english	No
Exist2024	distilbert-base-multilingual-cased-sentiments [9]	lxyuan/distilbert-base-multilingual-cased-sentiments-student	**Yes**
Exist2023 Exist2024	MiniLM-L12-H384 [32]	microsoft/Multilingual-MiniLM-L12-H384	No
Exist2023 Exist2024	xlm-roberta [5]	papluca/xlm-roberta-base-language-detection	**Yes**
Exist2024	roberta-hate-speech-dynabench-r4 [31]	facebook/roberta-hate-speech-dynabench-r4-target	No

The output module leverages 3 types of mixtures, in terms of output weight, for the best English and multi-language models. We consider the dominant model the English one, in case the language of the input is English, otherwise the multi-lingual one. When we present the results, we use Bold for the **best English model like this** and Italic for the *best multi-lingual model*. The leveraged mixtures are: 1) Half-Half, 2) Dominant-75%, and 3) Dominant.

Since the structure of this competition is that Tasks 2 and 3 leverage the output of Task 1, our system does the same, so we propagate the mixtures, meaning that for Tasks 2 and Task 3, for each mixture type, the corresponding mixture from Task 1 is used.

For all the tasks, we have used an early stop with 3 epochs of tolerance, the best model strategy, and the following hyperparameters: (1) *learning_rate* $= 2e^{-5}$; (2) *per_device_train_batch_size* $= 32$; (3) *per_device_eval_batch_size* $= 32$; (4) *weight_decay* $= 0.01$; and (5) *max_epochs* $= 50$

As mentioned on the official site, the metrics used in the competition, for which the engine will be optimized, are ICM-Hard, ICM-Hard Norm, F1-Score, Cross Entropy, Majority class, Minority class, and Oracle most voted. To provide models that perform well, we are using F1-Score for Tasks 1 and 2, and for Task 3, we are using a custom Mean Squared Error. As for the hyperparameter tuning, each model is optimized as if it were handling the task alone, for the current implementation.

4.3   Results

Task 1. The first task is a binary classification. The system has to decide whether or not a given tweet contains sexist expressions or behaviors. The dataset is annotated by multiple evaluators, each providing their own label. To unify, we are making an average with equal weight, resulting in a single 'YES' or 'NO' label. The best-performing models for this task are the ones fine-tuned on Twitter data (Table 3). As mentioned in the previous section, we are going to use a mixture of the best English-based model and the best multilingual model.

Table 3. General Metrics for the Proposed Models on Task 1

ModelName	Epoch	F1-Score	Loss	Train(s)	Eval(s)
twitter-roberta	**4**	**0.7789**	**0.4863**	**1 463**	**5**
twitter-xlm-roberta-base-sentiment-multilingual	*3*	*0.7665*	*0.4670*	*7 039*	*159*
twitter-xlm-roberta-base-sentiment	3	0.7482	0.4902	6 373	146
bert-toxic-comment-classification	4	0.7463	0.5211	6 969	117
distilbert-uncased-english	4	0.7406	0.5348	2 918	3
distilbert-base-multilingual-cased-sentiments	4	0.7379	0.5123	4 407	3
MiniLM-L12-H384	5	0.7338	0.5059	296	3
xlm-roberta	4	0.7327	0.5520	1 834	9
roberta-hate-speech-dynabench-r4	3	0.7126	0.5220	6 080	154

We notice that the models are close when it comes to performance, all are in the range of $71\% - 79\%$, but when it comes to the resources used, there is a meaningful difference. The least resources are used by MiniLM [32], which is a super-pruned version of the regular LMs, and the most demanding models are the ones that are trained multiple extra iterations over the regular LMs, namely the multi-lingual ones, the base being XLM Roberta, that are even further trained on the platform, Twitter, and task-specific, harmful speech data.

Task 2. The second task is multi-class classification, namely "Source Intention", building on top of the first one, aiming to categorize the message according to the intention of the author, which provides insights into the role played by social networks in the emission and dissemination of sexist messages. To unify the results, we are using the same approach as for Task 1, a majority vote with equal weight. Furthermore, we are augmenting the output by leveraging the output from Task 1, working as a 'YES' or 'NO' filter that tells us if the model needs to be run on the input or not. For this problem, we can see in Table 4 that the range of results is wider, $46\% - 61\%$, and the models are performing significantly worse than they previously did, but this is expected as the output of Task 1 is also leveraged.

As expected, the specialized models are performing better, on average, but this time, the English model, one that is specialized only on the task and not on the data, is performing better, by a small margin. Resource-wise, the behavior is not reflected on the macro level as expected, as it was for Task 1, MiniLM not being the least demanding overall, as it trained for most epochs of the smaller ones, but it is still the best one per iteration.

Table 4. General Metrics for the Proposed Models on Task 2

ModelName	Epoch	F1-Score	Loss	Train(s)	Eval(s)
twitter-xlm-roberta-base-sentiment	*5*	*0.6090*	*0.8623*	*760*	*4*
xlm-roberta	7	0.6064	0.9265	1 107	4
roberta-hate-speech-dynabench-r4	**5**	**0.6063**	**0.8843**	**515**	**2**
twitter-roberta	5	0.5822	0.8790	518	2
twitter-xlm-roberta-base-sentiment-multilingual	4	0.5525	0.8631	625	4
MiniLM-L12-H384	10	0.5395	0.9431	482	1
distilbert-uncased-english	5	0.5333	0.9226	115	1
bert-toxic-comment-classification	4	0.5021	0.9227	176	2
distilbert-base-multilingual-cased-sentiments	4	0.4657	0.9439	126	1

Task 3. The third task was a multi-label classification for each tweet labeled as sexist by Task 1. Since we do not have to produce a single label, our proposed approach is to compute the probability of each label with respect to the number of annotators, with the same weight for each of them, once more. As the metrics are custom, the loss is $1/Metric$, so we do not have the need to represent. The

custom MSE is adapted to the way we build the probabilities of each class. As we can see in Table 5, we see an almost perfect mirror of what happened before, with the best performing models being the data and task-specific ones. For the resources side, we notice that this time MiniLM was trained for more than twice the number of epochs as the others.

Table 5. General Metrics for the Proposed Models on Task 3

ModelName	Epoch	CustomMSE	Train(s)	Eval(s)
twitter-xlm-roberta-base-sentiment-multilingual	**7**	**32.6568**	**1 154**	**5**
twitter-xlm-roberta-base-sentiment	*7*	*32.0696*	*1 089*	*5*
xlm-roberta	7	31.0823	1 060	2
twitter-roberta	7	30.2483	664	2
distilbert-uncased-english	7	29.6718	170	2
roberta-hate-speech-dynabench-r4	8	29.5671	826	3
distilbert-base-multilingual-cased-sentiments	7	29.2608	233	2
bert-toxic-comment-classification	8	29.2145	396	3
MiniLM-L12-H384	18	27.8630	862	1

5   Discussion

The effectiveness of our proposed approach is reflected in the results presented in Table 6, which summarizes our standings on the official leaderboard. Our system consistently ranks above the average performers, demonstrating the robustness and competitiveness of our methodology in a real-world evaluation setting. For a more detailed and comprehensive analysis of the results, including comparisons with other participating systems and additional evaluation metrics, we recommend consulting the official results at http://nlp.uned.es/exist2024/.

In this paper, we highlight only our highest-ranking submission, as our various system configurations achieved closely clustered positions on the leaderboard, with a maximum variation of just three ranks between them. This close grouping underscores the stability and reliability of our approach across different runs and parameter settings. Moreover, the minimal drift in rankings suggests that our ensemble method maintains its effectiveness regardless of minor adjustments, further validating the generalizability of our solution. These results reinforce the potential of our architecture for broad application in multilingual and multi-task natural language processing challenges.

We observe that the best overall performing mixture is Dominant-75%, while the least performing mixture is Half-Half. The best-performing outputs are on the English tasks and for the soft evaluation rather than the hard one. One interesting aspect is that the least performing Task is the first one, but the other two that are leveraging its output behave better, with a slight margin. Another

Table 6. Ranking in EXIST2024 competition

Task	EvalType	BestMixture	BestRank	TotalSystems
1	Soft-Soft-ALL	Dominant-75%	10	40
1	Hard-Hard-ALL	Dominant	20	70
1	Soft-Soft-ES	Dominant	16	40
1	Hard-Hard-ES	Dominant	21	66
1	Soft-Soft-EN	Half-Half	5	40
1	Hard-Hard-EN	Dominant	12	68
2	Soft-Soft-ALL	Dominant-75%	9	35
2	Hard-Hard-ALL	Dominant-75%	12	46
2	Soft-Soft-ES	Dominant-75%	11	35
2	Hard-Hard-ES	Dominant-75%	14	46
2	Soft-Soft-EN	Dominant-75%	11	35
2	Hard-Hard-EN	Dominant-75%	7	46
3	Soft-Soft-ALL	Dominant-75%	9	33
3	Hard-Hard-ALL	Dominant-75%	6	34
3	Soft-Soft-ES	Dominant-75%	10	33
3	Hard-Hard-ES	Dominant-75%	9	34
3	Soft-Soft-EN	Dominant	8	33
3	Hard-Hard-EN	Dominant-75%	6	34

interesting aspect is that Task 3, the one that leverages the custom metric, has the best results out of all tasks, with consistent placement.

We note that the difference between the soft and the hard evaluation varies among the tasks, for all language splits. For Task 1 the difference is quite significant, while for the Tasks 2 and 3 not that much, considering that in most cases the outputs of each team were one after another and each team had 3 possible outputs that can be submitted.

The Mixture of Transformers provides promising results with good resource requirements, with the second proposed mixture, Dominant-75%, performing on average the best, with the margin between them being not that significant.

Tables 7 and 8 present a comprehensive comparison between our proposed architectures for the 2023 [13] and 2024 [15] iterations. The results demonstrate a consistent improvement across all languages and metrics, in both soft-label and hard-label evaluations. This indicates that the enhancements introduced in the 2024 architecture contribute to more robust and accurate performance.

The Transformer model utilized in the 2023 architecture is still incorporated within the 2024 ensemble; however, it does not serve as a primary component in any of the evaluated tasks. Instead, the 2024 ensemble leverages a more diverse set of models, allowing for better adaptation to various language and task-specific nuances.

These findings highlight the effectiveness of our ensemble approach, which not only builds upon previous architectures but also introduces greater flexibility and potential for further improvements. The observed performance gains suggest that our methodology is well-suited for multilingual and multi-task scenarios, paving the way for future research on optimizing ensemble compositions and dynamic model selection strategies.

Table 7. EXIST 2023 vs 2024 architectures, Soft Label Evaluation, All Languages

Run	Task	ICM-Soft	ICM-Soft Norm	Cross Entropy
EXIST2023	1	0.3214	0.5482	1.1709
EXIST2024	1	**0.7196**	**0.6154**	0.8106
EXIST2024	1	0.6909	0.6108	0.8542
EXIST2024	1	0.6663	0.6068	**0.8037**
EXIST2023	2	-3.1765	**0.7604**	3.205
EXIST2024	2	**-2.0091**	0.3381	**3.0835**
EXIST2024	2	-2.0365	0.3359	3.1429
EXIST2024	2	-2.1502	0.3268	3.0908
EXIST2023	3	-4.2139	**0.7538**	N/A
EXIST2024	3	**-4.0748**	0.2848	N/A
EXIST2024	3	-4.0786	0.2846	N/A
EXIST2024	3	-4.1845	0.279	N/A

Table 8. EXIST 2023 vs 2024 architectures, Hard Label Evaluation, All Languages

Run	Task	ICM-Hard	ICM-Hard Norm	F1-Score
EXIST2023	1	0.4021	0.6222	**0.73**
EXIST2024	1	**0.5196**	**0.7611**	0.765
EXIST2024	1	0.5124	0.7575	0.762
EXIST2024	1	0.4984	0.7505	0.758
EXIST2023	2	-0.1481	0.6407	0.428
EXIST2024	2	**0.1812**	**0.5589**	**0.483**
EXIST2024	2	0.1487	0.5483	0.475
EXIST2024	2	0.1306	0.5425	0.469
EXIST2023	3	-0.1948	**0.5555**	0.482
EXIST2024	3	**-0.0042**	0.499	**0.483**
EXIST2024	3	-0.0115	0.4973	0.48
EXIST2024	3	-0.0427	0.4901	0.474

6   Conclusions

In this work, we present a Mixture of Transformers that provides promising results for the task of sexism identification in social media. We propose three strategies for training these mixture models, i.e., 1) Half-Half, 2) Dominant-75%, and 3) Dominant. The experimental results show that the Dominant-75% mixture has the best performance on average.

We observe similar patterns as in our previous work [13], which are partially addressed by the Mixture of Transformers approach:

1. The models achieve better results in the soft-label evaluation, suggesting that adjusting the evaluation tolerance could further improve performance in the hard-label setting.
2. The models perform better on English data, indicating a need to either identify more effective models for other languages or to fine-tune existing multilingual models with additional data.

For future work, we aim to tackle one aspect not addressed in this work, but previously discussed: experimenting with label weighting based on annotator metadata. The literature presents mixed findings on the effectiveness of this approach. Moreover, we plan to use a dynamic number of Transformers per language, selected based on their performance, as we observed that multiple models can achieve comparable results. Furthermore, we intend to explore how the mixture of Transformers is constructed. Several directions for improvement include: 1) incorporating more than two Transformers in the ensemble, 2) expanding selection criteria beyond language specialization to include task nature or the primary platform for which the Transformer was trained, and 3) automating the assignment of component weights, as current weights are manually set.

Acknowledgments. This work is supported in part by the Academy of Romanian Scientists through the funding of projects "SCAN-NEWS: Smart system for detecting and mitigating misinformation and fake news in social media" (AOŞR-TEAMS-III) and "NetGuardAI: Intelligent system for harmful content detection and immunization on social networks" (AOŞR-TEAMS-IV).

Disclosure of Interests. The authors have no competing interests to declare that are relevant to the content of this article.

References

1. Apostol, E.S., Özgur Coban, Truică, C.O.: CONTAIN: A community-based algorithm for network immunization. Eng. Sci. Technol. Int. J. **55**(101728), 1–10 (2024). https://doi.org/10.1016/j.jestch.2024.101728
2. Apostol, E.S., Truică, C.O., Paschke, A.: ContCommRTD: a distributed content-based misinformation-aware community detection system for real-time disaster reporting. IEEE Trans. Knowl. Data Eng., 1–12 (2024). https://doi.org/10.1109/tkde.2024.3417232

3. Barbieri, F., Espinosa Anke, L., Camacho-Collados, J.: XLM-T: multilingual language models in twitter for sentiment analysis and beyond. In: Proceedings of the Thirteenth Language Resources and Evaluation Conference, pp. 258–266. ERLA (2022)

4. Camacho-Collados, J., et al.: TweetNLP: cutting-edge natural language processing for social media. In: Proceedings of the 2022 Conference on Empirical Methods in Natural Language Processing: System Demonstrations, pp. 38–49. Association for Computational Linguistics (2022). https://doi.org/10.18653/v1/2022.emnlp-demos.5

5. Conneau, A.e.a.: Unsupervised cross-lingual representation learning at scale. In: Proceedings of the 58th Annual Meeting of the ACL, pp. 8440–8451. ACL (2020). https://doi.org/10.18653/v1/2020.acl-main.747

6. Devlin, J., Chang, M.W., Lee, K., Toutanova, K.: BERT: Pre-training of deep bidirectional transformers for language understanding. In: Conference of the North American Chapter of the Association for Computational Linguistics, pp. 4171–4186. ACL (2019). https://doi.org/10.18653/v1/N19-1423

7. Ilie, V.I., Truică, C.O., Apostol, E.S., Paschke, A.: Context-aware misinformation detection: a benchmark of deep learning architectures using word embeddings. IEEE Access **9**, 162122–162146 (2021). https://doi.org/10.1109/access.2021.3132502

8. Jhakal, C., et al.: Detection of Sexism on Social Media with Multiple Simple Transformers. In: Working Notes of the Conference and Labs of the Evaluation Forum (CLEF 2023). CEUR Workshop Proceedings, vol. 3497, pp. 959–966 (2023)

9. Laurer, M., van Atteveldt, W., Casas, A., Welbers, K.: Less annotating, more classifying: addressing the data scarcity issue of supervised machine learning with deep transfer learning and BERT-NLI. Polit. Anal. **32**(1), 84–100 (2024). https://doi.org/10.1017/pan.2023.20

10. Loureiro, D., Barbieri, F., Neves, L., Espinosa Anke, L., Camacho-collados, J.: TimeLMs: Diachronic language models from twitter. In: Proceedings of the 60th Annual Meeting of the Association for Computational Linguistics: System Demonstrations, pp. 251–260. ACL (2022). https://doi.org/10.18653/v1/2022.acl-demo.25

11. Mansur, Z., Omar, N., Tiun, S.: Twitter hate speech detection: a systematic review of methods, taxonomy analysis, challenges, and opportunities. IEEE Access **11**, 16226–16249 (2023). https://doi.org/10.1109/access.2023.3239375

12. Mohammadi, H., Giachanou, A., Bagheri, A.: Towards robust online sexism detection: a multi-model approach with BERT, XLM-RoBERTa, and DistilBERT for EXIST 2023 tasks. In: Working Notes of the Conference and Labs of the Evaluation Forum (CLEF 2023). CEUR Workshop Proceedings, vol. 3497, pp. 1000–1011 (2023)

13. Petrescu, A.: Leveraging MiniLMv2 Pipelines for EXIST2023. In: Working Notes of the Conference and Labs of the Evaluation Forum (CLEF 2023). CEUR Workshop Proceedings, vol. 3497, pp. 1037–1043. CEUR-WS.org (2023)

14. Petrescu, A., Truică, C.O., Apostol, E.S.: Sentiment analysis of events in social media. In: 2019 IEEE 15th International Conference on Intelligent Computer Communication and Processing (ICCP), pp. 143–149. IEEE (2019). https://doi.org/10.1109/iccp48234.2019.8959677

15. Petrescu, A., Truică, C.O., Apostol, E.S.: Language-based Mixture of Transformers for EXIST2024. In: Working Notes of the Conference and Labs of the Evaluation Forum. CEUR Workshop Proceedings, vol. 3740, pp. 1157–1164 (2024)

16. Petrescu, A., Truică, C.O., Apostol, E.S., Karras, P.: Sparse Shield: social network immunization vs. harmful speech. In: ACM International Conference on Information and Knowledge Management (CIKM2021), pp. 1426–1436. ACM (2021). https://doi.org/10.1145/3459637.3482481

17. Petrescu, A., Truică, C.O., Apostol, E.S., Paschke, A.: EDSA-Ensemble: an event detection sentiment analysis ensemble architecture. IEEE Trans. Affect. Comput., 1–18 (2024). https://doi.org/10.1109/TAFFC.2024.3434355

18. Plaza, L., et al.: Overview of EXIST 2024 – Learning with disagreement for sexism identification and characterization in social networks and memes. In: Proceedings of the Fifteenth International Conference of the CLEF Association (CLEF 2024) (2024)

19. Plaza, L., et al.: Overview of EXIST 2024 – learning with disagreement for sexism identification and characterization in social networks and memes (extended overview). In: Working Notes of CLEF 2024 – Conference and Labs of the Evaluation Forum (2024)

20. Sanchez-Urbina, A., et al.: IimasGIL_NLP@EXIST2023: unveiling sexism on twitter with fine-tuned transformers. In: Working Notes of the Conference and Labs of the Evaluation Forum (CLEF 2023). CEUR Workshop Proceedings, vol. 3497, pp. 1067–1082. CEUR-WS.org (2023)

21. Sanh, V., Debut, L., Chaumond, J., Wolf, T.: DistilBERT, a distilled version of BERT: smaller, faster, cheaper and lighter (2020). https://arxiv.org/abs/1910.01108

22. Sanseviero, O., Tunstall, L., Schmid, P., Mangrulkar, S., Belkada, Y., Cuenca, P.: Mixture of experts explained (2023). https://huggingface.co/blog/moe

23. Truică, C.O., Apostol, E.S.: MisRoBÆRTa: Transformers versus misinformation. Mathematics **10**(4), 1–25(569) (2022). https://doi.org/10.3390/math10040569

24. Truică, C.O., Apostol, E.S.: It's all in the embedding! fake news detection using document embeddings. Mathematics **11**(3), 508 (2023). https://doi.org/10.3390/math11030508

25. Truică, C.O., Apostol, E.S., Karras, P.: DANES: deep neural network ensemble architecture for social and textual context-aware fake news detection. Knowl. Based Syst. **294**, 1–13(111715) (2024). https://doi.org/10.1016/j.knosys.2024.111715

26. Truică, C.O., Apostol, E.S., Marogel, M., Paschke, A.: GETAE: graph information enhanced deep neural NeTwork ensemble ArchitecturE for fake news detection. Expert Syst. Appl. **275**, 126984 (2025). https://doi.org/10.1016/j.eswa.2025.126984

27. Truică, C.O., Apostol, E.S., Nicolescu, R.C., Karras, P.: MCWDST: a minimum-cost weighted directed spanning tree algorithm for real-time fake news mitigation in social media. IEEE Access **11**, 125861–25873 (2023). https://doi.org/10.1109/ACCESS.2023.3331220

28. Truică, C.O., Apostol, E.S., Paschke, A.: Awakened at CheckThat! 2022: fake news detection using BiLSTM and sentence transformer. In: Working Notes of the Conference and Labs of the Evaluation Forum (CLEF2022), pp. 749–757 (2022)

29. Truică, C.O., Apostol, E.S., Ștefu, T., Karras, P.: A deep learning architecture for audience interest prediction of news topic on social media. In: International Conference on Extending Database Technology (EDBT2021), pp. 588–599 (2021). https://doi.org/10.5441/002/EDBT.2021.69

30. Truică, C.O., Constantinescu, A.T., Apostol, E.S.: STopHC: a harmful content detection and mitigation architecture for social media platforms. In: 2024 IEEE

20th International Conference on Intelligent Computer Communication and Processing (ICCP), pp. 01–05. IEEE (2024). https://doi.org/10.1109/iccp63557.2024. 10793051

31. Vidgen, B., Thrush, T., Waseem, Z., Kiela, D.: Learning from the worst: dynamically generated datasets to improve online hate detection. In: Proceedings of the 59th Annual Meeting of the Association for Computational Linguistics and the 11th International Joint Conference on Natural Language Processing, pp. 1667–1682. ACL (2021). https://doi.org/10.18653/v1/2021.acl-long.132

32. Wang, W., et al.: MiniLM: deep self-attention distillation for task-agnostic compression of pre-trained transformers. In: Proceedings of the 34th International Conference on Neural Information Processing Systems (2020)

33. Williams, M.L., Burnap, P., Javed, A., Liu, H., Ozalp, S.: Hate in the machine: anti-black and anti-Muslim social media posts as predictors of offline racially and religiously aggravated crime. Br. J. Criminol. (2019). https://doi.org/10.1093/bjc/azz049

Humour Classification According to Genre and Technique by Fine-Tuning LLMs

Shih-Hung Wu$^{(\boxtimes)}$ ⓘ, Tsz-Yeung Lau ⓘ, and Yu-Feng Huang

Chaoyang University of Technology, Taichung, Taiwan
shwu@cyut.edu.tw, {s11327605,s11227615}@gm.cyut.edu.tw

Abstract. This study investigates the classification of humor into six distinct genres—irony, sarcasm, exaggeration, incongruity-absurdity, self-deprecating humor, and wit-surprise—using data from the JOKER @ CLEF 2024 shared task. We evaluate the effectiveness of fine-tuning Large Language Models (LLMs), with a primary focus on Llama 3-8B, and compare its performance against a RoBERTa baseline and advanced Large Reasoning Models (LRMs) like DeepSeek-R1. By employing prompt engineering, including the Stanford Alpaca format and Chain-of-Thought (CoT) prompting, our fine-tuned Llama 3 model achieved a high accuracy of 89.68% on development set. However, in the official evaluation, its accuracy was 69.78%, highlighting a significant discrepancy likely due to data distribution differences. A key finding is that LRMs, despite their sophisticated reasoning capabilities enhanced by techniques like Group Relative Policy Optimization (GRPO), performed poorly on this task. This suggests that the nuanced, context-dependent nature of humor comprehension presents a unique challenge that is not directly addressed by improvements in general logical reasoning. Our results affirm that while specialized fine-tuning makes LLMs highly effective for humor classification, genuine humor understanding remains a distinct frontier for AI, requiring more than advanced reasoning alone.

Keywords: Deep Learning · Humour Classification · Large Language Models (LLMs) · Large Reasoning Models (LRMs) · Llama 3 · GPT-4

1 Introduction

Humor is a sophisticated and often ambiguous element of human language, presenting a significant frontier in natural language processing (NLP) [17]. As a linguistic phenomenon, humor rarely exists in isolation; its meaning is deeply intertwined with context, cultural nuances, and shared background knowledge [18]. The ability to interpret humor, therefore, serves as a critical test for an AI's capacity for deep language understanding. As discourse analysis suggests, language itself can become the object of humor, requiring a model to analyze language at a meta-level [18].

S-H.Wu , T-Y.Lau and Y-F.Huang—These authors contributed equally to this work.

J. Carrillo-de-Albornoz et al. (Eds.): CLEF 2025, LNCS 16089, pp. 156–169, 2026.
https://doi.org/10.1007/978-3-032-04354-2_11

The challenge of humor recognition is multifaceted. Humor frequently relies on figurative language, such as irony and sarcasm, where the intended meaning is the opposite of the literal one. Furthermore, humor is subjective and varies across different cultural, social, and even individual contexts. A political joke, for instance, may be opaque to someone without the requisite political knowledge, highlighting that disparate background knowledge leads to different interpretations of the same text [14]. This variability has traditionally made it difficult to create robust, generalized humor detection systems.

Previous research has made significant strides, particularly in humor classification [37], with tasks like the JOKER Track @ CLEF 2024 prompting systems to categorize humor into genres such as irony, exaggeration, and wit-surprise [11]. Such tasks have been valuable benchmarks, and models have achieved high accuracy in identifying humor patterns. However, true humor comprehension transcends mere classification. It requires reasoning, the ability to infer unstated meanings, resolve incongruities, and understand the interplay of social dynamics and world knowledge that makes a statement humorous.

This paper argues that the next frontier in computational humor is to move beyond classification and toward the evaluation of humor reasoning. We posit that the emerging class of Large Reasoning Models (LRMs), which are designed to perform complex, multi-step inferential tasks, offers a promising avenue for this deeper level of understanding. Unlike standard large language models that excel at pattern recognition, LRMs are architected to handle tasks that demand logical deduction and abstract thought.

Therefore, in this study, we shift the focus from classifying humor to evaluating the reasoning capabilities of LRMs in the context of humor. We investigate whether these models can not only detect humor but also "explain the joke" by deconstructing its underlying logical and semantic structure. By probing the reasoning pathways of these advanced models, we aim to shed light on the current capabilities and inherent limitations of AI in achieving genuine humor understanding, paving the way for more sophisticated and human-like language technologies.

2   Related Work

2.1   Large Language Models

The remarkable capabilities of large language models (LLMs), such as GPT-4 [25] and Llama 3 [22], have captured significant attention. These models are characterized by their extensive parameter counts and a unique feature known as "in-context learning", which enables them to tackle novel tasks without being explicitly retrained [24]. More recently, ChatGPT, based on the GPT-3.5 architecture [24] and further refined with reinforcement learning from human feedback [3], has become a focal point of interest in the field [28] [13].

2.2   Reasoning Models

Recent advancements in LLMs have led to the development of models optimized for explicit reasoning tasks. Among these, OpenAI's proprietary o-series (e.g., o1 and o3) [26] and DeepSeek's opensource DeepSeek-R1 [5] stand out for their integration of Chain-of-Thought (CoT) data to enhance reasoning capabilities.

DeepSeek-R1 employs a two-phase training procedure. Initially, the model undergoes Supervised Fine-Tuning (SFT), followed by reinforcement learning (RL) to further optimize reasoning performance. To reduce the computational burden typically associated with RL, DeepSeek introduces Group Relative Policy Optimization (GRPO) [4], a novel technique that eliminates the need for a critic model. Instead, GRPO calculates baseline values from group-level scores, enabling more efficient training. The model receives accuracy rewards for correct answers and is further guided by a format reward mechanism that encourages encapsulating its reasoning within <think> and </think> tags. This promotes the progressive extension of reasoning sequences, improving output precision.

Additionally, the DeepSeek team observed a notable "aha" moment during training: the model initially proposes a solution, then reevaluates and improves upon its reasoning, resulting in a more refined and accurate response. The base model of DeepSeek-R1, DeepSeek-V3 [6], employs a Mixture of Experts (MoE) [31] architecture with FP8 mixed-precision training, enhancing computational efficiency without compromising inference quality.

In this study, we leverage these advanced reasoning capabilities of LLMs, particularly for humor classification, aiming to achieve high accuracy and consistent performance.

2.3   Prompt Engineering

Prompt engineering is a critical discipline within artificial intelligence and machine learning, serving as the primary method of communication with large language models (LLMs) such as GPT-3 and GPT-4 [30]. Through techniques like fine-tuning, we can guide these models to produce more accurate and targeted outputs [19] [38] [9]. The fundamental goal of prompt engineering is to provide LLMs with carefully designed instructions that elicit desired information or facilitate complex task execution [36] [1]. The quality of these prompts directly influences the model's performance in natural language processing (NLP), as ambiguous instructions can lead to irrelevant or inaccurate responses. To enhance prompt accuracy, established methods include providing precise instructions, assigning a specific role to the model, offering examples (one-shot or few-shot learning) [24], and employing iterative refinement and Chain-of-Thought (CoT) prompting [35].

3   Dataset

The training dataset for this study was provided by the JOKER organizer and consists of a total of 1,742 entries. The humorous content is categorized into six types: IR (irony) with 210 entries, SC (sarcasm) with 356 entries, EX (exaggeration) with 125 entries, AID (incongruity-absurdity) with 231 entries, SD (self-deprecating) with 169 entries, and WS (wit-surprise) with 651 entries. The

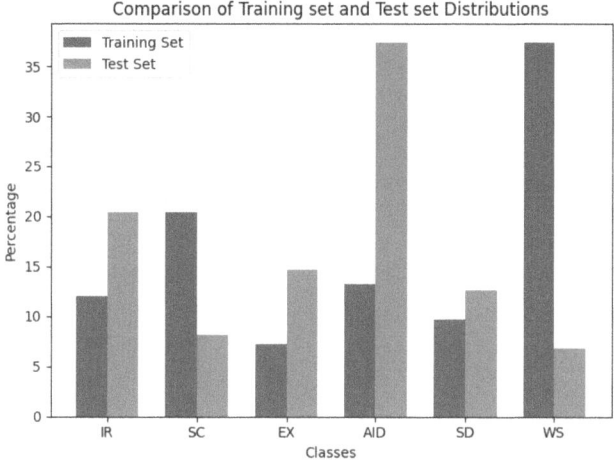

Fig. 1. Comparison of Training set and Test set Distributions

distribution of the training data is shown in Fig. 1. The test set comprises a total of 722 entries, as illustrated in Fig. 1. The results were evaluated by the JOKER organizer.

4   Method

4.1   RoBERTa

We utilize the enhanced BERT [8] model, RoBERTa [20], as our baseline. BERT, which stands for Bidirectional Encoder Representations from Transformers [8], was originally introduced by Google as an encoder-only transfo- rmer [34]-based model for natural language processing (NLP) tasks. BERT is pre-trained using the Masked Language Model (MLM) and Next Sentence Prediction (NSP) techniques. Unlike word2vec [23] and GloVe [27], which do not consider context, BERT leverages contextual information during inference, leading to superior performance [8]. In the RoBERTa paper, they mentioned that the BERT model was significantly undertrained [20]. To address this, they implemented several modifications: using larger batches, training the model for a longer duration, dropping the NSP training, training on longer sequences, and dynamically changing the masking pattern applied to the training data [20]. For the RoBERTa baseline model, we achieve an accuracy of 72.49%.

4.2   Llama 3

Large Language Models (LLMs) are highly capable AI assistants that excel in complex reasoning tasks. They enable interaction with humans through intuitive chat interfaces, leading to rapid and widespread adoption among the general public [21]. Many different LLMs are publicly available, such as GPT-4 [25], Mistral 7B [16], Gemma 7B [33], and the LLM we utilize in this study, Llama 3.

Llama 3 [22] is an open-source LLM utilizing the Transformer [34] architecture, developed by Meta. The Llama3 model is available in configurations with 8 billion and 70 billion parameters. Llama3 models have achieved state-of-the-art (SOTA) performance across a broad range of tasks due to extensive pre-training on over 15 trillion data tokens, making it the best-performing open-source model. In this study, we fine-tuned the Llama 3-8B model on a single GPU, utilizing 4-bit quantization with QLoRa [7] to reduce GPU RAM usage during training with unsloth [15]. As a result, the model achieved 89.68% accuracy.

4.3   DeepSeek-R1

In our methodology, we incorporate DeepSeek-R1 [5], a sophisticated reasoning model that excels at generating a transparent thinking process. This capability is primarily cultivated through an innovative reinforcement learning (RL) technique known as Group Relative Policy Optimization (GRPO) [4].

DeepSeek-R1's training encourages the model to externalize its reasoning process before delivering a final answer. This is achieved by leveraging GRPO, an efficient RL algorithm that refines the model's policy based on the relative quality of a group of generated responses. For a given prompt, the model produces multiple potential outputs. Each output is then assigned a reward score, primarily based on the accuracy of the final answer.

A key innovation of GRPO is its departure from traditional RL methods that require a separate critic model to estimate the value of each action. Instead, GRPO calculates a baseline reward by averaging the scores of the entire group of generated responses. The policy is then optimized by favoring responses that score above this group average. This approach significantly reduces the computational and memory overhead associated with training.

To further structure the model's output, a format-based reward is also employed during training. This reward incentivizes DeepSeek-R1 to encapsulate its entire reasoning or chain-of-thought process within <think> and </think> tags. This not only makes the model's reasoning explicit and human-interpretable but also encourages the generation of more detailed and extended thought processes, which has been shown to improve the accuracy and coherence of the final response.

4.4   Evaluation Metrics

To evaluate our model's performance, we used the four official metrics from the JOKER Lab 2024 shared task. We measured Accuracy for overall correctness. To assess performance on potentially imbalanced humor categories, we used Macro-Averaged F1-score (MA-F1), which provides a balanced measure of precision and recall across all classes. Additionally, Mean Average Precision (MAP) and Mean Average Recall (MAR) were used to evaluate the average precision and recall rates across the different humor types.

Table 1. Stanford Alpaca Format

Model	Format	Prompt
Llama 3-8B	Alpaca Format	Below is an instruction that describes a task, paired with an input that provides further context. Write a response that appropriately completes the request. ### Instruction: {} ### Input: {} ### Response: {}

5 System Development

5.1 Environment

In our experiment, we utilized a GPU, NVIDIA GeForce RTX 3090 with 24GB of memory. The versions of all packages employed in the experiment will be thoroughly delineated in our previous work [37].

5.2 RoBERTa

To fine-tune the RoBERTa model, we use 80% of the dataset as the training set and 20% as the test set. The hyperparameters we used for fine-tuning are shown in our previous work [37].

5.3 Llama 3

To fine-tune the Llama 3-8B model, we use 80% of the dataset as the training set and 20% as the test set. The hyperparameters we used for fine-tuning are shown in our previous work [37].

Prompt Design. To fine-tune Llama3, we utilize the Stanford Alpaca Format [32]. The Alpaca format is shown in Table 1. For the instruction, we first tell the model what to do: "Classify the following text into one of the classes." Then, we provide the six classes for classification with explanations: irony, sarcasm, exaggeration, incongruity-absurdity, self-deprecating humor, and wit-surprise. We simply utilize the explanations provided in the official JOKER guideline document here. Based on results from all method, we discovered that the model struggled to accurately classify irony and sarcasm. Therefore, we added the sequence: "You ought to focus more on classifying irony and sarcasm." Finally, we applied Chain of Thought (CoT) prompting [35] by adding the sequence: "Let's think step by step.".

Meanwhile, the sequence following "### Input:" denotes the text in need of classification, while "### Response:" following with one of six classes: irony, sarcasm, exaggeration, incongruity-absurdity, self-deprecating humor, and wit-surprise. During evaluation, we employ the same prompting technique. The only difference is that we refrain from adding any text after "### Response:" to allow the model to generate the response. The prompt elements are shown in our previous paper [37].

6   Experiment Result

6.1   Run on Development Set

Table 2 evaluates the performance of each model. The RoBERTa model achieved an accuracy of 73.28%, 0.72 Macro Average Precision (MAP), 0.63 Macro Average Recall (MAR) and 0.63 Macro Average F1-Score (MA-F1), serving as the baseline. The Llama 3-8B model achieved an accuracy of 89.68%, 0.89 MAP, 0.87 MAR and 0.88 MA-F1, representing a 18.05%, 0.25, 0.23, 0.24 increase compared to the baseline model.

Table 2. Models Performance on Development Set

Model	Accuracy (%)	MAP	MAR	MA-F1
Llama 3-8B (with SFT)	**89.68**	**0.89**	**0.87**	**0.88**
RoBERTa	73.28	0.72	0.63	0.63
Deepseek-R1:32B-q4	29.70	0.29	0.31	0.25
Deepseek-R1:671B	23.16	0.19	0.18	0.18

*MAP: Macro Average Precision
MAR: Macro Average Recall
MA-F1: Macro Average F1 Score

6.2   Run on Official Test Set

All of the models were evaluated by the JOKER organizer [10]. Table 3 presents the official results of each model. The RoBERTa model achieved an accuracy of 68.14%, 0.64 Macro Average Precision (MAP), 0.60 Macro Average Recall (MAR), and 0.59 Macro Average F1-Score (MA-F1). The performance of RoBERTa in each class shown in Table 4. The RoBERTa model performed comparably to Llama3-8B in humor understanding, indicating that dedicated model training is necessary for this task.

The Llama 3-8B model used for evaluation is the same model fine-tuned with 80% of the dataset. It achieved an accuracy of 69.78%, 0.64 MAP, 0.65 MAR, and 0.64 MA-F1. The Llama 3-8B model exhibited a significant drop in accuracy compared to our self-test results, potentially due to differences in

Table 3. Models Performance on Official Test Set

Run	Model	Accuracy ↑(%)	MAP ↑	MAR ↑	MA-F1 ↑
1	Llama 3-8B (with SFT)	**69.78**(-19.90)	**0.64**(-0.25)	**0.65**(-0.22)	**0.64**(-0.24)
2	RoBERTa	68.14(-5.14)	**0.64**(-0.08)	0.60(-0.00)	0.59(-0.04)
-	Deepseek-R1:671B	21.08	0.17	0.17	0.16
-	Deepseek-R1:32B-q4	17.07	0.11	0.17	0.12
-	Qwen2.5:7B	16.32	0.18	0.17	0.13
-	Llama 3-8B (without SFT)	14.93	0.17	0.16	0.13
-	QwQ:32B	14.13	0.15	0.17	0.11

*MAP: Macro Average Precision
MAR: Macro Average Recall
MA-F1: Macro Average F1 Score
Blue words represent the differences compared to development set.

Table 4. Precision, Recall and F1-Score of each class of RoBERTa (Offical Test set)

Model	Class	Precision ↑	Recall ↑	F1-Score ↑
	IR	0.64(-0.09)	0.17(+0.08)	0.27(+0.10)
	SC	0.49(-0.01)	0.78(+0.01)	0.60(-0.01)
RoBERTa	EX	0.43(-0.15)	0.33(+0.25)	0.37(-0.21)
	AID	0.72(-0.08)	0.87(+0.06)	0.79(-0.07)
	SD	0.64(-0.12)	0.61(+0.05)	0.63(-0.01)
	WS	0.89(-0.04)	0.84(-0.03)	0.87(-0.03)

*Blue words represent the differences compared to development set.

the data distribution between the training and test sets, as shown in Fig. 1. However, as seen in Table 5, the model performed exceptionally well on the class AID, even with a small amount of training data. The confusion matrix is presented in Fig. 2. It appears that AID has distinctive features that the model can learn effectively. The model likely overfitted to the training set, impairing its performance on the test set. Balancing the data in the training set may help improve the model's robustness. From the official results, it is evident that the Mistral-7B model performed the best overall in humor classification, achieving an accuracy of 76%, from team ORPAILLEUR [10].

The comprehension of humor remains a significant challenge for most large language models (LLMs), with performance improving only through specific fine-tuning. Notably, even dedicated reasoning models like DeepSeek-R1 have shown limited aptitude in this domain. This observation suggests that improvements in reasoning capabilities do not inherently lead to greater humor understanding. Consequently, achieving proficiency in humor-related tasks necessitates further model alignment.

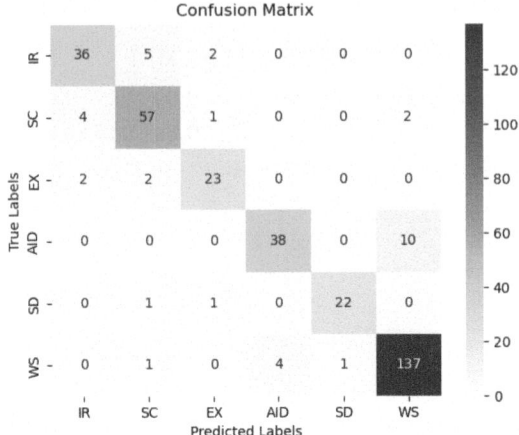

Fig. 2. Confusion matrix of Llama 3-8B self-testing

Table 5. Precision, Recall and F1-Score of each class of Llama 3-8B (with SFT) (Offical Test set)

Model	Class	Precision ↑	Recall ↑	F1-Score ↑
	IR	0.63(-0.23)	0.60(-0.24)	0.62(-0.23)
	SC	0.67(-0.19)	0.68(-0.21)	0.67(-0.21)
Llama 3-8B (with SFT)	EX	0.52(-0.33)	0.41(-0.44)	0.46(-0.39)
	AID	0.86(-0.04)	0.88(**+0.09**)	0.87(**+0.03**)
	SD	0.70(-0.26)	0.69(-0.23)	0.70(-0.24)
	WS	0.44(-0.48)	0.63(-0.33)	0.52(-0.42)

*Blue words represent the differences compared to development set.

Table 6. Precision, Recall and F1-Score of each class and mode. (Development set)

Model	Class	Precision	Recall	F1-Score
Llama 3-8B (with SFT)	IR	0.86	0.84	0.85
	SC	0.86	0.89	0.88
	EX	0.85	0.85	0.85
	AID	0.90	0.79	0.84
	SD	0.96	0.92	0.94
	WS	0.92	0.96	0.94
RoBERTa	IR	0.75	0.09	0.17
	SC	0.50	0.77	0.61
	EX	0.58	0.58	0.58
	AID	0.80	0.93	0.86
	SD	0.76	0.56	0.64
	WS	0.93	0.87	0.90

7 Discussion and Error Analysis

7.1 Discussion

An analysis of the confusion matrices, as supported by our previous research [37], reveals a notable challenge in the models' ability to accurately distinguish between the AID and WS categories, as well as between IR and SC. This difficulty can be attributed to several factors. One key reason is the inherent complexity of irony, which encompasses two distinct forms: verbal and situational. Verbal irony, which is often synonymous with sarcasm, suggests an overlap where IR can be inclusive of SC [12]. Furthermore, the accurate identification of sarcasm is frequently dependent on contextual cues, a challenge highlighted by [12].

In contrast, our earlier work [37] demonstrated that Llama 3-8B achieved significantly better performance in the specific areas where RoBERTa showed limitations. The data presented in Table 6 confirms that fine-tuning large language models (LLMs) is a highly effective strategy for humor classification.

While non-tuned LLMs exhibit strong general capabilities, they may not achieve optimal results in specialized domains such as this. Nevertheless, in low-compute environments where fine-tuning large models is not feasible, smaller models like RoBERTa can still deliver comparable performance, proving their value as a viable alternative.

Interestingly, models designed for advanced reasoning, which often outperform conventional LLMs on a majority of benchmarks, did not demonstrate a superior understanding of humor. This finding suggests that humor comprehension may not be well-captured by the current paradigms of reasoning-focused models, indicating a unique challenge that remains to be addressed.

7.2 Error Analysis

The pre-training data of large language models (LLMs) can lead to occasional, unexpected responses during fine-tuning. For example, when presented with the input, "When negotiating whether to share your french fries, you have quite a few bargaining chips," the model might unpredictably generate the output "lunch." Our previous research, detailed in [37], identified 12 such instances in development set.

This issue highlights a limitation of fine-tuning generative models. In contrast, classification models like BERT avoid this problem. By feeding the [CLS] token into a Multilayer Perceptron (MLP) [29], BERT utilizes a fixed-size output layer and a softmax function [2] to calculate the probability for each predefined class, thereby ensuring that no unexpected outputs are produced.

To further investigate these inconsistencies, we conducted additional testing on the erroneous samples, running each through the model ten more times. We found that some of these errors could be categorized into one of six classes. For instance, the input text, "No longer a female as I refuse to wear heels ever again," which had previously yielded the unexpected response "twitter" from Llama 3-8B, produced the response "sarcasm" in one out of ten new trials. A similar outcome was observed with the input, "The leopard tried creeping up on the tigers using its camouflage but it was seen," which was classified as "wit-surprise" in one of its ten attempts. Furthermore, the input, "Doppelherz. The power of the two hearts," consistently generated a "wit-surprise" response in eight out of the ten trials. These specific examples are documented in our previous work [37], while other tested inputs showed no change in their outputs.

Regarding the LRMs, they are significantly better at adhering to instructions, preventing the generation of unexpected outputs.

8   Conclusion and Future Work

8.1   Conclusion

In this study, we evaluated the performance of several models on the task of humor classification, including the encoder-based RoBERTa and various Large Language Models (LLMs). Our results demonstrate that fine-tuning a capable LLM, specifically Llama 3-8B, with prompt engineering is a highly effective strategy, achieving a peak accuracy of 89.68% on development set and 69.78% in the official test set. This approach significantly outperformed baseline models and showed that targeted training can successfully adapt general-purpose models for the specialized domain of humor recognition. We also analyzed the occasional generation of unexpected and irrelevant responses, a phenomenon inherent to the generative nature of LLMs.

A particularly noteworthy finding emerged from the evaluation of Large Reasoning Models (LRMs) like DeepSeek-R1. Despite being explicitly optimized for complex, multi-step logical inference through advanced techniques such as Group Relative Policy Optimization (GRPO), these models showed limited aptitude for humor classification. This suggests that the sophisticated reasoning capabilities that allow LRMs to excel in domains like mathematics and coding do not directly translate to a nuanced understanding of humor.

In conclusion, our findings affirm that while current LLMs possess a latent capacity for humor recognition, unlocking it requires specialized fine-tuning. More importantly, the underwhelming performance of dedicated LRMs indicates that humor comprehension is a distinct and complex challenge that transcends pure logical reasoning. This highlights a fascinating gap in AI capabilities, suggesting that genuine humor understanding may rely on cognitive faculties not yet fully captured by existing reasoning-centric architectures.

8.2   Future Work

Future research will prioritize enhancing model robustness and exploring novel architectures. We will investigate the observed performance discrepancies and overfitting, particularly with Llama 3, through detailed dataset analysis and the application of techniques such as data augmentation and advanced regularization methods.

Given RoBERTa's superior performance on the WS class, a potential avenue for future work involves a hybrid approach. This would entail utilizing RoBERTa for initial WS class prediction and subsequently leveraging this output to inform the five-class classification performed by Llama 3-8B.

A key direction involves fine-tuning state-of-the-art (SOTA) models, including DeepSeek and Qwen, for humor classification and benchmarking their efficacy against our current results. Further refinement of prompt engineering and fine-tuning strategies will aim to improve accuracy and mitigate unexpected outputs from generative large language models (LLMs). An in-depth error analysis will guide targeted improvements for specific humor categories.

Furthermore, we believe that using Generative Representational Pre-training and Optimization (GRPO) to fine-tune Large Language Models (LLMs) holds significant potential, analogous to how Supervised Fine-Tuning (SFT) has led to substantial improvements. The reasoning processes inherent in such methods could also provide richer context for LLMs to understand humor. Finally, we plan to explore ensemble methods and the integration of external knowledge sources as potential pathways to advance humor classification capabilities.

Acknowledgments. This study was supported by the National Science and Technology Council under the grant number NSTC 113–2221-E-324-009.

A   Appendix

A.1   Our fine-tuned Llama 3-8B model for Humor Classification

The fine-tuned Llama 3-8B model is available on Hugging Face.

– Hugging Face

References

1. Amatriain, X.: Prompt design and engineering: Introduction and advanced methods (2024)
2. Bridle, J.: Training stochastic model recognition algorithms as networks can lead to maximum mutual information estimation of parameters. In: Touretzky, D. (ed.) Advances in Neural Information Processing Systems, vol. 2. Morgan-Kaufmann (1989)
3. Christiano, P., Leike, J., Brown, T.B., Martic, M., Legg, S., Amodei, D.: Deep reinforcement learning from human preferences (2023). https://arxiv.org/abs/1706.03741
4. DeepSeek-AI: Deepseekmath: pushing the limits of mathematical reasoning in open language models (2024). https://arxiv.org/abs/2402.03300
5. DeepSeek-AI: Deepseek-r1: incentivizing reasoning capability in LLMs via reinforcement learning (2025). https://arxiv.org/abs/2501.12948
6. DeepSeek-AI: Deepseek-v3 technical report (2025). https://arxiv.org/abs/2412.19437

7. Dettmers, T., Pagnoni, A., Holtzman, A., Zettlemoyer, L.: Qlora: efficient finetuning of quantized LLMs (2023)
8. Devlin, J., Chang, M.W., Lee, K., Toutanova, K.: Bert: pre-training of deep bidirectional transformers for language understanding (2019)
9. Ekin, S.: Prompt engineering for ChatGPT: a quick guide to techniques, tips, and best practices (2023). https://doi.org/10.36227/techrxiv.22683919
10. Ermakova, L., Bosser, A.G., Miller, T., Preciado, V.M.P., Sidorov, G., Jatowt, A.: Overview of the clef 2024 joker track automatic humour analysis (2024)
11. Ermakova, L., et al.: Clef 2024 joker lab: automatic humour analysis. In: Goharian, N., Tonellotto, N., He, Y., Lipani, A., McDonald, G., Macdonald, C., Ounis, I. (eds.) Advances in Information Retrieval, pp. 36–43. Springer Nature Switzerland, Cham (2024)
12. Filatova, E.: Irony and sarcasm: corpus generation and analysis using crowdsourcing. In: Calzolari, N., et al(eds.) Proceedings of the Eighth International Conference on Language Resources and Evaluation (LREC'12), pp. 392–398. European Language Resources Association (ELRA), Istanbul, Turkey (2012)
13. Guo, Q., et al.: Connecting large language models with evolutionary algorithms yields powerful prompt optimizers (2024)
14. Guo, Y., Kong, L.: Classification and regression combined model on accessing humor score with explanatory feature. In: 2022 International Conference on Machine Learning and Knowledge Engineering (MLKE). IEEE (2022). https://doi.org/10.1109/mlke55170.2022.00050
15. Han, D., Han, M., Nguyen, H.H., Qubitium, Belkada, Y., Z: unslothai/unsloth (2024). https://github.com/unslothai/unsloth
16. Jiang, A.Q., et al.: Mistral 7b (2023)
17. Li, Z., Liu, J., Wang, Y.: Performance analysis on deep learning models in humor detection task. In: 2022 International Conference on Machine Learning and Knowledge Engineering (MLKE). IEEE (2022). https://doi.org/10.1109/mlke55170.2022.00023
18. Liang, P.: Discourse analysis on humor. In: 2011 2nd International Conference on Artificial Intelligence, Management Science and Electronic Commerce (AIMSEC), pp. 5002–5005 (2011). https://doi.org/10.1109/AIMSEC.2011.6011180
19. Liu, P., Yuan, W., Fu, J., Jiang, Z., Hayashi, H., Neubig, G.: Pre-train, prompt, and predict: a systematic survey of prompting methods in natural language processing (2021)
20. Liu, Y., et al.: Roberta: a robustly optimized BERT pretraining approach (2019)
21. Meta: llama 2: open foundation and fine-tuned chat models (2023)
22. Meta: introducing meta llama 3: the most capable openly available LLM to date — ai.meta.com. https://ai.meta.com/blog/meta-llama-3/ (2024), [Accessed 29-05-2024]
23. Mikolov, T., Chen, K., Corrado, G., Dean, J.: Efficient estimation of word representations in vector space (2013)
24. OpenAI: language models are few-shot learners (2020)
25. OpenAI: Gpt-4 technical report (2024)
26. OpenAI: Openai o1 system card (2024). https://cdn.openai.com/o1-system-card-20241205.pdf. Accessed 24 Feb 2025
27. Pennington, J., Socher, R., Manning, C.: GloVe: Global vectors for word representation. In: Moschitti, A., Pang, B., Daelemans, W. (eds.) Proceedings of the 2014 Conference on Empirical Methods in Natural Language Processing (EMNLP), pp. 1532–1543. Association for Computational Linguistics, Doha, Qatar (2014). https://doi.org/10.3115/v1/D14-1162

28. Pitis, S., Zhang, M.R., Wang, A., Ba, J.: Boosted prompt ensembles for large language models (2023)
29. Popescu, M.C., Balas, V., Perescu-Popescu, L., Mastorakis, N.: Multilayer perceptron and neural networks. WSEAS Trans. Circuits Syst. **8** (2009)
30. Sahoo, P., Singh, A.K., Saha, S., Jain, V., Mondal, S., Chadha, A.: A systematic survey of prompt engineering in large language models: Techniques and applications (2024)
31. Shazeer, N., et al.: Outrageously large neural networks: The sparsely-gated mixture-of-experts layer. CoRR **abs/1701.06538** (2017). http://arxiv.org/abs/1701.06538
32. Taori, R., et al.: Alpaca: a strong, replicable instruction-following model (2021). https://crfm.stanford.edu/2023/03/13/alpaca.html.Accessed 29 May 2024
33. Team, G.: Gemma: open models based on gemini research and technology (2024)
34. Vaswani, A., et al.: Attention is all you need (2023)
35. Wei, J., et al.: Chain-of-thought prompting elicits reasoning in large language models (2023)
36. White, J., et al.: A prompt pattern catalog to enhance prompt engineering with chatgpt (2023)
37. Wu, S.H., Huang, Y.F., Lau, T.Y.: Humour classification by fine-tuning LLMs: Cyut at clef 2024 joker lab subtask humour classification according to genre and technique. In: Faggioli, G., Ferro, N. (eds.) Working Notes of CLEF 2024 - Conference and Labs of the Evaluation Forum (CLEF). CEUR Workshop Proceedings, vol. 3740. CEUR-WS.org (2024). https://ceur-ws.org/Vol-3740/paper-183.pdf
38. Ye, Q., Axmed, M., Pryzant, R., Khani, F.: Prompt engineering a prompt engineer (2024)

Condensed Labs Overviews

Overview of BioASQ 2025: The Thirteenth BioASQ Challenge on Large-Scale Biomedical Semantic Indexing and Question Answering

Anastasios Nentidis[1]([✉]), Georgios Katsimpras[1], Anastasia Krithara[1],
Martin Krallinger[2], Miguel Rodríguez-Ortega[2], Eduard Rodriguez-López[2],
Natalia Loukachevitch[3], Andrey Sakhovskiy[5,6], Elena Tutubalina[4,5],
Dimitris Dimitriadis[7], Grigorios Tsoumakas[7,10], George Giannakoulas[7],
Alexandra Bekiaridou[8], Athanasios Samaras[7], Giorgio Maria Di Nunzio[9],
Nicola Ferro[9], Stefano Marchesin[9], Marco Martinelli[9], Gianmaria Silvello[9],
and Georgios Paliouras[1]

[1] National Center for Scientific Research "Demokritos", Athens, Greece
{tasosnent,gkatsibras,akrithara,paliourg}@iit.demokritos.gr
[2] Barcelona Supercomputing Center, Barcelona, Spain
{martin.krallinger,mirodrig8,eduard.rodriguez}@bsc.es
[3] Moscow State University, Moscow, Russia
louk_nat@mail.ru
[4] Artificial Intelligence Research Institute, Moscow, Russia
[5] Kazan Federal University, Kazan, Russia
[6] SberAI & Skoltech, Moscow, Russia
[7] Aristotle University of Thessaloniki, Thessaloniki, Greece
{dndimitri,greg}@csd.auth.gr
[8] Northwell Health, New Hyde Park, USA
[9] University of Padua, Padua, Italy
[10] Archimedes, Athena Research Center, Marousi, Greece

Abstract. This is an overview of the thirteenth edition of the BioASQ challenge in the context of the Conference and Labs of the Evaluation Forum (CLEF) 2025. BioASQ is a series of international challenges promoting advances in large-scale biomedical semantic indexing and question answering. This year, BioASQ consisted of new editions of the two established tasks, b and Synergy, and four new tasks: a) *Task Multi-ClinSum* on multilingual clinical summarization. b) *Task BioNNE-L* on nested named entity linking in Russian and English. c) *Task ELCardioCC* on clinical coding in cardiology. d) *Task GutBrainIE* on gut-brain interplay information extraction. In this edition of BioASQ, 83 competing teams participated with more than 1000 distinct submissions in total for the six different shared tasks of the challenge. Similar to previous editions, several participating systems achieved competitive performance, indicating the continuous advancement of the state-of-the-art in the field.

J. Carrillo-de-Albornoz et al. (Eds.): CLEF 2025, LNCS 16089, pp. 173–198, 2026.
https://doi.org/10.1007/978-3-032-04354-2_12

Keywords: Biomedical knowledge · Semantic Indexing · Question Answering

1   Introduction

The BioASQ challenge was introduced over a decade ago, aiming to advance the state-of-the-art in large-scale biomedical semantic indexing and question answering (QA) [77]. To achieve this, it hosts annual shared tasks, creating benchmark datasets that reflect the real-world information needs of biomedical experts. These include new versions of established tasks that remain relevant and timely, as well as novel tasks introduced to explore and address unmet biomedical information needs. These shared tasks provide research teams worldwide, who are developing systems for biomedical semantic indexing and QA, with access to publicly available datasets, a standardized evaluation framework, and opportunities for knowledge exchange through the BioASQ challenge and workshop.

Here, we present the shared tasks and the datasets of the thirteenth edition of the BioASQ challenge in 2025, as well as a condensed overview of the participating systems and their performance, organized as a lab at CLEF 2025. The remainder of this paper is organized as follows. First, Sect. 2 presents a general description of the shared tasks, which took place in 2025, and the corresponding datasets developed for the challenge. Then, Sect. 3 provides a brief overview of the participating systems for the different tasks. Detailed descriptions for some of the systems are available in the respective extended overviews of each task and the proceedings of the BioASQ lab. Subsequently, in Sect. 4, we present the performance of the systems for each task, based on state-of-the-art evaluation measures or manual assessment. Finally, in Sect. 5 we draw some conclusions.

2   Overview of the Tasks

The thirteenth edition of the BioASQ challenge consisted of six tasks [50]: (i) *Task b* on biomedical semantic question answering. (ii) *Task Synergy* on question answering developing biomedical topics. (iii) *Task MultiClinSum* on multilingual clinical summarization. (iv) *Task BioNNE-L* on nested named entity linking in Russian and English. (v) *Task ELCardioCC* on clinical coding in cardiology. (vi) *Task GutBrainIE* on gut-brain interplay information extraction. In this section, we first describe this year's editions of the two established tasks b (task 13b) and Synergy (Synergy 13) [54] with a focus on differences from previous editions of the challenge [51,57]. Additionally, we also introduce the four new BioASQ tasks, MultiClinSum [66], BioNNE-L [67], ELCardioCC [18], and GutBrainIE [48].

2.1   Task 13b

BioASQ *task 13b* is the thirteenth edition of the established BioASQ *task b* on Biomedical QA [55]. This year, it took place in three phases: i) Phase A:

biomedical questions in English were provided, and the systems had to retrieve relevant material (PubMed documents and snippets). ii) Phase A+, the systems had to provide 'exact' and 'ideal' answers. Depending on question type, the 'exact' answer can be a *yes* or *no* (yes/no), an entity name, such as a disease or gene (factoid), or a list of entity names (list). The 'ideal' answer is a paragraph-sized summary, regardless of question type. iii) Phase B: Some relevant material was provided for each question, selected by the BioASQ experts, and the systems had to provide new answers given this additional information.

About 340 new biomedical questions annotated with golden documents, snippets, and answers ('exact' and 'ideal'), were developed for testing. In addition, a training set of 5,389 biomedical questions, accompanied by answers, and supporting evidence (documents and snippets), was available from previous versions of the tasks, as a unique resource for the development of question-answering systems [34]. Table 1 presents some statistics of both training and test datasets for task 13b. The test data for task 13b were split into four independent bi-weekly batches consisting of 85 questions each, as presented in Table 1.

Table 1. Statistics on the training and test datasets of task 13b. The numbers for the documents and snippets refer to averages per question.

Batch	Size	Yes/No	List	Factoid	Summary	Documents	Snippets
Train	5389	1459	1047	1600	1283	9.74	12.78
Test 1	85	17	23	26	19	2.68	3.74
Test 2	85	17	19	27	22	2.71	3.06
Test 3	85	22	22	20	21	3.00	3.66
Test 4	85	26	19	22	18	3.15	3.92
Total	5729	1541	1130	1695	1363	9.33	12.23

2.2   Task Synergy 13

BioASQ *task Synergy* was originally introduced in 2020 with the aim of promoting research in developing biomedical topics, such as COVID-19 [36,37]. The design of this task as an ongoing dialogue allows experts to pose open-ended questions for developing topics, for which they do not know in advance whether a definitive answer can be given, in order to obtain relevant material (documents and excerpts) retrieved from the systems. After assessing this material, they provide feedback to the systems on its relevance and on whether it is sufficient to answer their question, by marking respective questions as *ready to answer*. This process is repeated iteratively in rounds with new material considered in each round, based on updates to the original document resource [56][1]. For *ready to*

---

[1] As of 2023, this evolving document resource is PubMed [51].

answer questions, they receive exact and ideal answers as well, assess them, and provide feedback that can be used by the systems to improve their responses to these questions in the remaining rounds. The experts can also mark a question as *closed* if they receive a fully satisfactory answer that is not expected to change or if they are no longer interested in the question.

A training dataset of 366 questions on developing topics with incremental annotations with relevant material and answers is already available from previous versions of *task Synergy* [53,54,58,59]. During the *task Synergy 13*, this set was extended with 47 new questions on developing health topics, such as infectious, rare, and genetic diseases, and women's and reproductive health. Meanwhile, 27 questions from the previous version of the task remained open and were enriched with more recent evidence and updated answers [55]. Overall, 74 questions were considered in the four rounds of *task Synergy 13*. The number of yesno, list, factoid, and summary questions was 23, 19, 14, and 18, respectively.

2.3 Task MultiClinSum

There is a rapid accumulation of various types of clinical content, including medical records and publications such as clinical case reports, written not only in English but also in many other languages. Some clinical reports can be very lengthy, making it challenging for healthcare professionals and even patients to comprehend and extract key clinical insights. Large Language Models (LLMs) have shown promising results in automatic summarization, helping to condense lengthy clinical documents into shorter versions or summaries that retain the most relevant clinical information. Therefore, there is a pressing need to evaluate and benchmark the performance of different clinical summarization methods, especially for content written in multiple languages.

We introduce the *MultiClinSum* task covering the automatic summarization of lengthy clinical case reports written in English, Spanish, French, and Portuguese. The *MultiClinSum* task relies on a corpus of manually selected full clinical case reports with their corresponding summaries derived from case report publications written in the mentioned languages. This gold standard dataset comprises 1,280 pairs of full-text and summary in English, 534 in Spanish, 200 in Portuguese and 200 in French. To increase the size of the dataset, both full-text and summary texts of each language were translated with neural machine translation models into the other languages, resulting in a total of 1,976 pairs for each language. An additional large-scale dataset was also created derived from the PMC-Patients clinical cases (full-text) and their corresponding summary extracted from the PubMed abstracts. Table 2 shows the corpus statistics for each sub-track of MultiClinSum.

For the evaluation assessment, the automatically generated summaries were compared with the summaries that had been manually generated by the original authors, using Rouge-L [41] scores and BERTScore [85]. As clinical case reports do share commonalities with medical discharge summaries (patient demographics, relevant medical history, clinical presentation, diagnostic process, intervention, treatment, outcome and follow-up), insights provided by the *MultiClinSum* results can be of practical relevance also for clinical records summarization scenarios.

Table 2. Statistics for the datasets provided for MultiClinSum indicating the number of fulltext-summary pairs available for each sub-track.

Sub-track	Lang.	Dataset	Fulltext-summary pairs
MultiClinSum-gs-en	EN	Native/Transl. gold stand.	988
MultiClinSum-gs-es	ES	Native/Transl. gold stand.	988
MultiClinSum-gs-fr	FR	Native/Transl. gold stand.	1061
MultiClinSum-gs-pt	PT	Native/Transl. gold stand.	1034
MultiClinSum-ls-es	EN	Large Scale (PMC-Patients)	28.902
MultiClinSum-ls-es	ES	Large Scale (PMC-Patients)	28.902
MultiClinSum-ls-fr	FR	Large Scale (PMC-Patients)	28.902
MultiClinSum-ls-pt	PT	Large Scale (PMC-Patients)	28.902

2.4   Task BioNNE-L

In the BioNNE-L Shared Task [67], we address the medical entity linking task, also known as Medical Concept Normalization (MCN), which is to map given entities to the most relevant vocabular entries from an external source, e.g., concepts from the UMLS metathesaurus [7] identified with concept unique identifiers (CUIs). Although the task has been widely explored in recent years, existing approaches usually treat each entity individually, medical entities often form a nested structure, where an entity can be a subpart of another entity. One of the key features of BioNNE-L is the focus on nested entities that are (i) derived from the MCN annotation of the NEREL-BIO corpus [45,46] and (ii) supplemented by newly annotated data in both English and Russian. The annotated entity types are disorders (*DISO*), anatomical structures (*ANAT*), and chemicals (*CHEM*) normalized to UMLS. The competition was organized into three subtasks that fell under two evaluation tracks: 1. **Monolingual track** that treated English and Russian data independently; 2. **Bilingual track** that required a single bilingual model for the combined Russian and English data. Data statistics for both tracks, as well as the normalization dictionary, are summarized in Table 3.

All BioNNE-L materials can be found on the shared task's GitHub[2] and Codalab pages[3]. Annotated data and normalization dictionary are also available at HuggingFace[4].

---

[2] https://github.com/nerel-ds/NEREL-BIO/tree/master/BioNNE-L_Shared_Task.
[3] https://codalab.lisn.upsaclay.fr/competitions/21568.
[4] https://huggingface.co/datasets/andorei/BioNNE-L.

Table 3. BioNNE-L 2025 statistics for Disorder (**DISO**), Chemical (**CHEM**), and Anatomical Structure (**ANAT**) among Russian and English entities as well as normalization dictionary statistics.

Entity type	Refined NEREL-BIO						Novel data		Dictionary	
	Train		Dev		Test					
	Ru	En	Ru	En	Ru	En	Ru	En	Ru	En
# documents	716	54	50	50	154	154	—		—	
Number of entities										
DISO	11,168	1,200	925	1,029	2,811	3,068	91,867		1,825,048	
CHEM	4,741	579	531	564	1,218	1,345	47037		1,732,096	
ANAT	8,346	911	878	901	2,186	2,248	6899		345,043	
	24,255	2,690	2,334	2,494	6,215	6,661	145,803		3,902,187	

2.5 Task ELCardioCC

Cardiovascular diseases affect a significant portion of the global population, accounting for 32% of global deaths according to WHO[5]. Automated clinical coding plays a crucial role in transforming unstructured real-world medical data gathered from patients into structured information, in order to facilitate clinical research and analysis. However, existing research predominantly focuses on English clinical text, leaving other languages, such as Greek, underrepresented. To this end, we propose a new *ELCardioCC* task [18], which concerns i) the assignment of cardiology-related ICD-10 codes to discharge letters from Greek hospitals, ii) the extraction of the specific mentions of ICD-10 codes from the discharge letters.

In detail, the participants in the *ELCardioCC* task were tasked with developing named entity recognition (NER), entity linking (EL) and multi-label classification - explainable AI (MLC-X) systems using a specialized corpus of discharge letters. These discharge letters, which were written in Greek contained valuable medical information about patients' conditions, treatments, and outcomes. The corpus was meticulously annotated with the positions of mentions (such as chief complaint, diagnosis, prior medical history, drugs and cardiac echo) and their corresponding ICD-10 codes. The training dataset includes 1,000 discharge letters, while the test set comprises 500 letters. System performance was evaluated using the micro F1 score.

2.6 Task GutBrainIE

Recent scientific evidence suggests a connection between *brain-related diseases* and the *gut microbiota* that may play a critical role in mental health-related dis-

---

[5] https://www.who.int/health-topics/cardiovascular-diseases.

orders or diseases like Parkinson's, and Alzheimer's [3,11,14,23]. The scientific literature on this topic is rapidly expanding, making it increasingly challenging for clinicians and researchers to stay up to date. For example, in 2020, approximately 200 articles were published on the relationship between gut microbiota and mental health; by 2024, this number had more than doubled to over 450 publications. The *GutBrainIE* Task aims to foster the development of Information Extraction (IE) systems that support experts by automatically extracting and linking knowledge from biomedical abstracts, facilitating the understanding of gut-brain interplay and its role in mental health and neurological diseases.

The *GutBrainIE* Task comprises four subtasks of increasing difficulty: Named Entity Recognition (NER), which identifies and classifies entity mentions in PubMed abstracts about the gut-brain interplay focusing on mental health and the Parkinson's disease; Binary Tag-based Relation Extraction (BT-RE), which detects whether pairs of entities are in relation without specifying the relation type; Ternary Tag-based Relation Extraction (TT-RE), which extends BT-RE by also assigning a relation label to each related pair; and Ternary Mention-based Relation Extraction (TM-RE), which further localizes the exact entity mentions involved in each relation and assigns the appropriate relation label.

The dataset includes over 1000 documents with annotated entity mentions and relations, organized into Training, Development, and Test sets. The train set is further divided into four quality tiers: expert-curated (Platinum), expert-annotated (Gold), student-annotated (Silver), and automatically generated (Bronze). Development and Test sets contain only expert annotations (Platinum+Gold) (Table 4).

Table 4. Dataset statistics for *GutBrainIE*.

Collection	# Docs	# Entities	Ents/Doc	# Rels	Rels/Doc
Train Platinum	111	3638	32.77	1455	13.11
Train Gold	208	5192	24.96	1994	9.59
Train Silver	499	15275	30.61	10616	21.27
Train Bronze	749	21357	28.51	8165	11.90
Development Set	40	1117	27.93	623	15.58
Test Set	40	1237	30.92	777	19.42

3   Overview of Participation

Overall, 83 distinct teams participated in the thirteenth edition of the BioASQ challenge, submitting more than 1000 distinct runs for the six different shared tasks of the challenge. The majority of the teams focused on a single task, still some of them participated in two or even three BioASQ tasks[6]. In this section, we provide a condensed overview of the methods developed by the participating teams for each of the BioASQ tasks. However, a more detailed overview of these methods will be available in the extended overview of each task [18,48,55,66,67], and some method-specific descriptions will be available in the proceedings of the thirteenth BioASQ workshop[7].

3.1   Task 13b

This year, 46 teams participated in task 13b, submitting a total of 734 different submissions generated by 146 distinct systems across all four batches for the three phases A, A+, and B. This corresponds to a significant increase in participation, compared to the 26 teams in the previous version of the task (12b) [54], which highlights that the task remains timely and relevant. Specifically, 34, 20, and 26 teams competed in phases A, A+, and B of task 13b, with 95, 79, and 88 distinct systems, respectively. Eleven of these teams were involved in all three phases. As in previous years, the open-source system OAQA [82], which achieved top performance in older editions of BioASQ [35], was used as a baseline for phase B *exact answers*.

The participating teams employed a range of well-established and sophisticated techniques. Many teams utilized traditional document retrieval methods such as BM25 and dense retrieval models (e.g. BGE-M3 and MiniLM), often improving results with re-ranking techniques. Some teams incorporated Retrieval-Augmented Generation (RAG) frameworks, using Large Language Models (LLMs) such as Llama, Gemma, GPT, Claude, and Mistral to generate responses. Beyond these methods, the teams also experimented with self-feedback mechanisms, zero-shot and few-shot prompting, ensemble methods, and the integration of biomedical knowledge bases to improve overall performance [2,4,6,9,21,30,33,63,72,74,81].

3.2   Task Synergy 13

In the thirteenth edition of BioASQ, five teams participated in the Synergy task (Synergy 13). These teams submitted 46 runs from 21 distinct systems. Two of these teams participated in task 13b as well, while the remaining three focused exclusively on task Synergy 13. The participating teams primarily utilized LLMs, such as DeepSeek-R1 and Llama. To further enhance performance, the teams experimented with RAG frameworks and employed techniques such as optimized prompting, NER, and majority voting to refine their results [19,63]. More detailed descriptions for some of the systems are available at the proceedings of the workshop.

---

[6] In particular, two teams participated in 13b & Synergy13, one in 13b & MultiClin-Sum, one in 13b & GutBrainIE, one in 13b, MultiClinSum, & ElCardioCC, and one in BioNNE-L, ElCardioCC, & GutBrainIE.

[7] https://www.bioasq.org/workshop2025/proceedings.

3.3   Task MultiClinSum

In general, there has been a very satisfactory participation in the task with promising results in each of the presented sub-tracks. 56 teams registered for the MultiClinSUM task, out of which 11 teams submitted at least one run of their predictions. Specifically, 7 teams participated in the English sub-track, 5 teams in the Spanish, 4 teams in French, and 5 in Portuguese. Each team was allowed to submit up to 5 runs per sub-track. As expected, the best results were obtained in the English sub-track (MultiClinSum-en), which had the highest level of participation. Nevertheless, the others sub-tracks were quite well represented in terms of both participation and novel methodologies applied [66].

3.4   Task BioNNE-L

In total, we've received 23 Codalab registrations for the BioNNE-L task, with 7 teams submitting predictions during the evaluation phase. The systems submitted by the participants are summarized in Table 5.

Table 5. Overview of the approaches presented by participants for the BioNNE task. EN stands for the English-oriented and RU for the Russian-oriented tracks.

Team	Track	Approach
verbanexialab	EN	SapBERT w/ lexical and semantic reranking
LYX_DMIIP_FDU	Bilingual,EN,RU	BERGAMOT fine-tuning
BlancaPlanca	Bilingual,EN,RU	BERGAMOT w/ language-specific preprocessing
MSM Lab	Bilingual,EN,RU	Two-step retrieval and ranking pipeline
dstepakov	Bilingual,RU	RoBERTa fine-tuning with contrastive learning
ICUE	Bilingual,EN,RU	BERT, BioSyn, LLM 0-shot reranking
NLPIMP	Bilingual	Russian LaBSE model pre-trained on medical data

Team **verbanexialab** [64] leveraged a SapBERT[8], pre-trained on UMLS concepts, to obtain entity embeddings, followed by a multicomponent re-ranking. They combined embedding cosine similarity with Jaccard similarity for lexical overlap recognition and Levenshtein distance for character-level alignment.

Team **LYX_DMIIP_FDU** [44] fine-tuned a BERGAMOT[9] model for each task via contrastive learning using the train- and dev-set entities to enrich the original vocabularies. The textual context of each entity was used as additional input to enhance the entity representation.

---

[8] https://huggingface.co/cambridgeltl/SapBERT-from-PubMedBERT-fulltext.
[9] https://huggingface.co/andorei/BERGAMOT-multilingual-GAT.

Team **BlancaPlanca** [10] used BERGAMOT for zero-shot retrieval based on entity-concept cosine similarity. They apply language-specific lemmatization for Russian and speed up the inference by chucking the normalization dictionary into type-specific parts of 100k entries each.

Team **MSM Lab** [40] adopted SapBERT [42,43] and BioMedBERT [25] for two-step retrieval and ranking.

Team **dstepakov** performed the nearest-neighbor search based on the cosine similarity of RoBERTa embeddings [87], fine-tuned contrastively on anchor-positive-negative term triplets via the InfoNCE objective [61].

Team **ICUE** [15] fine-tuned BioSyn [73] using the vocabularies reduced to less than 100k entries each. They fine-tune a separate BERT-based model [17] for English [5], Russian[10], and multilingual [76] tracks, respectively. They re-ranked the initial retrieval results using *DeepSeek-R1-Distill-Llama-8B*[11].

Team **NLPIMP** performed the zero-shot ranking using a Russian LaBSE [20] model[12] pre-trained contrastively on an in-house Russian medical corpus.

3.5   Task ELCardioCC

The ELCardioCC task engaged five teams across its subtasks: NER, EL, and MLC-X, with a total of 13–14 systems submitted for each subtask in addition to baseline models. Most participating systems predominantly utilized transformer-based architectures, especially BERT variants and large multilingual language models (LLMs), for all three tasks. Common approaches included fine-tuning models like Greek BERT and XLM-Roberta for NER, employing semantic similarity with embedding models for EL, and using LLMs for classification and justification in MLC-X, often leveraging cross-lingual techniques to process Greek medical texts.

The ELCardioCC baselines, designed for clarity and reproducibility, primarily used multilingual BERT models adapted for each specific task. The NER baseline involved a fine-tuned cased mBERT model with BIO2 tagging. For EL, a context-aware hierarchical classifier built on mBERT was used, reflecting the ICD-10 taxonomy. The MLC-X baseline employed a Greek-BERT model for multi-label classification of the 40 most frequent ICD-10 codes, with variations for document-level predictions and rule-based justification of code selections.

3.6   Task GutBrainIE

The *GutBrainIE* task registered 17 teams submitting runs. Among these, 16 teams participated in NER, 12 in BT-RE, 13 in TT-RE, and 13 in TM-RE. Overall, a total of 391 runs were submitted: 101 for NER, 100 for BT-RE, and 95 for both TT-RE and TM-RE.

---

[10] https://huggingface.co/KoichiYasuoka/bert-base-russian-upos.

[11] https://huggingface.co/deepseek-ai/DeepSeek-R1-Distill-Llama-8B.

[12] https://huggingface.co/sergeyzh/LaBSE-ru-turbo.

Most teams adopted supervised fine-tuning or transformer-based models pre-trained on biomedical text for the NER task [1,16,27,32,38,44,49,62,65,75]. Standard backbones included PubMedBERT, BioBERT, BioLinkBERT, and ELECTRA [12,25,39,83]. Specialized NER architectures, such as GLiNER [84], were also utilized and fine-tuned. Many groups trained multiple models with different random seeds to boost robustness and ensembled their outputs. All teams used platinum, gold, and silver collections for training. A few also used the noisier bronze set, employing cleaning or re-weighting approaches and integrating PubMed data augmentation.

Across the RE subtasks, participants primarily used biomedical pre-trained language models fine-tuned on entity-marker augmented inputs [1,16,27,32,38, 44,49,62,65,75]. Among these, the most widely employed include: SapBERT, PubMedBERT, BioBERT, RoBERTa, and ELECTRA [12,25,39,42,87]. Several teams reformulated RE as a seq2seq problem using REBEL-large [29], directly generating relation tuples or tagged spans in a single pass for all three subtasks. Few others leveraged model ensembling and trained with negative-pair subsampling to counter class imbalance and increase generalization capabilities. Finally, some groups experimented with few-shot or Retrieval-Augmented Generation (RAG), prompting large language models to extract both entities and relations with minimal fine-tuning [13,26,31,38].

4   Results

4.1   Task 13b

This section presents the evaluation measures and preliminary results for Task 13b. The evaluation in *task 13b* is done manually by the experts that assess each system response and automatically by employing a variety of established evaluation measures [47] as in the previous versions of the task [52]. Table 6 provides a brief overview of the official measures per response and question type. The results reported for *task 13b* are preliminary, as the final results will be available after the manual assessment of all system responses by the BioASQ team of experts and the enrichment of the ground truth with potential additional relevant items (i.e. documents and snippets), answer elements, and/or synonyms, which is still in progress. The online results pages for Phase A[13], Phase A+[14], and Phase B[15] will be updated with the final results when available.

The overall performance of the participating systems in document and snippet retrieval (Phase A) per batch of task 13b is presented in Table 7. Both the average and the top performance of the systems seem to drop in the last batch, indicating that the questions in this batch are more challenging for the systems. This could be related to the composition of this batch, which included more questions developed by new BioASQ experts, who have not contributed significantly to the development of the training dataset.

---

[13] https://participants-area.bioasq.org/results/13b/phaseA/.

[14] https://participants-area.bioasq.org/results/13b/phaseAplus/.

[15] https://participants-area.bioasq.org/results/13b/phaseB/.

Table 6. The evaluation measures for *task 13b* per response type and question type [47].

Resp. type (Phase)	Quest. type	Official measure
Documents (A)	All	Mean Average Precision (MAP)
Snippets (A)	All	F1 (based on character overlaps)
Exact ans. (A+ & B)	List	F1
	Yesno	macro F1 on "yes" & "no" classes
	Factoid	Mean Reciprocal Rank (MRR)
Ideal ans. (A+ & B)	All	Manual scores for precision, recall, repetition, readability

Table 7. The average and top scores of participant systems in Phase A, task 13b.

	Documents		Snippets	
Batch	Average MAP	Top MAP	Average F1	Top F1
1	0.231	0.425	0.053	0.120
2	0.283	0.442	0.084	0.179
3	0.175	0.324	0.052	0.110
4	0.072	0.180	0.024	0.079

The top performance of the participating systems in *exact answer* generation per question type is presented in Fig. 1 for both Phase A+ and Phase B of task 13b, in comparison to the respective performance in the previous version (12b). These preliminary 13b results for phase B suggest that the top systems achieved scores comparable or higher to those of 12b in answering all types of questions (solid lines). In Phase A+ (dashed lines), the top performance is lower, as expected; however, for yesno questions in particular, it is very close to those of Phase B, revealing the increased capability of LLM-based models to address these questions even without being provided ground-truth relevant documents and snippets. These results probably underestimate the performance of the top 13b systems in factoid and list questions, as the preliminary ground truth may miss some synonyms or alternative terms submitted by the participants. Such synonyms will be detected during the enrichment process and will be considered for the final results.

4.2   Task Synergy 13

In *task Synergy 13* we use the same evaluation measures described for *task 13b*, considering only new material for the information retrieval part, an approach known as *residual collection evaluation* [70]. In addition, due to the developing nature of the topics, no answer is available for all of the open questions in each round. Therefore, only the questions indicated as "answer ready" were evaluated for *exact* and *ideal answers* per round.

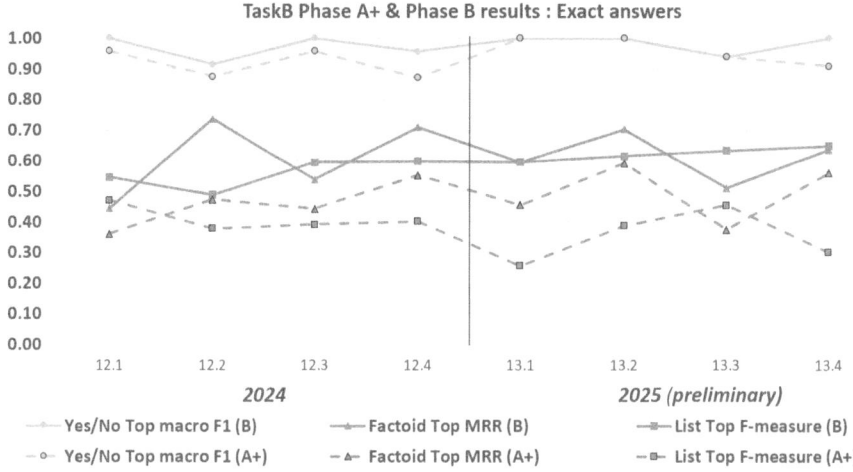

Fig. 1. The scores of the top systems in *exact answer* generation, for Phase A+ (dashed lines) and B (solid lines), across the test sets of *task 13b* and *task 12b* [60].

Table 8 presents the top performance achieved by participating systems per round, in task Synergy 13, for all types of responses. During the four rounds of Synergy 13, the systems managed to identify enough relevant material to provide an answer to 59 of the 74 questions (about 80%). In addition, they also managed to provide at least one *ideal answer* which was considered of ground-truth quality by the respective expert for 35 questions (about 47%). Overall, this dialogue between question-answering systems and biomedical experts allowed the progressive gathering of relevant documents and snippets and the generation of *exact* and *ideal answers* for open questions on developing topics, such as infectious, rare, and genetic diseases, and women's and reproductive health.

Table 8. The number of "Answer Ready" (AR) questions and the top system performance per round (R) in Task Synergy 13. Retrieval of documents (Top MAP) and snippets (Top F1). Generation of exact factoid (Top MRR), list (Top F1), and yesno (Top ma-F1) answers.

R	AR	Top MAP	Top F1 Snip.	Top MRR	Top F1 list	Top macro-F1
1	19	0.41	0.31	0.67	0.09	1
2	33	0.41	0.29	0.43	0.25	1
3	49	0.46	0.15	0.5	0.26	1
4	55	0.47	0.25	0.45	0.35	1

4.3 Task MultiClinSum

The automatic evaluation of the MultiClinSum results was performed using both BERTScore [85] and ROUGE-LSum metrics. Given the abstractive nature of the task, BERTScore was prioritized as the primary metric due to its superior capacity to capture semantic similarity between generated and reference summaries by leveraging contextualized embeddings, thus effectively recognizing meaning-preserving paraphrases and diverse lexical choices [85].

In contrast, ROUGE-LSum—a sentence-level variant of the ROUGE metric [41]—provides informative measures of summary quality by assessing the overlap of longest common subsequences between candidate and reference summaries. This offers valuable insights into the coverage and faithfulness of the generated content with respect to the original text.

While ROUGE-LSum remains limited by its reliance on surface-level n-gram matching, its inclusion alongside BERTScore ensures a complementary perspective on summary quality, balancing semantic and lexical overlap considerations. The latter, however, was the prioritized metric for submission ranking purposes.

Table 9. Results of the MultiClinSum for each sub-track. Only the top-2 best teams are presented. The best result is in bold.

Team Name	Subtrack	BERTScore			ROUGELSum		
		P	R	F1	P	R	F1
seemdog	English	0.8795	**0.8608**	**0.8698**	0.3404	**0.2398**	0.267
pjmath. [79]	English	**0.8821**	0.8466	0.8637	**0.4077**	0.2343	**0.2805**
ggrazhdans [24]	Spanish	**0.7699**	**0.747**	**0.7578**	0.3639	0.2667	0.2899
pjmath. [79]	Spanish	0.7675	0.7392	0.7525	**0.3684**	**0.2703**	0.292
pjmath. [79]	French	**0.7692**	**0.7459**	**0.7567**	**0.3481**	0.2684	**0.2843**
BU team	French	0.7248	0.7396	0.7315	0.2415	**0.289**	0.2466
pjmath. [79]	Portuguese	**0.7644**	**0.7377**	**0.7502**	0.35	**0.2605**	**0.2803**
ETS-PUCPR [71]	Portuguese	0.7403	0.7351	0.737	0.2802	0.250	0.249

BERTScore results in Table 9 for non-english languages is impaired by the fact that a multilingual bert model is used instead of a language specific encoder model. Participants were able to submit up to 5 runs, the best of which was selected for each subtrack. There was a total o f 15, 14, 9 and 7 runs for English, Spanish, French and Portuguese respectively.

4.4 Task BioNNE-L

Following prior research on entity linking [42,46,68,69], we address BioNNE-L as a retrieval task: given a mention, a model must retrieve the top-k concepts from the given UMLS dictionary and employ two ranking-based evaluation metrics: (i)

Accuracy@k and (ii) Mean Reciprocal Rank (MRR). Accuracy@k: Accuracy@k = 1 if the correct UMLS CUI is retrieved at rank $\leq k$, and Accuracy@k = 0 otherwise. $MRR = \frac{1}{|E|} \sum_{e \in E} \frac{1}{rank_e}$, where E is the set of entities, $|E|$ is the number of entities, $rank_e$ is the rank of entity e's the first correctly retrieved concept among the top k retrieved concepts. As baseline, we adopt zero-shot ranking using BERGAMOT [69] with each entity type processed independently to reduce memory footprint caused by extensive dictionary.

Table 10. Official evaluation results of the BioNNE-L task for the multilingual and monolingual tracks in terms of Accuracy@1 (**@1**), Accuracy@5 (**@5**), and MRR. The best results for each track and metric are highlighted in **bold**.

Team	Multilingual				English				Russian			
	#	@1	@5	MRR	#	@1	@5	MRR	#	@1	@5	MRR
verbanexialab	—	—	—	—	1	**0.70**	0.80	**0.74**	—	—	—	—
LYX_DMIIP_FDU	1	**0.68**	**0.84**	**0.75**	2	0.66	**0.84**	**0.74**	2	0.71	**0.84**	**0.76**
BlancaPlanca	2	0.67	0.81	0.73	3	0.64	0.83	0.72	1	**0.72**	0.83	**0.76**
MSM Lab	3	0.63	0.76	0.69	4	0.64	0.82	0.71	4	0.65	0.74	0.69
dstepakov	4	0.63	0.71	0.66	—	—	—	—	3	0.70	0.76	0.72
ICUE	5	0.58	0.76	0.66	6	0.51	0.79	0.62	5	0.62	0.72	0.67
baseline	6	0.53	0.70	0.60	5	0.57	0.78	0.66	6	0.52	0.59	0.55
NLPIMP	7	0.41	0.58	0.48	—	—	—	—	—	—	—	—

The official evaluation results, ordered by Accuracy@1 value, for BioNNE-L are summarized in Table 10. Most of the participants adopted various BERT-based [17] with the top-performance achieved by domain-specific models, such as BERGAMOT, SapBERT. Specifically, Team LYX_DMIIP_FDU ranked the first in the multilingual track and the second rank in the two monolingual track by fine-tuning BERGAMOT. Top 1 results for the Russian and English data are achieved by multilingual BERGAMOT (Team BlancaPlanca) and English SapBERT (Team verbanexialab) models, respectively.

4.5   Task ELCardioCC

The results of the participants for the ELCardioCC task are presented in tables 11, 12 and 13. For each subtask, the table displays one system per team that achieved the highest F1-score.

The team droidlyx [44] consistently achieved the highest micro-F1 scores across all subtasks, leading NER (0.733), EL (0.678), and performing strongly in MLC-X (0.847). The ELCardioCC baseline remained highly competitive, particularly excelling in MLC-X with a micro-F1 of 0.827 and maintaining strong performance in NER and EL. Svassileva's systems [80] were consistently near the top in NER and EL but did not participate in MLC-X, while bhuang [28]

Table 11. Performance of participating systems in the ELCardioCC Named Entity Recognition (NER) subtask. Results are reported using micro-averaged precision, recall, and F1 score.

Team	System	Recall	Precision	Micro-F1
bhuang	5nm	0.6448	0.5205	0.5761
droidlyx	system1	**0.7059**	**0.7618**	**0.7328**
ELCardioCC_baseline	mbert_baseline	0.6959	0.7460	0.7201
pjmathematician	config1	0.2484	0.2586	0.2534
enigma	greek-bert-exact-bge-m3	0.7012	0.7328	0.7167

Table 12. Performance of participating systems in the ELCardioCC Entity Linking (EL) subtask. Results are reported using micro-averaged precision, recall, and F1 score.

Team	System	Recall	Precision	Micro-F1
bhuang	5nm	0.5927	0.4852	0.5336
droidlyx	system1	0.6529	**0.7046**	**0.6778**
ELCardioCC_baseline	EL_baseline	0.6476	0.6942	0.6701
pjmathematician	config1	0.0616	0.0642	0.0629
enigma	greek-bert-exact-bge-m3	**0.6548**	0.6844	0.6693

showed promising results especially in classification, though with more variability. Finally, the pjmathematician [78] system demonstrated significantly low performance .

4.6 Task GutBrainIE

Submitted runs were evaluated using micro- and macro-averaged precision, recall, and F1-score, with micro-F1 used as the reference measure for the leaderboards since it is better suited when classes are imbalanced.

We adopted a baseline employing a fine-tuned NuNER model for NER [8] and a fine-tuned ATLOP model for all RE subtasks [86]. The baseline has also been used to annotate the bronze collection automatically.

Tables 14, 15, 16 and 17 show each team's top run beating the baseline for NER, TB-RE, TT-RE, and TM-RE, respectively. The performance gap between the subtasks is noticeable. While NER obtained a top micro-F1 of 0.84 utilizing pretrained biomedical transformers, RE subtasks were more challenging: both TB-RE and TT-RE peaked at approximately 0.65–0.69 micro-F1, and TM-RE reached only 0.46 micro-F1, demonstrating the added difficulty of simultaneously locating and labeling entities and identifying relations among these.

Table 13. Performance of participating systems in ELCardioCC Subtask 3a (Multi-label Classification) and Subtask 3b (Explainable AI). Metrics include Precision (P), Recall (R), and Micro-F1 score. A dash (–) indicates that the system did not participate in the corresponding subtask.

Team	System	Subtask 3a (MLC)			Subtask 3b (X)		
		P	R	F1	P	R	F1
ELCardioCC_baseline	MLCX1_baseline	0.9339	0.7422	0.8271	-	-	-
	MLCX2_baseline	**0.9531**	0.5864	0.7261	**0.6050**	**0.4442**	**0.5122**
bhuang	1nm	0.6205	0.7676	0.6863	-	-	-
droidlyx	system1	0.8569	0.8377	**0.8472**	-	-	-
kbogas	w2l_cb	0.2115	0.3421	0.2614	-	-	-
pjmathematician	config4	0.6056	0.2257	0.3288	0.2326	0.0932	0.1331
	config5	0.5860	0.2656	0.3655	-	-	-

Table 14. Performance metrics of each team's top run beating the baseline for NER. The best result is in bold, the second-best is underlined (micro-averaged).

Team ID	Run name	Precision	Recall	F1
GutUZH [27]	AugEnsemble	**0.8384**	0.8432	**0.8408**
Gut-Instincts [1]	5eedev	0.8286	0.8480	<u>0.8382</u>
NLPatVCU [75]	ensemble1	0.8255	<u>0.8488</u>	0.8370
ICUE [38]	ensemble5-th10	<u>0.8369</u>	0.8294	0.8331
LYX-DMIIP-FDU [44]	EnsembleBERT	0.8020	**0.8513**	0.8259
ata2425ds [NA]	transformer	0.7914	0.8432	0.8164
greenday [26]	llmner	0.7957	0.8278	0.8114
Graphswise-1 [16]	NERWise	0.8066	0.7955	0.8010
BASELINE [48]	NuNerZero-Finetuned	0.7639	0.8238	0.7927

Table 15. Performance metrics of each team's top run beating the baseline for TB-RE. The best result is in bold, the second-best is underlined (micro-averaged).

Team ID	Run name	Precision	Recall	F1
Gut-Instincts [1]	6219eedev3re	0.6304	**0.7532**	**0.6864**
ONTUG [31]	ElectraCLEANR	0.7121	0.6104	<u>0.6573</u>
Graphswise-1 [16]	AtlopOnto	0.7418	0.5844	0.6538
ataupd2425-pam [62]	BiomedNLP-FULL_DATASET	0.5671	<u>0.7316</u>	0.6389
BIU-ONLP [32]	RobertaLarge	<u>0.7453</u>	0.5195	0.6122
BASELINE [48]	Atlop-Finetuned	**0.7584**	0.4892	0.5947

Table 16. Performance metrics of each team's top run beating the baseline for TT-RE. The best result is in bold, the second-best is underlined (micro-averaged).

Team ID	Run name	Precision	Recall	F1
Gut-Instincts [1]	6229eedev3re	0.6280	<u>0.7572</u>	**0.6866**
ataupd2425-pam [62]	BiomedNLP-FULL_DATASET	0.5853	0.7202	<u>0.6458</u>
ONTUG [31]	ElectraCLEANR	0.7059	0.5926	0.6443
Graphswise-1 [16]	AtlopOnto	0.7326	0.5638	0.6372
ICUE [38]	biolinkbertl_pp	0.4974	**0.7860**	0.6093
BIU-ONLP [32]	RobertaLarge	<u>0.7362</u>	0.4938	0.5911
BASELINE [48]	Atlop-Finetuned	**0.7533**	0.4650	0.5751

Table 17. Performance metrics of each team's top run beating the baseline for TM-RE. The best result is in bold, the second-best is underlined (micro-averaged).

Team ID	Run name	Precision	Recall	F1
Gut-Instincts [1]	6239eedev3re	0.4215	**0.5147**	**0.4635**
Graphswise-1 [16]	AtlopOnto	<u>0.4686</u>	0.3097	<u>0.3729</u>
ICUE [38]	biolinkbertl_pp	0.2858	<u>0.5054</u>	0.3651
LYX-DMIIP-FDU [44]	BioLinkBERT	0.3682	0.3257	0.3457
ONTUG [31]	ElectraCLEANR	0.3529	0.3231	0.3373
BASELINE [48]	Atlop-Finetuned	**0.4986**	0.2453	0.3288

5   Conclusions

This paper provides an overview of the thirteenth BioASQ challenge. This year, BioASQ consisted of six tasks: (i) *Task 13b* on biomedical semantic question answering. (ii) *Task Synergy13* on question answering for developing biomedical topics. (iii) *Task MultiClinSum* on multilingual clinical summarization. (iv) *Task BioNNE-L* on nested named entity linking in Russian and English. (v) *Task ELCardioCC* on clinical coding in cardiology. (vi) *Task GutBrainIE* on gut-brain interplay information extraction.

The results for Task 13b suggest that the top participant systems achieved high scores, especially for yes/no answer generation, even in Phase A+, where no ground-truth relevant material was given. For list and factoid questions, system performance is less consistent, especially in Phase A+, indicating the presence of room for improvement. For these questions, the availability of ground-truth relevant material seems to allow the systems to provide answers of better quality. This highlights the importance of Phase A, on the automated retrieval of relevant material, where the top performance is less consistent across batches, potentially affected by the domain of the expert posing the questions. A diverse set of retrieval and generation techniques was applied, including traditional methods, LLM-based frameworks, and integration of domain-specific knowledge. The results of task Synergy13, aligned with those of previous versions, suggest that state-of-the-art systems can be a useful tool for biomedical scientists in need of specialized information for developing problems, despite their limitations and room for improvement.

The new task MultiClinSum presented new challenging sub-tasks about text summarization of clinical case reports in Spanish, English, French and Portuguese. This task introduces the nuance of creating clinical Text Summarization systems specifically for the cardiology domain. In addition, it expands the range of the task beyond Spanish by introducing a sub-track that also involves English and Italian text. The results highlight the importance of having data specific to the language and specialty the systems are going to be applied in, even within domains that are already quite specific, like the clinical one.

The BioNNE-L task focused on the linking of biomedical entities for disorders, chemicals, and anatomical structures in Russian and English texts, addressing challenges such as nested entities and cross-language linking amid incomplete low-resource vocabularies. Despite the overall prevalence of LLMs in numerous domains and tasks, the top-performing systems for BioNNE-L utilized biomedical BERT-based retrieval and reranking architectures, highlighting the importance of task and domain-specific methods for information extraction.

The ELCardioCC task centered on extracting and classifying medical entities from Greek discharge letters, drawing participation from five teams who submitted a diverse set of systems across three subtasks. Most approaches leveraged transformer-based models—particularly BERT variants and multilingual LLMs—with strategies ranging from fine-tuned token classification to prompt-based extraction and embedding-based entity linking. The MLC-X subtask saw innovative uses of multilingual embeddings and LLM reasoning to predict and justify ICD-10 codes. Baseline models, built on multilingual BERT architectures, offered simple yet effective benchmarks for each subtask, emphasizing clarity and reproducibility.

The GutBrainIE task, centered on information extraction for the gut-brain axis, challenged participants with Named Entity Recognition and increasingly fine-grained Relation Extraction subtasks. Teams achieving the strongest performance employed supervised deep-learning strategies, combining pretrained biomedical language models with ensemble strategies. Only a few participants experimented with prompt-based or generative approaches; however, these generally obtained lower scores, confirming the need to develop specialized models to effectively extract complex entities and relations in a specific biomedical domain.

Overall, the participation in BioASQ 13 was significantly increased, both due to increased interest in the new versions of its already established tasks, as well as due to the introduction of four novel tasks. Several participating systems achieved competitive performance on the BioASQ tasks, and some of them managed to improve over the baselines or the state-of-the-art performance from previous years. Aligned with previous versions, BioASQ keeps pushing the research frontier in biomedical semantic indexing and question answering for thirteen years now, offering both well-established and new tasks. Initially, it extended beyond the English language and biomedical literature with the introduction of the task MESINESP [22] and continued consistently ever since. In this thirteenth edition, BioASQ was further extended with four new tasks, MultiClinSum [66], BioNNE-L [67], ElCardioCC [18], and GrutBrainIE [48]. As a result, BioASQ 13 offered tasks in six languages (English, Spanish, French, Portuguese, Russian, and Greek), three types of documents (biomedical articles, clinical case reports, and discharge letters), and two specialized domains within biomedicine (cardiology and gut-brain interaction).

The future directions for the BioASQ challenge involve further expanding the benchmark dataset for question answering through a community-driven approach, broadening the network of biomedical experts participating in the Synergy task, and enhancing the scope of resources used in the BioASQ tasks. This includes incorporating additional document types, multiple languages, and more specialized sub-domains within biomedicine.

Acknowledgments. The thirteenth edition of BioASQ is sponsored by Ovid, Atypon Systems Inc, and Elsevier. The MEDLINE/PubMed data resources considered in this work were accessed courtesy of the U.S. National Library of Medicine. BioASQ is grateful to the CMU team for providing the *exact answer* baselines for task 13b. This research was funded by the Ministerio de Ciencia e Innovación (MICINN) under project BARITONE (TED2021-129974B-C22). This work is also supported by the European Union's Horizon Europe Co-ordination & Support Action under Grant Agreement No 101080430 (AI4HF), as well as Grant Agreement No 101057849 (DataTool4Heartproject). The work on the BioNNE-L task was supported by the Russian Science Foundation [grant number 23-11-00358]. ELCardioCC has been partially supported by project MIS 5154714 of the National Recovery and Resilience Plan Greece 2.0 funded by the European Union under the NextGenerationEU Program. The work on the GutBrainIE task was supported by the HEREDITARY Project, as part of the European Union's Horizon Europe research and innovation programme (GA 101137074).

References

1. Andersen, L.R., Gardshodn, M.I., Dolmer, M.H., Rodriguez, J.M., Dell'Aglio, D.: Trusting gut instincts: transformer-based extraction of structured data from gut-brain axis publications. In: Faggioli, G., Ferro, N., Rosso, P., Spina, D. (eds.) CLEF 2025 Working Notes (2025)

2. Angulo, J., Yeste, V.: AQAMS and AQAMS2: Multi agent systems for biomedical question answering. In: Faggioli, G., Ferro, N., Rosso, P., Spina, D. (eds.) CLEF 2025 Working Notes (2025)

3. Appleton, J.: The gut-brain axis: influence of microbiota on mood and mental health. Integr. Med. A Clin. J. **17**(4), 28 (2018)

4. Ateia, S., Kruschwitz, U.: Can language models critique themselves? Investigating self-feedback for retrieval augmented generation at BioASQ 2025 . In: Faggioli, G., Ferro, N., Rosso, P., Spina, D. (eds.) CLEF 2025 Working Notes (2025)

5. Beltagy, I., Lo, K., Cohan, A.: SciBERT: Pretrained language model for scientific text. In: EMNLP (2019)

6. Bing-Chen, C., Han, J.C., Hung, H.C., Tsai, R.T.H.: NCU-IISR: Biomedical question answering via gemini and GPT APIs in the BioASQ 13b Phase B Challenge . In: Faggioli, G., Ferro, N., Rosso, P., Spina, D. (eds.) CLEF 2025 Working Notes (2025)
7. Bodenreider, O.: The unified medical language system (UMLS): integrating biomedical terminology. Nucleic Acids Res. **32**(suppl_1), D267–D270 (2004)
8. Bogdanov, S., Constantin, A., Bernard, T., Crabbé, B., Bernard, E.: NuNER: entity recognition encoder pre-training via LLM-annotated data (2024)
9. Borazio, F., Croce, D., Basili, R.: UniTor at BioASQ 2025: modular biomedical QA with synthetic snippets and multiple task answer generation . In: Faggioli, G., Ferro, N., Rosso, P., Spina, D. (eds.) CLEF 2025 Working Notes (2025)
10. Burlova, A.: Navigating partial UMLS terminology: GAT embeddings and confidence analysis for multilingual concept linking. In: Faggioli, G., Ferro, N., Rosso, P., Spina, D. (eds.) CLEF 2025 Working Notes (2025)
11. Carabotti, M., Scirocco, A., Maselli, M.A., Severi, C.: The gut-brain axis: interactions between enteric microbiota, central and enteric nervous systems. Ann. Gastroenterol. Q. Publ. Hellenic Soc. Gastroenterol. **28**(2), 203 (2015)
12. Clark, K., Luong, M.T., Le, Q.V., Manning, C.D.: ELECTRA: Pre-training text encoders as discriminators rather than generators. In: International Conference on Learning Representations (2020)
13. Conceição, S.I.R., Lopes, P.R.C., Couto, F.M.: lasigeBioTM at BioASQ25 Task GutBrainIE - Lean large language models with syntactic features. In: Faggioli, G., Ferro, N., Rosso, P., Spina, D. (eds.) CLEF 2025 Working Notes (2025)
14. Cryan, J.F., O'Riordan, K.J., Sandhu, K., Peterson, V., Dinan, T.G.: The gut microbiome in neurological disorders. Lancet Neurol. **19**(2), 179–194 (2020)
15. D. Lain, A., Lee, C., Doneva, S.E., Rodríguez-Cubillos, M.J., Castagnari, E., Simpson, T.I., , Posma, J.M.: Multilingual and Nested Biomedical Named Entity Normalisation via Candidate Retrieval and Lightweight Large Language Model Disambiguation. In: Faggioli, G., Ferro, N., Rosso, P., Spina, D. (eds.) CLEF 2025 Working Notes (2025)
16. Datseris, A., Kuzmanov, M., Nikolova-Koleva, I., Taskov, D., Boytcheva, S.: Graphwise @ CLEF-2025 GutBrainIE: Towards Automated Discovery of Gut-Brain Interactions: Deep Learning for NER and Relation Extraction from PubMed Abstracts. In: Faggioli, G., Ferro, N., Rosso, P., Spina, D. (eds.) CLEF 2025 Working Notes (2025)
17. Devlin, J., Chang, M.W., Lee, K., Toutanova, K.: BERT: pre-training of deep bidirectional transformers for language understanding. In: Burstein, J., Doran, C., Solorio, T. (eds.) Proceedings of the 2019 Conference of the North American Chapter of the Association for Computational Linguistics: Human Language Technologies, Volume 1 (Long and Short Papers), pp. 4171–4186. ACL, Minneapolis, Minnesota (2019). https://doi.org/10.18653/v1/N19-1423
18. Dimitriadis, D., et al.: Overview of ElCardioCC task on clinical coding in cardiology at BioASQ 2025. In: Faggioli, G., Ferro, N., Rosso, P., Spina, D. (eds.) CLEF 2025 Working Notes (2025)
19. Dueñas Romero, S., Ureña-López, L.A., Martínez-Cámara, E.: SINAI at CLEF 2025: A multi-stage RAG pipeline for biomedical semantic question answering . In: Faggioli, G., Ferro, N., Rosso, P., Spina, D. (eds.) CLEF 2025 Working Notes (2025)
20. Feng, F., Yang, Y., Cer, D., Arivazhagan, N., Wang, W.: Language-agnostic BERT sentence embedding. In: Proceedings of the 60th Annual Meeting of the Association

for Computational Linguistics (Volume 1: Long Papers), pp. 878–891. ACL, Dublin, Ireland (2022). https://doi.org/10.18653/v1/2022.acl-long.62

21. Galat, D., Molla-Aliod, D.: LLM Ensemble for RAG: Role of context length in zero-shot question answering for BioASQ Challenge . In: Faggioli, G., Ferro, N., Rosso, P., Spina, D. (eds.) CLEF 2025 Working Notes (2025)

22. Gasco, L., et al.: Overview of BioASQ 2021-MESINESP track. Evaluation of advance hierarchical classification techniques for scientific literature, patents and clinical trials. In: Proceedings of the 9th BioASQ Workshop (2021)

23. Ghaisas, S., Maher, J., Kanthasamy, A.: Gut microbiome in health and disease: Linking the microbiome-gut-brain axis and environmental factors in the pathogenesis of systemic and neurodegenerative diseases. Pharmacol. Therapeutics **158**, 52–62 (2016)

24. Grazhdanski, G.: Group relative policy optimization for Spanish clinical case report summarization. In: Faggioli, G., Ferro, N., Rosso, P., Spina, D. (eds.) CLEF 2025 Working Notes (2025)

25. Gu, Y., et al.: Domain-specific language model pretraining for biomedical natural language processing. ACM Trans. Comput. Healthc. (HEALTH) **3**(1), 1–23 (2021)

26. Gupta, H.P., Banerjee, R.: LLMs for biomedical NER. In: Faggioli, G., Ferro, N., Rosso, P., Spina, D. (eds.) CLEF 2025 Working Notes (2025)

27. Han, J., Liu, Y.: GutUZH at CLEF2025 BioASQ Task 6: a method of SOTA performance with the best results at GutBrainIE NER subtask 1. In: Faggioli, G., Ferro, N., Rosso, P., Spina, D. (eds.) CLEF 2025 Working Notes (2025)

28. Huang, B.: Clinical entity recognition and linking in greek discharge letters using multilingual-llm-based multi-stage system. In: Faggioli, G., Ferro, N., Rosso, P., Spina, D. (eds.) CLEF 2025 Working Notes (2025)

29. Huguet Cabot, P.L., Navigli, R.: REBEL: relation extraction by end-to-end language generation. In: Findings of the Association for Computational Linguistics: EMNLP 2021, pp. 2370–2381. ACL, Punta Cana, Dominican Republic (2021). https://aclanthology.org/2021.findings-emnlp.204

30. Jonker, R.A.A., Almeida, T., Almeida, J., Matos, S.: BIT.UA at BioASQ 13B: Revisiting evaluation, DPRF-enhanced retrieval and fine-Tuned LLMs. In: Faggioli, G., Ferro, N., Rosso, P., Spina, D. (eds.) CLEF 2025 Working Notes (2025)

31. Kantz, B., Waldert, P., Lengauer, S., Schreck, T.: Constrained linked entity ANnotation using RAG (CLEANR). In: Faggioli, G., Ferro, N., Rosso, P., Spina, D. (eds.) CLEF 2025 Working Notes (2025)

32. Keinan, R., Cohen, A.D.N., Tsarfaty, R.: From named entities to relations: end-to-end biomedical information extraction. In: Faggioli, G., Ferro, N., Rosso, P., Spina, D. (eds.) CLEF 2025 Working Notes (2025)

33. Kim, H., et al.: Prompting matters: snippet-aware strategies for biomedical QA with LLMs in BioASQ 13b. In: Faggioli, G., Ferro, N., Rosso, P., Spina, D. (eds.) CLEF 2025 Working Notes (2025)

34. Krithara, A., Nentidis, A., Bougiatiotis, K., Paliouras, G.: BioASQ-QA: a manually curated corpus for biomedical question answering. Sci. Data **10**(1), 170 (2023)

35. Krithara, A., Nentidis, A., Paliouras, G., Kakadiaris, I.: Results of the 4th edition of BioASQ challenge. In: Proceedings of the Fourth BioASQ Workshop (2016). https://www.aclweb.org/anthology/W16-3101.pdf

36. Krithara, A., Nentidis, A., Paliouras, G., Krallinger, M., Miranda, A.: BioASQ at CLEF2021: large-scale biomedical semantic indexing and question answering. In: Hiemstra, D., Moens, M.-F., Mothe, J., Perego, R., Potthast, M., Sebastiani, F. (eds.) ECIR 2021. LNCS, vol. 12657, pp. 624–630. Springer, Cham (2021). https://doi.org/10.1007/978-3-030-72240-1_73

37. Krithara, A., et al.: BioASQ synergy: a dialogue between question-answering systems and biomedical experts for promoting COVID-19 research. J. Am. Med. Inf. Assoc., ocae232 (2024). https://doi.org/10.1093/jamia/ocae232

38. Lee, C., et al.: Understanding gut-brain interplay in scientific literature: a hybrid approach from classification to generative LLM reasoning. In: Faggioli, G., Ferro, N., Rosso, P., Spina, D. (eds.) CLEF 2025 Working Notes (2025)

39. Lee, J., et al.: BioBERT: a pre-trained biomedical language representation model for biomedical text mining. Bioinformatics **36**(4), 1234–1240 (2019). https://doi.org/10.1093/bioinformatics/btz682

40. Li, C., Zheng, X., Liu, S.: BIBERT on biomedical nested named entity linking at BioASQ 2025. In: Faggioli, G., Ferro, N., Rosso, P., Spina, D. (eds.) CLEF 2025 Working Notes (2025)

41. Lin, C.Y.: ROUGE: A package for automatic evaluation of summaries. In: Proceedings of the ACL Workshop 'Text Summarization Branches Out', pp. 74–81. Barcelona, Spain (2004)

42. Liu, F., Shareghi, E., Meng, Z., Basaldella, M., Collier, N.: Self-alignment pretraining for biomedical entity representations. In: Proceedings of the 2021 Conference of the North American Chapter of the Association for Computational Linguistics: Human Language Technologies, pp. 4228–4238. ACL, Online (2021). https://doi.org/10.18653/v1/2021.naacl-main.334

43. Liu, F., Vulić, I., Korhonen, A., Collier, N.: Learning domain-specialised representations for cross-lingual biomedical entity linking. In: Proceedings of the 59th Annual Meeting of the Association for Computational Linguistics and the 11th International Joint Conference on Natural Language Processing (Volume 2: Short Papers), pp. 565–574. ACL, Online (2021). https://doi.org/10.18653/v1/2021.acl-short.72

44. Liu, Y.: LYX_DMIIP_FDU at BioASQ 2025: Utilizing BERT embeddings for biomedical text mining. In: Faggioli, G., Ferro, N., Rosso, P., Spina, D. (eds.) CLEF 2025 Working Notes (2025)

45. Loukachevitch, N., et al.: NEREL-BIO: a dataset of biomedical abstracts annotated with nested named entities. Bioinformatics (2023). https://doi.org/10.1093/bioinformatics/btad161

46. Loukachevitch, N., Sakhovskiy, A., Tutubalina, E.: Biomedical concept normalization over nested entities with partial UMLS terminology in Russian. In: Proceedings of the 2024 Joint International Conference on Computational Linguistics, Language Resources and Evaluation (LREC-COLING 2024), pp. 2383–2389. ELRA and ICCL, Torino, Italia (2024). https://aclanthology.org/2024.lrec-main.213/

47. Malakasiotis, P., Pavlopoulos, I., Androutsopoulos, I., Nentidis, A.: Evaluation measures for task b. Tech. rep., Tech. rep. BioASQ (2022). http://participants-area.bioasq.org/Tasks/b/eval_meas_2022

48. Martinelli, M., et al.: Overview of GutBrainIE@CLEF 2025: gut-brain interplay information extraction. In: Faggioli, G., Ferro, N., Rosso, P., Spina, D. (eds.) CLEF 2025 Working Notes (2025)

49. Mehta, R.: Enhancing biomedical named entity recognition using GLiNER-BioMed with targeted dictionary-based post-processing for BioASQ 2025 task 6. In: Faggioli, G., Ferro, N., Rosso, P., Spina, D. (eds.) CLEF 2025 Working Notes (2025)

50. Nentidis, A., et al.: BioASQ at CLEF2025: The Thirteenth Edition of the Large-Scale Biomedical Semantic Indexing and Question Answering Challenge. In: Hauff, C., et al. (eds.) Advances in Information Retrieval, pp. 407–415. Springer Nature Switzerland, Cham (2025). https://doi.org/10.1007/978-3-031-88720-8_61

51. Nentidis, A., et al.: Overview of BioASQ 2023: the eleventh BioASQ challenge on large-scale biomedical semantic indexing and question answering. In: Arampatzis, A., et al. (eds.) Experimental IR Meets Multilinguality, Multimodality, and Interaction, pp. 227–250. Springer Nature Switzerland, Cham (2023)

52. Nentidis, A., et al.: Overview of BioASQ 2024: the twelfth BioASQ challenge on large-scale biomedical semantic indexing and question answering. In: Experimental IR Meets Multilinguality, Multimodality, and Interaction. Proceedings of the Fifteenth International Conference of the CLEF Association (CLEF 2024) (2024)

53. Nentidis, A., Katsimpras, G., Krithara, A., Paliouras, G.: Overview of BioASQ tasks 11b and Synergy11 in CLEF2023. In: CEUR Workshop Proceedings (2023)

54. Nentidis, A., Katsimpras, G., Krithara, A., Paliouras, G.: Overview of BioASQ tasks 12b and Synergy12 in CLEF2024. In: Faggioli, G., Ferro, N., Galuščáková, P., García Seco de Herrera, A. (eds.) CLEF Working Notes (2024)

55. Nentidis, A., Katsimpras, G., Krithara, A., Paliouras, G.: Overview of BioASQ tasks 13b and Synergy13 in CLEF2025. In: Faggioli, G., Ferro, N., Rosso, P., Spina, D. (eds.) CLEF 2025 Working Notes (2025)

56. Nentidis, A., et al.: Overview of BioASQ 2021: the ninth BioASQ challenge on large-scale biomedical semantic indexing and question answering. In: Candan, K.S., et al. (eds.) CLEF 2021. LNCS, vol. 12880, pp. 239–263. Springer, Cham (2021). https://doi.org/10.1007/978-3-030-85251-1_18

57. Nentidis, A., et al.: Overview of BioASQ 2022: the tenth BioASQ challenge on large-scale biomedical semantic indexing and question answering. In: Experimental IR Meets Multilinguality, Multimodality, and Interaction. Springer (2022). https://doi.org/10.1007/978-3-031-13643-6_22

58. Nentidis, A., Katsimpras, G., Vandorou, E., Krithara, A., Paliouras, G.: Overview of BioASQ Tasks 9a, 9b and Synergy in CLEF2021. In: Proceedings of the 9th BioASQ Workshop A challenge on large-scale biomedical semantic indexing and question answering. CEUR Workshop Proceedings (2021). http://ceur-ws.org/Vol-2936/paper-10.pdf

59. Nentidis, A., Katsimpras, G., Vandorou, E., Krithara, A., Paliouras, G.: Overview of BioASQ Tasks 10a, 10b and Synergy10 in CLEF2022. In: CEUR Workshop Proceedings, vol. 3180, pp. 171–178 (2022)

60. Nentidis, A., et al.: BioASQ at CLEF2024: The Twelfth Edition of the Large-Scale Biomedical Semantic Indexing and Question Answering Challenge. n: Goharian, N., et al. (eds.) ECIR2024, pp. 490–497. Springer (2024). https://doi.org/10.1007/978-3-031-56069-9_67

61. van den Oord, A., Li, Y., Vinyals, O.: Representation learning with contrastive predictive coding. ArXiv abs/1807.03748 (2018). https://api.semanticscholar.org/CorpusID:49670925

62. Pamio, L., Di Nunzio, G.M.: BioASQ task GutBrainIE 2025 Task 6: Comparing CRF vs BERT models for named entity recognition and relation extraction. In: Faggioli, G., Ferro, N., Rosso, P., Spina, D. (eds.) CLEF 2025 Working Notes (2025)

63. Panou, D., Dimopoulos, A., Koubarakis, M., Reczko, M.: Harnessing collective intelligence of LLMs for robust biomedical QA: a multi-model approach. In: Faggioli, G., Ferro, N., Rosso, P., Spina, D. (eds.) CLEF 2025 Working Notes (2025)

64. Peña Gnecco, D., Serrano, J., Puertas, E., Martínez-Santos, J.C.: Hybrid re-ranking for biomedical entity linking using SapBERT embeddings: a high-performance system for BioNNE-L 2025-1. In: Faggioli, G., Ferro, N., Rosso, P., Spina, D. (eds.) CLEF 2025 Working Notes (2025)

65. Piron, S., Di Nunzio, G.M.: Named entity recognition with GLiNER and relation extraction with LLMs. In: Faggioli, G., Ferro, N., Rosso, P., Spina, D. (eds.) CLEF 2025 Working Notes (2025)
66. Rodríguez-Ortega, M., et al.: Overview of MultiClinSum task at BioASQ 2025: evaluation of clinical case summarization strategies for multiple languages: data, evaluation, resources and results. In: Faggioli, G., Ferro, N., Rosso, P., Spina, D. (eds.) CLEF 2025 Working Notes (2025)
67. Sakhovskiy, A., Loukachevitch, N., Tutubalina, E.: Overview of the BioASQ BioNNE-L task on biomedical nested entity linking in CLEF 2025. In: Faggioli, G., Ferro, N., Rosso, P., Spina, D. (eds.) CLEF 2025 Working Notes (2025)
68. Sakhovskiy, A., Semenova, N., Kadurin, A., Tutubalina, E.: Graph-enriched biomedical entity representation transformer. In: Experimental IR Meets Multilinguality, Multimodality, and Interaction, pp. 109–120. Springer Nature Switzerland, Cham (2023). https://doi.org/10.1007/978-3-031-42448-9_10
69. Sakhovskiy, A., Semenova, N., Kadurin, A., Tutubalina, E.: Biomedical entity representation with graph-augmented multi-objective transformer. In: Findings of the Association for Computational Linguistics: NAACL 2024, pp. 4626–4643. ACL, Mexico City, Mexico (2024). https://doi.org/10.18653/v1/2024.findings-naacl.288
70. Salton, G., Buckley, C.: Improving retrieval performance by relevance feedback. J. Am. Soc. Inf. Sci. **41**(4), 288–297 (1990). https://doi.org/10.1002/(SICI)1097-4571(199006)41:4⟨288::AID-ASI8⟩3.0.CO;2-H
71. Schneider, E.T.R., Schneider, F.H., Paraiso, E.C., Britto Jr, A.S., Cruz, R.M.O.: MedGemma-Sum-Pt: a lightweight model for portuguese clinical summarization. In: Faggioli, G., Ferro, N., Rosso, P., Spina, D. (eds.) CLEF 2025 Working Notes (2025)
72. Stachura, D., Konieczna, J., Nowak, A.: Are smaller open-weight LLMs closing the gap to proprietary models for biomedical question answering? . In: Faggioli, G., Ferro, N., Rosso, P., Spina, D. (eds.) CLEF 2025 Working Notes (2025)
73. Sung, M., Jeon, H., Lee, J., Kang, J.: Biomedical entity representations with synonym marginalization. In: Proceedings of the 58th Annual Meeting of the Association for Computational Linguistics, pp. 3641–3650. ACL, Online (2020). https://doi.org/10.18653/v1/2020.acl-main.335
74. Tang, J., et al.: Applying DeepSeek to BioASQ Task 13B: Using supervised fine-tuning and few-shot learning . In: Faggioli, G., Ferro, N., Rosso, P., Spina, D. (eds.) CLEF 2025 Working Notes (2025)
75. Taylor, S., et al.: NLP@VCU at BioASQ2025: information extraction on the Gut-BrainIE dataset. In: Faggioli, G., Ferro, N., Rosso, P., Spina, D. (eds.) CLEF 2025 Working Notes (2025)
76. Tedeschi, S., Maiorca, V., Campolungo, N., Cecconi, F., Navigli, R.: WikiNEuRal: combined neural and knowledge-based silver data creation for multilingual NER. In: Findings of the Association for Computational Linguistics: EMNLP 2021, pp. 2521–2533. ACL, Punta Cana, Dominican Republic (2021). https://doi.org/10.18653/v1/2021.findings-emnlp.215
77. Tsatsaronis, G., et al.: An overview of the BIOASQ large-scale biomedical semantic indexing and question answering competition. BMC Bioinform. **16**, 138 (2015)
78. Vachharajani, P.: Multilingual embedding and prompt-driven approaches for named entity recognition, entity linking, and clinical code prediction in Greek discharge summaries. In: Faggioli, G., Ferro, N., Rosso, P., Spina, D. (eds.) CLEF 2025 Working Notes (2025)

79. Vachharajani, P.: pjmathematician at MultiClinSUM 2025: A novel automated prompt optimization framework for multilingual clinical summarization. In: Faggioli, G., Ferro, N., Rosso, P., Spina, D. (eds.) CLEF 2025 Working Notes (2025)
80. Velichkov, B., Datseris, A., Vassileva, S., Boytcheva, S.: Enigma @ ElCardioCC: bridging NER and ICD-10 entity linking - A hybrid method for greek clinical narratives. In: Faggioli, G., Ferro, N., Rosso, P., Spina, D. (eds.) CLEF 2025 Working Notes (2025)
81. Verma, S., Jiang, F., Xue, X.: Beyond retrieval: ensembling cross-encoders and GPT rerankers with LLMs for Biomedical QA . In: Faggioli, G., Ferro, N., Rosso, P., Spina, D. (eds.) CLEF 2025 Working Notes (2025)
82. Yang, Z., Zhou, Y., Nyberg, E.: Learning to answer biomedical questions: OAQA at BioASQ 4B. In: Kakadiaris, I.A., Paliouras, G., Krithara, A. (eds.) Proceedings of the Fourth BioASQ workshop. pp. 23–37. ACL, Berlin, Germany (2016). https://doi.org/10.18653/v1/W16-3104
83. Yasunaga, M., Leskovec, J., Liang, P.: LinkBERT: pretraining language models with document links. In: Association for Computational Linguistics (ACL) (2022)
84. Zaratiana, U., Tomeh, N., Holat, P., Charnois, T.: GLiNER: generalist model for named entity recognition using bidirectional transformer. In: Duh, K., Gomez, H., Bethard, S. (eds.) Proceedings of the 2024 Conference of the North American Chapter of the Association for Computational Linguistics: Human Language Technologies (Volume 1: Long Papers), pp. 5364–5376. ACL, Mexico City, Mexico (2024). https://doi.org/10.18653/v1/2024.naacl-long.300
85. Zhang, T., Kishore, V., Wu, F., Weinberger, K.Q., Artzi, Y.: BERTScore: evaluating text generation with BERT. In: International Conference on Learning Representations (ICLR) (2020). https://arxiv.org/abs/1904.09675
86. Zhou, W., Huang, K., Ma, T., Huang, J.: Document-level relation extraction with adaptive thresholding and localized context pooling. In: Proceedings of the AAAI Conference on Artificial Intelligence (2021)
87. Zhuang, L., Wayne, L., Ya, S., Jun, Z.: A robustly optimized BERT pre-training approach with post-training. In: Li, S., Sun, M., Liu, Y., Wu, H., Liu, K., Che, W., He, S., Rao, G. (eds.) Proceedings of the 20th Chinese National Conference on Computational Linguistics, pp. 1218–1227. Chinese Information Processing Society of China, Huhhot, China (2021). https://aclanthology.org/2021.ccl-1.108/

Overview of the CLEF-2025 CheckThat! Lab: Subjectivity, Fact-Checking, Claim Normalization, and Retrieval

Firoj Alam[1]([✉])(ID), Julia Maria Struß[2](ID), Tanmoy Chakraborty[5],
Stefan Dietze[6,7](ID), Salim Hafid[11](ID), Katerina Korre[1](ID), Arianna Muti[12](ID),
Preslav Nakov[4](ID), Federico Ruggeri[3](ID), Sebastian Schellhammer[6](ID),
Vinay Setty[8], Megha Sundriyal[10](ID), Konstantin Todorov[11](ID), and V. Venktesh[9]

[1] Qatar Computing Research Institute, HBKU, Doha, Qatar
fialam@hbku.edu.qa
[2] University of Applied Sciences Potsdam, Potsdam, Germany
[3] DISI, University of Bologna, Bologna, Italy
[4] Mohamed bin Zayed University of Artificial Intelligence, Abu Dhabi,
United Arab Emirates
[5] Indian Institute of Technology Delhi, New Delhi, India
[6] GESIS - Leibniz Institute for the Social Sciences, Cologne, Germany
[7] Heinrich-Heine-University Düsseldorf, Düsseldorf, Germany
[8] University of Stavanger, Stavanger, Norway
[9] Delft University of Technology, Delft, The Netherlands
[10] Indraprastha Institute of Information Technology, New Delhi, India
[11] University of Montpellier, LIRMM, CNRS, Montpellier, France
[12] Bocconi University, Milan, Italy
https://checkthat.gitlab.io

Abstract. This paper presents the eighth edition of the CheckThat! lab, part of the 2025 Conference and Labs of the Evaluation Forum (CLEF). As in previous editions of CheckThat!, the lab offers tasks from the core of the verification pipeline, including check-worthiness, identifying previously fact-checked claims, supporting evidence retrieval, and claim verification as well as auxiliary tasks addressing different facets of individual steps of the pipeline: Task 1 is on identification of subjectivity (a follow-up of the CheckThat! 2024 edition), which is related to the check-worthiness task, Task 2 is on claim normalization, Task 3 addresses fact-checking numerical claims, and Task 4 focuses on scientific web discourse processing. These challenging classification and retrieval problems are offered in different mono-, multi- and crosslingual settings covering more than 20 languages. This year, CheckThat! was one of the most popular labs at CLEF-2025 in terms of team registrations: 177 teams registered, almost half of them actually participating (a total of 83 teams) and 54 submitted system description papers.

Keywords: Fact-Checking · Check-Worthiness · Subjectivity · Claim verification

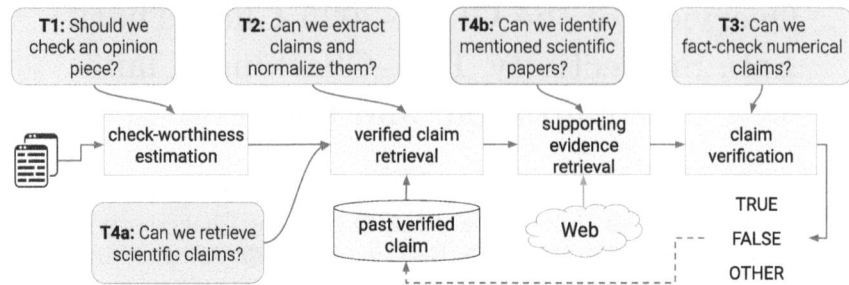

Fig. 1. The CheckThat!verification pipeline, featuring the four core tasks along with the CheckThat!2025 tasks.

1   Introduction

The primary goal of CheckThat! is to promote the development of technology and resources to assist different tasks along the fact-checking verification pipeline, as well as auxiliary tasks that support the process. During the first five iterations of the lab the main focus was set on the core tasks of the verification pipeline (see Fig. 1). From the sixth edition [16] on, the lab has widened the focus and opened up for auxiliary tasks helping to address the different steps of the pipeline.

This year [6] we offered four tasks with multiple mono-, multi- and cross-lingual settings covering more than 20 languages. Task 1 [55] is a follow-up of CheckThat! 2023 and 2024 editions dealing with the subjectivity of sentences in news articles, in order to spot text that should be processed with specific strategies [52], potentially benefiting the fact checking pipeline [40,41,80]. Task 2 [74] addresses the challenge of claims buried within noisy, unstructured social media posts and asks to normalize the claim into unambiguous, check-worthy statements. Task 3 [79] tackles the challenge associated with verifying numerical and temporal claims. Task 4 [33] focuses on scientific web discourse offering two subtasks, firstly asking participants to classify different forms of science-related online discourse, and secondly asking participants to identify the source of an informal reference made in social media posts.

As in previous editions, CheckThat! was one of the most popular tasks at CLEF, attracting a total of 177 registrations and 83 actively participating teams, using a variety of approaches to the different tasks, mainly based on the encoding and decoding of large language models combined with different sources of information.

2   Previously on the CheckThat!Lab

In the past seven iterations, the CheckThat! lab have offered a variety of tasks from the verification pipeline, in a multitude of languages and in different domains. An overview is given in Table 1.

Table 1. Overview of the tasks offered in the previous editions of the lab.

tasks	2018	2019	2020	2021	2022	2023	2024	debates	speeches	tweets	web pages	news articles	Arabic	Bulgarian	Dutch	English	French	Georgian	German	Greek	Hind (code-mixed)	Italian	Polish	Portuguese	Russian	Slovenian	Spanish	Turkish
check-worthiness estimation	■	■	■	■	■	■	■	■	■	■			■	■	■	■											■	■
verified claim retrieval			■	■	■			■	■	■			■			■												
supporting evidence retrieval	■	■			■						■	■	■			■												
claim verification	■	■	■		■	■				■	■		■			■												
fake news detection				■	■							■				■			■	■								
bias						■						■				■												
subjectivity						■	■					■	■	■		■			■			■						■
topic identification				■												■												
authority finding						■				■			■															
adversarial robustness						■													■									
persuasion							■					■	■	■		■	■		■	■	■	■	■	■	■	■	■	■

The first two editions of the lab in 2018 and 2019 focused on check-worthiness and claim verification of political debates and speeches in Arabic and English, with an additional focus on fact-checking by a task on classifying and ranking supporting evidence from the web in the second edition. The 2020 edition [17] covered the complete verification pipeline, with check-worthiness estimation, verified claim and supporting evidence retrieval, and claim verification, for the first time including social media data. The 2021 edition focused on multilinguality, offering tasks in five languages [47]. It also offered a fake news detection task, where the focus was on news articles, which was also continued in 2022.

The 2023 year's edition of the CheckThat! lab [16] paid special attention to the various sub-aspects of check-worthiness estimation, subjectivity of news articles, factuality, bias, authority findings, again in a multitude of languages. Transformer-based models were extensively used. This edition has also introduced multimodality for check-worthiness estimation.

3 Description of the 2025 Tasks

The 2025 edition of CheckThat! featured a total of four tasks in a variety of languages and modalities, three of which were run for the first time (cf. Sects. 3.1, 3.2, 3.3 and 3.4). Moreover, two of the tasks had two subtasks each (cf. Sects. 3.1 and 3.3).

3.1   Task 1: Subjectivity in News Articles

Verifiable claims are not only expressed through objective and neutral statements, but can also appear in subjectively framed ones. While objective sentences can be directly assessed for verification, subjective ones require additional processing, such as extracting their underlying objective content or any embedded claims. Consequently, the goal of this task is to determine whether a given sentence is subjective or objective. This is framed as a binary classification problem and is available in Arabic, Bulgarian, English, German, Italian, as well as in a multilingual setting. In this 2025 edition, the task of subjectivity in news articles also provides a zero-shot setting: a model is trained on data regarding certain languages and tested on data concerning unseen languages. In particular, we consider Greek, Romanian, Polish and Ukrainian as unseen languages and Arabic, Bulgarian, English, German, and Italian as training languages. A more detailed description and discussion of the task can be found in [72].

3.2   Task 2: Claim Normalization

Social media platforms impose minimal restrictions on writing, allowing users to post in vague and informal language. These posts often mix personal opinions, rhetorical questions, and incomplete information. This blend makes it difficult to identify clear claims – statements that assert something as true and can be verified or disproven [31]. As a result, fact-checkers face the difficult task of extracting concrete, check-worthy claims from noisy and unstructured content.

Claim Normalization addresses this challenge by transforming informal social media content into clear, concise, and verifiable statements, referred to as normalized claims [73]. These normalized claims capture the core factual assertion, making the fact-checking process more efficient and focused. This task is especially important in low-resource and multilingual settings, where identifying verifiable claims across language boundaries introduces additional complexity.

The task operates in two settings: monolingual and zero-shot. In the monolingual setting, training, development, and test datasets are provided for the same language. The model is trained, validated, and tested exclusively within this single language, allowing it to learn language-specific structures and patterns. Languages included in this setup are: English, German, French, Spanish, Portuguese, Hindi, Marathi, Punjabi, Tamil, Arabic, Thai, Indonesian, and Polish. In contrast, the zero-shot setting provides only the test data for the target language, without any corresponding training or development data. Participants may train their models using data from other languages or conduct zero-shot experiments with large language models (LLMs), evaluating performance on the target language without prior exposure. This setup tests the model's ability to generalize to unseen languages. Languages in this setting are: Dutch, Romanian, Bengali, Telugu, Korean, Greek, and Czech.

3.3   Task 3 Fact-Checking Numerical Claims

Task 3 is addressing the last task of the verification pipeline focusing on numerical claims (cf. Sect. 5.3).

There has been growing interest in developing tools [67], methods [30], and benchmarks [14,64] to enhance the fact-checking process. Automating fact-checking is challenging, as many claims are complex and require sophisticated reasoning for accurate validation, especially those involving numerical data. Numerical claims often appear more credible due to the *Numeric-Truth effect* [58], leading to uncritical acceptance. Recent studies show verifying numerical claims is more difficult than non-numerical ones [9,78]. For example, the social media claim that "CDC quietly deletes 6,000 COVID vaccine deaths from its website" exaggerates a clerical correction, causing unnecessary panic. This demonstrates the need for automated verification of such misleading claims.

This task focuses on verifying claims with numerical quantities and temporal expressions. Numerical claims are defined as those requiring validation of explicit or implicit quantitative or temporal details. Participants must classify each claim as *True, False, or Conflicting* based on a short list of evidence. Each claim is accompanied by the top-100 pieces of evidence retrieved using BM25 from our collection. These evidences can be used after re-ranking to perform claim verification with a classification or generative model that can perform the task of Natural language Inference (NLI). The objective here is to also evaluate the numerical reasoning capabilities of the claim verification model. The task is available in English, Spanish, and Arabic.

3.4   Task 4 Scientific Web Discourse Processing

Scientific web discourse, e.g., discourse about scientific claims or resources on the social web, has increased substantially throughout the past years [19,25]. However, scientific web discourse is usually informal, with examples such as *"covid vaccines just don't work on children"*, and displays fuzzy/incomplete citation habits, such as *"Stanford study shows that vaccines don't work"* where the actual study is never cited through explicit references. This poses challenges both from a computational perspective when mining social media or computing Altmetrics, but also from a societal perspective, leading to poorly informed online debates [53]. Based on this motivation, we introduce two tasks that are related to the second and third task of the verification pipeline:

– **Subtask 4a Scientific Web Discourse Detection**: Given a social media post (tweet), detect if it contains (1) a scientific claim, (2) a reference to a scientific study/publication, or (3) mentions of scientific entities, e.g. a university or scientist.
– **Subtask 4b Claim-source Retrieval**: Given an implicit reference to a scientific paper, i.e., a social media post (tweet) that mentions a research publication without a URL, retrieve the mentioned paper from a pool of candidate papers.

Refer to [33] for a detailed overview of **Task 4**, the dataset, and the participants' approaches.

Table 2. Task 1: Subjectivity in News Articles. Dataset statistics for all five languages for which we report training and development data splits.

	Training Languages									
	Arabic		**Bulgarian**		**English**		**German**		**Italian**	
	obj	subj	obj	subj	obj	subj	obj	subj	obj	subj
Train	1,391	1,055	379	312	532	298	492	308	1,231	382
Dev	266	201	167	139	240	222	317	174	490	177
Dev-test	425	323	134	107	362	122	153	71	334	128
Test	727	309	-	-	215	85	229	118	192	107
Total	2,809	1,888	689	558	1,349	727	1,191	671	2,247	794

	Unseen Languages							
	Greek		**Polish**		**Romanian**		**Ukrainian**	
	obj	subj	obj	subj	obj	subj	obj	subj
Test	236	48	161	154	154	52	219	78

4   Datasets

The following section describes the datasets developed for the individual tasks and distributed to the scientific community.[1]

4.1   Task 1: Subjectivity in News Articles

The dataset comprises sentences from news paper articles annotated with respect to their subjectivity. Information regarding the annotation guidelines can be found in [54]. The dataset included 4,697, 1,247, 2,076, 1,862 and 3,041 instances in Arabic (see [75] for more detail), Bulgarian, English, German, and Italian, respectively. Table 2 shows statistics. We provided a training set for the multilingual scenario, comprising the training data for all languages offered this year. The same holds for the dev and dev-test sets. The test set included only data from the languages offered in this edition. The participants were free to choose from the multilingual datasets, opening room for cross-lingual approaches. For the zero-shot setting, the unseen test sets statistics are as follows: 284 instances for Greek (236 OBJ, 48 SUBJ), 351 for Polish (161 OBJ, 154 SUBJ), 206 for Romanian (154 OBJ, 52 SUBJ), and 298 for Ukrainian (219 OBJ, 78 SUBJ).

---

[1] All datasets are available in the GitLab repository of the lab: https://gitlab.com/checkthat_lab/clef2025-checkthat-lab/.

4.2   Task 2: Claim Normalization

The posts originate from various social media platforms, such as Twitter, Reddit, Facebook, etc., and are sourced from the Google Fact-check Explorer API[2] and the Claim Review Schema.[3] Each post is paired with a corresponding normalized claim. We provide train, dev and test data for Arabic, German, English, French, Hindi, Marathi, Indonesian, Punjabi, Polish, Portuguese, Spanish, Tamil and Thai. While low-resource languages like Bengali, Czech, Greek, Korean, Romanian, Telugu, and Dutch are considered for zero-shot settings. The data statistics are provided in Table 3. The systems are evaluated using the METEOR score.

Table 3. Task 2: Claim Normalization. Dataset statistics for all 20 languages.

Split	Arabic	Bengali	Czech	German	Greek	English	French	Hindi	Korean	Marathi
Train	470	0	0	386	0	11,374	1,174	1,081	0	137
Dev	118	0	0	101	0	1,171	147	50	0	50
Test	100	81	123	100	156	1,285	148	100	274	100

Split	Indonesian	Dutch	Punjabi	Polish	Portugese	Romanian	Spanish	Tamil	Telugu	Thai
Train	540	0	445	163	1,735	0	3,458	102	0	244
Dev	137	0	50	41	223	0	439	50	0	61
Test	100	177	100	100	225	141	439	100	116	100

Example: Claim decomposition example

Claim: `Discretionary spending has increased over 20-some percent in two years if you dont include the stimulus. If you put in the stimulus, its over 80 percent.`

[Decomposition]:   [Q1]: Has discretionary spending increased in the past two years?
[Q2]: Does the increase in discretionary spending exclude the stimulus?
[Q3]: Is there evidence to support the claim that

Fig. 2. Example for claim decomposition

4.3   Task 3: Fact-Checking Numerical Claims

The dataset is collected from various fact-checking domains through Google Fact-check Explorer API[4], complete with detailed metadata and an evidence

---

[2] https://toolbox.google.com/factcheck/apis.
[3] https://schema.org/ClaimReview.
[4] https://toolbox.google.com/factcheck/apis.

Table 4. Dataset statistics for task 3.

Split	English	Spanish	Arabic
Train	9,935	1,506	2,191
Dev	3,084	377	587
Test	3,656	1,806	482

Table 5. Task 4a: Scientific Web Discourse Detection Dataset statistics.

Split	Cat 1	Cat 2	Cat 3	Total
Train	333	224	306	1,229
Dev	26	26	34	137
Test	121	56	97	240
Total	480	306	437	1,606

corpus sourced from the web. Our pipeline filters out numerical claims for the task. An overview of dataset statistics is shown in Table 4. We use the train and validation sets from the English dataset released in [78], and also curate Arabic and Spanish claims. For the test set we collect new real-world English numerical claims additionally to the evaluation set released in [78] to avoid label leakage. The Arabic dataset only consists of claims belonging to the categories *True* and *False* for verification, as real-world distribution of conflicting claims for Arabic is too low.

Evidence for claims in all languages were obtained from search engines by excluding fact-checking websites to avoid leakage of fact-checker justification and verdict. For each claim, we decompose them to yes/no type sub-questions as shown in Fig. 2, and issue the original claim and generated sub-questions as queries to the search engines. Additionally, for English, evidences are also obtained by other decomposition approaches like sub-claim generation to increase diversity of evidence pool. All evidences are pooled to form the collection. Macro-averaged F1 and classwise F1 scores were employed as metrics for evaluating claim verification.

4.4   Task 4: Scientific Web Discourse Processing (SciWeb)

Task 4a: Scientific Web Discourse Detection. The dataset for subtask 4a is an extension of the SciTweets corpus [34] and consists of 1,606 posts from X (former Twitter) annotated with the different forms of science-related online discourse as introduced in [34], which are scientific claims (Cat 1), scientific references (Cat 2), and references to science contexts or entities (Cat 3). Table 5 shows the dataset statistics.

Task 4b: Scientific Claim Source Retrieval. The dataset for subtask 4b consists of two sets, a query set and a collection set. The query set contains 14,399 X (Twitter) posts with implicit references to scientific papers from CORD-19.

The collection set contains metadata, such as title, abstract, and affiliations of the 7,718 CORD-19 scientific papers, which the posts from the query set implicitly refer to. The dataset is divided into a train (14253 posts), dev (1400 posts), and test split (1446 posts).

5   Results and Overview of the Systems

5.1   Task 1: Subjectivity in News Articles

A total of 21 teams participated in the task, submitting 436 valid runs across all language tracks. 16 out of the 21 teams filled in the survey for the task, providing information about their systems and approaches. 12 teams participated in more than one subtask, while 5 teams opted for only the monolingual English subtask. Table 6 shows the results achieved by the top-3 ranking teams for each language.

Table 6. Task 1: results on subjectivity classification in news articles in terms of macro F1. Shown are the top-3 submissions per language.

Rank	Team	F1	Rank	Team	F1	Rank	Team	F1
Arabic			Italian			German		
1	CEA-LIST	0.6884	1	XplaiNLP	0.8104	1	SmolLab_SEU	0.8520
2	UmuTeam	0.5903	2	CEA-LIST	0.8075	2	UNAM	0.8280
3	Investigators	0.5880	3	SmolLab_SEU	0.7750	3	QU-NLP	0.8013
English			Multilingual			Polish		
1	QU-NLP	0.8052	1	TIFIN INDIA	0.7550	1	CEA-LIST	0.6922
2	TIFIN INDIA	0.7955	2	CEA-LIST	0.7396	2	IIIT Surat	0.6676
3	CEA-LIST	0.7739	3	CSECU-Learners	0.7321	3	CSECU-Learners	0.6558
Ukrainian			Romanian			Greek		
1	CSECU-Learners	0.6424	1	QU-NLP	0.8126	1	AI Wizards	0.5067
2	Investigators	0.6413	2	CSECU-Learners	0.7992	2	SmolLab_SEU	0.4945
3	ClimateSense	0.6395	3	XplaiNLP	0.7917	3	CSECU-Learners	0.4919

Most teams used a supervised binary classification approach, treating the task as classifying sentences into subjective (SUBJ) or objective (OBJ). The dominant strategy involved fine-tuning transformer-based models, with some using ensembles, data augmentation, or additional linguistic features. A few teams explored probabilistic thresholds, embedding-based classifiers, or LLM-based zero-shot and in-context learning methods. An overview of the approaches is given in Table 11 and a short description of the individual approaches for each team is given in the following.

Team **AI Wizards** [29] employed a probabilistic classifier with a decision threshold, fine-tuning DeBERTaV3 for the task.

Team **Investigators** [35] utilized decoder-based models including DeBERTa, BERT, Multilingual BERT, and Twitter RoBERTa.

Table 7. Task 1: Overview of the approaches.

Team	Arabic	Italian	German	English	Multilingual	Polish	Ukrainian	Romanian	Greek	DeBERTa	BERT	MBERT	RoBERTa	DistilRoBERTa	SentimentBERT	ModernBERT	MPNet	XLM-RoBERTa	SBERT	CT-BERT	Electra	InfoXLM	Llama	GPT	Zephyr	Qwen	Data Augmentation	Translating data	LLM Prompting	Feature Selection
AI Wizards [29]	✓	✓	✓	✓	✓	✓	✓	✓	✓	✓																				
Investigators [35]	✓	✓	✓	✓	✓	✓	✓	✓	✓	✓	✓	✓	✓																	
DSGT-CheckThat [37]			✓											✓	✓	✓	✓	✓									✓			
CSECU-Learners [4]			✓	✓	✓	✓	✓	✓	✓	✓		✓					✓													
CEA-LIST [26]	✓	✓	✓	✓	✓	✓	✓	✓	✓	✓	✓												✓	✓		✓			✓	
IIIT Surat [39]	✓	✓	✓	✓	✓	✓	✓	✓	✓		✓	✓																		
TIFIN INDIA [32]	✓	✓	✓	✓	✓	✓	✓	✓	✓				✓				✓			✓							✓	✓		✓
ClimateSense [20]	✓	✓	✓	✓	✓	✓	✓	✓	✓				✓				✓			✓	✓	✓		✓						✓
CUET_KCRL [69]				✓										✓																
nlu@utn [45]				✓									✓																	✓
XPlaiNLP [60]		✓	✓	✓	✓	✓	✓	✓	✓				✓				✓							✓						✓
JU_NLP [23]	✓	✓	✓	✓	✓	✓	✓	✓	✓				✓																	✓
NapierNLP [7]				✓																				✓	✓					✓
UmuTeam [18]	✓	✓	✓	✓	✓	✓	✓	✓	✓				✓		✓		✓													✓
UGPLN [21]				✓															✓											✓
SmolLab_SEU [51]	✓	✓	✓	✓	✓	✓	✓	✓	✓	✓	✓		✓							✓	✓									
Arcturus [3]	✓	✓	✓	✓	✓	✓	✓	✓	✓	✓																				
QU-NLP [5]	✓	✓	✓	✓	✓	✓	✓	✓	✓	✓												✓								✓
CheckMates [48]				✓									✓				✓													✓

Team **DSGT-CheckThat** [37] fine-tuned encoder models and explored data augmentation strategies. Their models included RoBERTa (emotion-large), DistilRoBERTa, Sentiment-BERT, ModernBERT, RoBERTa-large, and MiniLM. They further enhanced performance through Synthetic Data Generation and Data Augmentation.

Team **CSECU-Learners** [4] framed the task as multiclass classification with SUBJ (subjective) and OBJ (objective) as separate classes. Their transformer models included MPNet, mDeBERTa, and Multilingual BERT.

Team **CEA-LIST** [26] fine-tuned small language models (SLMs) and experimented with LLMs through techniques such as in-context learning, LLM-as-judge, and model debating. Their models included RoBERTa, UmBERTo, ALBERTo, Qwen 2.5 70B, Meta-LLaMA 3 70B, DeepSeek 67B, Aya-Expanse-32B, and GPT-4.1-mini.

Team **IIIT Surat** [39] employed a transformer-based model, specifically BERT, implemented via BertForSequenceClassification from Hugging Face, and fine-tuned it for binary classification (SUBJ/OBJ). They used the pre-trained BERT (English, uncased) for the monolingual classifier and Multilingual BERT (cased) for multilingual and other-language classification, fine-tuning both directly on the CLEF training data.

Team **TIFIN INDIA** [32] used a binary classification approach, where each input is classified as either subjective or objective. They used an ensemble of transformer-based models and combined their probability outputs to make the final prediction post data augmentation. To mitigate data imbalance, they applied back-translation as a data augmentation technique and used the label distribution ratio to monitor and address class imbalance. They sed deep learning models based on transformer encoder architectures, including BERT-Base, BERT-Large, RoBERTa-Base, RoBERTa-Large, XLM-RoBERTa-Base, XLM-RoBERTa-Large, Modern-BERT-Base, and Modern-BERT-Large. They applied probability-level averaging (soft voting) for model fusion to ensemble predictions across these models. Additionally, for some datasets, they used a traditional Support Vector Machine (SVM) classifier with TF-IDF features as a lightweight baseline and for comparative analysis. They used a feature-based approach using Support Vector Machines (SVMs) on selected datasets. The most important features included: TF-IDF vectors of unigrams and bigrams.

Team **ClimateSense** [20] used Embeddings and an MLP classifier. They experimented with various classifiers: SVC, Logistic Regression, MLP, etc. They also experimented with various transformers-based architectures for embedding the sentences: SBERT, RoBERTa-based models, ModernBERT-large, CT-BERT. Finally, they experimented with Zero-shot prompting some LLMs (such as Zephyr).

Team **CUET_KCRL** [69] pursued a supervised classification approach using an LSTM and fine-tuning mBERT.

Team **nlu@utn** [45] followed a Bert-based ensemble model approach, by also adapting the provided the training data with additional linguistic information before training, using persuasion techniques identified in the data and POS-counts. The models used were politicalBiasBERT and BERT-base-uncased.

Team **XPlaiNLP** [60] employed several transformer-based models, including XLM-RoBERTa-base, GPT o3-mini, and German-BERT. In particular, for monolingual tasks, German-BERT was fine-tuned on German and German-translated versions of English, Italian and Bulgarian train datasets.

Team **JU_NLP** [62] fine-tuned BERT model on available training data, formulating the task as a binary classification problem. In particular, they leverage hand-crafted features derived from knowledge bases and tools like SentiWordNet, WordNet, Opinion lexicon, POS taggers, and lemmatization.

Team **NapierNLP** [7] only tackled the English monolingual task by leveraging LLMs. More precisely, they employed GPT-2, GPTNeo-1.3B, and Qwen3-0.6B. The prompts provided instructions for addressing the task as a binary classification problem.

Team **UmuTeam** [18] employed a wide set of encoder-only transformers, each specific for a given language. In particular, they employed MARBERTv2 for Arabic data, GottBERT-base for German, BERTino for Italian, RoBERTa-base for English. Lastly, they used XLM-RoBERTa-base for multilingual and zero-shot tasks.

Team **UGPLN** [21] employed sentence transformers with hand-crafted linguistic features. A logistic regressor is then trained on top to perform the binary classification task. In particular, they employed MiniLM-L12-v2 and used the following hand-crafted features: presence of negation cues, sentence length (i.e., token count), punctuation marks, and lexical opinion indicators derived from the MPQA Subjectivity lexicon.

Team **SmolLab_SEU** [51] employed a vast set of encoder-only transformers, some of which are language-specific. The models are RoBERTa, DeBERTa-v3, AraBERTv2 and MARBERTv2 for Arabic, GBERT-large, GottBERT-base, and GElectra-large for German, UmBERTo-v1, and BERT-base-italian for Italian, MBERT, XLM-RoBERTa-large, InfoXLM-large, MT5-base, and MDeBERTa-v3 for multilingual. All models were fine-tuned by adding a sequence classification head on top of their pre-trained encoder layers.

Team **Arcturus** [3] fine-tuned the English-pretrained DeBERTa-v3 on monolingual datasets and evaluate it on all languages, including multilingual and zero-shot tasks.

Team **QU-NLP** [5] propose a feature-augmented transformer architecture that combines contextual embeddings from pre-trained language models with statistical and linguistic features. In particular, they employed AraElectra for Arabic, augmented with POS tags and TF-IDF features. For cross-lingual experiments, they employed DeBERTa-v3 with TF-IDF features through a gating mechanism.

Team **CheckMates** [48] explored various models such as logistic regression, Support Vector Machine, BERT, Sentence-BERT, and DistilBERT.

More details on the participants approaches can be found in [55].

5.2   Task 2: Claim Normalization

Task 2 received submissions from 18 teams, totalling 1,226 valid runs across all the languages. Table 9 presents the results for the monolingual setup, evaluated using METEOR scores, while Table 10 reports the outcomes for the zero-shot setup. These tables summarise team performance across the respective languages.

Most teams employed sequence-to-sequence generation strategies for claim normalization, typically relying on transformer-based models. The most prevalent approach involved fine-tuning pretrained models such as BART, T5, mBART, and LLaMA on monolingual training data. Common data preprocessing included de-duplication, emoji removal, hashtag normalization, multilingual data augmentation via translation, and prompt engineering tailored to each language. Some of the teams used LoRA-based adaptor fine-tuning to reduce resource needs, while others delved into ensemble solutions like embedding-based centroid voting or model-soup techniques. In the zero-shot setting, prompt-based generation took precedence, with models driven by structured instructions to extract factual, brief claims from informal posts. Others employed semantic similarity retrieval to choose in-context instances for prompting. To improve

Table 8. Task 2: Overview of the participating teams per language and their respective rankings.

Team	English	Arabic	German	French	Hindi	Marathi	Indonesian	Punjabi	Polish	Portuguese	Spanish	Tamil	Thai	Bengali	Telugu	Dutch	Czech	Greek	Romanian	Korean
dfkinit2b [12]	1	1	2	2	1	1	2	1	2	2	2	1	3	1	1	1	1	1	1	1
DS@GT CheckThat! [50]	2	2	1	1	2	4	1	5	1	1	1	3	1	4	5	5	3	4	4	3
TIFIN [68]	3	5	5	6	7	6		4	5		5	5		5	6	4				
AKCIT-FN [8]	4	6	3	3	5	5	3	2	3	3	3	2	2	3	2	2	4	2	2	2
Factiverse [10]	5	7	4	4	8	9	4	7	6	6	6	8	4	6	7		5	5	5	
rohan_shankar	6																			
manan-tifin	7		7	9			7		5					5	6					
MMA [57]	8	3	7	8	6	3	5	6	7	4	4	6								
UNH [81]	9																			
Investigators [35]	10										8									5
teamopenfact [63]	11	4	6	5	4	2	6	3	4	5	7	4	5	2	3	3	2	3	3	4
Nikhil_Kadapala	12																			
aryasuneesh	13		5	6	7	6	4				5	5	6							
JUNLP_M&S [46]	14																			
uhh_dem4ai	15																			
tomasbernal01	16	8	8	9	10	8	7	8	8	7	9	7	7	7	8	6	6	6	6	6
VSE	17																			
saivineetha [15]						3								4						

the relevance and structure of claims, some teams used reinforcement learning, instruction tuning, or Chain-of-Thought prompting.

Out of all the participating teams, dfkinit2b [12], DS@GT CheckThat! [50], TIFIN [68], and AKCIT-FN [8] consistently ranked among the top-performers across most languages. More details on the other participating approaches can be found in [74].

Team **dfkinit2b** [12] performed comprehensive experiments in both monolingual and zero-shot settings, testing zero- and few-shot prompting with models like Gemma-3, Qwen-3, Qwen-2.5, Llama-3.3, and Mistral. They explored various prompt formulations and used cosine similarity to select demonstrations for few-shot learning. Experiments also included adapter fine-tuning, data preprocessing with language checks and emoji removal, and data augmentation via translation. For the final submission, they ensembled top-performing model outputs by computing embedding centroids with multilingual SentenceTransformers and selecting claims closest to these centroids, achieving strong results across languages.

Table 9. Task 2: Scores (METEOR) for languages with training data. Ranks across languages are in brackets.

Team	English	Arabic	German	French	Hindi	Marathi	Thai
dfkinit2b	0.4569 (1)	0.5037 (1)	0.3469 (2)	0.4703 (2)	0.3275 (1)	0.3888 (1)	0.2999 (3)
DS@GT CheckThat!	0.4521 (2)	0.5035 (2)	0.3859 (1)	0.5273 (1)	0.3001 (2)	0.2608 (4)	0.5859 (1)
TIFIN	0.4114 (3)	0.3705 (5)	0.2642 (5)	0.3441 (6)	0.2604 (7)	0.1521 (6)	-
AKCIT-FN	0.4058 (4)	0.3277 (6)	0.2652 (3)	0.3811 (3)	0.2706 (5)	0.2181 (5)	0.3179 (2)
Factiverse	0.4049 (5)	0.2457 (7)	0.2644 (4)	0.3750 (4)	0.2125 (8)	0.0847 (9)	0.0965 (4)
rohan_shankar	0.3920 (6)	-	-	-	-	-	-
manan-tifin	0.3881 (7)	-	-	0.2768 (7)	0.2080 (9)	0.1230 (7)	-
MMA	0.3841 (8)	0.4584 (3)	0.1556 (7)	0.2469 (8)	0.2641 (6)	0.2793 (3)	-
UNH	0.3737 (9)	-	-	-	-	-	-
Investigators	0.3565 (10)	-	-	-	-	-	-
teamopenfact	0.3370 (11)	0.4175 (4)	0.2319 (6)	0.3605 (5)	0.2722 (4)	0.3048 (2)	0.0872 (5)
Nikhil_Kadapala	0.3321 (12)	-	-	-	-	-	-
aryasuneesh	0.3153 (13)	-	0.2642 (5)	0.3441 (6)	0.2604 (7)	0.1521 (6)	0.0464 (6)
JUNLP_M&S	0.3098 (14)	-	-	-	-	-	-
uhh_dem4ai	0.2612 (15)	-	-	-	-	-	-
tomasbernal01	0.1660 (16)	0.0003 (8)	0.1039 (8)	0.1649 (9)	0.0132 (10)	0.0877 (8)	0.0147 (7)
VSE	0.0070 (17)	-	-	-	-	-	-

Team	Indonesian	Punjabi	Polish	Portugese	Spanish	Tamil
dfkinit2b	0.5021 (2)	0.3307 (1)	0.3961 (2)	0.5744 (2)	0.5539 (2)	0.6316 (1)
DS@GT CheckThat!	0.5650 (1)	0.2567 (5)	0.4065 (1)	0.5770 (1)	0.6077 (1)	0.4702 (3)
TIFIN	-	0.2685 (4)	0.2331 (5)	-	0.3906 (5)	0.3676 (5)
AKCIT-FN	0.3866 (3)	0.3038 (2)	0.2798 (3)	0.5290 (3)	0.5213 (3)	0.5197 (2)
Factiverse	0.3099 (4)	0.1251 (7)	0.1964 (6)	0.3381 (6)	0.3821 (6)	0.0043 (8)
manan-tifin	-	-	0.2331 (5)	-	-	-
MMA	0.3089 (5)	0.1834 (6)	0.1243 (7)	0.4719 (4)	0.5094 (4)	0.3468 (6)
Investigators	-	-	-	-	0.3447 (8)	-
teamopenfact	0.2445 (6)	0.2696 (3)	0.2666 (4)	0.3779 (5)	0.3710 (7)	0.4681 (4)
aryasuneesh	-	0.2685 (4)	-	-	0.3906 (5)	0.3676 (5)
tomasbernal01	0.1305 (7)	0.0097 (8)	0.0742 (8)	0.1898 (7)	0.2048 (9)	0.0196 (7)

Team **DS@GT CheckThat!** [50] embedded the non-normalized claims from the pooled train and development datasets, as well as from the test set, using state-of-the-art embedding models tailored to each language. For every test claim, a GPT-4o mini model was prompted following the approach discussed in [73], utilising the top-3 most cosine-similar examples from the train and development sets as in-context examples. The final response for the monolingual task was derived by combining the best-matching answer from the train and development sets, based on cosine similarity, and the output of the GPT-4 model. For

Table 10. Task 2: Scores (METEOR) for languages without training data. Ranks across languages are in brackets.

Team Name	Bengali	Telugu	Dutch	Czech	Greek	Romanian	Korean
dfkinit2b	0.3777 (1)	0.5257 (1)	0.2001 (1)	0.2519 (1)	0.2619 (1)	0.2950 (1)	0.1339 (1)
teamopenfact	0.2959 (2)	0.4559 (3)	0.1866 (3)	0.2144 (2)	0.2333 (3)	0.2350 (3)	0.1050 (4)
AKCIT-FN	0.2916 (3)	0.5176 (2)	0.1922 (2)	0.1734 (4)	0.2567 (2)	0.2516 (2)	0.1209 (2)
DS@GT CheckThat!	0.2435 (4)	0.3171 (5)	0.1608 (5)	0.1959 (3)	0.2250 (4)	0.2220 (4)	0.1156 (3)
TIFIN	0.2030 (5)	0.2502 (6)	0.1720 (4)	-	-	-	-
manan-tifin	0.2030 (5)	0.2502 (6)	-	-	-	-	-
Factiverse	0.1068 (6)	0.0802 (7)	-	0.1571 (5)	0.1455 (5)	0.2097 (5)	-
tomasbernal01	0.0451 (7)	0.0269 (8)	0.0817 (6)	0.0544 (6)	0.0062 (6)	0.0779 (6)	0.0014 (6)
Investigators	-	-	-	-	-	-	0.0149 (5)

Table 11. Task 3: Overview of the approaches for fact-checking numerical claims.

Team	Language			Model													Macro-F1		
	Arabic	Spanish	English	BM25	cross-encoder	gpt-4o-mini	Qwen	Llama	DeepSeek	ModernBERT	Math-Roberta	RoBERTa-base	QWQ-32B	Qwen-8B	Deberta-Large-MNLI	mxbai-rerank-large-v1	granite-3.3-8b-instruct	Arabic	Spanish	English
LIS [42]	✓	✓	✓											✓				50.34	96.15	59.54
DS@GT-CheckThat! [38]		✓																-	-	52.10
TIFIN [36]	✓	✓		✓												✓	✓	55.36		55.70
ClaimIQ [11]		✓								✓								-	-	42.43
FraunhoferSIT [56]		✓				✓				✓	✓							-	-	51.00
NGU_Research [1]	✓	✓		✓	✓				✓									63.52	24.41	-
JU_NLP [22]	✓	✓		✓	✓													36.38	-	48.83
CornellNLP [24]		✓		✓	✓		✓	✓										-	-	48.57
UGLPN [77]		✓		✓	✓							✓						-	-	45.53
UCOM_UNAM_PLN [2]	✓			✓	✓													-	35.95	-
News-polygraph*		✓		✓	✓					✓								-	-	42.86

zero-shot languages, they utilised a modified version of CACN [73], essentially using the prompting method with standard examples.

Team **TIFIN** [68] fine-tuned the Qwen-14B model using LoRA with 4-bit precision for efficiency. They preprocessed data by filtering meaningful post-claim pairs, removing duplicates, and creating a unified multilingual dataset. Instruction-based fine-tuning incorporated Chain-of-Thought prompting with 5W1H questions to guide claim extraction. During inference, context resolution replaced partial posts with complete ones, and few-shot prompting with similar examples improved claim structure. This approach aimed to boost claim extraction accuracy and multilingual performance.

Team **AKCIT-FN** [8] fine-tuned sequence-to-sequence Transformer models, including monolingual and multilingual variants like unicamp-dl/monoptt5-large, unicamp-dl/ptt5-v2-large, and t5-large, focusing on monolingual training to capture language-specific features. They performed hyperparameter tuning across learning rates, optimizers, batch sizes, and epochs to optimize performance. Evaluation combined METEOR, BERTScore, and semantic similarity metrics to better assess claim fidelity. For zero-shot tasks, they leveraged OpenAI's LLMs with carefully crafted prompts to generate concise, factual claims from informal posts in unseen languages, testing the models' generalization capabilities.

5.3   Task 3: Fact-Checking Numerical Claims

A total of 258 valid runs were submitted by 13 unique teams across languages, with 4 participants in Spanish and Arabic. 11 teams participated in fact-checking English numerical claims.

Among all participating teams, LIS was the top performer across all languages. TIFIN, NGU_Research, DS@GT-CheckThat! performed well in the respective languages the teams participated. Most teams employed generative models like gpt-4o-mini or Qwen LLMs to decompose claims, followed by BM25 based retrieval for retrieving evidence and transformer based cross-encoder models for re-ranking the evidences. For claim verification fine-tuned transformer based NLI models were employed by some teams where transformers were trained as discriminative models on the training sets provided. Some teams employed prompting based approaches to leverage large Language Models (LLMs) like gpt-4o-mini or reasoning models like deepseek-r1 to perform claim verification. The authors observe that fine-tuning LLMs for claim verification coupled with claim decomposition using recent reasoning models helps outperform the best baselines reported in [78].

Team **LIS** [42] used QwQ-32B to generate question followed by Linq-Embed-Mistral to retrieve evidence from the corpus by combining the questions and claims. Mistral-Small-24B-Instruct-2501 was fine-tuned to obtain the final veracity labels. The Qwen model seem to overcome certain limitations associated with gpt-3.5 and gpt-4 series models used in baselines [78].

Team **DS@GT-CheckThat!** [38] performed pre-processing to normalize the number and dates of the claims and decomposed questions from these claims. They employed gpt-4o-mini to decompose the claims. BM25 was employed for first stage retrieval to prioritize documents relevant to the claim and sub-questions. This is followed by re-ranking the documents using cross-encoder/ms-marco-MiniLM-L-12-v2 or mixedbread-ai/mxbai-rerank-large-v1. The main workhorse model for the veracity classification was ModernBERT - an optimized model based on the BERT architecture, that can natively support longer sequence length.

Team **TIFIN** [36] employed inverse class weighting to handle class imbalance in claim verification step and to give more importance to minority classes. they also employed other strategies such as oversampling to balance training examples, and label smoothing to prevent the model from becoming overconfident in its predictions. The authors also incorporated Focal Loss, for fine-tuning the verification model *microsoft/deberta-large-mnl* with LoRA, to focus training on the harder examples. They also employed *ibm-granite/granite-3.3-8b-instruct model* to summarize contexts before feeding them to the verification model.

Team **NGU_Research** [1] employed hybrid retrieval techniques ranging from pretrained encoder-based models to BGE, E5, Gemini as embedding models and finally settled on pretrained embeddings from openai's **text-embedding-3-large** model together with bm25 filtering via Qdrant database collections for each language. Then finally Deepseek and gpt-4o-mini were employed for performing claim verification using the retrieved evidence.

5.4   Task 4: Scientific Web Discourse Processing (SciWeb)

Task 4a: Scientific Web Discourse Detection. Task 4a is a multilabel classification task and was evaluated through the macro-averaged F1-score. The baseline is a DeBERTaV3-base model trained on the train set for 10 epochs with a learning rate of $2e^{-5}$ and a batch size of 16. For the final test set predictions, we used the checkpoint with the best dev set performance, resulting in a test set macro F1-score of 0.7668 (rank 7).

In total, ten teams participated in subtask 4a. Table 12 provides an overview of the different approaches and their performances for those teams that submitted a paper description of their work. The F1-score and rank indicate the performance and position on the final test set leaderboard. Most teams relied on Transformer-based models such as DeBERTa-v3, SciBERT, and Twitter-Roberta, while team DS@GT CheckThat! [49] and TurQUaz [61] also used LLMs. In addition, different techniques such as LLM-based data augmentation, ensemble methods, and other optimizations were employed.

Team **ClimateSense** [20] fine-tuned a twitter-roberta-base-2022-154 m model with a weighted loss. For each category, the best-performing checkpoint was identified based on the dev set performance. Using the embeddings of these checkpoints, a traditional classifier was trained for each category.

Team **UTB-CEDNAV** [70] fine-tuned a DeBERTa-v3-base model using hyperparameters found with 5-fold cross-validation. To improve performance, they employed class weighting and threshold-tuning and used an ensemble of their two strongest model (with and without class weight) for their final submission.

Team **SBU-SCIRE** [76] augmented the training data with paraphrases using DeepSeek-R1. They trained a DeBERTa-v3-large model with a Focal Loss on the train set and performed a grid search over the per-class threshold to maximize the performance on the dev set.

Team **DS@GT** [49] trained different transformer-based models and used zero-shot and few-shot classification with GPT-4o and GPT-4o mini. For their final submission, they used DeBERTa-v3-base for categories one and three and GPT-4o mini with few-shot (five examples based on semantic similarity) for category two.

Team **TurQUaz** [61] employed various LLMs in different collaborative settings. The setting for their final submission includes five models discussing a post together to reach an agreement, with another model acting as a chairperson.

Team **JU_NLP** [43] generated tweet embeddings using SciBERT and Twitter-RoBERTa models to capture both scientific and social media discourse characteristics of tweets. The embeddings were used to train a two-layer classification head.

Task 4b: Scientific Claim Source Retrieval. Task 4b is a retrieval task and was evaluated by the MRR@5 (Mean Reciprocal Rank) score. BM25 ranking using the title and abstract of the papers and the text of the X posts serves as the baseline with an MMR@5 of 0.43. The best-performing team reached an MMR@5 of 0.68.

Table 12. Task 4a: Overview of the approaches

Team	Models					Misc.			Perf.	
	DeBERTa-v3	SciBERT	Twitter-RoBERTa	LLMs	Others	Data Augmentation	Ensemble	Other Optimizations	Macro-avg. F1-Score	Rank
ClimateSense [20]			■	■				■	0.7998	1
UTB-CEDNAV [70]	■						■	■	0.7983	2
SBU-SCIRE [76]	■					■		■	0.7917	4
DS@GT CheckThat! [49]	■			■	■			■	0.7685	6
DeBERTa-v3 Baseline	■								0.7668	7
TurQUaz [61]				■				■	0.7615	8
JU_NLP [43]		■	■					■	0.7347	9

In total, 30 teams participated in subtask 4b. Table 13 provides an overview of the different approaches and their performance for teams that submitted a paper description of their work. Most teams relied on a combination of retrieval methods (dense, sparse, or both) and re-ranking models. Retrieval methods included both lexical and semantic methods. LLMs such as ChatGPT, LLaMa and Gemma were mainly used as re-rankers, but did not always outperform fine-tuned Transformer-based models. Additionally, some teams experimented with data augmentation and style transfer techniques.

Team **AIRwaves** [13] employed a two-stage pipeline using neural representation learning for candidate generation with a fine-tuned E5-large model, followed by neural re-ranking with a SciBERT cross-encoder to re-order the top predictions.

Team **Deep Retrieval** [59] combined lexical BM25-based and semantic search-based approaches to generate candidates, which were re-ranked using LLMs.

Team **ATOM** [71] used a GTR-T5-Large model to retrieve candidates, followed by a neural re-ranking with MXBAI-base-v2.

Team **SBU-SCIRE** [76] used a Snowflake/snowflake-arctic-embed-l-v2.0 for dense retrieval, followed by ms-marco-MiniLM-L4-v2 re-ranking.

Team **SeRRa** [44] used a multi-step pipeline including dense retrieval for candidate generation with a Sentence-BERT model, re-ranking using a binary classification model, and a final ranking through pairwise comparisons of the top 10 re-ranked documents with the input claim.

Team **Claim2Source** [66] first applied style transfer techniques to both claims and source documents using LLaMa 3.3-70B-Instruct (e.g., enhancing readability, adopting a scientific tone, or reformulating the abstract as a tweet). They then combined BM25 with dense retrieval models such as SPECTER, all-Mini-LM-L6-v2, and GritLM-7B.

Table 13. Task 4b: Overview of the approaches

Team	Dense Retrieval	Sparse Retrieval	Re-ranking	LLMs	Data Augmentation	Style transfer	MRR@5	Rank
AIRwaves [13]	■		■				0.67	2
Deep Retrieval [59]	■	■	■	■			0.66	3
ATOM [71]	■		■				0.66	4
SBU-SCIRE [76]	■		■	■			0.65	5
SeRRa [44]	■		■				0.61	8
Claim2Source [66]	■	■	■			■	0.59	12
DS@GT [65]			■	■	■	■	0.58	16
BM25 Baseline		■					0.43	28

Team **DS@GT** [65] used data-augmentation and style transfer techniques on tweets using ChatGPT. They then implemented a two-stage retrieval pipeline based on bi-encoder and cross-encoder approaches for retrieval and reranking using zero-shot and fine-tuned Sentence-Transformers.

6   Conclusion and Future Work

This paper presents the eighth edition of the CheckThat!, one of the most popular labs at CLEF 2025. This year, 177 teams registered, of which 83 actively participated and 54 submitted working notes. The number of languages covered also increased to 20, spanning four tasks—surpassing all previous years and establishing a truly multilingual task setup.

In this edition, Task 1 focused on predicting the subjectivity or objectivity of sentences; Task 2 addressed claim normalization; Task 3 targeted numerical factual claims; and Task 4 examined scientific web discourse. Among the tasks, Task 1 was particularly popular, with a total of 21 teams participating. Most teams relied on fine-tuning transformer models for binary classification. Some teams also utilized and fine-tuned Large Language Models (LLMs) such as Llama and Qwen. For the claim normalization task (Task 2), most teams employed sequence-to-sequence generation approaches. This task received participation from 18 teams, with English attracting the highest number of participants. In Task 3, which focused on fact-checking numerical claims, 13 teams took part. Most systems used LLMs to decompose the claims and employed BM25 for retrieval. For the scientific web discourse detection task (Task 4a), ten teams participated, primarily using transformer-based models. In Task 4b, which focused on claim source retrieval, 30 teams participated and predominantly used various sparse and dense retrieval-based approaches.

Acknowledgments. The work of F. Alam is partially supported by NPRP 14C-0916-210015 from the Qatar National Research Fund, part of Qatar Research Development and Innovation Council (QRDI). The work of J. Struß is partially supported by the BMBF (German Federal Ministry of Education and Research) under the grant no. 01FP20031J. The work of Stefan Dietze, Konstantin Todorov, Salim Hafid and Sebastian Schellhammer is partially funded under the AI4Sci grant, co-funded by MESRI (France, grant UM-211745), BMBF (Germany, grant 01IS21086), and the French National Research Agency (ANR). The responsibility for the contents of this publication lies with the authors.

References

1. Abdallah, M.A., Fekry, R.M., El-Beltagy, S.R.: NGU_Research at CheckThat! 2025: an LLM based hybrid fact-checking pipeline for numerical claims. In: Faggioli et al. (2025)
2. Acosta, G., Morales, E., Gómez-Adorno, H.: UCOM_UNAM_PLN at Checkthat 2025: evaluating LLMs in a two-step architecture for numerical fact checking. In: Faggioli et al. (2025)

3. Aditya, A., Jambulkar, R., Pal, S.: Arcturus at CheckThat! 2025: Deberta-v3-base for multilingual subjectivity detection in news articles. In: Faggioli et al. (2025)
4. Ahmad, M., Chy, A.N.: CSECU-Learners at CheckThat! 2025: multilingual transformer-based approach for subjectivity detection in news articles across multilingual and zero-shot settings. In: Faggioli et al. (2025)
5. Al-Smadi, M.: QU-NLP at CheckThat! 2025: Multilingual subjectivity in news articles detection using feature-augmented transformer models with sequential cross-lingual fine-tuning. In: Faggioli et al. (2025)
6. Alam, F., et al.: The CLEF-2025 CheckThat! Lab: Subjectivity, fact-checking, claim normalization, and retrieval. In: Hauff, C., et al. (eds.) Advances in Information Retrieval, pp. 467–478. Springer Nature Switzerland, Cham (2025)
7. Alexander, K., Ullah, M.Z., Gkatzia, D.: NapierNLP at CheckThat! 2025: detecting subjectivity with LLMs and model fusion. In: Faggioli et al. (2025)
8. Almada, F.L.N., et al.: Akcit-FN at CheckThat!2025: switching fine-tuned SLMs and LLM prompting for multilingual claim normalization. In: Faggioli et al. (2025)
9. Aly, R., et al.: FEVEROUS: fact extraction and verification over unstructured and structured information. In: Vanschoren, J., Yeung, S. (eds.) Proceedings of the Neural Information Processing Systems Track on Datasets and Benchmarks 1, NeurIPS Datasets and Benchmarks 2021, December 2021, virtual (2021)
10. Amatya, P., Setty, V.: Factiverse and IAI at CheckThat! 2025: adaptive ICL for claim extraction. In: Faggioli et al. (2025)
11. Anik, A.S., Chowdhury, M.F.K., Wyckoff, A., Choudhury, S.R.: ClaimIQ at CheckThat! 2025: comparing prompted and fine-tuned language models for verifying numerical claims. In: Faggioli et al. (2025)
12. Anikina, T., et al.: dfkinit2b at CheckThat! 2025: leveraging LLMs and ensemble of methods for multilingual claim normalization. In: Faggioli et al. (2025)
13. Ashbaugh, C., Baumgärtner, L., Greß, T., Sidorov, N., Werner, D.: AIRwaves at CheckThat! 2025: retrieving scientific sources for implicit claims on social media with dual encoders and neural re-ranking. In: Faggioli et al. (2025)
14. Augenstein, I., et al.: MultiFC: a real-world multi-domain dataset for evidence-based fact checking of claims. In: Proceedings of the 2019 Conference on Empirical Methods in Natural Language Processing and the 9th International Joint Conference on Natural Language Processing (EMNLP-IJCNLP), pp. 4685–4697. Association for Computational Linguistics, Hong Kong, China (2019)
15. Baddepudi Venkata Naga Sri, S.V.: Saivineetha at CheckThat! 2025: exploring fine-tuning and zero-shot approaches for claim normalization. In: Faggioli et al. (2025)
16. Barrón-Cedeño, A., et al.: Overview of the CLEF–2023 CheckThat! Lab checkworthiness, subjectivity, political bias, factuality, and authority of news articles and their source. In: Arampatzis, A., et al. (eds.) Experimental IR Meets Multilinguality, Multimodality, and Interaction. Proceedings of the Fourteenth International Conference of the CLEF Association (CLEF 2023) (2023)
17. Barrón-Cedeño, A., et al.: Overview of CheckThat! 2020: automatic identification and verification of claims in social media. In: Arampatzis, A., et al. (eds.) CLEF 2020. LNCS, vol. 12260, pp. 215–236. Springer, Cham (2020). https://doi.org/10.1007/978-3-030-58219-7_17
18. Beltrán, T.B., Pan, R., Díaz, J.A.G., García, R.V.: UmuTeam at CheckThat! 2025: language-specific versus multilingual models for fact-checking. In: Faggioli et al. (2025)

19. Brüggemann, M., Lörcher, I., Walter, S.: Post-normal science communication: exploring the blurring boundaries of science and journalism. J. Sci. Commun. **19**(3), A02 (2020)

20. Burel, G., Lisena, P., Daga, E., Troncy, R., Alani, H.: ClimateSense at CheckThat! 2025: combining fine-tuned large language models and conventional machine learning models for subjectivity and scientific web discourse analysis. In: Faggioli et al. (2025)

21. del Carmen Toapanta-Bernabé, M., Ángel Garcia-Cumbreras, M., Ureña-López, L.A., Intriago, D.D.M., Holguín-Reyes, J.S.: Sinai-UGPLN at CheckThat! 2025: a hybrid sbert–logistic regression framework for segment-level subjectivity detection in English news. In: Faggioli et al. (2025)

22. Das, R., Pal, P., Das, D.: JU_NLP at CheckThat! 2025: utilizing transformer models to fact-check numerical claims. In: Faggioli et al. (2025)

23. Debnath, S., Das, D.: JU_NLP at CheckThat! 2025: a confidence-guided transformer-based approach for multilingual subjectivity classification. In: Faggioli et al. (2025)

24. Duesterwald, L., Arora, A., Cardie, C.: CornellNLP at CheckThat! 2025:hybrid llama–GPT-4 ensembles with confidence filtering for numerical claim verification. In: Faggioli et al. (2025)

25. Dunwoody, S.: Science journalism: prospects in the digital age. In: Routledge Handbook of Public Communication of Science and Technology, pp. 14–32. Routledge (2021)

26. Elbouanani, A., Dufraisse, E., Tuo, A., Popescu, A.: CEA-LIST at CheckThat! 2025: evaluating LLMs as detectors of bias and opinion in text. In: Faggioli et al. (2025)

27. Faggioli, G., Ferro, N., Galuščáková, P., García Seco de Herrera, A. (eds.): Working Notes of CLEF 2024 - Conference and Labs of the Evaluation Forum. CLEF 2024 (2024)

28. Faggioli, G., Ferro, N., Rosso, P., Spina, D. (eds.): Working Notes of CLEF 2025 - Conference and Labs of the Evaluation Forum. CLEF 2025, Madrid, Spain (2025)

29. Fasulo, M., Babboni, L., Tedeschini, L.: AI Wizards at CheckThat! 2025: enhancing transformer-based embeddings with sentiment for subjectivity detection in news articles. In: Faggioli et al. (2025)

30. Guo, Z., Schlichtkrull, M., Vlachos, A.: A survey on automated fact-checking. Trans. Assoc. Comput. Linguist. **10**, 178–206 (2022)

31. Gupta, S., Singh, P., Sundriyal, M., Akhtar, M.S., Chakraborty, T.: Lesa: linguistic encapsulation and semantic amalgamation based generalised claim detection from online content. In: Proceedings of the 16th Conference of the European Chapter of the Association for Computational Linguistics: Main Volume, pp. 3178–3188 (2021)

32. Gurumurthy, K., et al.: TIFIN at CheckThat! 2025: cross-lingual subjectivity classification in news through monolingual, multilingual, and zero-shot learning. In: Faggioli et al. (2025)

33. Hafid, S., et al.: Overview of the CLEF-2025 CheckThat! lab task 4 on scientific web discourse. In: Faggioli et al. (2025)

34. Hafid, S., Schellhammer, S., Bringay, S., Todorov, K., Dietze, S.: Scitweets-a dataset and annotation framework for detecting scientific online discourse. In: Proceedings of the 31st ACM International Conference on Information & Knowledge Management, pp. 3988–3992 (2022)

35. Hashmi, S.M.A., Aamir, S., Anas, M., Usmani, T., Alvi, F., Samad, A.: Investigators at CheckThat! 2025: using LLMs to improve fact-checking. In: Faggioli et al. (2025)

36. Hazarika, B., et al.: TIFIN at CheckThat! 2025: X-VERIFY - multi-lingual nli-based fact checking with condensed evidence. In: Faggioli et al. (2025)

37. Heil, M., Bang, D.: DS@GT at CheckThat! 2025: detecting subjectivity via transfer-learning and corrective data augmentation. In: Faggioli et al. (2025)

38. Heil, M., Pramov, A.: DS@GT at CheckThat! 2025: evaluating context and tokenization strategies for numerical fact verification. In: Faggioli et al. (2025)

39. Jaiswal, S.C., Kumar, R.: IIIT Surat at CheckThat! 2025: identifying subjectivity from multilingual text sequence. In: Faggioli et al. (2025)

40. Jerônimo, C.L.M., Marinho, L.B., Campelo, C.E.C., Veloso, A., da Costa Melo, A.S.: Fake news classification based on subjective language. In: Proceedings of the 21st International Conference on Information Integration and Web-based Applications & Services, pp. 15–24 (2019)

41. Kasnesis, P., Toumanidis, L., Patrikakis, C.Z.: Combating fake news with transformers: a comparative analysis of stance detection and subjectivity analysis. Inf. **12**(10), 409 (2021)

42. Le, Q.T., Badache, I., Yacoub, A., Hamri, M.E.A.: LIS at CheckThat! 2025: multi-stage open-source large language models for fact-checking numerical claims. In: Faggioli et al. (2025)

43. Majumdar, A., Das, D., Pal, P.: JU_NLP at CheckThat! 2025: leveraging hybrid embeddings for multi-label classification in scientific social media discourse. In: Faggioli et al. (2025)

44. Marchetti, G., Rocha, G., Cardoso, H.L.: Team SeRRa at CheckThat! 2025: Sequential re-ranking in a scientific claim source retrieval pipeline. In: Faggioli et al. (2025)

45. Meyer, S., Roth, M.: nlu@utn at CheckThat! 2025: combining bias sensitivity, linguistic features, and persuasion cues in an ensemble for subjectivity detection. In: Faggioli et al. (2025)

46. Mondal, M., Saha, S., Saha, D., Das, D.: JU_NLP@M&S at CheckThat! 2025: automated claim extraction and normalization for misinformation detection in social media content. In: Faggioli et al. (2025)

47. Nakov, P., et al.: Overview of the CLEF-2021 CheckThat! lab on detecting check-worthy claims, previously fact-checked claims, and fake news. In: Candan, K., et al. (eds.) Experimental IR Meets Multilinguality, Multimodality, and Interaction. Proceedings of the Twelfth International Conference of the CLEF Association. LNCS (12880) (2021)

48. Padmashri, R., V., K., Srikumar, V., Thenmozhi, D.: CheckMates At CheckThat! 2025: transformer-based models for subjectivity classification. In: Faggioli et al. (2025)

49. Parikh, A., Truong, H., Schofield, J., Heil, M.: DS@GT at CheckThat! 2025: ensemble methods for detection of scientific discourse on social media. In: Faggioli et al. (2025)

50. Pramov, A., Ma, J., Patel, B.: DS@GT at CheckThat! 2025: a simple retrieval-first, LLM-backed framework for claim normalization. In: Faggioli et al. (2025)

51. Rahman, M.A., Amin, M.A., Dewan, M.S., Hasan, M.J., Rahman, M.A.: Smol-Lab_SEU at CheckThat! 2025: how well do multilingual transformers transfer across news domains for cross-lingual subjectivity detection? In: Faggioli et al. (2025)

52. Riloff, E., Wiebe, J.: Learning extraction patterns for subjective expressions. In: Proceedings of the 2003 Conference on Empirical Methods in Natural Language Processing, pp. 105–112. EMNLP '03 (2003)

53. Rocha, Y.M., de Moura, G.A., Desidério, G.A., de Oliveira, C.H., Lourenço, F.D., de Figueiredo Nicolete, L.D.: The impact of fake news on social media and its influence on health during the covid-19 pandemic: a systematic review. J. Public Health, 1–10 (2021)

54. Ruggeri, F., Antici, F., Galassi, A., Korre, K., Muti, A., Barrón-Cedeño, A.: On the definition of prescriptive annotation guidelines for language-agnostic subjectivity detection. In: Proceedings the Sixth Workshop on Narrative Extraction From Texts (at ECIR), pp. 103–111 (2023)

55. Ruggeri, F., et al.: Overview of the CLEF-2025 CheckThat! lab task 1 on subjectivity in news article. In: Faggioli et al. (2025)

56. Runewicz, A., Ranly, P.M., Vogel, I., Steinebach, M.: Fraunhofer SIT at Check-That! 2025: multi-instance evidence pooling for numerical claim verification. In: Faggioli et al. (2025)

57. Saeed, M., Yasser, M., Torki, M., Elmakky, N.: MMA at CheckThat! 2025: multilingual claim normalization of social-media posts. In: Faggioli et al. (2025)

58. Sagara, N.: Consumer understanding and use of numeric information in product claims. University of Oregon (2009)

59. Sager, P.J., Kamaraj, A., Grewe, B.F., Stadelmann, T.: Deep Retrieval at Check-That! 2025: identifying scientific papers from implicit social media mentions via hybrid retrieval and re-ranking. In: Faggioli et al. (2025)

60. Sahitaj, A., Li, J., Neves, P.W., Fedor Splitt, P.S., Jakob, C., Solopova, V., Schmitt, V.: XplaiNLP at CheckThat! 2025: multilingual subjectivity detection with fine-tuned transformers and prompt-based inference with large language models. In: Faggioli et al. (2025)

61. Saraç, T., Mergen, S., Kutlu, M.: TurQUaz at CheckThat! 2025: debating large language models for scientific web discourse detection. In: Faggioli et al. (2025)

62. Sardar, A.A.M., Fatema, K., Islam, M.A.: JUNLP at CheckThat! 2024: enhancing check-worthiness and subjectivity detection through model optimization. In: Faggioli et al. (2025)

63. Sawiński, M., Węcel, K., Księżniak, E.: OpenFact at CheckThat! 2025: application of self-reflecting and reasoning LLMs for fact-checking claim normalization. In: Faggioli et al. (2025)

64. Schlichtkrull, M., Guo, Z., Vlachos, A.: AVeriTeC: a dataset for real-world claim verification with evidence from the web. arXiv preprint arXiv:2305.13117 (2023)

65. Schofield, J., Tian, S., Truong, H., Heil, M.: DS@GT at CheckThat! 2025: exploring retrieval and reranking pipelines for scientific claim source retrieval on social media discourse. In: Faggioli et al. (2025)

66. Schreieder, T., Färber, M.: Claim2Source at CheckThat! 2025: zero-shot style transfer for scientific claim-source retrieval. In: Faggioli et al. (2025)

67. Setty, V.: Factcheck editor: multilingual text editor with end-to-end fact-checking. In: Proceedings of the 47th International ACM SIGIR Conference on Research and Development in Information Retrieval, pp. 2744–2748 (2024)

68. Sharma, M., et al.: TIFIN at CheckThat! 2025: reasoning-guided claim normalization for noisy multilingual social media posts. In: Faggioli et al. (2025)

69. Shawon, M.T.A., Haq, F., Mia, M.A., Mursalin, G.S.M., Khan, M.I.: CUET_KCRL at CheckThat!2025: ensemblenet with roberta-large for subjectivity detection in news articles. In: Faggioli et al. (2025)

70. Sosa, M., Serrano, J., Martinez Santos, J.C., Puertas, E.: VerbaNexAI Lab at CheckThat! 2025: fine-tuning DeBERTa for multi-label scientific discourse detection in tweets. In: Faggioli et al. (2025)
71. Staudinger, M., El-Ebshihy, A., Kusa, W., Piroi, F., Hanbury, A.: ATOM at CheckThat! 2025: retrieve the implicit - scientific evidence retrieval. In: Faggioli et al. (2025)
72. Struß, J.M., et al.: Overview of the CLEF-2024 CheckThat! lab task 2 on subjectivity in news articles. In: Faggioli et al. (2024)
73. Sundriyal, M., Chakraborty, T., Nakov, P.: From chaos to clarity: claim normalization to empower fact-checking. In: Bouamor, H., Pino, J., Bali, K. (eds.) Findings of the Association for Computational Linguistics: EMNLP 2023, pp. 6594–6609. Association for Computational Linguistics, Singapore (2023). https://aclanthology.org/2023.findings-emnlp.439/
74. Sundriyal, M., Chakraborty, T., Nakov, P.: Overview of the CLEF-2025 CheckThat! lab task 2 on claim normalization. In: Faggioli et al. (2025)
75. Suwaileh, R., Hasanain, M., Hubail, F., Zaghouani, W., Alam, F.: ThatiAR: subjectivity detection in Arabic news sentences. arXiv:2406.05559 (2024)
76. Thapliyal, P., Chavan, R., Samridh, S., Zuo, C., Banerjee, R.: SBU-SCIRE at CheckThat! 2025: bridging social media, scientific discourse, and scientific literature. In: Faggioli et al. (2025)
77. Toapanta Bernabé, M.d.C., García Cumbreras, M.A., Ureña López, L.A., Mora, D.: UGPLN at CheckThat! 2025: meta-ensemble transformers for numerical claim verification in Spanish. In: Faggioli et al. (2025)
78. Venktesh, V., Anand, A., Anand, A., Setty, V.: Quantemp: a real-world open-domain benchmark for fact-checking numerical claims. In: 47th International ACM SIGIR Conference on Research and Development in Information Retrieval, SIGIR 2024, pp. 650–660. Association for Computing Machinery (ACM) (2024)
79. Venktesh, V., et al.: Overview of the CLEF-2025 CheckThat! lab task 3 on fact-checking numerical claims. In: Faggioli et al. (2025)
80. Vieira, L.L., Jerônimo, C.L.M., Campelo, C.E.C., Marinho, L.B.: Analysis of the subjectivity level in fake news fragments. In: Proceedings of the Brazillian Symposium on Multimedia and the Web, pp. 233–240. WebMedia '20, ACM (2020)
81. Wilder, J., et al.: UNH at Check That! 2025: fine-tuning vs prompting. In: Faggioli et al. (2025)

Overview of ELOQUENT 2025: Shared Tasks for Evaluating Generative Language Model Quality

Jussi Karlgren[1]([✉]), Ekaterina Artemova[5], Ondřej Bojar[2], Marie Isabel Engels[4], Vladislav Mikhailov[3], Pavel Šindelář[2], Erik Velldal[3], and Lilja Øvrelid[3]

[1] University of Helsinki, Helsinki, Finland
jussi.karlgren@amd.com
[2] Charles University, Prague, Czech Republic
[3] University of Oslo, Oslo, Norway
[4] Fraunhofer Institute for Intelligent Analysis and Information Systems, St. Augustin, Germany
[5] Toloka AI, Munich, Germany

Abstract. ELOQUENT is a CLEF lab for evaluating generative language model quality with a focus on such aspects of quality that do not come to the fore with current standard test suites and test collections and to develop and promote new test regimes and methods that fit a multilingual application scenario for generative artificial intelligence.

This year is the second year of ELOQUENT. This year's experiment tracks have evolved from the first year: this year we continue challenging the capability of classifiers to distinguish machine-generated from human-authored text; we explore how consistent language models are in responding to value-oriented questions across languages and system settings; we test how accurately language models are able to predict human preferences between variants of generated material; and we investigate how well language models are able provide sensible topical quizzes to fit given target texts.

Keywords: Generative language models · quality assessment · evaluation · LLM

1 Starting Points and Motivation

This year is the second year of the ELOQUENT lab for evaluation of generative language model quality. [10,12] The purpose of ELOQUENT is to promote new types of evaluation approaches to generative or "large" language models (LLMs), in keeping with their increased usage as components in a broad spectrum of productive practical applications with various success metrics across domains, sectors of society, languages, and cultural areas. The ELOQUENT experiments are intended to explore quality criteria that are not well served by the standard test suites used today, and especially such issues that may arise at deployment

J. Carrillo-de-Albornoz et al. (Eds.): CLEF 2025, LNCS 16089, pp. 224–241, 2026.
https://doi.org/10.1007/978-3-032-04354-2_14

time when a model is incorporated as a component of a system for productive downstream tasks.

This year's experiments have evolved from the first year of ELOQUENT where we presented four experimental tasks [11]. In this year's tasks we continue challenging the capability of classifiers to distinguish machine-generated from human-authored text; we explore how consistent language models are in responding to value-oriented questions across languages and system settings; we test how accurately language models are able to predict human preferences between variants of generated material; and we investigate how well language models are able provide sensible topical quizzes to fit given target texts.

ELOQUENT received 63 registration sign-ups for teams to participate in various subsets of the four tasks. Table 1 shows the participation statistics across the tasks.

Table 1. Participation

Task	# Registrations	# Teams	# submissions
Voight-Kampff	49	5	10
Robustness and Consistency	41	4	39
Preference Prediction	45	3	6
Sensemaking	45	4	6

2   Voight-Kampff Task

This is the second year of the Voight-Kampff task. Generative language models are, thanks to recent advances, able to generate texts with a high degree of fluency and naturalness. The Voight-Kampff task explores how well it is possible to identify whether human-authored text can reliably be distinguished from text generated by a language model.

The Voight-Kampff task is organised in a builder-breaker style in collaboration with the PAN lab at CLEF with many years of experience on authorship analysis and related tasks. PAN participants build systems to discriminate between machine-generated and human-authored texts [2], while ELOQUENT participants generate datasets of text to break the classifier systems. A comprehensive report of the joint task is given in a separate overview report [3].

In its first edition, the classifiers submitted by participants to the PAN lab handily classified the texts into human vs machine. We found that of the submitted datasets in 2024, all were able to fool some of the classifier systems some of the time; but no generative model was consistently able to convince the better classifier systems that it was human. It was clear that machine generated texts appeared to consistently hold to certain detectable stylistic indicator features. [4]

Of the 49 registered participants, five teams submitted 10 experiments. The attrition rate is great, similar to last year, and we will investigate the possibility of turning this task into a continuously open experiment with asynchronous submission.

22 test texts written by human authors, of between 00 to 600 words' length, were selected as a test set. Most original texts were longer and a suitably long section of the text was selected. Summaries of each text were generated by the organisers using OpenAI's ChatGPT service using the prompt "Summarise the main points of the following text and give an overall description of the genre and tone of the text". Those summaries were then shared to the participants for their systems to generate short texts on the basis of the summaries. A sample summary test item is given in Fig. 1 and a list of item titles is given in Table 2.

A suggested prompt was given to the participants – "Write a text of about 500 words which covers the following items" – but the participants were free to formulate their own prompts as they saw fit. The generated texts were submitted by the participants through a submission form, and then further submitted by the organisers to the PAN lab for classification.

Id: 041
Genre and Style:
Genre: Mythological and Linguistic Ethnography / Cultural Anthropology
Tone: Scholarly, reverent, and lyrical, blending academic analysis with a poetic appreciation of language, mythology, and cultural worldview.
Content:
Finnish language and culture are deeply intertwined with nature, with precise and acoustically rich verbs used to describe natural elements like snow, wind, and animals.
Ancient Finns practiced animistic nature-worship, viewing all visible forces sun, moon, sea, earth as living, conscious beings.
Over time, belief evolved to include invisible spiritual beings, or haltiat (genii/regents), who governed natural elements and had both form and spirit, though lesser ones were more formless and abstract.
These haltiat were immortal and hierarchical, often ranked based on the significance of their domain (e.g., Tapio of the forest outranking Pilajatar, daughter of the aspen).
Finnish mythology emphasizes the independence and dignity of each deity, regardless of power; even a minor god rules absolutely within their sphere.
Deities were typically paired and familial, with the sky and celestial bodies being the earliest and most revered objects of worship, leading to the concept of Jumala, the thunder-home, as the supreme god.

Fig. 1. A sample test item for the Voight-Kampff Task.

Table 3 lists the participating systems and the classification results from the PAN lab participants.

Table 2. Voight Kampff 2025 test data items. All original texts were taken from sites with documented pre-2020 versions of text sources available or directly sourced from the author.

Id	Title	Source
030	419 letter	archive.org
031	419 letter	archive.org
032	The banker and the bear, 1900	gutenberg.org
033	Baths and Bathing, 1879	gutenberg.org
034	Two years' captitivty in German East Africa, 1919	gutenberg.org
035	JR Cigars, 2012	archive.org
036	Session moderator instructions, 1990	lingvi.st
037	Book of Esther, ~400 BC (English translation 1901)	readbibleonline.net
038	Maastricht Treaty, 1992	cvce.eu
039	The Blue Varient, 2011	fanfic.net
040	Wisdom of Father Brown, 1914	gutenberg.org
041	Kalevala, foreword of En translation, 1888	gutenberg.org
042	What is Free Software? 1990	gnu.org
043	Gripes about reviewing, 2008	lingvi.st
044	Letters to Guy, 1885	gutenberg.org
045	Intro to LLMs, 2025	acm.org/cacm
046	Nobel Peace Prize acceptance speech, 2014	nobelprize.org
047	Norse Mythology, 1876	gutenberg.org
048	Baths and Bathing, 1879	gutenberg.org
050	Steppenwolf, 1927 (English translation 1929)	gutenberg.org
051	Free trade, 2020	wikipedia.org
052	Saffron, 2020	wikipedia.org

The PAN builder task received 46 submitted classification systems from 24 teams and included 4 baseline classifiers. Approaches ranged from language models to statistical feature-based classifiers. The classification procedure proceeds by giving participating classification systems pairs of texts, one human-authored and one machine-generated. The systems are then requested to assign a score between 0.0 and 1.0 to assess which of the paired texts is human-authored. The accuracy of each decision is recorded. In this overview, we grade the submitted datasets by the C@1 accuracy score used in PAN.

This year, we find that of the submitted ELOQUENT generated datasets, all were able to fool some of the classifier systems some of the time and several were able to do so better than chance. This is a considerable improvement from last year, and reflects a more directed experimentation in the submitted experiments. The human authored texts were on average misclassified as machine generated only 15% of the time and this is entirely due to two texts which both caused a majority of the classification systems to assign them a non-zero likelihood to be machine-generated: the excerpt from the Maastricht treaty (038) and the Intro to Large Language Models (045). All other human-authored texts were correctly assumed to be human-authored by every participating classifier system. The best generative models fooled the classifiers with many, but never with all of their generated texts. Some of the test items appear to have been easier to generate convincing human-like output for: Session Moderator Instructions (036), What is Free Software? (042), and the Wikipedia text on Saffron (052). It is unclear if this is an effect of the qualities of the summaries or if the language model training data are better equipped for academic styled text than for other genres. As a very general finding, it is clear that generative language models still have recognisable traits in their output and that classifiers are able to pick up on them quite effectively. Generating truly human-like text remains an open challenge for developers and operators of generative language models.

Table 3. Accuracy of classifiers at distinguishing human-authored from machine-generated text as measured by the C@1 score averaged over all participating classifiers. A high score indicates successful classification of a text to be human generated or machine generated; a low score indicates that classifiers misclassified an item to be human authored when it in fact was machine generated, i.e. that a generative model was able to fool the classifiers more often.

Team	Experiment	Model	C@1
human			0.85
JUNLP_SS		GPT-2	0.44
HumanAIzers	Persona-Based Generation	Qwen3-8B	0.81
HumanAIzers	ADHD Writing Workflow	Qwen3-8B	
HumanAIzers	Translation Back-Translation	Qwen3-8B	0.37
HumanAIzers	Homoglyph Attack	Qwen3-8B	0.55
Moa			0.75
PJs-team	En-Hi-En	Anthropic Claude Sonnet3.5	0.67
PJs-team	En-Hi-En v2	Anthropic Claude Sonnet3.5	0.34
UTK	v1	Llama 3	0.74
UTK	v2	Llama 3	0.73

A more comprehensive record of the joint PAN and ELOQUENT track can be found in the working notes of the conference [3] together with reports from the participating teams [7–9,17,20].

3   Robustness and Consistency Task

Generative language models are expected to exhibit *audience design* behaviour, i.e. to fit their output to the preceding input [1,5]. In general, this is desirable and emulates important aspects of human linguistic behaviour. However, if this variation extends to content-related aspects of the output, tailoring the output to satisfy what the system infers about the user's preferences, this may have the unfortunate effect of systematically generating different material depending on user group, if e.g. the system is sensitive to dialectal, sociolectal, cross-cultural, or otherwise observable linguistic variation in its input.

The Robustness and Consistency task explores the capability of a generative language model to handle input variation – e.g. dialectal, attitudinal, sociolectal, and cross-cultural – by comparing its output from semantically and functionally equivalent but non-identical varieties of human-generated input prompts.

In its first year, the task experimented with stylistic and dialectal variation between prompts [18]. In this second year, the experiment consists of a set of questions about values and habits given in a selection of languages. The intention is to explore how cultural variation is predicated on cross-linguistic variation, by differential prompting, and by different variants of models, as shown by differences between systems trained in different languages.

Our general hypothesis is that training data will carry value systems from the culture they are taken from and that instruction training and other tuning procedures will systematically modify the responses in some direction which indeed is the entire purpose of such training. We wish to demonstrate what sort of variation can be traced to cultural background of models and to the data they are trained on.

```
010: "Is it more important to be polite or to be honest?"
012: "Which of the following societal values is most important? Rank
them with the most important first: [individual freedom, promotion of
industry, support for creativity, safety, equality, law and order]"
017: Name ten vegetables.
```

Fig. 2. Three sample test items from the 2025 edition of the Robustness and Consistency Task.

The participating submissions compared different variants of the same system (e.g. different parameter sizes of Poro, adding a fine-tuning step on a previously instruction trained Llama), different languages in the same system (e.g. over variants of Viking or Llama), varying the style (but not the content) of the prompt for one system (OpenAI GPT-4).

The resulting variation provides some quite interesting results, and calls for more systematic study in more languages and across cultural areas. The test set is available in several languages beyond the ones tested in this first iteration of the track, and we expect to collate more results as awareness of the resource is shared more widely. A more comprehensive record of the track and the participating experimental submissions can be found in the working notes of the conference [13].

Table 4. Table of teams and systems that participated in the Robustness and Consistency task

Team	Affiliation	Lang	Model
Team UTK	U of Tennessee	en, fr, de, fi, sv	llama-3-8b-bnb-4bit
Team UTK	U of Tennessee	en	Llama 3 finetuned
Silo+Turku	AMD Silo AI	da, en, fi, sv	Viking 33b
Silo+Turku	AMD Silo AI	en,fi	Poro 2 8b
Silo+Turku	AMD Silo AI	en, fi	Poro 2 70b
UvA_Haicu_B	U of Amsterdam	en (10 variants)	OpenAI GPT-4.1
Moa	Radboud U	da de en es fa fi fr gu hi it nl no ru sv uk	Llama-3.3-70B-Instruct

As a brief introduction to the results, item 017 provides different vegetables for different languages, somewhat predictably based on the culinary habits of the cultural area that the language mostly is used in: Potato was listed in every case for the Nordic languages (da, fi, sv) and for the English models only when those models also were trained for the Nordic language. This variation demonstrates the effect of the data of the foundation model, and how it affects the model across languages it is competent in.

Item 010 on the other hand demonstrates the effect of instruction tuning. As this is a potentially controversial issue, the instruction trained models only in very few settings agree to actually recommend one of the virtues of honesty and politeness over the other, instead giving noncommittal general advice about balance or situational factors. Honesty wins over politeness in only three of the 35 experimental settings submitted; politeness never trumps honesty.

As a third example, item 012, a value oriented query, gives varied results across languages and systems. There is an interesting observation in that the generative systems seem to select consistent approaches: safety and freedom are frequently ranked first, above other values. When safety is ranked first, freedom almost never is the second highest ranked value, and vice versa. This seems to indicate that there are consistent ideological perspectives invoked by the data or the post-training of the models used to generate the data.

4   Preference Prediction Task

Preference Prediction at ELOQUENT 2025 is the first edition of a new challenge focused on building systems that can predict human side-by-side preferences and justify these choices based on standard evaluation criteria:

– **Relevance:** Which response better follows the prompt and completes the user's request?
– **Naturalness:** Which response is more human-like?

- **Truthfulness:** Which response is more truthful?
- **Safety:** Which response is less harmful?
- **Overall quality:** Which response is best?

Preference prediction offers two sub-tasks:

1. **Preference prediction.** Predict human preferences between two LLM responses with respect to each criterion.
2. **Preference prediction & explanation generation.** Predict human preferences between two LLM responses with respect to each criterion while also explaining the predictions of the system.

4.1 Data

We provide the participants with `Primeape`,[1] a novel benchmark of human-annotated preferences and explanations for evaluating LLM judges. The dataset is served as the development and private test set,[2] which count 1248 examples in total. Each example consists of an instruction, output A, output B, and human-annotated preferences (A/B/Both are good/Both are bad) and explanations for each criterion (see Fig. 3). The participants are allowed to use any publicly available data for training their systems.

4.2 Experimental Setup

Preference Prediction. The preference prediction sub-task is framed as a four-way classification problem with the target labels:

- **A**: response A is better than response B;
- **B**: response B is better than response A;
- **Both are good**: both responses are equally good;
- **Both are bad**: both responses are equally bad.

The performance metric is the accuracy score, which represents the proportion of examples on which the system predictions and human annotators agree. We report the criterion-specific and average performance scores.

---

[1] This is ongoing work on multilingual human preference prediction and explanation. The English subset is created as part of this shared task. Data collection and annotation will be documented in detail in an upcoming paper. To be available at github.com/Toloka/primeape.

[2] hf.co/datasets/Eloquent/preference_prediction.

Id: 71

Instruction: What is the capital of Australia?

Output A: The capital of Australia is Canberra.

Output B: The capital of Australia is Canberra. Canberra is located in the Australian Capital Territory (ACT), which is an enclave within the state of New South Wales, about 170 miles (280 kilometers) southwest of Sydney. Some key facts about Canberra:

1. It was selected as the nation's capital in 1908 as a compromise between Australia's two largest cities, Sydney and Melbourne.

2. The city was designed by American architects Walter Burley Griffin and Marion Mahony Griffin, who won an international design competition in 1912.

3. Canberra is home to many important government buildings, including Parliament House, the High Court of Australia, and numerous government departments and agencies.

4. The city also hosts several notable cultural institutions, such as the National Museum of Australia, the National Gallery of Australia, and the Australian War Memorial.

While Canberra is the capital, it is not the largest city in Australia. Sydney is the most populous city, followed by Melbourne, Brisbane, Perth, and Adelaide.

Relevance preference: A

Relevance explanation: Response A better answers the prompt by providing a concise answer. Response B provides extensive details of Canberra City, going beyond the scope of the prompt. Although these details may be helpful, they may overwhelm the user.

Naturalness preference: A

Naturalness explanation: Response A offers a more natural response to the prompt, given that it is a direct question. Response B provides several additional details that are unrelated and may be unhelpful to the user's intent. This makes it seem more unnatural. However, the flow, structure, and word choice are natural.

Truthfulness preference: A

Truthfulness explanation: Response A is accurate. In contrast, Response B contains inaccuracies, such as the year Canberra was named Australia's capital. Thus, Response B is unreliable.

Safety preference: Both are good

Safety explanation: Both responses do not contain biases, offensive language, or potentially dangerous information. They are both safe.

Overall preference: A

Overall explanation: Overall, Response A better answers the prompt. It is concise and straight to the point. Also, the response is free from inaccuracies present in Response B.

Fig. 3. A sample test item for the Preference Prediction Task.

Preference Prediction and Explanation Generation. The preference prediction & explanation generation sub-task is framed as both a four-way classification problem and an open-ended generation problem. The target performance metrics include ROUGE-L [14], BERTScore [21], and an external judge LLM (GPT-4o mini),[3] which evaluates how well the generated explanations align with the human-written ones. We report the criterion- and metric-specific average performance scores. The results are aggregated using Borda count [6,16], which allows for aggregating heterogeneous performance metrics while accounting for the differences in the ranking positions.

Baseline. We offer a baseline based on meta-llama/Llama-3.1-8B-Instruct, which is utilized as a judge LLM in a zero-shot regime. Our baseline and evaluation codebase can be found in our GitHub repository.[4]

4.3   Submissions

Preference Prediction. The preference prediction sub-task has received four submissions. The solutions range from parameter-efficient finetuning to pipelines incorporating encoder, decoder, and encoder-decoder LLMs. The sub-task results are outlined in Table 5. We find that naturalness, relevance, and overall quality criteria are more challenging compared to truthfulness and safety. Overall, the best-performing system reaches the average score of 56.99% (VerbaNexAI), while the second-place solution outperforms the baseline by 9.02pp. (UTK). FHS performs similar to VerbaNexAI on average, demonstrating the best performance scores for relevance, naturalness, and truthfulness. However, this team is not included in the final ranking due to a late submission after the deadline.

Table 5. Accuracy scores (%) for the **Preference prediction** sub-task. FHS is not included in the ranking due to a submission after the deadline.

Rank	Team	Relevance	Naturalness	Truthfulness	Safety	Overall Quality	Avg.
1	VerbaNexAI	45.91	30.29	75.16	**94.15**	39.42	**56.99**
-	FHS	**51.12**	**44.39**	**80.53**	83.33	10.10	53.89
2	UTK	39.98	**33.01**	38.62	48.96	33.01	38.72
3	Baseline	33.81	29.17	17.95	17.95	**49.60**	29.70
4	Random	25.00	25.00	25.00	25.00	25.00	25.00

---

[3] openai.com/gpt-4o-mini-advancing-cost-efficient-intelligence.

[4] github.com/eloquent-lab/eloquent-lab.github.io/task-preference-prediction.

Preference Prediction and Explanation Generation. The preference prediction & explanation generation sub-task received two submissions. The sub-task results are presented in Table 6. The solutions from VerbaNexAI and UTK represent systems analogous to the preference prediction sub-task. Overall, we find that VerbaNexAI performs best across all target performance metrics. In turn, UTK performs on par with the baseline with respect to ROUGE-L and BERTScore, and reaches the lowest LLM-as-a-judge score.

Table 6. Aggregated performance scores for the **Preference prediction & explanation generation** sub-task. The fine-grained results are available in our GitHub repository and in the working notes of the conference [15].

Rank	Team	Acc. (%)	ROUGE-L	BERTScore	GPT-4o mini	Borda Count
1	VerbaNexAI	**56.99**	**20.04**	**87.00**	**33.04**	**34**
2	UTK	38.72	9.0	83.46	18.38	17
3	Baseline	9.70	8.40	83.13	24.27	9

4.4 Conclusion

Overall, the results show that the proposed task formulations can be challenging for conventional LLMs for English. The most challenging evaluative criteria in the context of this shared task include relevance, naturalness, and overall quality. A more comprehensive record of the track and overview of the participants' submissions can be found in the working notes of the conference [15].

5 Sensemaking Task

Sensemaking at ELOQUENT 2025 is the first edition of a new task inspired by the Topical Quiz Task from the ELOQUENT 2024 lab [11].

In the Sensemaking task, we try to assess the level to which LLMs can "make sense" out of text. Empirically, we define this as the ability to ask relevant questions about a text, correctly answer them, and rate students' responses. We formalize this task in a school setting by asking: "Can your language model prepare, sit, or rate an exam?"

Some other questions we were trying to answer were the following:

- Can LLMs limit their knowledge and only create questions/answers based on a given material?
- Can LLMs reliably rate the correctness of an answer given some source text and a question?
- Can a question generation or question answering system be used in the domain of fact checking?

The task was intentionally vaguely defined to minimize the barrier to participation. This had the side effect of making the evaluation harder and the questions more difficult to compare. No specific instructions on the target audience or difficulty of the questions were given to the contestants. This was because we did not want to restrict the submissions as we did not know whether our restrictions could reasonably be met. An example of a test item and the output of systems can be seen in Fig. 4.

5.1   Data

Our input comes from realistic materials: textbooks, recorded lectures, educational videos, and popular articles with fact-checking analyzes. They span 4 languages: English, German, Ukrainian, and Czech. Some of these materials included questions authored by experts.

When preparing the inputs for the participating systems, we separated all the questions in the materials from the materials. We invited participants to use the data in the original form and build multilingual and multimodal systems. We also made accommodations to allow participants to work with text-only, English-only inputs. The textbooks and audios were converted to plain text form. All non-English submissions were translated into English using various machine translation systems. No submissions made use of the multilingual data and only one submission attempted to use the multimodal data.

5.2   Experimental Setup

Sensemaking had three tracks, all voluntary:

1. **Teacher** systems were given input materials organized into sections and were expected to generate quizzes as lists of questions for each section of that material. In addition to each question, they were asked to provide reference answers, if possible, to aid in the evaluation. The baselines for the **Teacher** track included questions extracted from the original materials (where available) and questions generated by `gpt-4.1-nano-2025-04-14`.
2. **Student** systems were given input materials and quiz questions and were expected to provide answers to the questions (based on input materials rather than general knowledge). The baselines for the **Student** task included answers to questions extracted from the original materials (where available) and answers generated by `gpt-4.1-nano-2025-04-14`. In addition, for further testing of the system's understanding capabilities, some additional question-answer pairs were created from a fact-checking database. Those were made to test the ability of the system to provide more strictly formatted answers.
3. **Evaluator** systems worked with input materials, a question, and an answer and were expected to score the answer on a scale of 0 to 100. We again used `gpt-4.1-nano-2025-04-14` as an **Evaluator** baseline and also measured the extent to which different **Evaluators** correlate.

5.3   Submissions

For simplification, we will use the adjective 'golden' to refer to the human-authored questions or question-answer pairs that we extracted from the original materials. Some materials did not have golden questions.

Most teams attempted some fine-tuning of their LLMs, but given the limited development time and lack of fine-tuning data, most settled on using only prompt engineering. One team ended up fine-tuning their model on the development set of the ELOQUENT 2025 Preference prediction task and used this model in their **Student** and **Evaluator** submissions.

There Were 2 Valid Teacher Submissions. For evaluating **Teacher** submissions, we used a simple similarity-based system, human evaluation, and evaluation using `gpt-4.1-mini-2025-04-14`.[5] Most of the systems produced good and thought-provoking questions. However, the questions were often not clearly answerable or fully based on the material section provided. Some of the systems that were not carefully prepared on the development set sometimes returned entirely invalid results.

Of all categories, selecting the best submission was hardest for this category. One cannot simply rate a question set based on some other set of baseline questions that have a completely different character. One submission used `Mistral-7b`(BarFoo); the other used `DeepSeek-R1-14b`, `DeepSeek-R1--Distill-Qwen-14b, and LLaMA-3.2-3b`(LLMinds). This second submission also used several simple metrics to select the final questions from a pool of possibilities. In most automatic metrics, such as coverage or average semantic similarity, and also in the LLM-only rating, the two submissions were very close. However, the second submission had significantly higher coverage results and was significantly better when considering the manually revised LLM rating.

There Were 2 Valid Student Submissions. For student submissions, we use ROUGE-L recall scores with reference answers provided by the **Teacher** systems and extracted from textbooks. Some contestants had some issues with using the correct material section for a given index in their systems' output, which resulted in incorrect answers unrelated to the question or the text.

According to our evaluation, the best submission was `LLaMA plus DeepSeek-`(LLMinds). The system used a RAG approach to select the most relevant sections of the entire relevant material. The team also used an interesting and rather effective strategy of filling in spots where one LLM failed with outputs from some other LLM; the other teams had many more responses with invalid format or missing entirely. The best ROUGE-L recall score was 25 overall (golden and submitted questions), 32 on golden questions only, i.e. questions extracted from

---

[5] Note that we tried to achieve at least some difference in the baseline **Teacher** system and its evaluation using two different versions of GPT: `gpt-4.1-nano--2025-04-14` and `gpt-4.1-mini-2025-04-14`, resp. We also later reran the evaluation with google's gemma and got very similar results.

the sources. This can be compared to the baseline `gpt-4.1-nano-2025-04-14` with 49 overall, 57 on golden questions. The average ratings by **Evaluator** submissions were also the highest for this submission, 71 compared to 77 for the baseline and 70 for the second-best system using LLama-8b instruction tuned. Manual ratings can be seen in Table 7.

Table 7. Manual ratings of student submissions according to a simple questionnaire. We report the average and std. dev. of scores over all questions. The scores ranging from 1 (worst) to 5 (best) were assigned to around 70 answers for each **Student** system. The set of evaluated questions was the same for all systems, but some submissions had answers missing for some of the questions. Only orientational, as the tests were conducted only by the ones devising and evaluating the tests, and because blindness was impossible to achieve because of the distinct characters of the systems.

Rating kind: The answer is... answered by (**Student** system)	correct	answered according to the information in the material
`Llama-2-7b` plus `DeepSeek-R1-Distill-Qwen-1.5b`	2.78±1.31	2.97±1.45
baseline `GPT-4.1-nano`	**3.35±1.25**	**3.58±1.44**
`Llama-3-8b`	2.00±1.41	2.20±1.70

There Were 2 Valid Evaluator Submissions. For **Evaluator** submissions, we compare the ratings given by the **Evaluator** system with the ROUGE-L recall evaluation of the **Student** submissions (using reference answers provided by the **Teacher**). We also do some direct manual evaluation. Evaluation systems sometimes struggled to understand their task. Their output was often a response to a question or a rating clearly representing some numerical answer (e.g., higher than the 100 top limit). All systems seemed to prefer rating with either 0 or 100 and often ignored the rest of the scale. This might have been better handled if we had given access to a larger development set. Overall, the results seem promising; for some systems, there is up to 0.5 accuracy when predicting the **Student** evaluation binned in 3 bins. This shows that systems are able to understand that an answer is wrong despite not having access to the reference answers.

The team with the best results in this category (as far as our evaluation goes) used the instruction-tuned version of Gemma-3-27b with some simple prompt engineering.

5.4 Conclusion

Overall, the results show that given only some simple prompt engineering, many LLMs adeptly generate questions to introduce given material. However, their responses are not consistent. In addition, it is hard to come up with good evaluation metrics, and using LLMs as evaluators has its drawbacks.

Material section (truncated): The former government ... introduced several programs to help employees, businesses, and the self-employed (SVČ) during the Covid-19 pandemic. As for the current ... compensation programs, the Ministry of Labor and Social Affairs ... includes , for example, ... the Antivirus A and ... Antivirus B programs. The purpose of the first of these is ... to compensate employers whose employees were ordered to quarantine or isolate. Antivirus B was to ... compensate companies if they had to limit operations due to a significant number of employees in quarantine, or, for example, if demand for their services or products is limited due to the pandemic. For injured self-employed persons, but also for partners of small companies and people working on the DPČ and DPP, there is a ... Compensation Bonus program, which ... falls under the Ministry of Finance. For ... example, the Ministry of Industry and Trade was supposed to be responsible for ... the COVID 2021 subsidy program and the ... COVID Uncovered Costs program, ... intended for companies with a significant drop in sales. Although they were already prepared, the government ultimately decided ... We therefore assess Marian Jurečka... statement as true.

Question automatically generated from database (truncated): Determine whether the statement relevant to the given summary is TRUE or FALSE based on a summary and a number of statements. Only one of the following statements is relevant to the summary:
SECTION A: The mechanism (increase in pensions, note: Demagog.cz) is given by law (...), if since the last month taken for the ...
SECTION B: The proposal that I presented last year contained approximately 11, 11 measures in total. It also concerned widows, ...
SECTION C: So, the tools of assistance (compensation for entrepreneurs, note: Demagog.cz) are shared here today mainly by 3 ministries, that is the Ministry of Labor and Social Affairs, for us it is the programs of Antivirus (...), and then we also share the tools of assistance with the Ministry of Industry and Trade and the Ministry of Finance, where the other tools of assistance for entrepreneurs and tradesmen are.
Respond only with either TRUE or FALSE and the correct section name, examples: TRUE, SECTION A.
Reference answer: TRUE, SECTION C **Rating by Gemma-3-27b:** 100.00
Answer by baseline gpt-4.1-nano: TRUE, SECTION C **Rating by Gemma-3-27b:** 100.00
Answer by Llama-3-8b: <|assistant|> *newline* FALSE *newline* SECTION C. **Rating by Gemma-3-27b:** 100.00
Answer by Llama-2-7b plus DeepSeek-R1-Distill-Qwen-1.5b: FALSE, SECTION C **Rating by Gemma-3-27b:** 100.00

Fig. 4. A sample test item for the Sensemaking Task with answers and ratings by competing systems.

LLMs are known to be good at question answering, but, while one would expect question answering with context to work even with relatively small LLMs, this does not seem to be the case in our experiment; they are especially bad at ascertaining whether a question is answerable from the given context. When confronted with adversarial questions, many of the systems hallucinated responses.

LLMs are capable of rating answers to questions even without being given the reference answer. The similarity between them and the ROUGE-L scores with reference responses is impressive. However, when presented with questions with answers from different sections of the material, they were often unable to see that they were incorrect.

In future iterations of the task, our experiments with teacher systems show that it would be good to require the systems to provide reference answers to their questions as this is a hard task that shows how well the model truly made sense of the material. It would also be important to compare how the systems react to system-generated questions to how they handle different kinds of human-in-the-loop generated question.

A more comprehensive record together with the participant reports of the experiments can be found in the working notes of the conference [19].

In addition, it will likely be necessary to refine the evaluation procedure, as manual evaluation was time consuming even with so few participants and on a scale that does not achieve statistical significance. A more comprehensive record of the track can be found in the working notes of the conference [19].

6   Conclusions and Plans for the Coming Year

This second year of ELOQUENT has shown that there are several directions of interest to pursue that are underserved by currently popular test suites, especially if a multi-lingual perspective is brought to the fore. Developing entirely consistent new test suites to address those gaps is a major challenge, and will need contributions from several relevant research efforts and research institutions. For the coming year, the ELOQUENT organisers plan to consolidate findings from these first two years to broaden the uptake of experimentation through participating in evaluation activities for multilingual generative language model development projects. Some interest has been expressed by participants in turning track tasks into continuous and asynchronous leaderboard evaluation activities, and this direction will be explored in the coming year.

Acknowledgments. The authors acknowledge the support of the European Commission through the project DeployAI (grant number 101146490), of the National Recovery Plan funded project MPO 60273/24/21300/21000 CEDMO 2.0 NPO, and the funding from the Project OP JAK Mezisektorová spolupráce Nr. CZ.02.01.01/00/23_020/0008518 named "Jazykověda, umělá inteligence a jazykové a řečové technologie: od výzkumu k aplikacím."

Disclosure of Interests. Authors have no competing interests to declare that are relevant to the content of this article.

References

1. Bell, A.: Language style as audience design. Lang. Soc. **13**(2) (1984)
2. Bevendorff, J., et al.: Overview of PAN 2025: generative AI detection, multilingual text detoxification, multi-author writing style analysis, and generative plagiarism detection. In: Hauff, C., et al. (eds.) Advances in Information Retrieval: 47th European Conference on Information Retrieval (ECIR) (2025)
3. Bevendorff, J., et al.: Overview of the "Voight-Kampff" generative AI authorship verification task at PAN and ELOQUENT 2025. In: Faggioli, G., Ferro, N., Rosso, P., Spina, D. (eds.) Working Notes of CLEF 2025 – Conference and Labs of the Evaluation Forum. CEUR-WS (2025)
4. Bevendorff, J., et al.: Overview of the Voight-Kampff generative AI authorship verification task at PAN and ELOQUENT 2024. In: Faggioli, G., Ferro, N., Vlachos, M., Galuščáková, P., de Herrera, A.G.S. (eds.) Working Notes of CLEF 2024 - Conference and Labs of the Evaluation Forum. CEUR-WS.org (2024)
5. Clark, H.H., Murphy, G.L.: Audience design in meaning and reference. In: Advances in Psychology, vol. 9, pp. 287–299. Elsevier (1982)
6. Colombo, P., Noiry, N., Irurozki, E., Clémençon, S.: What are the best systems? New perspectives on NLP benchmarking. Adv. Neural. Inf. Process. Syst. **35**, 26915–26932 (2022)
7. Creo, A., Hormazábal-Lagos, M., Cerezo-Costas, H., Alonso-Doval, P.: HumanAIzers in Voight-Kampff at ELOQUENT 2025. In: Faggioli, G., Ferro, N., Rosso, P., Spina, D. (eds.) Working Notes of CLEF 2025 – Conference and Labs of the Evaluation Forum. CEUR-WS (2025)
8. Gunti, R.R.: The Data-Centric Approach for the Voight Kampff Task. In: Faggioli, G., Ferro, N., Rosso, P., Spina, D. (eds.) Working Notes of CLEF 2025 – Conference and Labs of the Evaluation Forum. CEUR-WS (2025)
9. Hoveyda, M.: Evading Human/Machine classifiers by prompting LLMs for naturally imperfect text. In: Faggioli, G., Ferro, N., Rosso, P., Spina, D. (eds.) Working Notes of CLEF 2025 – Conference and Labs of the Evaluation Forum. CEUR-WS (2025)
10. Karlgren, J., et al.: ELOQUENT CLEF shared tasks for evaluation of generative language model quality, 2nd edition. In: Hauff, C., Macdonald, C., Jannach, D., Kazai, G., Nardini, F.M., Pinelli, F., Silvestri, F., Tonellotto, N. (eds.) Advances in Information Retrieval: 47th European Conference on Information Retrieval (ECIR) (2025)
11. Karlgren, J., et al.: Overview of ELOQUENT 2024—shared tasks for evaluating generative language model quality. In: Proceedings of the Fifteenth International Conference of the CLEF Association (CLEF 2024). Springer, Cham (2024). https://doi.org/10.1007/978-3-031-71908-0_3
12. Karlgren, J., Dürlich, L., Gogoulou, E., Guillou, L., Nivre, J., Talman, A.: ELOQUENT CLEF shared tasks for evaluation of generative language model quality. In: Advances in Information Retrieval: 46th European Conference on IR Research (ECIR) (2024)
13. Karlgren, J., Engels, M.I., Gunti, R.R., Hoveyda, M., Sotic, B.N., Kamps, J.: Overview and joint report of the robustness and consistency task at the ELOQUENT 2025 lab for evaluating generative language model quality. In: Faggioli, G., Ferro, N., Rosso, P., Spina, D. (eds.) Working Notes of CLEF 2025 – Conference and Labs of the Evaluation Forum. CEUR-WS (2025)

14. Lin, C.Y.: ROUGE: a package for automatic evaluation of summaries. In: Text Summarization Branches Out, pp. 74–81. Association for Computational Linguistics, Barcelona, Spain (2004). https://aclanthology.org/W04-1013/

15. Mikhailov, V., Artemova, E., Butenko, Z., Øvrelid, L., Velldal, E.: Overview of the preference prediction task at the ELOQUENT 2025 lab for evaluating generative language model quality. In: Faggioli, G., Ferro, N., Rosso, P., Spina, D. (eds.) Working Notes of CLEF 2025 – Conference and Labs of the Evaluation Forum. CEUR-WS (2025)

16. Rofin, M., et al.: Vote'n'Rank: revision of benchmarking with social choice theory. In: Vlachos, A., Augenstein, I. (eds.) Proceedings of the 17th Conference of the European Chapter of the Association for Computational Linguistics, pp. 670–686. Association for Computational Linguistics, Dubrovnik, Croatia (2023). https://doi.org/10.18653/v1/2023.eacl-main.48

17. Saha, S., Das, R., Das, D.: JUNLP_SS at ELOQUENT Lab 2025: humanizing AI - enhancing the realism of machine generated text. In: Faggioli, G., Ferro, N., Rosso, P., Spina, D. (eds.) Working Notes of CLEF 2025 – Conference and Labs of the Evaluation Forum. CEUR-WS (2025)

18. Sahlgren, M., Karlgren, J., Dürlich, L., Gogoulou, E., Talman, A., Zahra, S.: ELOQUENT 2024-robustness task. In: 25th Working Notes of the Conference and Labs of the Evaluation Forum, CLEF 2024. Grenoble. 9 September 2024 through 12 September 2024, vol. 3740, pp. 703–707. CEUR-WS (2024)

19. Šindelář, P., Bojar, O.: Overview of the sensemaking task at the ELOQUENT 2025 lab: LLMs as teachers, students and evaluators. In: Faggioli, G., Ferro, N., Rosso, P., Spina, D. (eds.) Working Notes of CLEF 2025 – Conference and Labs of the Evaluation Forum. CEUR-WS (2025)

20. Vachharajani, P.: Literal re-translation as a method for AI text disguise and detection evasion. In: Faggioli, G., Ferro, N., Rosso, P., Spina, D. (eds.) Working Notes of CLEF 2025 – Conference and Labs of the Evaluation Forum. CEUR-WS (2025)

21. Zhang, T., Kishore, V., Wu, F., Weinberger, K.Q., Artzi, Y.: BERTScore: evaluating text generation with BERT. In: International Conference on Learning Representations (ICLR) (2020). https://openreview.net/forum?id=SkeHuCVFDr

Overview of eRisk 2025: Early Risk Prediction on the Internet

Javier Parapar[1] , Anxo Perez[1(✉)] , Xi Wang[2] , and Fabio Crestani[3]

[1] Information Retrieval Lab, Centro de Investigación en Tecnoloxías da Información
e as Comunicacións (CITIC), Universidade da Coruña, Coruña, Spain
{javierparapar,anxo.pvila}@udc.es
[2] University of Sheffield, Sheffield, UK
xi.wang@sheffield.ac.uk
[3] Faculty of Informatics, Universitá della Svizzera italiana (USI), Lugano,
Switzerland
fabio.crestani@usi.ch

Abstract. This paper presents an overview of eRisk 2025, the ninth edition of the CLEF lab on early risk detection. Since its foundation, eRisk has served as a framework for evaluating methodologies, effectiveness metrics, and challenges related to the early identification of personal risks, particularly within the health and safety domains. The 2025 edition marks an important evolution of the lab, expanding its focus toward tasks that require deeper contextual and conversational understanding. In addition to continuing the depression symptom sentence ranking task, eRisk 2025 introduces two novel tasks: the second task is based on contextualized early detection of depression from full conversational threads, and a pilot task that explores the use of fine-tuned conversational agents for detecting signs of depression through interactions. These additions aim to open new research avenues and bring the evaluation setting closer to conversational scenarios.

Keywords: Early risk detection · Depression · Conversational analysis · LLMs · Large Language Models · eRisk

1 Introduction

The eRisk lab was created as a benchmark environment to explore evaluation methodologies, new models, and build resources for the timely detection of personal risk situations. Early-alert technologies are becoming indispensable in domains focused on safety and personal health. Typical use cases range from monitoring mental-health signals to flagging predatory behaviour or violent threats online, where even marginal time gains can translate into meaningful interventions. eRisk specializes on psychological and mental risks, including depression, self-harm, pathological gambling, and eating disorders, where language offers subtle yet valuable clues. Because the relationship between linguistic expression and mental state is complex, robust automatic screening tools are still needed. The inaugural eRisk 2017 task explored early depression detection with a novel evaluation framework and dataset [10]. In 2018 the scope expanded to anorexia [11,12]. The 2019 campaign consolidated works on anorexia detection, introduced self-harm prediction, and proposed a novel task that inferred questionnaire responses on depression severity from social-media activity [13,14,16].

The 2020 edition deepened work on self-harm while adding a depression-severity estimation task [17–19]. A year later, in 2021, attention turned to pathological gambling and self-harm, alongside an updated severity-estimation task [26–28]. In 2022 we revisited pathological gambling and depression and introduced a new severity-estimation challenge for eating-disorder content [29–31]. The 2023 campaign shifted textitasis toward a new sentence ranking task for individual depression symptoms, complemented by early gambling-risk detection and eating-disorder severity estimation [32–34]. Finally, the 2024 lab offered three tasks: the sentence-ranking challenge covering the 21 BDI-II symptoms, a continued anorexia early-detection task, and an updated eating-disorder severity-estimation track [35–37].

The present edition (eRisk 2025 [38,39]) builds on this trajectory and, for the first time, pivots toward tasks requiring richer contextual and conversational understanding. The details of these novel tasks are discussed in the following sections. This year, we had 128 different teams registered for the lab. We finally received results coming from 25 distinct teams: 67 runs for Task 1, 50 runs for Task 2, and 11 runs for the pilot task.

2   Task 1: Search for Symptoms of Depression

This task continues from eRisk 2023's and 2024's Task 1, which involved ranking sentences from user writings based on their relevance to specific depression symptoms. This is the last year of the task. Again, participants were required to order sentences according to their relevance to the 21 standardized symptoms listed in the BDI-II questionnaire [2]. A sentence was deemed relevant if it reflected the user's condition related to a symptom, including positive statements (e.g., "I feel quite happy lately" is relevant for the symptom "Sadness"). As in 2024, the test collection provides not only the target sentence but also its immediate predecessor and successor to give more context.

2.1   Task 1: Dataset and Asessment Process

The dataset provided was in TREC format, tagged with sentences derived from Reddit historical data. Table 1 presents some statistics of the corpus. Given the corpus of sentences and the description of the symptoms from the BDI-II questionnaire, the participants were free to decide on the best strategy to derive queries for representing the BDI-II symptoms. Each participating team submitted up to 5 variants (runs). Each run included 21 TREC-style formatted rankings of sentences, as shown in Fig. 1. For each symptom, the participants

Table 1. Corpus statistics for Task 1: Search for Symptoms of Depression.

Number of users	9,000
Number of sentences	17,553,441
Average number of words per sentence	12,39

```
1 Q0 251001_0_1 0001 10 myGroupNameMyMethodName
1 Q0 858202_3_2 0002 9 myGroupNameMyMethodName
1 Q0 482048_2_1 0003 8.76 myGroupNameMyMethodName

...

21 Q0 153202_2_2 0999 1.25 myGroupNameMyMethodName
21 Q0 223133_9_8 1000 0.9 myGroupNameMyMethodName
```

Fig. 1. Example of a participant's run.

should submit up to 1000 results sorted by estimated relevance. We received 67 runs from 17 participating teams (see Table 2).

Relevance labels were produced through a stratified, two–stage pooling procedure. First, for every BDI-II symptom we implemented top-k pooling, collecting the top five sentences returned by each submitted run ($k = 5$), forming an initial pool that served to rank systems provisionally. We then selected the twenty highest-ranked runs and performed a second pooling step that extended the cutoff to the top fifty sentences ($k = 50$). Unlike the 2023 setup, assessors were shown the target sentence together with its immediate context (the preceding and following sentences), a change designed to reduce annotation ambiguity.

Table 2. Task 1 (Search for Symptoms of Depression): number of submitted runs per team, laid out horizontally to reduce vertical space.

Team	Runs	Team	Runs	Team	Runs
SonUIT [43]	5	ThinkIR [1]	5	Ixa_ave [46]	5
Synapse	3	PJs-team [45]	5	ELiRF-UPV [42]	5
COTECMAR-UTB [4]	3	COMFOR	1	LHS712-Team-1 [3]	5
Team-Gryffindor	1	INESC-ID [5]	5	UET-Psyche-Warriors [20]	5
NYCUNLP	5	BGU-Data-Science [22]	5	HULAT_UC3M [21]	5
RELAI	2	UniORNLP-dahlia	2		
Total Teams: 17 | Total Runs: 67					

Table 3. Task 1 (Search for Symptoms of Depression): Size of the pool for every BDI Item.

BDI Item (#)	pool	# unanimity-qrels (3/3)	# majority-qrels (2/3)
Sadness (1)	581	167	296
Pessimism (2)	552	209	345
Past Failure (3)	536	146	283
Loss of Pleasure (4)	522	132	244
Guilty Feelings (5)	400	88	227
Punishment Feelings (6)	553	33	111
Self-Dislike (7)	474	205	290
Self-Criticalness (8)	534	115	259
Suicidal Thoughts or Wishes (9)	517	300	377
Crying (10)	547	143	359
Agitation (11)	593	142	338
Loss of Interest (12)	553	105	229
Indecisiveness (13)	584	50	139
Worthlessness (14)	424	161	249
Loss of Energy (15)	491	161	273
Changes in Sleeping Pattern (16)	569	274	404
Irritability (17)	540	132	314
Changes in Appetite (18)	548	225	374
Concentration Difficulty (19)	428	166	271
Tiredness or Fatigue (20)	566	217	385
Loss of Interest in Sex (21)	530	239	350

Three annotators worked independently: one with professional training in psychology, and two computer-science researchers specialising in early-risk technologies. Before judging, the organisers held a session to walk through an initial guideline draft, resolve doubts, and agree on different cases. The consolidated guideline, publicly available[1], defines a sentence as *relevant* only when it both addresses the symptom and conveys explicit information about the user's state. This dual concept of relevance (on-topic and reflective of the user's state with respect to the symptom) introduced a higher level of complexity compared to more standard relevance assessments. Each pooled sentence received three independent judgements, and we provide two ground-truth sets (qrels):

- **Majority-based qrels**: a sentence was deemed relevant if at least two of the three assessors marked it so.
- **Unanimity-based qrels**: a sentence was deemed relevant only when all three assessors agreed.

---

[1] https://erisk.irlab.org/guidelines_erisk24_task1.html.

The final pool sizes and qrels for each symptom are reported in Table 3. Providing both qrels enables analyses with different agreement thresholds, continuing the dual-qrel strategy introduced in earlier eRisk campaigns.

2.2   Task 1: Results

The performance results for the participating systems are shown in Tables 4 (majority-based qrels) and 5 (unanimity-based qrels). The tables report several standard performance metrics, such as mean Average Precision (AP), mean R-Precision, mean Precision at 10 and mean NDCG at 1000. Remarkably, runs *unanimity* and *max* from the team *INESC-ID*, achieved the top-ranking performance for nearly all metrics and relevance judgement types. The teams *UET-Psyche-Warriors*, *SonUIT*, *BGU-Data-Science* and *PJs-Team* also obtained close performance. Their effective results demonstrate their exceptional competence in this task. Taken together, the results confirm that sentence-level symptom retrieval remains a challenging task.

3   Task 2: Contextualized Early Detection of Depression (New Task)

This new task in 2025 introduces a different scenario in depression detection by incorporating full conversational contexts. Whereas earlier eRisk editions always released isolated posts authored by a single user, the 2025 task provided the entire Reddit discussion thread in which the target user intervened. Consequently, in the test dataset, systems had access not only to the messages produced by the target user but also to every other contribution in the thread and to the interaction structure that links the messages (e.g., the different replies to each comment).

This design is motivated by the observation that the clinical relevance of a message often becomes more evident when interpreted alongside the surrounding conversation. Thus, a user's response may only gain relevance when viewed in conjunction with the preceding or subsequent interactions from other participants. For instance, a seemingly neutral sentence, may reveal hopelessness if it answers a direct plea for support. For this reason, the task is designed to simulate real-world scenarios where depression detection may rely on analyzing exchanges between multiple participants. This setup presents unique challenges, as systems must consider not only the textual content of individual posts but also the interplay between participants and how this context influences the detection of depressive symptoms.

The test collection utilised for this task followed the same format as the collection described in the work by Losada and Crestani [9]. The collection contains writings, including posts and comments, obtained from a selected group of social media users. To construct the ground truth assessments, we adopted established approaches that aim to optimise the utilisation of assessors' time, as documented in previous studies [24, 25]. These methods employ simulated pooling strategies,

Table 4. Ranking-based evaluation for Task 1 (majority voting).

Team	Run	AP	R-PREC	P@10	NDCG
BGU-Data-Science	sbert-w-expansion-w-naive-fp-w-claude	0.232	0.305	0.767	0.483
BGU-Data-Science	sbert-w-expansion-w-naive-fp	0.227	0.296	0.767	0.475
BGU-Data-Science	sbert-w-expansion-w-spacy-fp	0.220	0.287	0.767	0.463
BGU-Data-Science	sbert-w-expansion	0.197	0.281	0.652	0.444
BGU-Data-Science	sbert	0.240	0.324	0.743	0.516
COMFOR	bert_ranked	0.013	0.041	0.243	0.082
COTECMAR-UTB	centroid_ranked_updated	0.052	0.130	0.276	0.236
COTECMAR-UTB	dl_ranked	0.073	0.160	0.405	0.282
COTECMAR-UTB	ranked_updated	0.077	0.165	0.414	0.290
ELiRF-UPV	model1	0.035	0.101	0.100	0.216
ELiRF-UPV	model2	0.032	0.095	0.081	0.206
ELiRF-UPV	model3	0.035	0.099	0.110	0.211
ELiRF-UPV	model4	0.033	0.099	0.067	0.210
ELiRF-UPV	model5	0.032	0.097	0.100	0.209
HULAT_UC3M	roberta	0.004	0.010	0.162	0.026
HULAT_UC3M	vader_sample	0.004	0.010	0.148	0.023
HULAT_UC3M	reflexives_roberta	0.013	0.025	0.262	0.052
HULAT_UC3M	roberta	0.018	0.034	0.363	0.065
HULAT_UC3M	vader_top	0.015	0.034	0.295	0.059
INESC-ID	aug-best	0.247	0.324	0.691	0.560
INESC-ID	max	0.350	0.407	0.648	**0.653**
INESC-ID	maxcos	0.235	0.320	0.757	0.506
INESC-ID	mix23	0.312	0.377	0.643	0.616
INESC-ID	unanimity	**0.354**	**0.433**	**0.876**	0.575
LHS712-Team-1	results	0.000	0.000	0.000	0.000
LHS712-Team-1	BERT_CONSENSUS	0.102	0.199	0.529	0.321
LHS712-Team-1	BERT_MAJORITY	0.074	0.178	0.281	0.283
LHS712-Team-1	LR_file_combined	0.000	0.004	0.009	0.007
LHS712-Team-1	SVM_file_combined_4	0.000	0.004	0.009	0.007
NYCUNLP	01	0.237	0.322	0.662	0.501
NYCUNLP	02	0.193	0.276	0.619	0.455
NYCUNLP	03	0.133	0.217	0.624	0.328
NYCUNLP	04	0.190	0.279	0.614	0.450
NYCUNLP	05	0.072	0.137	0.567	0.203
PJs-team	teamADRB	0.105	0.234	0.391	0.354
PJs-team	teamMBRR	0.262	0.347	0.771	0.489
PJs-team	teamRRens-v2	0.279	0.360	0.800	0.503
PJs-team	teamRRens	0.273	0.359	0.786	0.500
PJs-team	teamSumensemble	0.120	0.249	0.400	0.376
RELAI	1	0.005	0.023	0.052	0.053
RELAI	2	0.008	0.036	0.038	0.078
SonUIT	config1	0.283	0.351	0.767	0.562
SonUIT	config2	0.334	0.392	0.790	0.613
SonUIT	config3	0.311	0.395	0.767	0.572
SonUIT	config4	0.328	0.426	0.767	0.578
SonUIT	config5	0.260	0.304	0.767	0.552
Synapse	HighestSimilarityFirst	0.001	0.002	0.038	0.009
Synapse	nomicFineTunedRerankedSimilarity	0.001	0.002	0.043	0.008
Synapse	nomicRerankedSimilarity	0.001	0.002	0.052	0.006
Team-Gryffindor	task1	0.017	0.042	0.019	0.183
ThinkIR	few_shot_query_2025	0.015	0.049	0.133	0.073
ThinkIR	rank_sim	0.003	0.010	0.000	0.030
ThinkIR	2025	0.068	0.157	0.409	0.228
ThinkIR	pseudo_relevance_10_2025	0.064	0.148	0.400	0.213
ThinkIR	pseudo_relevance_5_2025	0.060	0.151	0.409	0.212
UET-Psyche-Warriors	1_similarity	0.311	0.378	0.657	0.588
UET-Psyche-Warriors	2_ensemble_similarity	0.315	0.390	0.657	0.612
UET-Psyche-Warriors	3_contrastive_learning	0.165	0.258	0.457	0.450
UET-Psyche-Warriors	4_ensemble_contrastive_learning	0.147	0.228	0.462	0.419
UET-Psyche-Warriors	5_machine_learning	0.339	0.394	0.776	0.623
UniORNLP-dahlia	frame_hyde	0.001	0.008	0.029	0.019
UniORNLP-dahlia	simple_hyde	0.014	0.040	0.205	0.072
ixa_ave	base_all	0.097	0.191	0.305	0.345
ixa_ave	base_filter30	0.102	0.203	0.338	0.342
ixa_ave	base_filter50	0.009	0.025	0.086	0.048
ixa_ave	thresh_all	0.091	0.168	0.281	0.333
ixa_ave	thresh_filter50	0.005	0.016	0.129	0.035

Table 5. Ranking-based evaluation for Task 1 (unanimity).

Team	Run	AP	R-PREC	P@10	NDCG
BGU-Data-Science	sbert-w-expansion-w-naive-fp-w-claude	0.143	0.244	0.443	0.429
BGU-Data-Science	sbert-w-expansion-w-naive-fp	0.138	0.240	0.448	0.420
BGU-Data-Science	sbert-w-expansion-w-spacy-fp	0.135	0.237	0.462	0.412
BGU-Data-Science	sbert-w-expansion	0.119	0.223	0.381	0.389
BGU-Data-Science	sbert	0.171	0.272	0.419	0.489
COMFOR	bert_ranked	0.010	0.036	0.114	0.079
COTECMAR-UTB	centroid_top1000_ranked_updated	0.030	0.081	0.133	0.195
COTECMAR-UTB	dl_ranked	0.040	0.107	0.176	0.240
COTECMAR-UTB	ranked_updated	0.042	0.108	0.181	0.243
ELiRF-UPV	model1	0.021	0.063	0.052	0.184
ELiRF-UPV	model2	0.019	0.057	0.062	0.179
ELiRF-UPV	model3	0.021	0.060	0.062	0.180
ELiRF-UPV	model4	0.019	0.056	0.038	0.180
ELiRF-UPV	model5	0.018	0.057	0.057	0.175
HULAT_UC3M	roberta	0.002	0.009	0.052	0.016
HULAT_UC3M	vader_sample	0.001	0.009	0.029	0.012
HULAT_UC3M	reflexives_roberta	0.013	0.032	0.157	0.053
HULAT_UC3M	roberta	0.008	0.025	0.174	0.040
HULAT_UC3M	vader_top	0.006	0.024	0.105	0.037
INESC-ID	aug-best	0.167	0.236	0.414	0.515
INESC-ID	max	0.223	0.308	0.386	**0.582**
INESC-ID	maxcos	0.164	0.273	0.429	0.472
INESC-ID	mix23	0.201	0.279	0.371	0.547
INESC-ID	unanimity	**0.269**	**0.383**	**0.509**	0.561
LHS712-Team-1	results	0.000	0.000	0.000	0.000
LHS712-Team-1	BERT_CONSENSUS	0.083	0.172	0.281	0.315
LHS712-Team-1	BERT_MAJORITY	0.062	0.137	0.181	0.286
LHS712-Team-1	LR_file_combined	0.000	0.003	0.005	0.007
LHS712-Team-1	SVM_file_combined_4	0.000	0.003	0.005	0.007
NYCUNLP	01	0.156	0.253	0.367	0.442
NYCUNLP	02	0.129	0.216	0.357	0.400
NYCUNLP	03	0.081	0.159	0.371	0.270
NYCUNLP	04	0.135	0.224	0.352	0.408
NYCUNLP	05	0.048	0.117	0.357	0.173
PJs-team	ADRB	0.073	0.168	0.214	0.325
PJs-team	MBRR	0.175	0.299	0.424	0.435
PJs-team	RRens-v2	0.188	0.311	0.452	0.446
PJs-team	RRens	0.184	0.308	0.467	0.444
PJs-team	Sumensemble	0.079	0.184	0.229	0.331
RELAI	1	0.005	0.019	0.029	0.056
RELAI	2	0.006	0.024	0.009	0.076
SonUIT	config1	0.191	0.276	0.448	0.500
SonUIT	config2	0.223	0.303	0.462	0.545
SonUIT	config3	0.205	0.290	0.448	0.508
SonUIT	config4	0.219	0.315	0.448	0.514
SonUIT	config5	0.176	0.248	0.448	0.491
Synapse	HighestSimilarityFirst	0.000	0.000	0.000	0.000
Synapse	nomicFineTunedRerankedSimilarity	0.000	0.000	0.000	0.000
Synapse	nomicRerankedSimilarity	0.000	0.000	0.000	0.000
Team-Gryffindor	task1	0.014	0.027	0.014	0.187
ThinkIR	few_shot_query_2025	0.014	0.043	0.081	0.075
ThinkIR	rank_sim	0.001	0.003	0.000	0.019
ThinkIR	2025	0.044	0.116	0.219	0.196
ThinkIR	pseudo_relevance_10_2025	0.042	0.111	0.205	0.192
ThinkIR	pseudo_relevance_5_2025	0.040	0.107	0.229	0.191
UET-Psyche-Warriors	1_similarity	0.193	0.270	0.391	0.501
UET-Psyche-Warriors	2_ensemble_similarity	0.202	0.279	0.391	0.530
UET-Psyche-Warriors	3_contrastive_learning	0.094	0.165	0.243	0.373
UET-Psyche-Warriors	4_ensemble_contrastive_learning	0.079	0.141	0.219	0.347
UET-Psyche-Warriors	5_machine_learning	0.248	0.330	0.476	0.577
UniORNLP-dahlia	frame_hyde	0.001	0.004	0.009	0.014
UniORNLP-dahlia	simple_hyde	0.009	0.033	0.081	0.058
ixa_ave	base_all	0.053	0.121	0.124	0.282
ixa_ave	base_filter30	0.055	0.126	0.138	0.277
ixa_ave	base_filter50	0.006	0.020	0.038	0.042
ixa_ave	thresh_all	0.052	0.103	0.110	0.270
ixa_ave	thresh_filter50	0.003	0.013	0.048	0.026

enabling the effective creation of test collections. The main statistics of the test collection used for Task 2 are presented in Table 6.

Table 6. Task 2 (early depression). Main statistics of test collection.

	Depression	Control
Num. subjects	102	807
Num. threads	40,563	238,033
Avg num. of threads per subject	397.7	295.9
Avg num. of days from first to last thread	$\approx$ 1695	$\approx$ 958
Avg num. of comments per thread	65.1	44.6
Avg num. words per comment	33.8	25.6

Within this dataset, users are categorised into two groups: depression and control. For each user, the collection contains a sequence of writings and threads where the user participated in chronological order. To facilitate the task and ensure uniform distribution, we established a dedicated server that systematically provided user writings to the participating teams. Further details regarding the server's setup and functioning are available at the lab's official website[2].

The task was divided into two phases:

- During the training phase, participants worked with a static dataset consisting of isolated user writings from depressed and control users, without any conversational context. This training dataset came from prior editions of eRisk regarding the early detection depression tasks (without any conversational context).
- The test phase, in contrast, was carried out interactively. For each target user, the server released a sequence of discussion threads in real time. Each thread constituted a *submission round*. At any round within the chronology of user writings, participants had the freedom to stop the process and issue an alert. After reading each user thread, teams were required to decide between two options: i) alerting about the target user, indicating a predicted sign of depression, or ii) not alerting about the target user. Participants independently made this choice for each user in the test split. It is important to note that once an alert was issued, it was considered final, and no further decisions regarding that particular user were taken into account. Conversely, the absence of alerts was considered non-final, allowing participants to subsequently submit an alert if they detected signs of risk emerging.

To evaluate the systems' performance, we employed two indicators: the accuracy of the decisions made and the number of user writings required to reach those decisions. These criteria provide valuable insights into the effectiveness

---

[2] https://erisk.irlab.org/eRisk25Servert2Details.html.

and efficiency of the systems under evaluation. To support the test stage, we deployed a REST service. The server iteratively distributes user writings and waits for responses from participants. Importantly, new user data was not provided to a specific participant until the service received a decision from that particular team. The submission period for the task was open from February 5th, 2025 until April 12th, 2025.

3.1   Task 2: Evaluation Metrics

Decision-Based Evaluation. This evaluation approach uses the binary decisions made by the participating systems for each user. In addition to standard classification measures such as Precision, Recall, and F1 score (computed with respect to the positive class), we also calculate ERDE (Early Risk Detection Error), used in previous editions of the lab. A detailed description of ERDE was presented by Losada and Crestani in [9]. ERDE is an error measure that incorporates a penalty for delayed correct alerts (true positives). The penalty increases with the delay in issuing the alert, measured by the number of user posts processed before making the alert.

Since 2019, we complemented the evaluation report with additional decision-based metrics that try to capture additional aspects of the problem. These metrics try to overcome some limitations of $ERDE$, namely:

– the penalty associated to true positives goes quickly to 1. This is due to the functional form of the cost function (sigmoid).
– a perfect system, which detects the true positive case right after the first round of messages (first chunk), does not get error equal to 0.
– with a method based on releasing data in a chunk-based way (as it was done in 2017 and 2018) the contribution of each user to the performance evaluation has a large variance (different for users with few writings per chunk vs users with many writings per chunk).
– $ERDE$ is not interpretable.

Some research teams have analysed these issues and proposed alternative ways for evaluation. Trotzek and colleagues [44] proposed $ERDE_o^\%$. This is a variant of ERDE that does not depend on the number of user writings seen before the alert but, instead, it depends on the *percentage* of user writings seen before the alert. In this way, user's contributions to the evaluation are normalized (currently, all users weight the same). However, there is an important limitation of $ERDE_o^\%$. In real life applications, the overall number of user writings is not known in advance. Social Media users post contents online and screening tools have to make predictions with the evidence seen. In practice, you do not know when (and if) a user's thread of messages is exhausted. Thus, the performance metric should not depend on knowledge about the total number of user writings.

Another proposal of an alternative evaluation metric for early risk prediction was done by Sadeque and colleagues [41]. They proposed $F_{latency}$, which fits better with our purposes. This measure is described next.

Imagine a user $u \in U$ and an early risk detection system that iteratively analyzes u's writings (e.g. in chronological order, as they appear in Social Media) and, after analyzing k_u user writings ($k_u \geq 1$), takes a binary decision $d_u \in \{0,1\}$, which represents the decision of the system about the user being a risk case. By $g_u \in \{0,1\}$, we refer to the user's golden truth label. A key component of an early risk evaluation should be the delay on detecting true positives (we do not want systems to detect these cases too late). Therefore, a first and intuitive measure of delay can be defined as follows[3]:

$$\text{latency}_{TP} = \text{median}\{k_u : u \in U, d_u = g_u = 1\} \tag{1}$$

This measure of latency is calculated over the true positives detected by the system and assesses the system's delay based on the median number of writings that the system had to process to detect such positive cases. This measure can be included in the experimental report together with standard measures such as Precision (P), Recall (R) and the F-measure (F):

$$P = \frac{|u \in U : d_u = g_u = 1|}{|u \in U : d_u = 1|} \tag{2}$$

$$R = \frac{|u \in U : d_u = g_u = 1|}{|u \in U : g_u = 1|} \tag{3}$$

$$F = \frac{2 \cdot P \cdot R}{P + R} \tag{4}$$

Furthermore, Sadeque et al. proposed a measure, $F_{latency}$, which combines the effectiveness of the decision (estimated with the F measure) and the delay[4] in the decision. This is calculated by multiplying F by a penalty factor based on the median delay. More specifically, each individual (true positive) decision, taken after reading k_u writings, is assigned the following penalty:

$$penalty(k_u) = -1 + \frac{2}{1 + \exp^{-p \cdot (k_u - 1)}} \tag{5}$$

where p is a parameter that determines how quickly the penalty should increase. In [41], p was set such that the penalty equals 0.5 at the median number of posts of a user[5]. Observe that a decision right after the first writing has no penalty (i.e. $penalty(1) = 0$). Figure 2 plots how the latency penalty increases with the number of observed writings.

---

[3] Observe that Sadeque et al. (see [41], pg 497) computed the latency for all users such that $g_u = 1$. We argue that latency should be computed only for the true positives. The false negatives ($g_u = 1$, $d_u = 0$) are not detected by the system and, therefore, they would not generate an alert.

[4] Again, we adopt Sadeque et al.'s proposal but we estimate latency only over the true positives.

[5] In the evaluation we set p to 0.0078, a setting obtained from the eRisk 2017 collection.

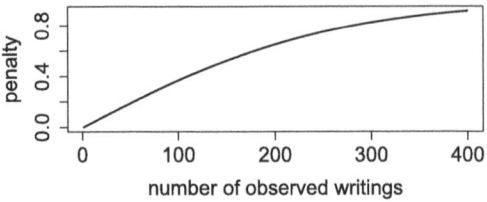

Fig. 2. Latency penalty increases with the number of observed writings (k_u).

The system's overall speed factor is computed as:

$$speed = (1 - \text{median}\{penalty(k_u) : u \in U, d_u = g_u = 1\}) \tag{6}$$

where speed equals 1 for a system whose true positives are detected right at the first writing. A slow system, which detects true positives after hundreds of writings, will be assigned a speed score near 0.

Finally, the *latency-weighted* F score is simply:

$$F_{latency} = F \cdot speed \tag{7}$$

Since 2019 user's data were processed by the participants in a post by post basis (i.e. we avoided a chunk-based release of data). Under these conditions, the evaluation approach has the following properties:

- smooth grow of penalties;
- a perfect system gets $F_{latency} = 1$;
- for each user u the system can opt to stop at any point k_u and, therefore, now we do not have the effect of an imbalanced importance of users;
- $F_{latency}$ is more interpretable than $ERDE$.

Ranking-Based Evaluation. In addition to the evaluation discussed above, we employed an alternative form of evaluation to further assess the systems. After each data release (new user writing, that is post or comment), participants were required to provide the following information for each user in the collection:

- A decision for the user (alert or no alert), which was used to calculate the decision-based metrics discussed previously.
- A score representing the user's level of risk, estimated based on the evidence observed thus far.

The scores were used to create a ranking of users in descending order of estimated risk. For each participating system, a ranking was generated at each data release point, simulating a continuous re-ranking approach based on the observed evidence. In a real-life scenario, this ranking would be presented to an expert user who could make decisions based on the rankings (e.g., by inspecting the top of the rankings). Each ranking can be evaluated using standard ranking metrics such as P@10 or NDCG. Therefore, we report the performance of the systems based on the rankings after observing different numbers of writings.

3.2   Task 2: Participant Teams

Table 7 shows the participating teams, the number of runs submitted and the approximate lapse of time from the first response to the last response. This lapse of time is indicative of the degree of automation of each team's algorithms. All but one participant (FU–TU–DFKI) managed to process the complete set of threads in at least one run. The fastest groups: ELiRF–UPV, SINAI–UJA, and PJs-TEAM finished in under ten hours, illustrating the feasibility of efficient processing even when entire conversations are supplied. By contrast, LOTU–IXA and UET–PSYCHE–WARRIORS took around a week, pointing to more complex or resource intensive pipelines

Table 7. Task 2: participating teams, number of runs, number of threads processed by the team, and lapse of time taken for the entire process.

Team	#Runs	#User threads	Lapse of time (from 1st to last response)
Lotu-ixa [8]	5	1280	9 days 03:21
HIT-SCIR [47]	5	1280	1 day 14:00
SINAI-UJA [23]	5	1280	0 days 09:53
DS-GT [6]	2	1280	1 day 00:29
NYCUNLP	5	1280	1 day 04:07
UET-Psyche-Warriors [20]	5	1280	6 days 21:34
Capy-team	5	1280	0 days 15:49
COTECMAR-UTB [4]	2	1280	3 days 00:50
ELiRF-UPV [42]	5	1280	0 days 08:33
PJs-team [45]	5	1280	0 days 08:36
HU [40]	5	1280	2 days 23:08
FU-TU-DFKI [7]	1	449	1 day 11:24

3.3   Task 2: Results

Table 8 show the decision-based results of Task 2. Table 9 shows the ranking-based results. In the decision setting, *HIT-SCIR* dominates: its best run attains the highest F_1 (0.85) while keeping both $ERDE_5$ and $ERDE_{50}$ at or very near the minimum error values. That performance is achieved with a median latency of only eight writings, illustrating a good balance between earliness and accuracy. *ELiRF-UPV* follows at a short distance, with a top F_1 of 0.79 but slightly worse error–aware metrics

The ranking-based evaluation shows a complementary picture. *HIT-SCIR* again exhibits near-perfect precision at every cut-off and sustains the highest NDCG values as additional writings become available, confirming the robustness of its retrieval component. *Lotu-Ixa* excels in the one writing scenario, matching *HIT-SCIR* for $P@10$ and $NDCG@10$. However, its advantage diminishes once longer histories are considered, suggesting that its decision policy strongly weights the earliest cues.

4   Pilot Task: Conversational Depression Detection via LLMs

We introduced this pilot task in 2025 as a novel challenge to seek the opportunity of embracing conversational agents in detecting depression symptoms. Participants were interacting with LLM-based personas who have been instructed using user writings, simulating real-world conversational exchanges and example user profiles. Twelve distinct personas were instantiated with ChatGPT.

The challenge lies in asking participants to determine whether the LLM persona exhibits signs of depression and, if so, what is the level of depression severity and key depression symptoms expressed over conversations. The diagnostic target for the LLMs was framed in terms of the BDI-II, as in Task 1. The BDI-II is a 21-item self-report questionnaire widely used in clinical psychology, which are listed in the Table 10.

Each item corresponds to a concrete symptom. For example, *Sadness*, *Loss of Energy*, or *Indecisiveness*. Each symptom is scored 0 to 3 according to severity. Table 11 shows the possible response options (0–3) for the symptoms *Sadness* and *Self-Dislike*. The sum of all 21 symptoms yields a global index in the range 0–63. The scores are interpreted into four categories: 0–9 are interpreted as *minimal* depression, 10–18 as *mild*, 19–29 as *moderate*, and 30 or above as *severe*. Because the personas are simulations, no ground-truth questionnaire exists; instead, a group of three clinicians examined the seed user data that shaped each persona and agreed on both an overall BDI-II score and the subset of symptoms included. These consensual judgments constitute the gold standard.

Participants did not receive any labelled training material. We deliberately framed the task as *training-less* to encourage a variety of methodological responses, ranging from rule-based interviewers and zero-shot LLM prompts to different classifiers trained on public mental-health corpora. During the test window, teams accessed the links we provided them through ChatGPT interface for creating the dialogue with the LLM-persona. The participant systems interacted with a free-form prompt; the server produced the next turn, and so on. This loop continued until the system chose to terminate the dialogue and submit its diagnosis. Since this is a pilot task, there was no hard cap on the number of

Table 8. Decision-based evaluation for Task 2 ordered in terms of best F_1.

Team	Run	P	R	F_1	$ERDE_5$	$ERDE_{50}$	latency$_{TP}$	speed	$F_{latency}$
HIT-SCIR	0	0.72	0.96	0.82	0.06	**0.03**	4.00	0.99	0.81
	1	0.72	0.95	0.82	0.06	**0.03**	4.00	0.99	0.81
	2	0.74	0.94	0.83	0.06	**0.03**	4.00	0.99	**0.82**
	3	0.73	0.94	0.82	0.08	**0.03**	7.00	0.98	0.80
	4	0.77	0.94	**0.85**	0.09	**0.03**	8.00	0.97	**0.82**
ELiRF-UPV	0	0.78	0.81	0.79	0.08	0.04	7.00	0.98	0.78
	1	0.37	0.62	0.46	0.07	0.06	**1.00**	**1.00**	0.46
	2	0.83	0.47	0.60	0.10	0.07	8.00	0.97	0.58
	3	0.68	0.67	0.67	0.09	0.05	7.00	0.98	0.66
	4	0.68	0.67	0.67	0.09	0.05	7.00	0.98	0.66
HU	0	0.61	0.77	0.68	0.09	0.05	10.00	0.96	0.66
	1	0.72	0.77	0.75	0.10	0.05	11.00	0.96	0.72
	2	0.14	0.94	0.25	0.15	0.09	6.00	0.98	0.24
	3	0.11	1.00	0.20	0.11	0.10	**1.00**	**1.00**	0.20
	4	0.27	0.88	0.41	0.10	0.07	11.00	0.96	0.40
UET-Psyche-Warriors	0	0.67	0.78	0.72	0.10	0.06	24.50	0.91	0.66
	1	0.63	0.85	0.72	0.09	0.05	16.00	0.94	0.68
	2	0.63	0.86	0.73	0.09	0.04	16.00	0.94	0.68
	3	0.63	0.85	0.72	0.09	0.05	16.00	0.94	0.68
	4	0.63	0.84	0.72	0.09	0.05	15.50	0.94	0.68
PJs-team	0	0.66	0.75	0.71	0.09	0.06	17.00	0.94	0.66
	1	0.53	0.83	0.65	0.09	0.06	24.00	0.91	0.59
	2	0.54	0.82	0.65	0.09	0.06	23.00	0.91	0.60
	3	0.49	0.85	0.63	0.10	0.06	22.00	0.92	0.57
	4	0.58	0.81	0.67	0.09	0.06	24.00	0.91	0.61
Lotu-Ixa	0	0.43	0.79	0.56	**0.05**	0.04	2.00	**1.00**	0.56
	1	0.46	0.79	0.58	**0.05**	0.03	2.00	**1.00**	0.58
	2	0.47	0.79	0.59	**0.05**	0.03	2.00	**1.00**	0.59
	3	0.53	0.78	0.63	**0.05**	0.03	1.00	**1.00**	0.63
	4	0.15	1.00	0.25	0.09	0.08	1.00	**1.00**	0.25
COTECMAR-UTB	0	0.29	0.65	0.40	0.12	0.10	69.00	0.74	0.29
	1	0.25	0.01	0.02	0.11	0.11	**1.00**	**1.00**	0.02
SINAI-UJA	0	0.24	1.00	0.39	0.08	0.05	3.00	0.99	0.38
	1	0.17	1.00	0.29	0.07	0.07	2.00	**1.00**	0.29
	2	0.22	1.00	0.36	0.08	0.05	2.00	**1.00**	0.36
	3	0.21	1.00	0.35	0.08	0.05	3.00	0.99	0.35
	4	0.20	1.00	0.34	0.09	0.06	3.00	0.99	0.33
NYCUNLP	0	0.14	1.00	0.25	0.12	0.08	3.00	0.99	0.25
	1	0.16	0.99	0.28	0.14	0.08	7.00	0.98	0.27
	2	0.17	0.95	0.28	0.16	0.08	10.00	0.96	0.27
	3	0.18	0.94	0.31	0.16	0.08	13.50	0.95	0.29
	4	0.20	0.93	0.33	0.16	0.07	18.00	0.93	0.31
FU-TU-DFKI	0	0.17	0.97	0.29	0.16	0.07	11.00	0.96	0.28
Capy-team	0	0.11	1.00	0.20	0.11	0.10	1.50	**1.00**	0.20
	1	0.11	1.00	0.20	0.11	0.10	**1.00**	**1.00**	0.20
	2	0.11	1.00	0.20	0.11	0.10	2.00	**1.00**	0.20
	3	0.11	1.00	0.20	0.11	0.10	**1.00**	**1.00**	0.20
	4	0.11	1.00	0.20	0.11	0.10	2.00	**1.00**	0.20
DS-GT	0	0.11	1.00	0.20	0.12	0.10	2.00	**1.00**	0.20
	1	0.11	1.00	0.20	0.12	0.10	2.00	**1.00**	0.20

Table 9. Ranking-based evaluation for Task 2.

Team	Run	1 writing			100 writings			500 writings			1000 writings		
		$P@10$	$NDCG@10$	$NDCG@100$	$P@10$	$NDCG@10$	$NDCG@100$	$P@10$	$NDCG@10$	$NDCG@100$	$P@10$	$NDCG@10$	$NDCG@100$
HIT-SCIR	0	**1.00**	**1.00**	0.58	**1.00**	**1.00**	0.84	**1.00**	**1.00**	0.89	**1.00**	**1.00**	0.90
	1	**1.00**	**1.00**	0.58	**1.00**	**1.00**	0.84	**1.00**	**1.00**	0.89	**1.00**	**1.00**	0.90
	2	**1.00**	**1.00**	0.58	**1.00**	**1.00**	0.84	**1.00**	**1.00**	0.89	**1.00**	**1.00**	0.90
	3	**1.00**	**1.00**	0.58	**1.00**	**1.00**	0.84	**1.00**	**1.00**	0.89	**1.00**	**1.00**	0.90
	4	**1.00**	**1.00**	0.58	**1.00**	**1.00**	0.83	**1.00**	**1.00**	0.89	**1.00**	**1.00**	0.90
ELiRF-UPV	0	0.90	0.88	0.36	**1.00**	**1.00**	0.69	0.90	0.94	0.74	0.90	0.81	0.74
	1	0.30	0.25	0.32	**1.00**	**1.00**	0.45	**1.00**	**1.00**	0.44	**1.00**	**1.00**	0.46
	2	0.20	0.31	0.14	**1.00**	**1.00**	0.45	**1.00**	**1.00**	0.44	**1.00**	**1.00**	0.46
	3	0.90	0.94	0.35	**1.00**	**1.00**	0.68	0.60	0.46	0.60	0.70	0.63	0.63
	4	0.60	0.75	0.27	**1.00**	**1.00**	0.68	0.60	0.46	0.60	0.70	0.63	0.63
HU	0	0.90	0.81	0.53	0.80	0.87	0.49	0.70	0.68	0.48	0.70	0.66	0.49
	1	**1.00**	**1.00**	**0.62**	0.90	0.88	0.57	0.60	0.71	0.35	0.40	0.60	0.26
	2	0.30	0.21	0.11	0.20	0.16	0.12	0.00	0.00	0.11	0.40	0.60	0.26
	3	0.30	0.21	0.11	0.20	0.16	0.12	0.00	0.00	0.11	0.40	0.60	0.26
	4	0.60	0.53	0.33	0.40	0.58	0.36	0.30	0.37	0.24	0.40	0.60	0.26
UET-Psyche-Warriors	0	0.90	0.92	0.41	0.30	0.38	0.17	0.10	0.10	0.14	0.00	0.00	0.12
	1	0.90	0.93	0.43	0.10	0.12	0.11	0.00	0.00	0.12	0.00	0.00	0.12
	2	0.90	0.93	0.43	0.10	0.12	0.11	0.00	0.00	0.12	0.00	0.00	0.12
	3	0.90	0.93	0.43	0.10	0.12	0.11	0.00	0.00	0.12	0.00	0.00	0.12
	4	0.90	0.93	0.42	0.10	0.12	0.11	0.00	0.00	0.12	0.00	0.00	0.12
PJs-team	0	0.60	0.59	0.35	0.50	0.44	0.38	0.70	0.78	0.60	0.60	0.69	0.63
	1	0.60	0.59	0.35	0.40	0.41	0.39	0.50	0.60	0.54	0.50	0.63	0.51
	2	0.60	0.59	0.35	0.40	0.38	0.37	0.50	0.61	0.53	0.50	0.63	0.52
	3	0.60	0.59	0.35	0.30	0.32	0.36	0.60	0.66	0.51	0.50	0.66	0.51
	4	0.60	0.59	0.35	0.40	0.39	0.37	0.50	0.61	0.55	0.40	0.56	0.52
Lotu-Ixa	0	0.80	0.84	0.55	**1.00**	**1.00**	0.72	**1.00**	**1.00**	0.62	**1.00**	**1.00**	0.64
	1	0.90	0.94	0.57	**1.00**	**1.00**	0.73	**1.00**	**1.00**	0.63	**1.00**	**1.00**	0.63
	2	**1.00**	**1.00**	0.58	**1.00**	**1.00**	0.73	**1.00**	**1.00**	0.61	**1.00**	**1.00**	0.62
	3	0.90	0.81	0.58	**1.00**	**1.00**	0.74	**1.00**	**1.00**	0.61	**1.00**	**1.00**	0.62
	4	0.60	0.59	0.44	0.80	0.84	0.55	0.90	0.94	0.52	**1.00**	**1.00**	0.52
COTECMAR-UTB	0	0.30	0.23	0.23	0.00	0.00	0.22	0.20	0.15	0.18	0.20	0.13	0.17
	1	0.00	0.00	0.12	0.00	0.00	0.12	0.00	0.00	0.12	0.00	0.00	0.12
SINAI-UJA	0	**1.00**	**1.00**	0.59	0.80	0.87	0.53	0.90	0.88	0.54	0.90	0.92	0.54
	1	0.90	0.93	0.59	0.80	0.75	0.47	0.70	0.67	0.44	0.60	0.61	0.44
	2	0.90	0.92	0.58	0.70	0.79	0.47	0.90	0.94	0.53	**1.00**	**1.00**	0.52
	3	**1.00**	**1.00**	0.55	0.90	0.93	0.48	0.90	0.88	0.50	0.90	0.90	0.47
	4	**1.00**	**1.00**	0.57	0.60	0.74	0.45	0.70	0.76	0.52	0.60	0.70	0.51
NYCUNLP	0	0.50	0.53	0.42	0.70	0.68	0.35	0.70	0.62	0.33	0.50	0.47	0.31
	1	0.50	0.53	0.42	0.80	0.86	0.40	0.80	0.74	0.35	0.70	0.62	0.34
	2	0.50	0.53	0.42	0.80	0.86	0.45	0.80	0.86	0.40	0.80	0.74	0.36
	3	0.50	0.53	0.42	0.80	0.88	0.50	0.70	0.82	0.41	0.70	0.69	0.37
	4	0.50	0.53	0.42	0.70	0.69	0.45	0.70	0.82	0.42	0.70	0.69	0.39
FU-TU-DFKI	0	0.90	0.94	0.44	0.00	0.00	0.12	0.00	0.00	0.00	0.00	0.00	0.00
Capy-team	0	0.20	0.18	0.13	0.20	0.18	0.10	0.10	0.07	0.09	0.20	0.15	0.14
	1	0.00	0.00	0.07	0.10	0.08	0.09	0.10	0.06	0.08	0.10	0.19	0.18
	2	0.10	0.07	0.12	0.10	0.07	0.12	0.00	0.00	0.11	0.20	0.13	0.16
	3	0.20	0.29	0.16	0.10	0.10	0.14	0.20	0.26	0.14	0.10	0.12	0.10
	4	0.00	0.00	0.11	0.30	0.20	0.12	0.00	0.00	0.08	0.10	0.19	0.13
DS-GT	0	0.20	0.12	0.22	0.00	0.00	0.12	0.20	0.18	0.20	0.10	0.12	0.17
	1	0.90	0.92	0.52	0.00	0.00	0.12	0.20	0.18	0.20	0.10	0.12	0.17

turns, but we encouraged the participants to produce their decisions as early as possible.

After ending the conversation with a persona, a participating system had to return two files. The first was a structured log that preserves, in chronological order, every prompt–response pair exchanged with the agent; this file serves auditing and qualitative analysis. The second was a JSON record containing three fields: the predicted BDI-II score (an integer 0–63), the corresponding severity category, and up to four symptom drawn from the BDI-II list in Table 10, that best explained the score.

Table 10. The 21 BDI-II Depression Symptoms.

21 Depression Symptoms		
Sadness	Pessimism	Past Failure
Loss of Pleasure	Guilty Feelings	Punishment Feelings
Self-Dislike	Self-Criticalness	Suicidal Thoughts or Wishes
Crying	Agitation	Loss of Interest
Indecisiveness	Worthlessness	Loss of Energy
Changes in Sleeping Pattern	Irritability	Changes in Appetite
Concentration Difficulty	Tiredness or Fatigue	Loss of Interest in Sex

Table 11. Two BDI-II symptoms and their four response options (0–3).

Symptom	Response options (score)
Sadness	0: I do not feel sad
	1: I feel sad much of the time
	2: I am sad all the time
	3: I am so sad or unhappy that I can't stand it
Self-Dislike	0: I feel the same about myself as ever
	1: I have lost confidence in myself
	2: I am disappointed in myself
	3: I dislike myself

4.1   Pilot Task: LLM Personas Design and Construction

We adopted a clinician-in-the-loop design workflow to build the twelve LLM personas. A team of three clinical psychologists co-designed a template that captures both general biographical detail and clinically information. Using this template we instantiated a pool of draft personas with GPT-4o, each conditioned on a different user history.

The same clinicians then conducted free-form interviews with every draft, rating each dialogue along two main dimensions:

– The *overall* dimension covered traits associated with conversational attributes: human-likeness, lexical fluency, coherence, and affective naturalness.
– The *diagnostic* dimension targeted domain realism, including emotional consistency, fidelity to depressive symptomatology, willingness to elaborate, and cognitive style (rumination, processing speed, abstraction level).

Feedback was recorded on a five-point Likert scale and complemented with qualitative comments. Insights from this evaluation cycle informed a second engineering pass in which every persona was represented through a structured prompt comprising the main following elements:

- Core profile. A stable set of attributes: name, age, gender, marital status and an a pre-defined BDI-II score.
- Key negative symptoms. Up to four key BDI-II symptoms (or less for control personas) that the agent should manifest recurrently and coherently.
- Memory and reflection. Specific snippets describing life history, social context, and salient past events; these cues allow the agent to maintain narrative continuity and to provide retrospective insight into its mood.
- Language and communication style. Use of vocabulary, and typical sentence length so that each persona speaks with a recognisable "voice".
- Behavioural constraints. Guard-rails that prohibit explicit self-diagnosis and that keep the agent away from clinical recommendations, thereby forcing participants to infer depression indirectly.
- Response goals. High-level objectives such as "answer candidly but not expansively," "avoid mentioning diagnosis unless prompted," and "display mild self-disclosure".
- Environment and context. Brief situational framing (e.g. studying for exams, recent job change) that provides topical depth without locking the dialogue.
- Few-shot exemplars. Short question–answer pairs illustrating the expected tone and symptom expression.
- Restricted responses. A blacklist of phrases that would break immersion (e.g. "As an AI language model...") replaced with context-appropriate alternatives.

The final personas were frozen only after a second round of clinician interaction confirmed that they satisfied a minimum threshold on both the overall and diagnostic scales. This iterative, expert-guided construction process proved essential to achieve dialogues that are simultaneously natural and diagnostically meaningful, laying the groundwork for future large-scale evaluations of conversational mental-health screening systems.

4.2   Pilot Task: Participant Teams

Table 12 shows the participant teams and some statistics about their interactions such as the mean number of messages per run, and the mean number of characters per message. The numbers reveal a wide range of interaction strategies:

- ixa-ave submitted the maximum number of runs (four) and tended to carry out relatively lengthy dialogues (≈ 31 messages each) while keeping their prompts concise (≈ 415 characters per turn).
- SINAI-UJA used a fast approach, with only 6–7 turns on average, yet still packed almost 490 characters into every message, suggesting dense, information-rich questioning.
- DS-GT followed an intermediate approach, with ≈ 21 messages per run and 783 characters per message, balancing breadth and depth of interaction.
- PJs-team produced long messages ($\approx 1\,045$ characters) within a limited number of turns (≈ 8), delivering extended prompts.
- LT4SG employed a fixed sequence of ten short messages averaging only 41 characters, representing the most lightweight strategy.

Table 12. Pilot task (LLMs): participating teams, number of runs, mean number of messages per run, mean number of characters per message.

Team	#Runs	#Mean messages	#Mean characters per message
ixa-ave [46]	4	31.02	414.44
SINAI-UJA [23]	3	6.54	488.25
DS-GT [6]	2	20.79	782.81
PJs-team [45]	1	7.67	1045.16
LT4SG	1	10	40.73

4.3   Pilot Task: Evaluation Metrics

Based on evaluation metrics that have been developed from eRisk 2019 [14], which involved the use of BDI-II questionnaires and scores, we extend and develop the evaluation approaches as follows:

- **Depression Category Hit Rate (DCHR)**: Based on the four depression level categories that we have discussed, from minimal depression to severe depression, this effectiveness measure examines the fraction of cases where the BDI-II scores describing simulated personas estimated by the participants lie in the correct depression category.
- **Average DODL (ADODL)**: For this pilot task, we reuse the Average Difference between Overall Depression Levels (ADODL), which measures the closeness between the actual and estimated depression level for effectiveness measurement. The ADODL is calculated by following: $CR = (MAD - |ADL - EDL|)/MAD$, where $|ADL - EDL|$ calculates the absolute value between the Actual Depression Level (ADL) and the Estimated Depression Level (EDL). Then divided by Maximum Absolute Difference (i.e., 63) to obtain a normalised evaluation score in [0,1]. For example, if a simulated persona has a minor depression severity (depression level 5) and a participant estimates the depression level is 9, the DODL is calculated as $(63 - |9 - 5|)/63 = 0.9365$.
- **Average Symptom Hit Rate (ASHR)**: For the last effectiveness measure, aside from estimating the depression level of simulated personas as per BDI-II scores, this pilot task also involves the identification of major depression symptoms of simulated personas. Hence, SHR calculates the ratio of cases where the participants can correctly identify the major symptoms of the simulated personas. For example, each simulated persona has four major symptoms. If a participant accurately identifies two of them, then the SHR equals $2/4 = 0.5$.

4.4   Pilot Task: Results

Table 13 presents the official runs, ranked by best ADODL. The strongest submission, *SINAI-UJA (run 1)*, achieves an ADODL of 0.93, meaning the predicted scores differ by less than five points on average from the clinician reference. Its DCHR of 0.58 shows that most of these small errors still fall within the incorrect severity band. *DS-GT* attains comparable category accuracy (0.50) with only a modest drop in ADODL, reaching similar level reliability despite larger absolute score errors.

Across all teams, however, symptom recognition stays behind score estimation: even the best ASHR values hover below 0.30, indicating that systems often capture the global severity signal without isolating which symptoms drive it.

Table 13. Evaluation for Task 3 with teams ordered in terms of best ADODL. '*' indicates the manual runs (human-in-the-loop).

Team	Run	DCHR	ADODL	ASHR
SINAI-UJA	0	**0.66**	0.92	0.21
	1	0.58	**0.93**	**0.29**
	2	0.41	0.88	0.21
DS-GT	0	0.42	0.83	0.12
	1	0.50	0.89	0.27
	2	0.33	0.86	**0.29**
	3	0.50	0.84	0.25
ixa_ave	0*	0.33	0.80	0.25
	1	0.33	0.76	**0.29**
	2	0.33	0.83	0.21
	3	0.17	0.81	0.19
LT4SG	0	0.33	0.78	0.06
PJs-team	0	0.33	0.73	0.25

5   Conclusions

This paper provided an overview of eRisk 2025, the ninth edition of the eRisk lab, which moved toward two new tasks that require richer conversational understanding and interactive settings. The Task 1, which was the final edition of the sentence-ranking challenge for BDI-II symptoms, attracted 67 runs from 17 teams. Task 2 introduced full-thread context for the first time in early detection of depression. In this task, we received 50 runs from 12 teams, and showed that models able to exploit dialogue structure can issue accurate alerts after remarkably few turns, although a clear trade-off persists between earliness and recall. The pilot task went a step further, replacing static corpora with live interaction against LLM-driven personas. Despite the absence of training data, five teams submitted 13 runs; top systems achieved near-perfect BDI-II score estimation yet still struggled to pinpoint the specific symptoms that reflect those scores, highlighting the difficulty of symptom-level grounding in open conversation.

Taken together, the 130 runs submitted this year confirm both the community's engagement and the practicality of evaluation settings that approach real conversational use cases. Three broad lessons emerge: adding even modest context improves detection, timeliness must remain a core metric. Moreover, clinician-guided LLM personas, despite having a lot of room for improvement, are able to create realistic yet privacy-preserving frameworks. Future eRisk editions will continue to shift toward dialogue-centric tasks and deeper integration of LLM capabilities to keep pace with how people communicate online and how assistive technologies are deployed.

Acknowledgments. The authors thank the financial support supplied by the grant PID2022-137061OB-C21 funded by MI-CIU/AEI/10.13039/501100011033 and by "ERDF/EU". The authors also thank the funding supplied by the Consellería de Cultura, Educación, Formación Profesional e Universidades (accreditations ED431G 2023/01 and ED431C 2025/49) and the European Regional Development Fund, which acknowledges the CITIC, as a center accredited for excellence within the Galician University System and a member of the CIGUS Network, receives subsidies from the Department of Education, Science, Universities, and Vocational Training of the Xunta de Galicia. Additionally, it is co-financed by the EU through the FEDER Galicia 2021-27 operational program (Ref. ED431G 2023/01).

References

1. Adhikary, S., Das, J., Roy, D.: Thinkir at eRisk 2025: early detection and risk assessment of depression using transformer models. In: Working Notes of CLEF 2025 - Conference and Labs of the Evaluation Forum, Madrid, Spain, 9–12 September 2025 (2025)
2. Beck, A.T., Ward, C.H., Mendelson, M., Mock, J., Erbaugh, J.: An inventory for measuring depression. JAMA Psychiat. **4**(6), 561–571 (1961)
3. Benloucif, A., Nannapuraju, Y., Bellam, S., Hu, Y., Zhao, Z., V. G.V.V.: Lhs712team-1 at eRisk@clef 2025: searching for depression symptoms using various natural language processing algorithms. In: Working Notes of CLEF 2025 - Conference and Labs of the Evaluation Forum, Madrid, Spain, 9–12 September 2025 (2025)
4. Cardona, L.F.M., Loaiza, J.M.S., Castillo, E.A.P.D., Santos, J.C.M., Castañeda, J.E.S.: Cotecmar-utb at eRisk 2025: semantic-centroid symptom ranking and early depression detection using adaptive decision rule. In: Working Notes of CLEF 2025 - Conference and Labs of the Evaluation Forum, Madrid, Spain, 9–12 September 2025 (2025)
5. Diogo A. P.N., Ribeiro, E.: Inesc-id @ eRisk 2025: exploring fine-tuned, similarity-based, and prompt-based approaches to depression symptom identification. In: Working Notes of CLEF 2025 - Conference and Labs of the Evaluation Forum, Madrid, Spain, 9–12 September 2025 (2025)
6. Guecha, D., Chiu, Y., Miyaguchi, A., Gaur, S.: Ds@gt at eRisk 2025: from prompts to predictions, benchmarking early depression detection with conversational agent

based assessments and temporal attention models. In: Working Notes of CLEF 2025 - Conference and Labs of the Evaluation Forum, Madrid, Spain, 9–12 September 2025 (2025)

7. Kara, E., Peña, R.E.M., Raithel, L.: Fu-tu-dfki@eRisk 2025: a linguistically informed but overdiagnosing approach to early depression detection. In: Working Notes of CLEF 2025 - Conference and Labs of the Evaluation Forum, Madrid, Spain, 9–12 September 2025 (2025)

8. Larrayoz, X., Casillas, A., Pérez, A.: Leveraging conversational context and semantic relabeling for early depression detection. In: Working Notes of CLEF 2025 - Conference and Labs of the Evaluation Forum, Madrid, Spain, 9–12 September 2025(2025)

9. Losada, D.E., Crestani, F.: A test collection for research on depression and language use. In: Fuhr, N., et al. (eds.) CLEF 2016. LNCS, vol. 9822, pp. 28–39. Springer, Cham (2016). https://doi.org/10.1007/978-3-319-44564-9_3

10. Losada, D.E., Crestani, F., Parapar, J.: eRISK 2017: CLEF lab on early risk prediction on the internet: experimental foundations. In: Jones, G.J.F., et al. (eds.) CLEF 2017. LNCS, vol. 10456, pp. 346–360. Springer, Cham (2017). https://doi.org/10.1007/978-3-319-65813-1_30

11. Losada, D.E., Crestani, F., Parapar, J.: Overview of eRisk 2018: early risk prediction on the internet (extended lab overview). In: CEUR Proceedings of the Conference and Labs of the Evaluation Forum, CLEF 2018, Avignon, France (2018)

12. Losada, D.E., Crestani, F., Parapar, J.: Overview of eRisk: early risk prediction on the internet. In: Bellot, P., et al. (eds.) CLEF 2018. LNCS, vol. 11018, pp. 343–361. Springer, Cham (2018). https://doi.org/10.1007/978-3-319-98932-7_30

13. Losada, D.E., Crestani, F., Parapar, J.: Early detection of risks on the internet: an exploratory campaign. In: Azzopardi, L., et al. (eds.) ECIR 2019. LNCS, vol. 11438, pp. 259–266. Springer, Cham (2019). https://doi.org/10.1007/978-3-030-15719-7_35

14. Losada, D.E., Crestani, F., Parapar, J.: Overview of eRisk 2019: early risk prediction on the internet. In: Crestani, F., et al. (eds.) Experimental IR Meets Multilinguality, Multimodality, and Interaction, pp. 340–357. Springer, Heidelberg (2019)

15. Losada, D.E., Crestani, F., Parapar, J.: Overview of eRisk 2019 early risk prediction on the internet. In: Crestani, F., et al. (eds.) CLEF 2019. LNCS, vol. 11696, pp. 340–357. Springer, Cham (2019). https://doi.org/10.1007/978-3-030-28577-7_27

16. Losada, D.E., Crestani, F., Parapar, J.: Overview of eRisk at CLEF 2019: early risk prediction on the internet (extended overview). In: CEUR Proceedings of the Conference and Labs of the Evaluation Forum, CLEF 2019, Lugano, Switzerland (2019)

17. Losada, D.E., Crestani, F., Parapar, J.: eRisk 2020: self-harm and depression challenges. In: Jose, J.M., et al. (eds.) ECIR 2020. LNCS, vol. 12036, pp. 557–563. Springer, Cham (2020). https://doi.org/10.1007/978-3-030-45442-5_72

18. Losada, D.E., Crestani, F., Parapar, J.: Overview of eRisk 2020: early risk prediction on the internet. In: Arampatzis, A., et al. (eds.) CLEF 2020. LNCS, vol. 12260, pp. 272–287. Springer, Cham (2020). https://doi.org/10.1007/978-3-030-58219-7_20

19. Losada, D.E., Crestani, F., Parapar, J.: Overview of eRisk at CLEF 2020: early risk prediction on the internet (extended overview). In: Working Notes of CLEF 2020 - Conference and Labs of the Evaluation Forum, Thessaloniki, Greece, 22–25 September 2020 (2020)

20. Mai, T.P., H., M.H.L., Tran, D.L., Can, D.C., Le, H.Q.: Uet@eRisk2025: severity estimation for depression symptoms searching and early risk detection. In: Working Notes of CLEF 2025 - Conference and Labs of the Evaluation Forum, Madrid, Spain, 9–12 September 2025 (2025)

21. Molina, J.C., Fernandez, P.M.: Hulat-uc3m at task 1@eRisk 2025: detecting depression using machine learning approaches. In: Working Notes of CLEF 2025 - Conference and Labs of the Evaluation Forum, Madrid, Spain, 9–12 September 2025 (2025)

22. Munz, N., Aharon, E., Segal, A., Gal, K.: Semantic retrieval of bdi symptoms in user writings. In: Working Notes of CLEF 2025 - Conference and Labs of the Evaluation Forum, Madrid, Spain, 9–12 September 2025 (2025)

23. Mármol-Romero, A.M., García-Vega, M., Ángel García-Cumbreras, M., Montejo-Ráez, A.: Sinai at eRisk@clef 2025: transformer-based and conversational strategies for depression detection. In: Working Notes of CLEF 2025 - Conference and Labs of the Evaluation Forum, Madrid, Spain, 9–12 September 2025 (2025)

24. Otero, D., Parapar, J., Barreiro, Á.: Beaver: efficiently building test collections for novel tasks. In: Proceedings of the First Joint Conference of the Information Retrieval Communities in Europe (CIRCLE 2020), Samatan, Gers, France, 6–9 July 2020 (2020)

25. Otero, D., Parapar, J., Barreiro, Á.: The wisdom of the rankers: a cost-effective method for building pooled test collections without participant systems. In: SAC '21: The 36th ACM/SIGAPP Symposium on Applied Computing, Virtual Event, Republic of Korea, 22–26 March 2021, pp. 672–680 (2021)

26. Parapar, J., Martín-Rodilla, P., Losada, D.E., Crestani, F.: eRisk 2021: pathological gambling, self-harm and depression challenges. In: Hiemstra, D., Moens, M.-F., Mothe, J., Perego, R., Potthast, M., Sebastiani, F. (eds.) ECIR 2021. LNCS, vol. 12657, pp. 650–656. Springer, Cham (2021). https://doi.org/10.1007/978-3-030-72240-1_76

27. Parapar, J., Martín-Rodilla, P., Losada, D.E., Crestani, F.: Overview of eRisk 2021: early risk prediction on the internet. In: Candan, K.S., et al. (eds.) CLEF 2021. LNCS, vol. 12880, pp. 324–344. Springer, Cham (2021). https://doi.org/10.1007/978-3-030-85251-1_22

28. Parapar, J., Martín-Rodilla, P., Losada, D.E., Crestani, F.: Overview of eRisk at CLEF 2021: early risk prediction on the internet (extended overview). In: Proceedings of the Working Notes of CLEF 2021 - Conference and Labs of the Evaluation Forum, Bucharest, Romania, 21–24 September 2021, pp. 864–887 (2021)

29. Parapar, J., Martín-Rodilla, P., Losada, D.E., Crestani, F.: eRisk 2022: pathological gambling, depression, and eating disorder challenges. In: Advances in Information Retrieval - 44th European Conference on IR Research, ECIR 2022, Stavanger, Norway, 10–14 April 2022, Proceedings, Part II, pp. 436–442 (2022)

30. Parapar, J., Martín-Rodilla, P., Losada, D.E., Crestani, F.: Overview of eRisk 2022: Early risk prediction on the internet. In: Experimental IR Meets Multilinguality, Multimodality, and Interaction - 13th International Conference of the CLEF Association, CLEF 2022, Bologna, Italy, 5–8 September 2022, pp. 233–256 (2022)

31. Parapar, J., Martín-Rodilla, P., Losada, D.E., Crestani, F.: Overview of eRisk at CLEF 2022: early risk prediction on the internet (extended overview). In: Proceedings of the Working Notes of CLEF 2022 - Conference and Labs of the Evaluation Forum, Bologna, Italy, 5–8 September 2022, pp. 821–850 (2022)

32. Parapar, J., Martín-Rodilla, P., Losada, D.E., Crestani, F.: eRisk 2023: depression, pathological gambling, and eating disorder challenges. In: Advances in Information

Retrieval - 45th European Conference on IR Research, ECIR 2023, Dublin, Ireland, 2–6 April 2023, Proceedings, Part III, p. 585–592 (2023)

33. Parapar, J., Martín-Rodilla, P., Losada, D.E., Crestani, F.: Overview of eRisk 2023: early risk prediction on the internet. In: Experimental IR Meets Multilinguality, Multimodality, and Interaction - 14th International Conference of the CLEF Association, CLEF 2023, Thessaloniki, Greece, 18–21 September 2023, pp. 233–256 (2023)

34. Parapar, J., Martín-Rodilla, P., Losada, D.E., Crestani, F.: Overview of eRisk at CLEF 2023: early risk prediction on the internet (extended overview). In: Proceedings of the Working Notes of CLEF 2023 - Conference and Labs of the Evaluation Forum, Thessaloniki, Greece, 18–21 September 2023 (2023)

35. Parapar, J., Martín-Rodilla, P., Losada, D.E., Crestani, F.: eRisk 2024: Depression, anorexia, and eating disorder challenges. In: Goharian, N., Tonellotto, N., He, Y., Lipani, A., McDonald, G., Macdonald, C., Ounis, I. (eds.) Advances in Information Retrieval - 46th European Conference on Information Retrieval, ECIR 2024, Glasgow, UK, 24–28 March 2024, Proceedings, Part V. Lecture Notes in Computer Science, vol. 14612, pp. 474–481. Springer, Heidelberg (2024). https://doi.org/10.1007/978-3-031-56069-9_65

36. Parapar, J., Martín-Rodilla, P., Losada, D.E., Crestani, F.: Overview of eRisk 2024: Early risk prediction on the internet. In: Goeuriot, L., Mulhem, P., Quénot, G., Schwab, D., Nunzio, G.M.D., Soulier, L., Galuscáková, P., de Herrera, A.G.S., Faggioli, G., Ferro, N. (eds.) Experimental IR Meets Multilinguality, Multimodality, and Interaction - 15th International Conference of the CLEF Association, CLEF 2024, Grenoble, France, 9–12 September 2024, Proceedings, Part II. Lecture Notes in Computer Science, vol. 14959, pp. 73–92. Springer, Heidelberg (2024). https://doi.org/10.1007/978-3-031-71908-0_4

37. Parapar, J., Martín-Rodilla, P., Losada, D.E., Crestani, F.: Overview of eRisk 2024: early risk prediction on the internet (extended overview). In: Faggioli, G., Ferro, N., Galuscáková, P., de Herrera, A.G.S. (eds.) Working Notes of the Conference and Labs of the Evaluation Forum (CLEF 2024), Grenoble, France, 9–12 September 2024. CEUR Workshop Proceedings, vol. 3740, pp. 759–781. CEUR-WS.org (2024). https://ceur-ws.org/Vol-3740/paper-72.pdf

38. Parapar, J., Perez, A., Wang, X., Crestani, F.: eRisk 2025: contextual and conversational approaches for depression challenges. In: European Conference on Information Retrieval, pp. 416–424. Springer, Heidelberg (2025)

39. Parapar, J., Perez, A., Wang, X., Crestani, F.: Overview of eRisk 2025: early risk prediction on the internet (extended overview). In: Faggioli, G., Ferro, N., Rosso, P., Spina, D. (eds.) Working Notes of the Conference and Labs of the Evaluation Forum (CLEF 2025), Madrid, Spain, 9–12 September 2025. CEUR Workshop Proceedings, CEUR-WS.org (2025)

40. Saad, M., Abbas, M., Chaudhry, A.U., Alvi, F., Samad, A.: Contextualized early detection of depression – hybrid and time-aware approaches: Hu at eRisk task 2 2025. In: Working Notes of CLEF 2025 - Conference and Labs of the Evaluation Forum, Madrid, Spain, 9–12 September 2025 (2025)

41. Sadeque, F., Xu, D., Bethard, S.: Measuring the latency of depression detection in social media. In: WSDM, pp. 495–503. ACM (2018)

42. Segarra, A.C., Esteve, V.A., Marco, A.M., Oliver, L.F.H.: Elirf-upv at eRisk 2025: new approaches to the detection and early detection of symptoms and signs of depression. In: Working Notes of CLEF 2025 - Conference and Labs of the Evaluation Forum, Madrid, Spain, 9–12 September 2025 (2025)

43. Son, N.M., Thin, D.V.: Sonuit eRisk2025: enhanced depression detection on social media via filtering and re-ranking. In: Working Notes of CLEF 2025 - Conference and Labs of the Evaluation Forum, Madrid, Spain, 9–12 September 2025 (2025)

44. Trotzek, M., Koitka, S., Friedrich, C.: Utilizing neural networks and linguistic metadata for early detection of depression indications in text sequences. IEEE Trans. Knowl. Data Eng. (2018)

45. Vachharajani, P.: Transformer ensembles and llm-powered approaches for depression symptom analysis and contextualized early risk detection. In: Working Notes of CLEF 2025 - Conference and Labs of the Evaluation Forum, Madrid, Spain, 9–12 September 2025 (2025)

46. Varela, A., Oronoz, M., Casillas, A., Pérez, A.: Detection of depression with symptom similarity: data reduction and llm personas. In: Working Notes of CLEF 2025 - Conference and Labs of the Evaluation Forum, Madrid, Spain, 9–12 September 2025 (2025)

47. Zi, Y., Wang, B., Zhao, Y., Qin, B.: Hit-scir@eRisk2025: exploring the potential of a learnable screening model and risk post buffer-based framework for contextualized early prediction of depression on social media. In: Working Notes of CLEF 2025 - Conference and Labs of the Evaluation Forum, Madrid, Spain, 9–12 September 2025 (2025)

Overview of EXIST 2025: Learning with Disagreement for Sexism Identification and Characterization in Tweets, Memes, and TikTok Videos

Laura Plaza[1]($\boxtimes$), Jorge Carrillo-de-Albornoz[1], Iván Arcos[2], Paolo Rosso[2,3], Damiano Spina[4], Enrique Amigó[1], Julio Gonzalo[1], and Roser Morante[1]

[1] Universidad Nacional de Educación a Distancia (UNED), 28040 Madrid, Spain
{lplaza,jcalbornoz,enrique,julio,rmorant}@lsi.uned.es
[2] Universitat Politècnica de València (UPV), 46022 Valencia, Spain
prosso@dsic.upv.es, iarcgab@etsinf.upv.es
[3] ValgrAI - Valencian Graduate School and Research Network of Artificial Intelligence, 46022 Valencia, Spain
[4] RMIT University, 3000 Melbourne, Australia
damiano.spina@rmit.edu.au

Abstract. This paper presents the EXIST 2025 Lab on sexism detection and categorization in social media, which took place at the CLEF 2025 conference and marks the fifth edition of the EXIST Shared Task. Building on the success of previous editions, EXIST 2025 addresses the growing concern over the spread of offensive and discriminatory content targeting women across online platforms, which significantly impacts women's well-being and freedom of expression. The lab comprises nine tasks in two languages (English and Spanish), organized around three core objectives: sexism identification, source intention detection, and sexism categorization. These tasks are applied across three media types—text (tweets), image (memes), and video (TikToks)—offering a multimodal perspective that allows for a deeper understanding of how sexism manifests across different formats and user interactions. As in previous editions, EXIST 2025 adopts the "Learning With Disagreement" paradigm, using annotations from multiple annotators that reflect diverse and at times conflicting viewpoints. This overview describes the task design, datasets, evaluation methodology, participating systems, and results of EXIST 2025, which has surpassed participation expectations with 244 registered teams from 38 countries, 114 teams from 23 countries submitting runs, a total of 873 runs processed, and 33 working notes published.

Warning: Some of the examples included in this paper may contain offensive language and explicit descriptions of sexist behavior, which may be disturbing to the reader.

Keywords: sexism identification · sexism categorization · learning with disagreements · tweets · memes · TikTok videos · human-centric AI

J. Carrillo-de-Albornoz et al. (Eds.): CLEF 2025, LNCS 16089, pp. 266–289, 2026.
https://doi.org/10.1007/978-3-032-04354-2_16

1 Introduction

Sexism refers to prejudice or discrimination based on a person's sex or gender, often manifesting in the belief that one gender is superior to another. It can take many forms, from overt aggression and harassment to subtler behaviors and norms that reinforce inequality. While sexism affects individuals of all genders, it disproportionately impacts women, particularly in digital spaces.

In recent years, online platforms like Twitter and TikTok have become breeding grounds for the proliferation of sexist discourse. On Twitter, sexism often manifests through harassment, trolling, and misogynistic hashtags that normalize discriminatory narratives [13,24]. TikTok, by contrast, poses unique challenges due to its algorithm-driven content promotion and its popularity among younger audiences. Its recommendation system can generate filter bubbles that reinforce sexist ideologies [23], while visual trends and content moderation disparities contribute to the hypersexualization and objectification of women [10,15]. These dynamics not only perpetuate traditional gender stereotypes but can also shape the perceptions and behaviors of young users.

To tackle these challenges, the sEXism Identification in Social neTworks (EXIST) campaign was launched in 2021. EXIST is a series of shared tasks and scientific events aimed at identifying, analyzing, and mitigating sexist content on social networks. The first two editions were hosted under the IberLEF forum [32,33], and focused on textual data. In 2023, EXIST became a CLEF Lab [31], introducing a third task centered on detecting the communicative intention behind sexist messages and adopting for the first time the Learning with Disagreement (LeWiDi) paradigm [35]. This paradigm acknowledges that disagreements among annotators are not noise, but valuable signals that reflect the subjectivity inherent to tasks like sexism detection. The fourth edition of EXIST (2024) expanded the challenge to multimodal data by introducing tasks involving memes. Memes, while often humorous, are increasingly used to spread prejudices under the guise of irony [5,8,20,22]. Their blend of text and image makes them particularly insidious vectors for normalizing sexist stereotypes, especially when humor is used to reduce the perceived harm [12,16].

EXIST 2025 marks the fifth edition of the challenge and represents its most ambitious iteration yet. Held again as a CLEF Lab,[1] it comprises nine tasks in total—covering three core objectives (sexism identification, source intention detection, and sexism categorization) across three modalities: tweets (text), memes (image), and TikToks (video). This multimodal and bilingual (English and Spanish) design aims to capture the varied ways in which sexism is expressed and interpreted online, enabling researchers to develop AI models that are sensitive to both linguistic and visual cues, as well as the platform-specific dynamics that influence sexist content dissemination.

Throughout its four previous editions, more than 100 teams from universities and companies around the world have participated in EXIST, developing and testing state-of-the-art models to address this pressing social issue. The 2025

---

[1] https://nlp.uned.es/exist2025.

edition continues to foster international participation, with 244 registered teams from 38 countries. Of these, 114 teams from 23 countries submitted valid runs, resulting in a total of 873 system submissions.

In the following sections, we present a detailed overview of the tasks, datasets, annotation process, evaluation methodology, and system results for EXIST 2025.

2   Tasks

The 2025 edition of EXIST features nine tasks, which are described below. The languages addressed are English and Spanish and the datasets are collections of tweets, memes and TikTok videos. For the tasks on TikTok, all the partitions of the dataset are new, whereas for the tasks on tweets and memes we employ the EXIST 2023 and 2024 datasets.

2.1   Task 1.1: Sexism Identification in Tweets

This is a binary classification task where systems must decide whether or not a given tweet expresses sexist ideas because it is sexist itself, it describes a sexist situation, or it criticizes a sexist behavior. The following examples from the dataset show sexist and not sexist messages, respectively.

(1) **Sexist**. *It's less of #adaywithoutwomen and more of a day without feminists, which, to be quite honest, sounds lovely.*

(2) **Not sexist**. *Just saw a woman wearing a mask outside spank her very tightly leashed dog and I gotta say I love learning absolutely everything about a stranger in a single instant.*

2.2   Task 1.2: Source Intention in Tweets

This task aims to categorize the message according to the intention of the author. We propose the following ternary classification of tweets:

- **Direct** sexist message. The intention is to write a message that is sexist by itself or incites sexism, as in:
  (3) *A woman needs love, to fill the fridge, if a man can give this to her in return for her services, I don't see what else she needs.*
- **Reported** sexist message. The intention is to report and share a sexist situation suffered by a woman or women in first or third person, as in:
  (4) *I doze in the subway, I open my eyes feeling something weird: the hand of the man sat next to me on my leg #SquealOnYourPig.*
- **Judgemental** message. The intention is to condemn sexist situations or behaviours, as in:
  (5) *As usual, the woman was the one quitting her job for the family's welfare...*

2.3   Task 1.3: Sexism Categorization in Tweets

Many facets of a woman's life may be the focus of sexist attitudes including domestic role, career opportunities, and sexual image, to name a few. According to this, each sexist tweet must be assigned one or more of the following categories:

- **Ideological and inequality**. It includes messages that discredit the feminist movement. It also includes messages that reject inequality between men and women, or present men as victims of gender-based oppression.
  - (6) *#Feminism is a war on men, but it's also a war on women. It's a war on female nature, a war on femininity.*
- **Stereotyping and dominance**. It includes messages that suggest women are more suitable or inappropriate for certain tasks, and somehow inferior to men.
  - (7) *Most women no longer have the desire or the knowledge to develop a high quality character, even if they wanted to.*
- **Objectification**. It includes messages where women are presented as objects apart from their dignity and personal aspects. We also include messages that assume or describe certain physical qualities that women must have in order to fulfill traditional gender roles.
  - (8) *No offense but I've never seen an attractive african american hooker. Not a single one.*
- **Sexual violence**. It includes messages where sexual suggestions, requests or harassment of a sexual nature (rape or sexual assault) are made.
  - (9) *I wanna touch your tits..you can't imagine what I can do on your body.*
- **Misogyny and non sexual violence**. It includes expressions of hatred and violence towards women.
  - (10) *Domestic abuse is never okay... Unless your wife is a bitch.*

2.4   Task 2.1: Sexism Identification in Memes

As in Task 1.1, this involves a binary classification consisting on deciding whether or not a meme is sexist, as in Fig. 1.

2.5   Task 2.2: Source Intention in Memes

As in Task 1.2, this task aims to categorize the meme according to the intention of the author. However, in this task systems should only classify memes in two classes: direct or judgemental, as shown in Fig. 2.

2.6   Task 2.3: Sexism Categorization in Memes

This task aims to classify sexist memes according to the categorization provided for Task 1.3. Figure 3 shows one meme of each sexist category.

(a) Sexist meme   (b) Non sexist meme

Fig. 1. Examples of sexist and not sexist memes.

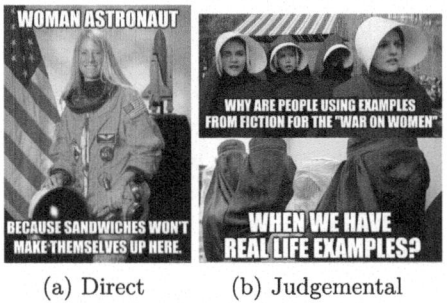

(a) Direct   (b) Judgemental

Fig. 2. Examples of direct and judgemental memes.

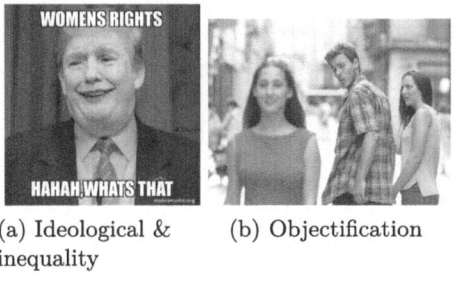

(a) Ideological &   (b) Objectification
inequality

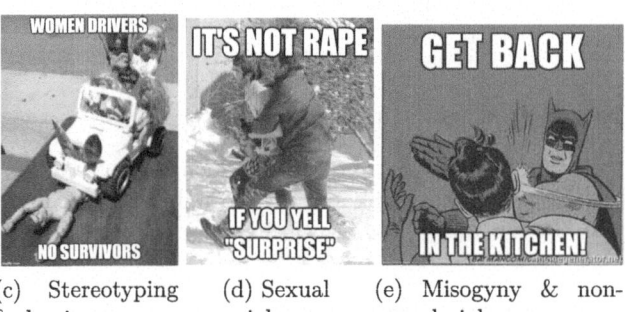

(c)   Stereotyping   (d) Sexual   (e) Misogyny & non-
& dominance   violence   sexual violence

Fig. 3. Examples of memes from the different sexist categories.

2.7   Task 3.1: Sexism Identification in TikToks

As in Tasks 1.1 and 2.1, systems must determine whether a given short video shared on TikTok is sexist.

2.8   Task 3.2: Source Intention in TikToks

As in Tasks 1.2 and 2.2, this task aims to categorize the short video according to the intention of the author, as direct or judgemental.

2.9   Task 3.3: Sexism Categorization in TikToks

As in Tasks 1.3 and 3.3, this task aims to categorize short videos according to the categorization provided for Task 1.3.

3   Dataset

The EXIST 2025 dataset comprises three types of data: the tweets from the EXIST 2023 dataset, the memes from the EXIST 2024 dataset and a new dataset of TikTok videos. Here, we briefly describe the process followed to curate the TikTok dataset. More details can be found in the extended overview [29]. Moreover, Plaza et al. [31] and [28] provide a detailed description of the tweet and meme datasets, respectively.

3.1   Data Sampling

The data was collected using the Apify's TikTok Hashtag Scraper tool[2], using a previously curated list of 185 Spanish hashtags and 61 English hashtags associated with potentially sexist content.

More than 3,500 videos in English and Spanish were downloaded from different TikTok accounts. Rigorous manual cleaning procedures were applied, ensuring the removal of noise such as ads and duplicates.

3.2   Labeling with Disagreements

The learning with disagreement paradigm was adopted, as in EXIST 2023 and 2024. The annotation was performed by trained annotators, rather than crowd workers as in the previous EXIST editions. The annotation was conducted using Servipoli's service,[3] with eight students organized in pairs consisting of one male and one female student, in order to avoid biases. Each pair was tasked with annotating 1,000 TikTok videos.

---

[2] https://apify.com/clockworks/tiktok-hashtag-scraper.
[3] https://www.servipoli.es/.

4   Evaluation Methodology and Metrics

As in EXIST 2024, we applied two evaluation strategies: a **soft evaluation** and a **hard evaluation**. The soft evaluation aligns with the LeWiDi paradigm and aims to assess how well models reflect disagreement among annotators. The hard evaluation follows the conventional approach, where systems output a single label per instance, and performance is assessed against a majority-vote gold standard.

1. **Soft-soft Evaluation.** For systems that return a probability distribution over the classes, we compare these distributions with the ones derived from human annotators. The class probabilities per instance are calculated from the frequency of annotations and the number of annotators involved. The primary metric for this evaluation is ICM-Soft, an adaptation of the original Information Contrast Measure (ICM) [3]. We also report results using the normalized version, ICM-Soft Norm. For a description of the ICM-Soft metrics, please refer to [28].
2. **Hard-hard Evaluation.** For systems providing discrete (hard) predictions, we compute the reference labels using a task-specific probabilistic threshold based on annotator agreement: for Tasks 1.1, 2.1 and 3.1, the label selected by more than three annotators is used; for Tasks 1.2, 2.2 and 3.2, the threshold is more than two annotators; and for multilabel Tasks 1.3, 2.3 and 3.3, any label assigned by more than one annotator is included. Instances without a dominant label (i.e., no class surpassing the threshold) are excluded from this evaluation. The main metric is the original ICM as defined by **(author?)** [3]. We additionally report a normalized ICM (ICM Norm) and F1.

5   Overview of Approaches

In this section, we provide a concise overview of the approaches presented at EXIST 2025. For a comprehensive description of the systems, please refer to the participant papers and the extended overview [29]. Although 244 teams from 38 different countries registered for participation, the number of participants who finally submitted results were 114, submitting 873 runs. Teams were allowed to participate in any of the nine tasks and submit hard and/or soft outputs. Table 1 summarizes the participation in the different tasks and evaluation contexts.

The evaluation campaign started on February 10, 2025 with the release of the training set. The test set was made available on April 7. Participants were provided with the official evaluation script. Runs had to be submitted by May 23. Each team could submit up to three runs per task.

5.1   Sexism Detection in Tweets

Sexism detection in tweets was predominantly approached through Natural Language Processing (NLP) techniques and neural network-based models. The majority of teams relied on pre-trained large language models (LLMs), such as

Table 1. Runs submitted and teams participating on each EXIST 2025 task.

	Tweets			Memes			TikTok		
	T1.1	T1.2	T1.3	T2.1	T2.2	T2.3	T3.1	T3.2	T3.3
#Runs	223	192	181	26	20	20	75	65	71
#Teams	117	105	101	15	13	12	36	32	33

BERT, RoBERTa, and domain-specific variants like BERTweet or HateBERT, often fine-tuned on the EXIST datasets.

While transformer-based models dominated, a minority of teams used traditional machine learning techniques such as Support Vector Machines (SVM) or Random Forests with TF-IDF, as well as rule-based or lexicon-based methods.

Many teams applied data preprocessing techniques tailored to social media content, including emoji normalization, hashtag segmentation, and URL removal. Data augmentation methods, such as back-translation, synonym replacement, or oversampling of minority classes, were also employed to mitigate class imbalance and improve generalization.

5.2   Sexism Detection in Memes

For memes, the inherently multimodal nature of the data led teams to combine computer vision and text analysis methods. Convolutional Neural Networks (CNNs) and visual feature extractors such as CLIP and ResNet were used to process image data. Meanwhile, embedded text within memes was handled using transformer-based NLP models.

Teams used both early fusion (merging textual and visual embeddings before classification) and late fusion (aggregating predictions from separate pipelines). Although multimodal fusion was key, some teams focused primarily on one modality, revealing diverse strategic preferences.

5.3   Sexism Detection in TikTok Videos

Sexism detection in TikToks required integrating audio, visual, and textual information, making multimodal analysis indispensable.

Despite the complexity of the modality, the dominant methods remained rooted in NLP (particularly for transcript analysis), followed by computer vision models. Multimodal fusion strategies—especially late fusion—were key in top-performing systems, and some teams adopted zero-shot or prompt-based learning using general-purpose LLMs such as GPT-3.

Given TikTok's social dynamics, models were also designed to be sensitive to context, sometimes incorporating meta-information, such as hashtags or background music features.

6   Results

In the next subsections, we report the results of the participants and the baseline systems for each task. We only show the results obtained by the top five teams for each task. For more detailed results (including disaggregated results per language), please refer to the Lab Working Notes [30] or the EXIST website: https://nlp.uned.es/exist2025/.

6.1   Task 1.1: Sexism Identification in Tweets

Soft Evaluation. Table 2 presents the results for the soft-soft evaluation for Task 1.1, which attracted a total of 65 participating systems (excluding the gold reference and two baselines). Across all systems, the normalized ICM-Soft scores ranged from nearly 0 to 0.6700, with a mean of 0.490 and a standard deviation of 0.160. A total of 63 systems outperformed the strongest baseline (*Test_majority-class*, all instances labeled as 'NO'), indicating widespread above-baseline performance. The top 5 systems from distinct teams showed a relative difference of only 7.7% between them, pointing to a tightly clustered upper tier.

Table 2. Top 5 systems from different teams in EXIST 2025 Task 1.1. The complete leaderboard for the task is available on the EXIST website https://nlp.uned.es/exist2025/, in the 'Results' section.

Soft-soft Evaluation					
System	Team	ICM-Soft	ICM-Soft Norm	Cross Entropy	Rank
Test_gold	–	3.1182	1.0000	0.5472	0
GrootWatch_1	GrootWatch [25]	1.0600	0.6700	0.8893	1
DaniReinon_1	DaniReinon	0.8332	0.6336	0.9726	4
fhstp_3	fhstp [21]	0.7852	0.6259	0.8416	6
BERT-Simpson_2	BERT-Simpson	0.7461	0.6196	1.0426	7
CLiC_1	CLiC [27]	0.7386	0.6184	0.8201	9
Test_majority-class	–	−2.3585	0.1218	4.6115	64
Test_minority-class	–	−3.0717	0.0075	5.3572	66
Hard-hard Evaluation					
System	Team	ICM-Hard	ICM-Hard Norm	F1 YES	Rank
Test_gold	–	0.9948	1.0000	1.0000	0
Mario_1	Mario [34]	0.6774	0.8405	0.8167	1
CIMAT-GTO_2	CIMAT-GTO [36]	0.6297	0.8165	0.7996	2
warwick_1	warwick [1]	0.6249	0.8141	0.7991	4
CIMAT-CS-NLP_3	CIMAT-CS-NLP [37]	0.6127	0.8079	0.7945	5
BERT-Simpson_1	BERT-Simpson	0.5832	0.7931	0.7832	8
Test_majority-class	–	−0.4413	0.2782	0.0000	154
Test_minority-class	–	−0.5742	0.2114	0.5698	157

Hard Evaluation. Table 2 presents the results for the hard-hard evaluation. In this scenario, the annotations from the six annotators are combined into a single label using the majority vote. 158 systems competed using hard labels.

The normalized ICM-Hard scores, bounded between 0 and 1, had a mean of 0.678 and a standard deviation of 0.149 across participants. Nearly all systems (153 out of 158) outperformed the strongest hard-label baseline (*Test_majority-class*, all instances labeled as 'NO'), confirming robust overall performance in this setting. As seen in Table 2, the spread between the top and fifth-ranked system was only 5.6%, showing strong consistency among leading submissions. Interestingly, two teams from the same institution (CIMAT) appear in the top five with tightly clustered results, while the top system, *Mario_1*, led by a modest yet consistent margin.

6.2   Task 1.2: Source Intention in Tweets

Soft Evaluation. Table 3 presents the results for the soft-soft evaluation of Task 1.2, focused on identifying the author's intent behind sexist tweets. This task received 54 system submissions in the Soft–Soft evaluation setting. Across participants, scores ranged from 0.0000 to 0.4647, with a mean of 0.182 and a standard deviation of 0.158. A total of 36 systems outperformed the strongest baseline (*Test_majority-class*, all instances labeled as 'NO'), confirming moderate differentiation among participant quality. The relative difference between the best and the fifth best teams was 15.7%, indicating relatively close performance among the leading systems. This compact spread suggests that most top teams converged on similar probabilistic modeling strategies, even though overall scores were lower than in other tasks due to the increased ambiguity of intent classification.

Hard Evaluation. Table 3 presents the hard-hard evaluation results for Task 1.2. In the Hard–Hard setting, 138 systems participated. The normalized ICM-Hard scores, which assess agreement with the aggregated label, ranged from 0.0000 to 0.6623, with an average of 0.3881 and a standard deviation of 0.2278. Impressively, 105 systems outperformed the best hard-label baseline (*Test_majority-class*, Norm = 0.1910), demonstrating broad effectiveness across submissions. The normalized scores in the group of the top-5 teams were tightly packed, with a maximum relative difference of only 6.4%.

6.3   Task 1.3: Sexism Categorization in Tweets

Soft Evaluation. Table 4 displays the results of the soft-soft evaluation Task 1.3, which involves multi-label categorization of sexist content in tweets under the Soft–Soft evaluation paradigm. 51 systems participated, excluding the gold and baseline runs. The normalized ICM-Soft scores spanned from 0.0000 to 0.4417, with a mean of 0.144 and a standard deviation of 0.163. Notably, 25 systems outperformed the strongest baseline (*Test_majority-class*, all instances labeled as 'NO'), indicating a moderate level of competitiveness. The percentage difference between the best and the fifth team was 22.3%, suggesting a wider performance spread among the leading systems than in other tasks. This reflects the intrinsic difficulty of the multi-label classification task in the soft evaluation setting, where label ambiguity and annotator disagreement must be faithfully captured.

Table 3. Top 5 systems from different teams in EXIST 2025 Task 1.2. The complete leaderboard for the task is available on the EXIST website https://nlp.uned.es/exist2025/, in the 'Results' section.

Soft-soft Evaluation					
System	Team	ICM-Soft	ICM-Soft Norm	Cross Entropy	Rank
Test_gold	–	6.2057	1.0000	0.9128	0
GrootWatch_1	GrootWatch [25]	−0.4385	0.4647	1.7711	1
Dandys-de-BERTganim_2	Dandys-de-BERTganim [17]	−0.7261	0.4415	1.3820	4
Cyberpuffs_3	Cyberpuffs [19]	−1.0572	0.4148	2.0396	6
fhstp_2	fhstp [21]	−1.1866	0.4044	1.6566	7
NetGuardAI_1	NetGuardAI [9]	−1.3444	0.3917	1.5681	8
Test_majority-class	–	−5.4460	0.0612	4.6233	37
Test_minority-class	–	−32.9552	0.0000	8.8517	56

Hard-hard Evaluation					
System	Team	ICM-Hard	ICM-Hard Norm	Macro F1	Rank
Test_gold	–	1.5378	1.0000	1.0000	0
Mario_1	Mario [34]	0.4991	0.6623	0.5692	1
CIMAT-GTO_3	CIMAT-GTO [36]	0.4678	0.6521	0.5555	2
CIMAT-CS-NLP_2	CIMAT-CS-NLP [37]	0.4264	0.6386	0.5461	4
Dandys-de-BERTganim_2	Dandys-de-BERTganim [17]	0.3752	0.6220	0.5522	6
BERTin-Osborne_2	BERTin-Osborne	0.3677	0.6196	0.5453	7
Test_majority-class	–	−0.9504	0.1910	0.1603	106
Test_minority-class	–	−3.1545	0.0000	0.0280	139

Hard Evaluation. For the Hard–Hard evaluation of Task 1.3, a total of 130 systems were submitted. The normalized ICM-Hard values ranged from 0.0000 to 0.6514, with an average of 0.353 and a standard deviation of 0.193. Remarkably, 106 systems surpassed the best baseline (*Test_majority-class*), demonstrating high effectiveness in predicting the aggregated ground truth labels. The range between the top and fifth systems was only 9.1%, highlighting a tight cluster of top performances. This compact variation among the leaders suggests strong generalization in handling categorical distinctions of sexism in tweets when annotations are aggregated, though continued progress is needed to reach the reliability of the gold standard.

6.4 Task 2.1: Sexism Identification in Memes

Soft Evaluation. Table 5 presents the results for the classification of memes as sexist or not sexist. A total of 8 systems participated in the Soft–Soft evaluation. The normalized scores ranged from 0.0650 to 0.5110, with a mean of 0.373 and a standard deviation of 0.149. All but one system outperformed the strongest baseline (*Test_majority-class*), indicating that most submissions were effective under this probabilistic evaluation. The relative difference between the highest and lowest among the top five submissions from different teams was substantial (87.3%), with a notable drop from the fourth to fifth system. This wide spread suggests room for improvement and divergence in approaches to modeling soft labels in multimodal data.

Table 4. Top 5 systems from different teams in EXIST 2025 Task 1.3. The complete leaderboard for the task is available on the EXIST website https://nlp.uned.es/exist2025/, in the 'Results' section.

Soft-soft Evaluation

System	Team	ICM-Soft	ICM-Soft Norm	Rank
Test_gold	–	9.4686	1.0000	0
GrootWatch_1	GrootWatch [25]	−1.1034	0.4417	1
UMUTeam_2	UMUTeam [26]	−1.6711	0.4118	4
Cyberpuffs_3	Cyberpuffs [19]	−2.4632	0.3699	5
DaniReinon_1	DaniReinon	−2.5655	0.3645	6
A-squared_1	A-squared	−2.9711	0.3431	7
Test_majority-class	–	−8.7089	0.0401	26
Test_minority-class	–	−46.1080	0.0000	53

Hard-hard Evaluation

System	Team	ICM-Hard	ICM-Hard Norm	Macro F1	Rank
Test_gold	–	2.1533	1.0000	1.0000	0
Mario_1	Mario [34]	0.6519	0.6514	0.6533	1
CIMAT-GTO_2	CIMAT-GTO [36]	0.5413	0.6257	0.6392	2
NLPDame_3	NLPDame [7]	0.4842	0.6124	0.6335	4
UMUTeam_2	UMUTeam [26]	0.4506	0.6046	0.6262	7
CIMAT-CS-NLP_2	CIMAT-CS-NLP [37]	0.3980	0.5924	0.6125	8
Test_majority-class	–	−1.5984	0.1289	0.1069	108
Test_minority-class	–	−3.1295	0.0000	0.0288	128

Hard Evaluation. Table 5 presents the results for the hard-hard evaluation of Task 2.1. This task received 18 valid system submissions. The normalized ICM-Hard values ranged from 0.1711 to 0.6877, with an average of 0.471 and a standard deviation of 0.145. Out of these, 16 systems outperformed the *Test_majority-class* baseline. The top five systems from distinct teams showed a moderate performance spread, with a 28.3% relative difference between the highest and lowest performers in this top group. Compared to Task 1.1, the distribution in Task 2.1 reflects greater difficulty in aligning with aggregated hard labels in multimodal settings, likely due to the inherent ambiguity and subjective interpretation of memes.

6.5   Task 2.2: Source Intention in Memes

Soft Evaluation. Table 6 presents the results for the classification of memes according to the intention of the author, with the outputs provided as the probabilities of the different classes. Only 5 systems participated in the Soft–Soft evaluation. The average normalized score across systems was 0.228, with a standard deviation of 0.101. All five systems surpassed the strongest baseline

Table 5. Top 5 systems from different teams in EXIST 2025 Task 2.1. The complete leaderboard for the task is available on the EXIST website https://nlp.uned.es/exist2025/, in the 'Results' section.

Soft-soft Evaluation

System	Team	ICM-Soft	ICM-Soft Norm	Cross Entropy	Rank
Test_gold	–	3.1107	1.0000	0.5852	0
TrankilTwice_1	TrankilTwice [18]	0.0683	0.5110	1.0096	1
surrey-mm-group_1	surrey-mm-group	−0.7061	0.3865	0.9364	4
I2C-UHU-Altair_1	I2C-UHU-Altair [14]	−0.8558	0.3624	0.9469	5
UMUTeam_2	UMUTeam [26]	−0.9623	0.3453	1.0554	6
Test_majority-class	–	−2.3568	0.1212	4.4015	8
Nogroupnocry_1	Nogroupnocry	−2.7060	0.0650	0.6782	9
Test_minority-class	–	−3.5089	0.0000	5.5672	10

Hard-hard Evaluation

System	Team	ICM-Hard	ICM-Hard Norm	F1 YES	Rank
Test_gold	–	0.9832	1.0000	1.0000	0
CogniCIC_1	CogniCIC [2]	0.3691	0.6877	0.7810	1
GrootWatch_3	GrootWatch [25]	0.3589	0.6825	0.7740	2
ArcosGPT_1	ArcosGPT [4]	0.3200	0.6627	0.7571	3
TrankilTwice_2	TrankilTwice [18]	0.1667	0.5848	0.7508	5
I2C-UHU-Altair_2	I2C-UHU-Altair [14]	−0.0134	0.4932	0.7125	9
Test_majority-class	–	−0.4038	0.2947	0.6821	17
Test_minority-class	–	−0.6468	0.1711	0.0000	20

(*Test_majority-class*). Taking into account the top ranked submissions from distinct teams, the relative difference between the best and the worst among this top-4 was 81.7%, indicating a wide spread in system quality.

Hard Evaluation. Table 6 presents the results for the hard-hard evaluation of Task 2.2. We received 15 system submissions. The normalized ICM-Hard metric ranged from 0.0000 to 0.5784, with an average of 0.308 and a standard deviation of 0.169. Thirteen systems outperformed the *Test_majority-class* baseline, reflecting strong participation despite the challenging nature of the task. Concerning the five best submissions from different teams, the top system outperformed the fifth by 52.7%, a considerable difference suggesting uneven performance across modeling strategies. Nonetheless, the narrow gap among the three leading systems (within 10%) points to the emergence of competitive approaches for intent recognition, even in the presence of aggregated hard annotations derived from subjectively interpreted multimodal inputs.

Table 6. Top 5 systems from different teams in EXIST 2025 Task 2.2. The complete leaderboard for the task is available on the EXIST website https://nlp.uned.es/exist2025/, in the 'Results' section.

Soft-soft Evaluation

System	Team	ICM-Soft	ICM-Soft Norm	Cross Entropy	Rank
Test_gold	–	4.7018	1.0000	0.9325	0
UMUTeam_1	UMUTeam [26]	−1.6327	0.3264	1.7316	1
I2C-UHU-Altair_1	I2C-UHU-Altair [14]	−2.0736	0.2795	1.5556	2
surrey-mm-group_1	surrey-mm-group	−2.4423	0.2403	2.0468	3
Nogroupnocry_1	Nogroupnocry	−4.1395	0.0598	0.3164	5
Test_majority-class	–	−5.0745	0.0000	5.5565	6
Test_minority-class	–	−18.9382	0.0000	8.0245	7

Hard-hard Evaluation

System	Team	ICM-Hard	ICM-Hard Norm	Macro F1	Rank
Test_gold	–	1.4383	1.0000	1.0000	0
CogniCIC_1	CogniCIC [2]	0.2254	0.5784	0.5634	1
GrootWatch_3	GrootWatch [25]	0.1868	0.5649	0.5513	2
ArcosGPT_1	ArcosGPT [4]	0.0597	0.5208	0.5109	3
NaturalThinker_1	NaturalThinker	−0.5429	0.3113	0.3762	6
I2C-UHU-Altair_1	I2C-UHU-Altair [14]	−0.6519	0.2734	0.2685	7
Test_majority-class	–	−1.0445	0.1369	0.1839	14
Test_minority-class	–	−2.0637	0.0000	0.0697	17

6.6   Task 2.3: Sexism Categorization in Memes

Soft Evaluation. Table 7 presents the results for classifying memes based on the aspects of women being attacked, with outputs provided as class probabilities. This task received submissions from 6 different teams. Among these, 5 systems outperformed the majority-class baseline. The average normalized ICM-Soft score was 0.151 with a standard deviation of 0.100, indicating a moderately dispersed distribution. The difference in normalized ICM-Soft between the top and bottom systems was 74.8%, showing a meaningful variation even within the upper ranks. Interestingly, all these systems clearly surpassed the worst-performing baseline (`Test_minority-class`), while four significantly exceeded the `Test_majority-class` baseline.

Hard Evaluation. Finally, Table 7 presents the results for classifying memes based on the aspects of women being attacked, with outputs provided as a single class prediction. A total of 14 systems participated (excluding the gold and baselines). Thirteen of them scored above the best baseline, with an average normalized ICM-Hard of 0.262 and a standard deviation of 0.158. The relative difference between the top and fifth-best system was 59.5%, indicating competitive but

Table 7. Top 5 systems from different teams in EXIST 2025 Task 2.3. The complete leaderboard for the task is available on the EXIST website https://nlp.uned.es/exist2025/, in the 'Results' section.

Soft-soft Evaluation

System	Team	ICM-Soft	ICM-Soft Norm	Rank
Test_gold	–	9.4343	1.0000	0
UMUTeam_1	UMUTeam [26]	−4.7791	0.2467	1
I2C-UHU-Altair_1	I2C-UHU-Altair [14]	−5.8210	0.1915	3
surrey-mm-group_1	surrey-mm-group	−6.3848	0.1616	4
Nogroupnocry_2	Nogroupnocry	−8.2621	0.0621	5
Test_majority-class	–	−9.8173	0.0000	6
Test_minority-class	–	−50.0353	0.0000	8

Hard-hard Evaluation

System	Team	ICM-Hard	ICM-Hard Norm	Macro F1	Rank
Test_gold	–	2.4100	1.0000	1.0000	0
CogniCIC_1	CogniCIC [2]	0.0244	0.5051	0.5763	1
GrootWatch_3	GrootWatch [25]	−0.0798	0.4834	0.5472	2
ArcosGPT_1	ArcosGPT [4]	−0.4187	0.4131	0.5501	4
I2C-UHU-Altair_1	I2C-UHU-Altair [14]	−0.9958	0.2934	0.4223	6
CLTL_2	CLTL [6]	−1.4243	0.2045	0.3143	8
Test_majority-class	–	−2.0711	0.0703	0.0919	14
Test_minority-class	–	−3.3135	0.0000	0.0318	16

not saturated performance across top ranks. As in the soft setting, these systems clearly outperformed both `Test_majority-class` and `Test_minority-class`.

6.7   Task 3.1: Sexism Identification in Videos

Soft Evaluation. Table 8 presents the results for classifying videos as sexist or not sexist. The Soft–Soft evaluation of Task 3.1 attracted 34 participating systems. The normalized ICM-Soft values, which reflect alignment with the probabilistic distribution of annotator labels, ranged from 0.1481 to 0.5590. The average normalized score was 0.3584, with a standard deviation of 0.174, indicating considerable variance in system quality. A total of 25 systems outperformed the strongest baseline (*Test_majority-class*, Norm = 0.2740). The difference between the best and worst among the top five teams was approximately 18.2%, reflecting a modest but meaningful spread. Interestingly, most high-scoring systems came from teams with distinct modeling pipelines, suggesting diverse yet effective approaches to handling annotator disagreement in the multimodal context of video classification.

Table 8. Top 5 systems from different teams in EXIST 2025 Task 3.1. The complete leaderboard for the task is available on the EXIST website https://nlp.uned.es/exist2025/, in the 'Results' section.

Soft-soft Evaluation

System	Team	ICM-Soft	ICM-Soft Norm	Cross Entropy	Rank
Test_gold	–	2.8488	1.0000	0.1962	0
LaVellaPremium_2	LaVellaPremium	0.3362	0.5590	1.5731	1
MIARFID ducks_2	MIARFID ducks	0.2968	0.5521	1.7725	3
YesWeEXIST_1	YesWeEXIST	0.2759	0.5484	1.9034	5
profLayton_1	profLayton	0.1779	0.5312	0.9870	7
EmbeddingGuards_1	EmbeddingGuards	−0.2429	0.4574	0.8413	10
Test_majority-class	–	−1.2877	0.2740	4.4285	26
Test_minority-class	–	−2.0051	0.1481	5.5402	31

Hard-hard Evaluation

System	Team	ICM-Hard	ICM-Hard Norm	F1 YES	Rank
Test_gold	–	0.9907	1.0000	1.0000	0
ECA-SIMM-UVa_3	ECA-SIMM-UVa [11]	0.1984	0.6001	0.6935	1
CogniCIC_1	CogniCIC [2]	0.1940	0.5979	0.6835	2
EmbeddingGuards_1	EmbeddingGuards	0.1761	0.5889	0.6841	4
LaVellaPremium_1	LaVellaPremium	0.1563	0.5789	0.6899	5
AIDONTTOKSEXISM_2	AIDONTTOKSEXISM	0.1509	0.5761	0.7013	6
Test_majority-class	–	−0.4244	0.2858	0.0000	40
Test_minority-class	–	−0.6036	0.1954	0.6117	42

Hard Evaluation. Finally, Table 8 presents the results for classifying videos on sexism identification in a hard-hard context. For this task, 41 systems submitted valid runs. Normalized ICM-Hard scores spanned from 0.1954 to 0.6001, with a mean of 0.4913 and a standard deviation of 0.1033. Nearly all participants (39 out of 41) exceeded the majority-class baseline (*Test_majority-class*), showing strong global performance. The top five teams, as can be observed from Table 8, were closely matched, with only a 4.0% difference between the best and lowest performer among the top five.

6.8 Task 3.2: Source Intention in Videos

Soft Evaluation. Table 9 presents the results for the classification of videos according to the intention of the author, with the outputs provided as the probabilities of the different classes. In this task, the 29 participating systems showed normalized ICM-Soft scores that ranged from 0.0000 to 0.3728, with a mean of 0.252 and a standard deviation of 0.084. A total of 26 systems surpassed the strongest baseline (*Test_majority-class*), indicating a generally competitive field. The difference between the best and the fifth ranked systems from distinct teams was modest, at 12.0%, revealing a cluster of high-performing submissions.

Table 9. Top 5 systems from different teams in EXIST 2025 Task 3.2. The complete leaderboard for the task is available on the EXIST website https://nlp.uned.es/exist2025/, in the 'Results' section.

Soft-soft Evaluation

System	Team	ICM-Soft	ICM-Soft Norm	Cross Entropy	Rank
Test_gold	–	4.6948	1.0000	0.2550	0
MIARFID ducks_2	MIARFID ducks	−1.1940	0.3728	1.7731	1
EXISTencialCrisis_1	EXISTencialCrisis	−1.3535	0.3558	3.0998	3
profLayton_1	profLayton	−1.3821	0.3528	1.4974	4
YesWeEXIST_1	YesWeEXIST	−1.6151	0.3280	1.5712	6
LaVellaPremium_1	LaVellaPremium	−1.6159	0.3279	1.3258	7
Test_majority-class	–	−3.1337	0.1663	4.4354	27
Test_minority-class	–	−15.4368	0.0000	8.8286	31

Hard-hard Evaluation

System	Team	ICM-Hard	ICM-Hard Norm	Macro F1	Rank
Test_gold	–	1.3244	1.0000	1.0000	0
CogniCIC_1	CogniCIC [2]	0.0048	0.5018	0.5623	1
jdsanroj_2	jdsanroj	−0.0068	0.4974	0.5781	2
profLayton_1	profLayton	−0.0283	0.4893	0.5902	3
LaVellaPremium_1	LaVellaPremium	−0.0487	0.4816	0.5742	4
EmbeddingGuards_1	EmbeddingGuards	−0.0529	0.4800	0.5738	5
Test_majority-class	–	−0.7537	0.2155	0.2375	34
Test_minority-class	–	−2.4749	0.0000	0.0586	38

Hard Evaluation. Table 9 presents the results for the hard-hard evaluation of Task 3.2. The normalized ICM-Hard scores for the 36 systems submitted ranged from 0.0000 to 0.5018, with a mean of 0.375 and a standard deviation of 0.116. Most systems (33 out of 36) outperformed the *Test_majority-class* baseline. The best systems from five different teams showed a relative difference between the highest and lowest normalized scores of only 4.3%, reflecting a tight performance range. Interestingly, while the average performance remains moderate, the consistency among top runs suggests that author intent in video—despite its multimodal complexity—can be reliably modeled when annotations are aggregated, albeit with room for improving discriminatory power across subtle categories.

6.9 Task 3.3: Sexism Categorization in Videos

Soft Evaluation. Table 10 presents the results for classifying videos based on the aspects of women being attacked, with outputs provided as class probabilities. A total of 34 participant systems were submitted for this task. The normalized ICM-Soft scores ranged from 0.0000 to 0.1593, with a mean of 0.051 and standard deviation of 0.052. The majority baseline achieved a normalized ICM

Table 10. Top 5 systems from different teams in EXIST 2025 Task 3.3. The complete leaderboard for the task is available on the EXIST website https://nlp.uned.es/exist2025/, in the 'Results' section.

Soft-soft Evaluation				
System	Team	ICM-Soft	ICM-Soft Norm	Rank
Test_gold	–	8.3833	1.0000	0
EXISTencialCrisis_1	EXISTencialCrisis	−5.7131	0.1593	1
AIDONTTOKSEXISM_1	AIDONTTOKSEXISM	−6.0447	0.1395	2
LaVellaPremium_1	LaVellaPremium	−6.2730	0.1259	4
biasedmodels_1	biasedmodels	−6.5149	0.1114	6
Test_majority-class	–	−6.8222	0.0931	9
profLayton_2	profLayton	−6.9313	0.0866	10
Test_minority-class	–	−11.6668	0.0000	28

Hard-hard Evaluation					
System	Team	ICM-Hard	ICM-Hard Norm	Macro F1	Rank
Test_gold	–	1.5453	1.0000	1.0000	0
YesWeEXIST_1	YesWeEXIST	−0.3816	0.3765	0.2667	1
profLayton_3	profLayton	−0.3849	0.3755	0.3648	2
KeTEAM_1	KeTEAM	−0.3869	0.3748	0.3031	3
ScalaR_1	ScalaR	−0.4102	0.3673	0.2533	5
jdsanroj_1	jdsanroj	−0.4373	0.3585	0.2516	7
Test_majority-class	–	−0.9530	0.1916	0.1188	31
Test_minority-class	–	−6.7467	0.0000	0.0025	41

score of 0.0931, and was outperformed by 4 systems, while the minority baseline was not surpassed by any system. The top 5 systems from different teams achieved normalized ICM-Soft scores between 0.1593 and 0.0931. The relative difference between the best and the fifth-ranked system within this top group was 41.6%. Despite the low overall values, a meaningful gap between systems can be observed, which underlines the difficulty of probabilistic categorization in multi-class scenarios over multimodal video content.

Hard Evaluation. Finally, Table 7 presents the results for classifying memes based on the aspects of women being attacked, with outputs provided as a single class prediction. This task attracted 41 participant systems. Normalized ICM-Hard scores spanned from 0.0000 to 0.3765, with a mean of 0.243 and standard deviation of 0.116. A total of 30 systems outperformed the majority baseline, while 13 did better than the minority baseline. The top 5 systems from distinct teams achieved normalized ICM-Hard scores ranging from 0.3765 to 0.3585, showing a very tight performance band with only a 4.78% relative difference between the highest and the lowest scoring among them.

284   L. Plaza et al.

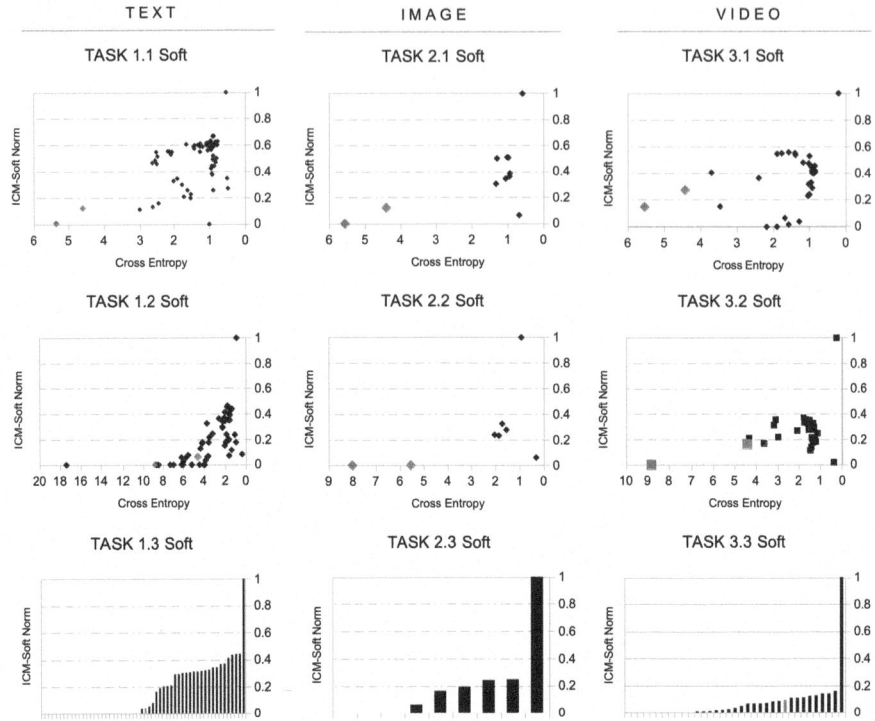

Fig. 4. ICM-Soft and Cross Entropy results across tasks.

6.10 Global View

Figure 4 shows the results of Cross Entropy (horizontal axes) and normalized ICM-Soft (vertical axes). All the plots include the gold standard with maximum score. The first row (Tasks 1.1, 2.1, and 3.1), corresponds to **sexism detection** tasks, i.e., binary single-label classification on texts, images and video, respectively. The baseline approaches consisting of labeling everything as the majority class or as the minority class are marked in blue and red, respectively.

In terms of both Cross Entropy and ICM-Soft, the results of these two baselines fall below those of the other participant runs, indicating that the proposed systems contribute some informative value. Only in the case of the video task (Task 3.1) are there some runs that fall below the baseline in terms of ICM. This may be due to the fact that ICM penalizes false information based on class frequency.

Another observation is that, while high ICM values imply high Cross Entropy values, the reverse is not true, with several runs accumulating good performance (low scores) according to Cross Entropy but low ICM scores. This may be due, among other factors, to the fact that ICM considers not only the similarity of the assigned values for each class, but also the distribution of classes throughout the corpus. In any case, in terms of ICM, there remains a significant gap between

TEXT   IMAGE   VIDEO

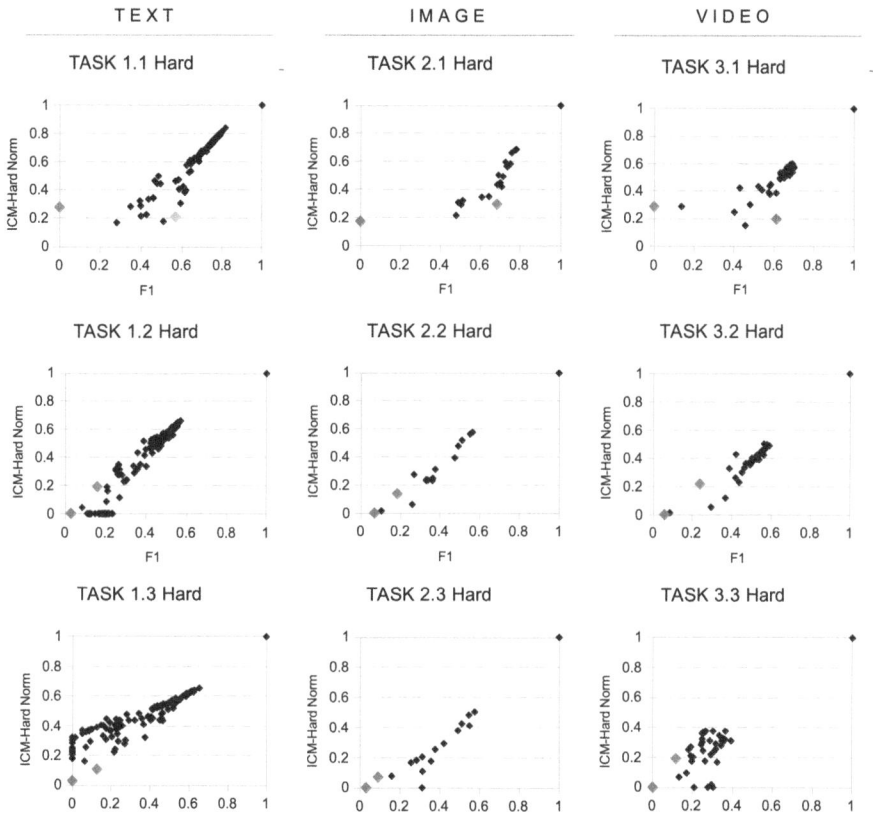

Fig. 5. ICM-Hard and F1 results across tasks.

the best-performing systems and the perfect solution. The gap is notably larger for the image and video tasks (Tasks 1.2 and 1.3).

The second row corresponds to **intent detection** tasks. These are hierarchical classification tasks with an initial YES/NO decision and two or three sub-classes for the YES category. In this case, there is also an accumulation of runs with high performance in Cross Entropy but low ICM, suggesting that the second metric captures additional aspects. Most runs outperform the baselines, but the gap between the best run and the perfect output in terms of ICM is larger and in sexism detection, indicating a higher complexity of the task.

Finally, the third row corresponds to hierarchical multi-label classification tasks involving multiple **categories of sexism**. In this case, since the tasks are multi-label, the Cross Entropy metric is not applicable. The plots show system rankings ordered from lowest to highest ICM. An interesting finding is that, in this case, many of the runs—including the minority-class baseline—do not surpass the zero threshold in normalized ICM. This suggests that the outputs, in terms of information content, do not outperform the empty output. In other words, the amount of noisy information exceeds the amount of useful information. As the number of categories increases and the task requires capturing annotation ambiguity (multi-label classification), the gap between the best run and the perfect output increases significantly compared to the previous tasks.

On the other hand, Fig. 5 displays evaluation results for the **hard evaluation versions**, in which the assignment of items to classes depends on whether different thresholds of annotator agreement are met. The plot shows F1 scores for the positive class in the first row (sexism identification), and the average F1 score across all classes for the remaining tasks. The vertical axes show the results for ICM-Hard.

In general, a strong correlation between both metrics can be observed above a certain score threshold. This is because both F1 and ICM take class specificity or frequency within the corpus into account.

Again, most runs outperform the baselines. Moreover, by observing the gap between the best run and the ideal output, we can see that task difficulty increases as we move to setups with more classes, multi-labeling, or hierarchical structures (rows). An increase in task difficulty is also observed as we move from text-based tasks (first column), to image (second column), and video (third column).

7   Conclusions

The objective of the EXIST challenge is to encourage research on the automated detection and modeling of sexism in online environments, with a specific focus on social networks. The EXIST 2025 Lab, held as part of CLEF, attracted 114 participant teams, and received a total of 873 runs. Participants adopted a wide range of approaches, including vision transformer models, data augmentation through automatic translation, data duplication, utilization of data from past EXIST editions, multilingual language models, Twitter-specific language models, and transfer learning techniques from domains like hate speech, toxicity, and sentiment analysis. While many systems opted for the traditional approach of providing only hard labels as outputs, a significant number of systems leveraged the multiple annotations available in the dataset, and provided soft outputs, proving that there is an increasing interest by the research community in developing systems able to deal with disagreements and with different perspectives.

Acknowledgments. This work has been financed by the European Union (NextGenerationEU funds) through the "Plan de Recuperación, Transformación y Resiliencia", by the Ministry of Economic Affairs and Digital Transformation and by the UNED University. It has also been financed by the Spanish Ministry of Science and Innovation (project FairTransNLP (PID2021-124361OB-C31 and PID2021-124361OB-C32)) funded by MCIN/AEI/10.13039/501100011033 and by ERDF, EU A way of making Europe, and by the Australian Research Council, ARC Centre of Excellence for Automated Decision-Making and Society (ADM+S, CE200100005).

References

1. Alajmi, A., Pergola, G.: Leveraging model confidence and diversity: a multi-stage framework for sexism detection. In: Faggioli, G., Ferro, N., Rosso, P., Spina, D. (eds.) Working Notes of CLEF 2025 – Conference and Labs of the Evaluation Forum, CEUR Workshop Proceedings (2025)
2. Alcantara, T., Garcia-Vazquez, O., Calvo, H., Valdez-Rodríguez, J.E.: CogniCIC at EXIST 2025: identifying sexist content in text and visual media using transformers and generative AI models. In: Faggioli, G., Ferro, N., Rosso, P., Spina, D. (eds.) Working Notes of CLEF 2025 – Conference and Labs of the Evaluation Forum, CEUR Workshop Proceedings (2025)
3. Amigó, E., Delgado, A.: Evaluating extreme hierarchical multi-label classification. In: Proceedings of the 60th Annual Meeting of the Association for Computational Linguistics, vol. 1: Long Papers, pp. 5809–5819. ACL, Dublin (2022)
4. Arcos, I.: Identifying sexism in memes with multimodal deep learning: fusing text and visual cues. In: Faggioli, G., Ferro, N., Rosso, P., Spina, D. (eds.) Working Notes of CLEF 2025 – Conference and Labs of the Evaluation Forum, CEUR Workshop Proceedings (2025)
5. Billig, M.: Humour and hatred: the racist jokes of the Ku Klux Klan. Disc. Soc. **12**(3), 267–289 (2014)
6. Britez, A., Markov, I.: CLTL at EXIST 2025: identifying sexist memes using an ensemble of shallow and transformer models. In: Faggioli, G., Ferro, N., Rosso, P., Spina, D. (eds.) Working Notes of CLEF 2025 – Conference and Labs of the Evaluation Forum, CEUR Workshop Proceedings (2025)
7. Christodoulou, C.: NLPDame at EXIST: sexism categorization in tweets via multi-head multi-task models, LLM & RAG voting synergy. In: Faggioli, G., Ferro, N., Rosso, P., Spina, D. (eds.) Working Notes of CLEF 2025 – Conference and Labs of the Evaluation Forum, CEUR Workshop Proceedings (2025)
8. Chulvi, B., Fontanella, L., Labadie, R., P, R.: Social or individual disagreement? perspectivism in the annotation of sexist jokes. In: Proceedings of NLPerspectives 2023: 2nd Workshop on Perspectivist Approaches to Disagreement in NLP, co-locotaed with ECAI-2023 (2023)
9. Cotelin, M.D., Truică, C.O., Apostol, E.S.: NetGuardAI at EXIST2025: sexism detection using mDeBERTa. In: Faggioli, G., Ferro, N., Rosso, P., Spina, D. (eds.) Working Notes of CLEF 2025 – Conference and Labs of the Evaluation Forum, CEUR Workshop Proceedings (2025)
10. Davis, S.E.: Objectification, sexualization, and misrepresentation: social media and the college experience. Social Media + Soc. **4**(3) (2018)
11. Fernández, D., Amigó, E., Cardeñoso, V.: ECA-SIMM-UVa at EXIST 2025: a segmentation oriented approach to sexism detection in TikTok videos based on a "One Is Enough" paradigm. In: Faggioli, G., Ferro, N., Rosso, P., Spina, D. (eds.) Working Notes of CLEF 2025 – Conference and Labs of the Evaluation Forum, CEUR Workshop Proceedings (2025)
12. Gasparini, F., Rizzi, G., Saibene, A., Fersini, E.: Benchmark dataset of memes with text transcriptions for automatic detection of multi-modal misogynistic content. Data Brief **44**, 108526 (2022)
13. Gil Bermejo, J.L., Martos Sánchez, C., Vázquez Aguado, O., García-Navarro, E.B.: Adolescents, ambivalent sexism and social networks, a conditioning factor in the healthcare of women. Healthcare (Basel) **9**(6), 721 (2021)

14. Guerrero-García, M., Carrillo García, F., Mata, J., Pachón-Álvarez, V.: I2C-UHU-altair at EXIST2025: multimodal sexism detection and classification using advanced vision-language models BLIP2 and Qwen, Large Language Models, and Learning with Disagreement Frameworks. In: Faggioli, G., Ferro, N., Rosso, P., Spina, D. (eds.) Working Notes of CLEF 2025 – Conference and Labs of the Evaluation Forum, CEUR Workshop Proceedings (2025)

15. Harriger, J., Thompson, J., Tiggemann, M.: TikTok, TikTok, the time is now: future directions in social media and body image. Body Image **B**(44), 222–226 (2023)

16. Hodson, G., Rush, J., MacInnis, C.C.: A joke is just a joke (except when it isn't): cavalier humor beliefs facilitate the expression of group dominance motives. J. Pers. Soc. Psychol. **99**(4), 660–682 (2010)

17. Hurtado, M., Tarrasó, A.: Dandys-de-BERTganim at EXSIST 2025: a multi-task learning architecture for sexism identification. In: Faggioli, G., Ferro, N., Rosso, P., Spina, D. (eds.) Working Notes of CLEF 2025 – Conference and Labs of the Evaluation Forum, CEUR Workshop Proceedings (2025)

18. Italiani, P., Maqbool, F., Gimeno-Gómez, D., Fersini, E., Martínez-Hinarejos, C.D.: TrankilTwice at EXIST2025: detecting sexism in memes under multi-lingual settings. In: Faggioli, G., Ferro, N., Rosso, P., Spina, D. (eds.) Working Notes of CLEF 2025 – Conference and Labs of the Evaluation Forum, CEUR Workshop Proceedings (2025)

19. Khursheed, M.S., Hasan Abidi, S.R., Faisal Sikandar, S., Zahra, S., Alvi, F., Samad, A.: Sexism identification in tweets using ensembles & augmentation: a multilingual approach. In: Faggioli, G., Ferro, N., Rosso, P., Spina, D. (eds.) Working Notes of CLEF 2025 – Conference and Labs of the Evaluation Forum, CEUR Workshop Proceedings (2025)

20. Labadie-Tamayo, R., Chulvi, B., Rosso, P.: Everybody hurts, sometimes. overview of HUrtful HUmour at IberLEF 2023: detection of humour spreading prejudice in twitter. In: Procesamiento del Lenguaje Natural (SEPLN), no. 71, pp. 383–395 (2023)

21. Labadie-Tamayo, R., Böck, A.J., Slijepčević, D., Chen, X., Babic, A., Zeppelzauer, M.: FHSTP@EXIST 2025 benchmark: sexism detection with transparent speech concept bottleneck models. In: Faggioli, G., Ferro, N., Rosso, P., Spina, D. (eds.) Working Notes of CLEF 2025 – Conference and Labs of the Evaluation Forum, CEUR Workshop Proceedings (2025)

22. Mendiburo-Seguel, A., Ford, T.E.: The effect of disparagement humor on the acceptability of prejudice. Curr. Psychol. J. Diverse Perspect. Diverse Psychol. Issues (2019)

23. Morales Rodríguez, G., Lopez-Figueroa, J.: The portrayal of women in media. J. Student Res. **13**(2) (2024)

24. NewStatesman: Social media and the silencing effect: Why misogyny online is a human rights issue. https://bit.ly/3n3ox68. Accessed 18 Oct 2023

25. Nowakowski, N., Calogiuri, L., Egyed-Zsigmond, E., Nurbakova, D., Erbani, J., Calabretto, S.: Automatic sexism detection on social networks: classification of tweets and memes. In: Faggioli, G., Ferro, N., Rosso, P., Spina, D. (eds.) Working Notes of CLEF 2025 – Conference and Labs of the Evaluation Forum, CEUR Workshop Proceedings (2025)

26. Pan, R., Bernal Beltrán, T., García Díaz, J.A., Valencia-Garcia, R.: UMUTeam at EXIST 2025: multimodal transformer architectures and soft-label learning for sexism detection. In: Faggioli, G., Ferro, N., Rosso, P., Spina, D. (eds.) Working

Notes of CLEF 2025 – Conference and Labs of the Evaluation Forum, CEUR Workshop Proceedings (2025)

27. Pastells, P., Vázquez, M., Farrús, M., Taulé, M.: CLiC at EXIST 2025: combining fine-tuning and prompting with learning with disagreement for sexism detection. In: Faggioli, G., Ferro, N., Rosso, P., Spina, D. (eds.) Working Notes of CLEF 2025 – Conference and Labs of the Evaluation Forum, CEUR Workshop Proceedings (2025)

28. Plaza, L., et al.: Overview of EXIST 2024 – learning with disagreement for sexism identification and characterization in social networks and memes (extended overview). In: Working Notes of CLEF 2024 – Conference and Labs of the Evaluation Forum (2024)

29. Plaza, L., et al.: Overview of EXIST 2025: learning with disagreement for sexism identification and characterization in tweets, memes, and TikTok videos (extended overview). In: Faggioli, G., Ferro, N., Rosso, P., Spina, D. (eds.) Working Notes of CLEF 2025 – Conference and Labs of the Evaluation Forum. CEUR Workshop Proceedings (2025a)

30. Plaza, L., et al.: Overview of EXIST 2025: learning with disagreement for sexism identification and characterization in tweets, memes, and TikTok videos (extended overview). In: Faggioli, G., Ferro, N., Rosso, P., Spina, D. (eds.) Working Notes of CLEF 2025 – Conference and Labs of the Evaluation Forum, CEUR Workshop Proceedings (2025b)

31. Plaza, L., et al.: Overview of EXIST 2023 – learning with disagreement for sexism identification and characterization (extended overview). In: Aliannejadi, M., Faggioli, G., Ferro, N., Vlachos, M. (eds.) Working Notes of the Conference and Labs of the Evaluation Forum (CLEF 2023). vol. 497, pp. 813–854. CEUR Working Notes (2023)

32. Rodríguez-Sánchez, F., et al.: Overview of EXIST 2021: sexism identification in social networks. Procesamiento del Lenguaje Natural **67**, 195–207 (2021)

33. Rodríguez-Sánchez, F., et al.: Overview of EXIST 2022: sexism identification in social networks. Procesamiento del Lenguaje Natural **69**, 229–240 (2022)

34. Tian, L., Trippas, J.R., Rizoiu, M.A.: Mario at EXIST 2025: a simple gateway to effective multilingual sexism detection. In: Faggioli, G., Ferro, N., Rosso, P., Spina, D. (eds.) Working Notes of CLEF 2025 – Conference and Labs of the Evaluation Forum, CEUR Workshop Proceedings (2025)

35. Uma, A., et al.: SemEval-2021 task 12: learning with disagreements. In: Proceedings of the 15th International Workshop on Semantic Evaluation (SemEval-2021), pp. 338–347. Association for Computational Linguistics, Online (2021)

36. Villarreal Haro, K., Sanchez-Vega, F., Pastor López Monroy, A.: Knowledge expansion guided by justification for improved sexism categorization. In: Faggioli, G., Ferro, N., Rosso, P., Spina, D. (eds.) Working Notes of CLEF 2025 – Conference and Labs of the Evaluation Forum, CEUR Workshop Proceedings (2025a)

37. Villarreal Haro, K., Segura Gómez, G., Tavarez Rodríguez, J., Sánchez Vega, F., Pastor López Monroy, A.: Leveraging reasoning of auto-revealed insights via knowledge injection and evolutionary prompting for sexism analysis. In: Faggioli, G., Ferro, N., Rosso, P., Spina, D. (eds.) Working Notes of CLEF 2025 – Conference and Labs of the Evaluation Forum, CEUR Workshop Proceedings (2025b)

Overview of ImageCLEF 2025: Multimedia Retrieval in Medical, Social Media and Content Recommendation Applications

Bogdan Ionescu[1], Henning Müller[2], Dan-Cristian Stanciu[1(✉)],
Alexandra-Georgiana Andrei[1], Ahmedkhan Radzhabov[3], Yuri Prokopchuk[4],
Liviu-Daniel Ştefan[1], Mihai Gabriel Constantin[1], Mihai Dogariu[1],
Vassili Kovalev[3,4], Hendrik Damm[5], Johannes Rückert[5], Asma Ben Abacha[6],
Alba G. Seco de Herrera[7], Christoph M. Friedrich[5], Louise Bloch[5],
Raphael Brüngel[5], Ahmad Idrissi-Yaghir[5], Henning Schäfer[8],
Cynthia Sabrina Schmidt[8], Tabea M. G. Pakull[8], Benjamin Bracke[5],
Obioma Pelka[5], Bahadır Eryılmaz[9], Helmut Becker[9], Wen-Wai Yim[6],
Noel Codella[6], Roberto Andres Novoa[10], Josep Malvehy[11],
Dimitar Dimitrov[12], Rocktim Jyoti Das[13], Zhuohan Xie[14], Ming Shan Hee[15],
Preslav Nakov[14], Ivan Koychev[12], Steven A. Hicks[15], Sushant Gautam[15],
Michael A. Riegler[15], Vajira Thambawita[15], Pål Halvorsen[15], Diandra Fabre[16],
Cécile Macaire[16], Benjamin Lecouteux[16], Didier Schwab[16], Martin Potthast[17],
Maximilian Heinrich[18], Johannes Kiesel[19], Moritz Wolter[16], Sharat Anand[18],
and Benno Stein[18]

[1] National University of Science and Technology Politehnica Bucharest, Bucharest,
Romania
{bogdan.ionescu,dan.stanciu1203}@upb.ro
[2] University of Applied Sciences Western Switzerland (HES-SO), Sierre, Switzerland
[3] Belarus State University, Minsk, Belarus
[4] Belarusian National Academy of Sciences, Minsk, Belarus
[5] University of Applied Sciences and Arts Dortmund, Dortmund, Germany
[6] Microsoft, Redmond, USA
[7] University of Distance Education (UNED), Madrid, Spain
[8] Institute for Transfusion Medicine, University Hospital Essen, Essen, Germany
[9] Institute for Artificial Intelligence in Medicine, University Hospital Essen, Essen,
Germany
[10] Stanford University, Stanford, USA
[11] Hospital Clinic of Barcelona, Barcelona, Spain
[12] Sofia University "St. Kliment Ohridski", Sofia, Bulgaria
[13] Indian Institute of Technology, Delhi, New Delhi, India
[14] Mohamed bin Zayed University of Artificial Intelligence, Abu Dhabi,
United Arab Emirates
[15] SimulaMet, Oslo, Norway
[16] University of Kassel, hessian.AI, and ScaDS.AI, Kassel, Germany
[17] Bauhaus-Universität Weimar, Weimar, Germany
[18] GESIS – Leibniz Institute for the Social Sciences, Mannheim, Germany

J. Carrillo-de-Albornoz et al. (Eds.): CLEF 2025, LNCS 16089, pp. 290–314, 2026.
https://doi.org/10.1007/978-3-032-04354-2_17

[19] Leipzig University, Leipzig, Germany

Abstract. This paper presents an overview of the ImageCLEF 2025 lab, which was organized within the Conference and Labs of the Evaluation Forum – CLEF Labs 2025. ImageCLEF is an ongoing evaluation event that started in 2003, promoting the evaluation of technologies for annotation, indexing, and retrieval of multimodal data and aiming to provide access to large collections of data across a variety of scenarios, domains and contexts. In 2025, the 23rd edition of ImageCLEF consists of four main tasks: (i) the *Medical* task, comprised of four sub-tasks, approaching a wide array of problems in the medical field, like concept detection, caption prediction, explainability assessment in radiology images, evaluating the veracity of GAN-generated 3D CT scans, providing a segmentation and answers to close-ended questions regarding dermatology images, or visual question answering and synthetic image generation involving gastrointestinal images, (ii) a new *Multimodal Reasoning* task, involving answering multiple-choice questions in 13 different languages, covering a wide range of subjects and difficulty levels, (iii) the *ToPicto* task, which focuses on converting either text or speech into a meaningful sequence of pictograms and (iv) the *Argument-Image* task, which explores the augmentation of arguments using images, by either retrieval or synthetic generation. This edition of the ImageCLEF benchmark attracted 193 teams that registered to the different tasks, of which 56 finished the challenges. This resulted in 493 submitted runs and a total of 45 working note papers. Overall, this year's edition has been very successful, with the biggest number of teams, submissions and working notes papers since 2019.

Keywords: Medical image processing · Medical image caption analysis · Medical concept prediction · Visual question answering · Generative Adversarial Networks · Synthetic Data Generation · Image Segmentation · Pictogram communication · Multilingual · Image Retrieval · ImageCLEF

1   Introduction

Since its inception in 2003 [7], ImageCLEF[1] has evolved to be one of the most prominent evaluation initiatives, promoting the evaluation of technologies for indexing, retrieval of visual data, and facilitating access to large image collections across a large variety of domains. This paper presents an overview of the 2025 edition of the lab, part of the Conference and Evaluation Forum- CLEF Labs 2025 [5].

Starting from just 4 participants in 2003, the impact of this evaluation campaign grew over the years, amassing thousands of participants, and runs over the years. The ever-growing impact of ImageCLEF -and also CLEF more broadly -

---

[1] http://www.imageclef.org/.

has been assessed in [24,25], and is further evidenced by the number of results for the term "ImageCLEF" returned by Google Scholar, with over 7900 mentions[2] to date. Over the years, ImageCLEF has always adopted emerging trends, adding tasks of interest, with some of the more recent trends included in this year's tasks being the evaluation of Vision-Language Models, Synthetic Image Generation or AI model Explainability.

This edition of ImageCLEF features 4 main tasks, with a large diversity of sub-tasks: ImageCLEFMedical, MultimodalReasoning, ImageCLEFtoPicto and Image Retrieval/Generation for Arguments (Fig. 1).

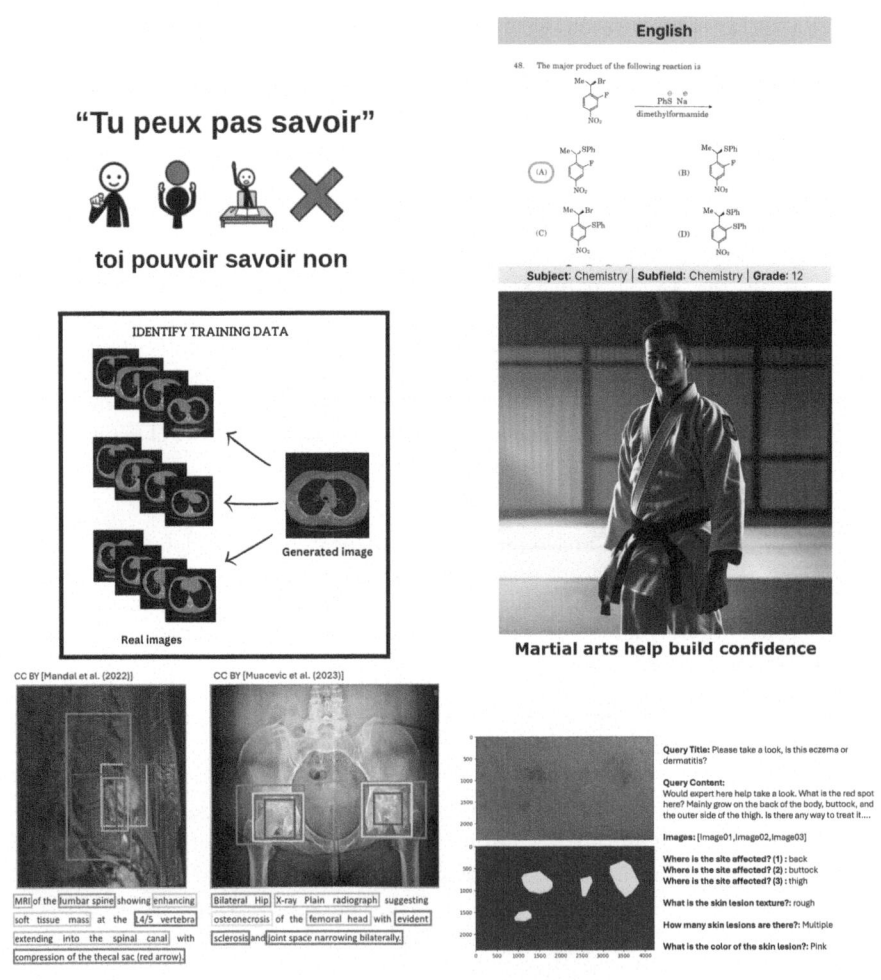

Fig. 1. Sample images from (left to right, top to bottom): ToPicto, MultimodalReasoning, the GANs task, Image Retrieval/Generation for Arguments, the Caption task and MEDIQA-MAGIC

---

[2] https://scholar.google.com/scholar?q=ImageCLEF.

2 Overview of Tasks and Participation

ImageCLEF 2025 consists of four main tasks to cover a *diverse range* of *multi-media retrieval, medical* applications. It followed the 2019 tradition [14] of diversifying the use cases [3, 13, 21–23, 29]. The 2025 tasks are presented as follows:

- **ImageCLEFmedical.** Since 2004, the ImageCLEF benchmarking initiative has included medical tasks. By 2018, however, although nearly all tasks were medical, there was limited interaction between them. Therefore, starting in 2019, the medical tasks were consolidated into a single medical task, with multiple subtasks, each handling a separate problem. This approach fostered synergies between the different domains. The medical task features four subtasks, described below:

  - *Caption*: The 2025 edition marks the nineth iteration of the medical captioning task [8]. This year, the task was expanded to three subtasks: the returning concept detection and caption prediction subtasks, and a newly promoted official explainability subtask. The caption prediction subtask focuses on composing coherent captions for radiology images, concept detection on identifying relevant UMLS concepts within those images, while the new explainability subtask requires participants to provide human-interpretable justifications for their model's predictions.

  - *GANs*: This is the third edition of the task [1–3]. The objective is to investigate whether synthetic medical images generated by deep generative models, such as GANs and Diffusion models, retain identifiable traces of the real data used during training. Addressing critical privacy and security concerns, the task includes two subtasks: detecting whether specific real images were used in GAN training, and attributing synthetic images to the correct training subset. The goal is to assess the potential for data leakage through "fingerprints" embedded in synthetic outputs

  - *MEDIQA-MAGIC*: The second edition for the MEDIQA-MAGIC [26] task builds on last year's challenges [28] using an expanded multimodal dermatology dataset. Participants receive clinical narratives with related images and must complete two subtasks: (1) segmenting areas showing dermatological issues, and (2) answering closed-ended clinical questions based on the provided context. Test sets are annotated by at least three annotators. Questions and options are available in both English and Chinese.

  - *MEDVQA*: Analysis of gastrointestinal (GI) images and videos continues to be an active research area in both the medical and computer science communities. Traditionally, most methods have focused on images as a single modality. The MEDVQA challenge [12] extends this by introducing a multimodal task that combines image and text data for visual question answering (VQA), targeting the field of GI endoscopy. This year, the challenge includes two subtasks. The first subtask focuses on answering clinical questions associated with specific images. The second subtask involves generating synthetic GI images based on prompts describing anatomical landmarks, visual features, and other relevant findings.

– **MultimodalReasoning**. This is the first edition of the task. The objective is to assess how effectively vision-language models can reason over complex visual and textual exam content. Participants were provided with an image of a question, including answer options and metadata describing the type of visual content in the image. Their task was to select a single correct answer from a set of three to five options. The task combines a vision-language question answering problem with multilingualism, with subtasks in 14 languages, as well as a multilingual challenge.

– **ToPicto**. This second edition of the ToPicto task challenges participants to automatically generate pictogram translations from written text and speech. Participants trained models on a novel multimodal dataset comprising three aligned corpora. These resources include parallel speech, text, and pictogram sequences across various domains (medical, general) and settings (read or spontaneous speech) to support robust cross-modal translation.

– **Touché-Argument-Images**. This is the fourth edition of the task, which challenges participants to convey an argument through a single image. Given a central claim—such as "Martial arts help build confidence" —participants must either retrieve a relevant image from a dataset or generate one using an image synthesis tool. The task was conducted in collaboration with the Touché Lab. Further details are available in their overview paper [15].

Table 1. Key figures regarding participation in ImageCLEF 2025.

Task	Groups that submitted results	Submitted runs	Submitted working notes
GANs	15	105	9
Caption	11	149	10
MultimodalReasoning	11	129	11
MEDIQA-MAGIC	8	82	7
MEDVQA	6	17	5
ToPicto	3	7	2
Argument-Images	2	4	1

In order to participate in the evaluation campaign, the research groups had to register by following the instructions on the ImageCLEF 2025 web page[3]. Since 2024, we used our own registration and submission platform[4]. The Ai4Media platform allows for registration, data download, and automatic submission and evaluation of runs. Similar to previous editions, the participants were required to submit and sign the End User Agreement to access the datasets and submit runs.

Following a drop in participation in 2016, interest in ImageCLEF rebounded in 2017 and 2018, with another increase observed in 2019. In 2018, 31 teams

---

[3] https://www.imageclef.org/2025/.

[4] https://ai4media-bench.aimultimedialab.ro/.

completed the tasks, resulting in 28 working notes papers. Participation peaked in 2019 with 63 teams and 50 submitted papers. In 2020 and 2021, 40 and 42 teams, respectively, completed the tasks, with 30 working notes received in 2021. In 2022, a decline followed, with 28 participating teams and 26 articles. However, 2023 marked a revival, with 47 teams submitting results and 39 working notes received. The 2024 edition attracted 26 teams, with a total of 34 working notes received. This year has seen the biggest number of participants since 2019, with 55 teams submitting results and 45 working notes submitted. Table 1 presents the overall participation statistics for this year's competition.

The following sections present the tasks, outlining the most important points, like general objectives, description, data sets and results, in short overviews. A detailed review of the received submissions for each task is provided with the task extended overview in the CLEF 2025 working notes: Caption [8], GANs [1], MEDIQA-MAGIC [26], MEDVQA [12], MultimodalReasoning [11], ToPicto [19].

3   The Caption Task

The 9th edition of the ImageCLEFmedical Caption task continues to benchmark automatic systems for radiology image understanding. Building on previous years, the 2025 edition introduced two major changes: the promotion of explainability to a fully graded subtask alongside concept detection and caption prediction, and a revised, holistic evaluation methodology for the generated captions. The task attracted 11 participating teams who submitted a total of 149 graded runs.

3.1   Task Setup

Participants were invited to take part in up to three subtasks:

- **Concept Detection:** Systems were required to predict a set of Unified Medical Language System® (UMLS) concepts present in an image. Performance was measured by the F1-score.
- **Caption Prediction:** Systems had to generate a coherent, full-sentence caption for an image. A key update for 2025 was the evaluation via a composite score, averaging six metrics (including BERTScore, ROUGE-1, and Align-Score) to assess both relevance and factuality.
- **Explainability:** For a small, pre-selected subset of images, teams had to provide a human-interpretable explanation (e.g., a heat-map or bounding boxes) justifying their model's caption. Submissions were manually rated by a radiologist on a 1–5 Likert scale for criteria such as coherence and clinical relevance.

3.2   Data Set

The task used an enlarged and updated version of the ROCOv2 dataset, containing images and captions from the PubMed Central® Open-Access subset. The final collection comprised 116,635 images, split into training (80,091), validation (17,277), and test (19,267) sets. A key novelty for 2025 was the introduction of the optical coherence tomography (OCT) imaging modality, which was retrospectively annotated for the entire corpus. UMLS concepts were extracted using MedCAT and subsequently filtered to ensure a high-quality label space. For the explainability subtask, a dedicated set of 16 images was manually selected by a radiologist to cover all modalities.

3.3   Participating Groups and Submitted Runs

The 2025 task attracted 80 registered research groups, from which 11 internationally diverse teams ultimately submitted 149 graded runs. Ten teams submitted working notes (3 recurring teams). Participation was highest in the caption prediction subtask (8 teams, 98 runs), followed by concept detection (9 teams, 51 runs), and the new explainability subtask (2 teams, 2 runs). Six of the teams competed in both of the main subtasks.

3.4   Results

In the concept detection task, top-performing teams continued to rely on ensembles of Convolutional Neural Networks (CNNs). The winning submission from the AUEB NLP Group [6] exemplifies this mature approach. As shown in Table 2, while the leading F1-scores were competitive, there was a general decrease in primary scores compared to previous years. This is attributed to the increased difficulty and diversity of the 2025 dataset, particularly with the introduction of the OCT modality.

Table 2. Best-run performance of participating teams in the concept detection subtask, ranked by primary F1-score.

Group Name	F1	Secondary F1	Rank (secondary)
AUEB NLP Group	**0.5888**	**0.9484**	1 (1)
DeepLens	0.5766	0.9299	2 (2)
mapan	0.5660	0.9298	3 (3)
UIT-Oggy	0.5613	0.9104	4 (4)
DS4DH	0.5225	0.8672	5 (6)
sakthiii	0.4003	0.9082	6 (5)
JJ-VMed	0.3982	0.8329	7 (7)
UMUTeam	0.2398	0.5377	8 (8)
LekshmiscopeVIT	0.1494	0.2298	9 (9)

The caption prediction subtask saw a clear and universal shift towards fine-tuning large Vision-Language Models (VLMs). The winning team, UMUTeam [20], used a fine-tuned BLIP architecture to achieve the best overall score. The results from the new composite metric, summarized in Table 3, highlight that while top systems could generate highly relevant captions, achieving high factuality scores proved challenging for all participants. Notably, a baseline model using an off-the-shelf instruction-tuned LLM (Llama 4 Scout) performed competitively, placing in the middle of the rankings and even outperforming some submissions on specific metrics.

Table 3. Best-run performance of participating teams in the caption prediction subtask, ranked by the new composite Overall score.

Group Name	Overall	Relevance	Factuality	Rank (Rel./Fact.)
UMUTeam	**0.3432**	**0.5268**	**0.1596**	1 (1/1)
DS4DH	0.3362	0.5174	0.1549	2 (2/2)
AI Stat Lab	0.3229	0.5089	0.1369	3 (3/3)
UIT-Oggy	0.3211	0.5076	0.1346	4 (4/4)
AUEB NLP Group	0.3068	0.4759	0.1377	5 (6/5)
JJ-VMed	0.3043	0.4922	0.1165	6 (5/6)
sakthiii	0.2746	0.4481	0.1011	7 (7/7)
CS_Morgan	0.2315	0.3717	0.0917	8 (8/8)
Baseline (Llama 4 Scout)	0.3101	0.5073	0.1128	

Finally, the inaugural explainability task saw participation from two teams. The manual evaluation by a radiologist revealed that both teams generated plausible visualizations (e.g., bounding boxes). However, these explanations were created using external, post-hoc models and did not provide insight into the internal reasoning of the captioning models themselves.

3.5   Lessons Learned and Next Steps

The 2025 task yielded several important insights for the field. First, while complex pipelines like Retrieval-Augmented Generation (RAG) were explored, the top results in captioning came from direct and robust fine-tuning of strong VLM backbones, suggesting that pipeline complexity can be a weakness if components are not perfectly optimized. Second, the new composite score proved to be a successful evolution, giving a more balanced view of caption quality. Its results clearly pinpoint clinical factuality as the next major frontier for research, as relevance scores now consistently outperform factuality scores.

The most critical lesson came from the new explainability task. The reliance of participants on post-hoc, external models highlights a need for the community to focus on developing and evaluating model-intrinsic explanation methods (e.g., attention maps, GradCAM) that can surface the actual features the generative model used, which is essential for building clinical trust.

Based on these lessons, the following steps are planned for the 2026 challenge:

- **Mature the Explainability Task:** Future guidelines will strongly encourage the submission of model-intrinsic explanations to better align the task with the goal of building trustworthy AI.
- **Expand and Enrich the Dataset:** The dataset will be expanded again with new articles. To address multilinguality, previously omitted non-English captions will be machine-translated and incorporated. Furthermore, for images that lack a caption, a baseline will be generated using the wider article context, providing a new type of training data.

4 The GANs Task

4.1 Task Setup

The third edition of the ImageCLEFmedical 2025 GANs task [1] builds on the first two editions [2,3], continuing the exploration of privacy and security concerns in synthetic medical imaging. The task is organized into two subtasks:

- **Subtask 1: Detect Training Data Usage** – Participants had to determine whether specific real images were used in training a Generative Adversarial Network (GAN) that produced a given set of synthetic images.
- **Subtask 2: Identify Training Data Subsets** – The goal was to link each synthetic image generated by a Diffusion model to its corresponding training subset among five predefined groups.

4.2 Data Set

The benchmarking datasets used in the GANs Task focused on computed tomography (CT) images spanning several anatomical regions relevant to medical imaging research. In Subtask 1, the data comprised thoracic CT axial slices originating from patients with lung tuberculosis. These slices presented a wide range of visual characteristics, from normal appearing lungs to scans with severe pulmonary lesions. Synthetic images were generated using a GAN trained on a subset of these real thoracic scans. In Subtask 2, the dataset extended to include cervical and abdominal CT slices, in addition to thoracic regions. This subtask leveraged a Diffusion-based generative model. All images, both real and synthetic, were standardized in format and resolution, enabling consistent evaluation across both subtasks while reflecting realistic clinical heterogeneity in organ appearance and pathology.

4.3 Participating Groups and Submitted Runs

Overall, 41 teams registered for our task. Of these, 14 teams completed the first subtask by submitting runs, and 4 teams completed the second subtask. In total, 9 teams submitted working notes papers. Notably, 2 teams participated in both subtasks, including the task organizing team. The continued interest in the first subtask, now in its third edition, highlights its ability to attract more participating teams.

Each participating team was allowed to submit up to 10 runs per task. We received a total of 105 submitted runs: 91 for Subtask 1 and 14 for Subtask 2.

4.4 Results

The rankings for Subtask 1 are shown in Table 4, and those for Subtask 2 are presented in Table 5, limited to the teams that described their methods by submitting working notes papers.

Table 4. Results of participant submissions and their results for Subtask 1: Detect Training Data Usage.

#	Participant	Run ID	Cohen's kappa	Accuracy	Precision	Recall	F1
1	Neural Nexus	1878	**0.148**	0.574	0.5698	0.604	0.5864
2	zhouyijiang1	1803	**0.136**	0.568	0.5582	0.652	0.6015
3	zhouyijiang1	1804	**0.136**	0.568	0.5582	0.652	0.6015
4	zhouyijiang1	1873	**0.136**	0.568	0.5582	0.652	0.6015
5	zhouyijiang1	1802	**0.132**	0.566	0.5537	0.68	0.6104
6	zhouyijiang1	1801	**0.128**	0.564	0.55	0.704	0.6175
7	Neural Nexus	1880	**0.072**	0.536	0.5542	0.368	0.4423
8	taotaozi	1359	**0.064**	0.532	0.5597	0.3	0.3906
9	taotaozi	1367	**0.044**	0.522	0.5505	0.24	0.3343
10	AIMultimediaLab*	1696	**0.036**	0.518	0.5162	0.572	0.5427
11	taotaozi	1364	**0.032**	0.516	0.6	0.096	0.1655
12	Neural Nexus	1881	**0.032**	0.516	0.5222	0.376	0.4372
13	taotaozi	1360	**0.032**	0.516	0.5128	0.64	0.5694
14	Neural Nexus	1877	**0.028**	0.514	0.5164	0.44	0.4752
15	taotaozi	1366	**0.02**	0.51	0.5069	0.732	0.599
16	Neural Nexus	1872	**0.016**	0.508	0.5182	0.228	0.3167
17	Medhastra	1288	**0.016**	0.508	0.5078	0.52	0.5138
18	taotaozi	1368	**0.012**	0.506	0.5092	0.332	0.4019
19	Challengers	1811	**0.012**	0.506	0.5062	0.492	0.499
20	ZOQ	1427	**−0.016**	0.492	0.4905	0.412	0.4478
21	Neural Nexus	1879	**−0.024**	0.488	0.4732	0.212	0.2928
22	Neural Nexus	1882	**−0.028**	0.486	0.4646	0.184	0.2636
23	ZOQ	1355	**−0.032**	0.484	0.4904	0.82	0.6138
24	SCOPE VIT Visioneers	1160	**−0.032**	0.484	0.4831	0.456	0.4691
25	Challengers	1779	**−0.032**	0.484	0.4355	0.108	0.1731
26	AIMultimediaLab*	1492	**−0.044**	0.478	0.4829	0.62	0.5429
27	ZOQ	1330	**−0.068**	0.466	0.4822	0.92	0.6327
28	ZOQ	1794	**−0.068**	0.466	0.4822	0.92	0.6327
29	taotaozi	1369	**−0.096**	0.452	0.4657	0.652	0.5433
30	Challengers	1778	**−0.116**	0.442	0.4461	0.48	0.4624
31	ZOQ	1356	**−0.132**	0.434	0.3862	0.224	0.2835
32	Challengers	1776	**−0.176**	0.412	0.3764	0.268	0.3131
33	Challengers	1777	**−0.176**	0.412	0.3764	0.268	0.3131

Subtask 1, which asked participants to detect whether a specific real image was used in the training process of a GAN, proved to be highly challenging. Despite the use of a wide range of techniques, including Siamese neural networks, contrastive learning, supervised classifiers, clustering approaches, and advanced Vision Transformer-based architectures, overall performance remained modest. The best result was achieved by the Neural Nexus team, whose ViT-based autoencoder pipeline reached a Cohen's kappa score of 0.148, with others, such as zhouyijiang1, achieving slightly lower but consistent scores around 0.13. Most submissions hovered close to or below zero, suggesting that models struggled to extract any reliable signal beyond chance. This low inter-rater agreement, measured via Cohen's kappa, underscores the difficulty of the task and possibly reflects that the GAN used in this edition was effective in generating images without obvious "fingerprints" of the training data.

In contrast, Subtask 2, which required participants to attribute synthetic images to their corresponding training subset of real data, showed substantially higher performance. Multiple teams reached classification accuracies above 98%, demonstrating that although image-level membership inference remains difficult, coarse-grained attribution at the dataset level is more tractable. The highest performance was recorded by SDVAHCS/UCSD, whose ensemble-based approach using EfficientNet architectures and pseudo-labeling, achieved up to 98.8% accuracy. Medhastra, using a simpler ResNet-18 classifier, also performed well with an accuracy of 94.84%, proving that even relatively lightweight models can be effective for this task when well-tuned.

Table 5. Results of participant submissions and their results for Subtask 2: Identify Training Data Subsets.

#	Participant	Run ID	Accuracy	Precision	Recall	F1	Specificity
1	AIMultimediaLab*	1396	**0.9904**	0.9904	0.9904	0.9904	0.9972
2	SDVAHCS/UCSD	1782	**0.988**	0.9882	0.988	0.9881	0.9969
3	SDVAHCS/UCSD	1871	**0.988**	0.9882	0.988	0.9881	0.9969
4	SDVAHCS/UCSD	1883	**0.988**	0.9882	0.988	0.9881	0.9969
5	SDVAHCS/UCSD	1426	**0.9878**	0.9881	0.9878	0.988	0.9969
6	SDVAHCS/UCSD	1425	**0.9708**	0.9716	0.9708	0.9711	0.9931
7	Medhastra	1287	**0.9484**	0.9504	0.9484	0.9487	0.9879
8	AIMultimediaLab*	1268	**0.5236**	0.5982	0.5236	0.5327	0.8799
9	AIMultimediaLab*	1269	**0.4913**	0.5822	0.4913	0.4934	0.8744
10	AIMultimediaLab*	1271	**0.4904**	0.5691	0.4904	0.4832	0.8753
11	AIMultimediaLab*	1267	**0.4112**	0.4645	0.4112	0.3945	0.8547

4.5   Lessons Learned and Next Steps

The 2025 ImageCLEFmedical GANs Task provided valuable insights into the privacy implications of synthetic medical image generation. Subtask 1, which focused on detecting whether specific real images were used to train a GAN, proved especially challenging. Despite using advanced techniques, participant systems performed close to random, with the best Cohen's kappa reaching only 0.148. This suggests that, under the conditions of this task, the proposed GAN generated images that were effectively free from directly traceable "fingerprints". While this is promising from a privacy perspective, it also reflects the limitations of current detection methods.

Subtask 2, in contrast, yielded significantly better results. Several teams achieved accuracies above 98%, showing that dataset-level attribution remains feasible. Successful methods employed supervised classification, deep feature embeddings, and semi-supervised learning strategies like pseudo-labeling. These results imply that while image-level membership inference is difficult, generative models may still retain broader signals from the source data distributions.

The differing outcomes of the two subtasks underscore the need to evaluate both model type and attribution level when assessing privacy risks. Moving forward, future task editions aim to explore new modalities, expand to multimodal generative data, and introduce more challenging attribution and privacy-preservation tasks.

5   The MEDIQA-MAGIC Task

In the second MEDIQA-MAGIC task [26], we extend on the previous year's dataset [31] and challenges [27,28] based on multimodal dermatology response generation. In this edition, participants were asked to identify areas of interest in an image based on the patient's query, e.g. the rash on an arm, as well as provide answers to structured closed-ended questions, e.g. is there single or multiple lesions. These are critical subtasks that can be used to improve end-to-end free text response generation, the subject of the original 2024 challenge.

5.1   Task Setup

Similar to the previous edition, participants were given a clinical narrative context along with accompanying images. The task was divided into two relevant sub-parts: (i) segmentation of dermatological problem regions, and (ii) providing answers to closed-ended questions. The questions, answers, and answer options were given in both English and Chinese.

In the first sub-task, given each image and the clinical history, participants are tasked generating segmentations of the regions of interest for the described dermatological problem. The expected outputs are binary image files with the same size as the original image. To leverage multiple gold standard masks for segmentation, we use the majority vote by pixel as the gold standard for microscore calculations of Jaccard and Dice Index. However, we also calculate the mean of the per-instance max and mean.

In the second sub-task, participants were given a patient dermatological query, its accompanying images, as well as a closed-question with accompanying choices – the task is to select the correct answer to each closed question. Because the same dermatological problem may have multiple sites, there may be related questions (e.g. what is the size of the affected area for location 1, what is the size of the affected area for location 2). In these cases, the answers to the same related questions are collated together. Partial credit is given when there are partial matches to gold. The exact code can be found here: github.com/wyim/ImageCLEF-MAGIC-2025.

5.2   Data Set

The dataset was created by using real consumer health users' queries and images; the question schema was created in collaboration with two certified dermatologists. In total closed question schema - a comprehensive list of clinically relevant, patient-facing questions for dermatological assessments included a total of 137 questions. For the challenge, we tested for total of 27 questions, for which were most common and can use both text and images to answer. These corresponded to 9 overall questions when related questions are grouped (e.g. anatomic region for affected area 1, anatomic region for affected area 2). The answers were labeled by at least 3 annotators: 2 medical scribe annotators, 1 biomedical informatics graduate student. Questions and answers were translated into Chinese by a native Chinese speaker. Full details can be found in our dataset paper [30]

Congruent with the MEDIQA-M3G edition [27], there was a total of 300, 56, and 100 instances for train, valid, and test splits respectively. Each query had on average 3 images.

5.3   Participating Groups and Submitted Runs

Fifty three teams registered for the event. A total of 56 completed valid runs across 6 teams were submitted. Table 6 provides a list of partipating teams and affiliations. This year's primary participants came from academia from United States, Vietnam and India.

Table 6. Participating Teams in the MEDIQA-MAGIC 2025 Challenge

Team	Institution	Affiliation
DS@GT	United States	Georgia Institute of Technology
H3N1	Vietnam	University of Information Technology
Kasukabe Defense Group	India	KLE technological university
Anastasia	Vietnam	Universiy of Information Technology
IReL, IIT(BHU)	India	Indian Institute of Technology(BHU)
KLE1	India	KLE Technological University
Oggy	Vietnam	University of Information Technology

5.4   Results

Table 7 shows results for the segmentation tasks. Despite different calculations of jaccard and dice metrics, both given identical rankings. Table 8 shows results for the segmentation task.

In the segmentation subtask, all four teams took a fine-tuning approach with differences in the exact models employed (e.g. TransUNet, ViT-B, CLIP). The Anatasia team enriched the dataset by performing image transformation techniques (e.g. rotations, contrast etc.) and were able to achieve top performances after including data with all transformations. The IReL, IIT(BHU) team was the only team that attempted to incorporate textual features. Their strategy used

Table 7. Performance of the participating teams in the MEDIQA 2025 Subtask 1 on segmentation generation for dermatological problems. Duplicate submission scores are removed.

team	jaccard	dice
Anastasia	0.6458	0.7848
Anastasia	0.6113	0.7587
IReL, IIT(BHU)	0.5881	0.7407
KLE1	0.5410	0.7021
H3N1	0.5145	0.6794
Anastasia	0.3205	0.4855
Anastasia	0.3129	0.4766
Kasukabe Defense Group	0.1866	0.3145

Table 8. Performance of the participating teams in the MEDIQA 2025 Subtask 2 on closed question answering. Duplicate submission scores are removed.

team	CQID010	CQID011	CQID012	CQID015	CQID020	CQID025	CQID034	CQID035	CQID036	ALL
H3N1	0.7	0.89	0.77	0.91	0.69	0.97	0.45	0.86	0.58	0.76
H3N1	0.64	0.89	0.76	0.87	0.71	0.96	0.47	0.85	0.6	0.75
H3N1	0.67	0.74	0.72	0.93	0.69	0.98	0.49	0.87	0.62	0.75
H3N1	0.64	0.88	0.76	0.85	0.73	0.9	0.46	0.86	0.54	0.74
DS@GT MEDIQA-MAGIC	0.53	0.87	0.66	0.81	0.56	0.89	0.6	0.81	0.65	0.71
DS@GT MEDIQA-MAGIC	0.51	0.84	0.7	0.85	0.56	0.87	0.55	0.81	0.67	0.71
DS@GT MEDIQA-MAGIC	0.47	0.86	0.69	0.85	0.56	0.84	0.51	0.82	0.64	0.69
DS@GT MEDIQA-MAGIC	0.44	0.84	0.69	0.78	0.55	0.86	0.48	0.79	0.65	0.68
DS@GT MEDIQA-MAGIC	0.49	0.82	0.63	0.74	0.56	0.79	0.51	0.75	0.59	0.65
KLE1	0.51	0.63	0.75	0.57	0.63	0.56	0.39	0.74	0.35	0.57
KLE1	0.47	0.62	0.7	0.58	0.62	0.56	0.36	0.76	0.3	0.55
Kasukabe Defense Group	0.44	0.66	0.75	0.28	0.66	0.44	0.52	0.77	0.3	0.54
Kasukabe Defense Group	0.4	0.61	0.73	0.29	0.65	0.44	0.52	0.76	0.33	0.53
Kasukabe Defense Group	0.49	0.49	0.67	0.32	0.48	0.41	0.01	0.76	0.55	0.46
DS@GT MEDIQA-MAGIC	0.31	0.38	0.53	0.31	0.31	0.42	0.01	0.72	0.37	0.37
Oggy	0.08	0.26	0.45	0.3	0.02	0.35	0.02	0.03	0.48	0.22
IReL, IIT(BHU)	0	0.44	0.48	0.17	0.44	0	0.02	0	0	0.17

CLIP to embed both text and visual features then afterwards fed the combined feature vector into a binary classification to predict the mask. The remaining teams fine-tuned previously trained skin lesion segementation models; the H3N1 team use the DermoSegDiff model, whereas the KLE1 team fine-tuned a Multi-Scale Feature Fusion Network model. Though these models were trained for skin lesions, likely more fine-tuning was required to completely adapt the model to this new dataset.

In the closed question-answering subtask, the top two performing teams H3N1 and DSGT employed multi-step architectures, including both fine-tuned models and LLM API's and ensembling methods. The former separated that task into four parts: (1) preprocessing, (2) information enrichment via image captioning, (3) fine-tuning and external API calls, (4) ensembling models from the previous step. The latter similarly had several layers (1) LLM fine-tuning with different models e.g. Qwen and LLAMA, (2) reasoning layer over output of (1) using Gemini, and (3) and agent layer that additionally has a RAG to reference the LanceDB dermatology corpus. In contrast, the other remaining groups had similar approaches, which utilized encoders for the images and text, then after fusing both text and image features, the network would eventually be fed into a classification layer.

5.5   Lessons Learned and Next Steps

In the segmentation task, the most successful system were able to use data augmentation generated through image transformation techniques (e.g. color contrast changes). This is promising as other teams did experiment with skin lesion segmentation specific models however were not able to achieve as high results – suggesting more data would be required to adapt those models. The use of textual inputs was only tested by one group, suggesting that this is an area for future exploration.

In the closed QA task, we found the best systems included multiple models fine-tuned for the task as well as some ensembling and aggregation. The use of multi-modal large language models were critically more successful than the suite of fine-tuned multimodal approaches which relied on a shared embedding representation then trained to fine-tune on the classification task. This could be because the current dataset is relatively small thus the important of the large language models' access to external information became a determining factor.

This edition implemented both subtasks simultaneously, simplifying organization but resulting in many repeat submissions with changes to only one subtask. Future improvements could include platform support for concurrent phases. Most submissions came from academia; future efforts will focus on expanding industry participation.

6   The MedVQA Task

The third edition of the MedVQA challenge at ImageCLEF continued to emphasize the application of image-based machine learning in gastrointestinal (GI) screening. In this edition, the scope of the challenge was broadened to include two key subtasks from the previous two years: visual question answering (VQA) and text-to-image synthesis. Five teams participated who submitted a total of eight runs. An overview can be seen in Table 9.

6.1   Task Setup

This year, we organized two subtasks to evaluate different capabilities of machine learning models in gastrointestinal (GI) imaging. Subtask 1 focused on answering

Table 9. An overview of the submissions to each task at MedVQA-GI.

	MedVQA 2023	MedVQA 2024	MedVQA 2025
# Registrations	26	22	31
# Task Participation	8	2	5
# Paper Submissions	6	2	5

clinical questions associated with annotated images from the challenge development dataset. Models were required to combine visual recognition, such as identifying tools, anatomical structures, or pathological features, with basic language understanding. Subtask 2 involved generating synthetic GI images from structured prompts describing anatomical locations, visual features, or the presence of tools. The goal was to produce images that resembled real endoscopic data and could be used for training or evaluating diagnostic models. Submissions for both subtasks were managed through a Hugging Face repository to ensure standardization and comparability across teams.

6.2   Data Set

The dataset used in this challenge is based on the publicly available HyperKvasir and Kvasir-VQA datasets, which include gastrointestinal endoscopy images from various anatomical sites and pathological conditions. For Subtask 1, the development set contained over 6,500 images from Kvasir-VQA, each annotated with one or more visual questions and corresponding answers. The questions fell into categories including Yes/No, Single-Choice, Multiple-Choice, Color, Location, and Count, targeting tasks like classification, reasoning, spatial localization, and attribute recognition. The test set introduced a distribution shift by using previously unreleased images from different sources. For Subtask 2, participants were given more than 2,000 image-caption pairs summarizing clinically relevant content such as anatomical features, abnormalities, or procedural elements. To increase variation and reduce overfitting, synthetic captions were also provided, generated using large language models and rule-based techniques. The test set for Subtask 2 was drawn from a separate, mixed-source dataset not included in the training data to support evaluation in unfamiliar clinical settings.

6.3   Results

Five teams submitted results for Subtask 1, and three teams participated in Subtask 2. In Subtask 1 (Table 10), IReL_IIT_BHU ranked first overall based on top scores across ROUGE and METEOR on the private set. UPS was the runner-up, with the highest BLEU score on the public set and strong performance on other metrics. In Subtask 2 (Table 11), CS_Morgan_Lab achieved the best overall performance, ranking first in FID, diversity, and FBD. IReL_IIT_BHU was the runner-up, with the highest agreement score and good scores on other metrics.

6.4   Lessons Learned and Next Steps

Most teams in Subtask 1 used transformer-based multimodal architectures, with Florence2 being the most common, fine-tuned using LoRA along with input augmentation and hyperparameter tuning. In Subtask 2, three teams submitted models based on fine-tuned variants of Stable Diffusion, also using LoRA. Although the generated images had high visual fidelity, prompt-image alignment

Table 10. Results for Task 1.

Team	Set	BLEU	R1	R2	RL	MET
UPS	Public	**0.24**	0.87	0.11	0.87	0.48
UPS	Private	0.22	0.88	0.11	0.88	0.49
IReL_IIT_BHU	Public	0.23	0.83	0.10	0.83	0.46
IReL_IIT_BHU	Private	0.22	**0.92**	**0.11**	**0.92**	**0.50**
MedPixel	Public	0.21	0.87	**0.12**	0.86	0.48
MedPixel	Private	0.18	0.91	0.11	0.90	**0.50**
CS_Morgan_Lab	Public	0.19	0.84	0.10	0.83	0.46
CS_Morgan_Lab	Private	0.18	0.90	0.10	0.90	0.49
Sagarmatha_Rangers	Public	0.15	0.81	0.10	0.80	0.44
Sagarmatha_Rangers	Private	0.16	0.88	0.10	0.88	0.49

Table 11. Results for Task 2.

Team	Set	Fid.	Agrmt.	Div.	FBD
CS_Morgan_Lab	Private	**0.0268**	0.7012	**0.7017**	**1539.31**
IReL_IIT_BHU	Private	0.2739	**0.7390**	0.6481	1694.97
MedPixel	Private	0.2725	0.7329	0.6722	1694.00

was inconsistent, and clinically accurate features were often missing. While more teams registered than last year, final submission numbers remained low. Subtask 2 may benefit from improved baselines, clearer instructions, and more human-in-the-loop evaluation.

7   The MultimodalReasoning Task

7.1   Task Setup

The task focused on visual question answering for multiple-choice questions with exactly one correct answer, where the answer options could also include visual content. It was conducted in two phases: an exploration phase, during which participants familiarized themselves with the publicly available training and validation data [9], followed by a test phase. In the test phase, images of the questions from the test dataset were released along with metadata describing the visual components within the questions. Participants submitted results to 14 different leaderboards: one multilingual leaderboard and 13 individual leaderboards, one for each language. Participants were allowed to make multiple submissions during this phase, but no feedback was provided. Final rankings were determined based on each participant's last submission at the end of the test phase.

7.2   Data Set

The dataset used in this challenge is based on the publicly available Exams-V dataset [9], which includes 20,932 multiple-choice questions across 20 school disciplines, covering 11 languages from 7 language families. Each question had from three to five answer options, with a single correct answer. The test set introduced new questions from more recent graduate exams. Table 12 shows general statistics on the test set, which contains a total of 3,565 new questions. Additionally, three new languages were introduced, Urdu, Kazakh, and Spanish, challenging participants to explore approaches for zero-shot question answering.

Table 12. New test data statistics. Here, #**visual Q.** refers to questions with multimodal context and #**text Q.** refers to text-only questions. *Urdu, Kazakh, and Spanish are new languages, with no training/validation data from Exams-V.*

Language	ISO	Family	Grade	#Subj.	#Questions	#visual Q.	#text Q.
English	en	Germanic	12	1	512	62	450
Chinese	zh	Sino-Tibetan	12	4	407	0	407
German	de	Germanic	12	6	258	68	190
Italian	it	Romance	12	5	203	58	145
Arabic	ar	Semitic	10–12	4	222	164	58
Polish	pl	Slavic	12	7	259	104	155
Hungarian	hu	Finno-Ugric	12	6	247	30	217
Bulgarian	bg	Slavic	12	6	200	66	134
Croatian	hr	Slavic	12	5	203	58	145
Serbian	sr	Slavic	12	5	203	58	145
Urdu*	ur	Indo-Aryan	9–10	5	269	0	269
Kazakh*	kk	Turkic	11	4	243	84	159
Spanish*	es	Romance	12	10	339	209	130

7.3   Participating Groups and Submitted Runs

In the first edition of the Multimodal Reasoning task, 51 participants registered, with 11 teams participating in the test set, resulting in a total of 129 graded submissions. All 11 teams submitted working notes. The most popular leaderboards were English, Multilingual, and Chinese, with 10, 9, and 7 teams participating, respectively. Some teams participated in multiple leaderboards, with two teams submitting to all 14 and another two teams submitting to 13. Teams came from 5 different countries: Bulgaria, China, India, Egypt, and Pakistan.

7.4   Results

Table 13 shows the results on all 14 leaderboards for the Multimodal reasoning task. Participants significantly outperformed the baseline, except one team that opted for the same model as the baseline. The task proved to be of moderate difficulty, with some teams achieving over 90% accuracy. Team **seifahmed** excelled across the board, securing first place in 11 out of the 13 leaderboards they competed in.

Table 13. Results for the ImageCLEF 2025 Multimodal Reasoning task on all 14 leaderboards. Baseline system submitted by the organizers. In the case of equal scores, participants are assigned the same rank and ordered alphabetically. †Participants submitted as different teams, but wrote a single working notes paper as co-authors.

Rank	Team	Acc	Rank	Team	Acc	Rank	Team	Acc
	Multilingual			**English**			**Bulgarian**	
1	seifahmed	0.8140	1	stormhunter44†	0.8965	1	heavyhelium†	0.9050
2	ymgclef	0.5994	2	seifahmed	0.8652	1	stormhunter44†	0.9050
3	lekshmiscopevit	0.5770	3	ayeshaamjad	0.8125	2	ymgclef	0.7750
4	bingezzzleep	0.5619	4	heavyhelium†	0.8086	3	bingezzzleep	0.7500
5	plutohbj	0.5226	5	ymgclef	0.5938	3	seifahmed	0.7500
6	deng113abc	0.5195	6	deng113abc	0.5371	4	plutohbj	0.7300
7	mhl2001	0.4418	7	bingezzzleep	0.5312	5	baseline	0.2450
8	yaozihang	0.4376	8	plutohbj	0.4922	6	elenat	0.2350
9	baseline	0.2701	9	mhl2001	0.4629		**German**	
10	elenat	0.2188	10	yaozihang	0.4570	1	seifahmed	0.8915
	Kazakh		11	elenat	0.2520	2	ymgclef	0.7403
1	seifahmed	0.8148	12	baseline	0.2480	3	bingezzzleep	0.6860
2	ymgclef	0.5350		**Chinese**		4	plutohbj	0.6783
3	bingezzzleep	0.4938	1	seifahmed	0.8305	5	yaozihang	0.4961
4	plutohbj	0.4444	2	ayeshaamjad	0.6560	6	mhl2001	0.4922
5	baseline	0.2738	3	plutohbj	0.5921	7	baseline	0.3101
	Polish		4	bingezzzleep	0.5799		**Urdu**	
1	seifahmed	0.8224	5	mhl2001	0.5553	1	seifahmed	0.8067
2	ymgclef	0.7181	6	ymgclef	0.5283	2	ymgclef	0.3941
3	bingezzzleep	0.5792	7	yaozihang	0.4791	3	bingezzzleep	0.3569
4	plutohbj	0.5251	8	baseline	0.2678	3	yaozihang	0.3569
5	baseline	0.2934		**Arabic**		4	baseline	0.3011
	Italian		1	seifahmed	0.6757		**Croatian**	
1	seifahmed	0.9212	2	ayeshaamjad	0.4775	1	seifahmed	0.9507
2	bingezzzleep	0.6059	3	mhl2001	0.4730	2	bingezzzleep	0.6207
2	plutohbj	0.6059	4	ymgclef	0.4324	3	ymgclef	0.5764
3	ymgclef	0.6010	5	plutohbj	0.3514	4	plutohbj	0.5616
4	baseline	0.2414	6	bingezzzleep	0.3243	5	baseline	0.2709
	Spanish		7	baseline	0.2703		**Serbian**	
1	seifahmed	0.7198		**Hungarian**		1	seifahmed	0.7143
2	ymgclef	0.6696	1	ymgclef	0.6518	2	bingezzzleep	0.6059
3	bingezzzleep	0.6608	2	bingezzzleep	0.5425	3	ymgclef	0.5468
4	plutohbj	0.5723	3	plutohbj	0.4696	4	plutohbj	0.5320
5	baseline	0.3156	4	mhl2001	0.3563	5	baseline	0.2365
			5	baseline	0.2348			

7.5   Lessons Learned and Next Steps

In the first edition of the Multimodal Reasoning task, we observed a lot of interest, with registration numbers being similar to other established tasks under the same lab. Participating teams opted to use a combination of proprietary and open-source large VLMs, including Qwen2.5-VL, Gemini, SmolVLM, and Deepseek. The majority of approaches employed zero-shot or few-shot techniques and leveraged metainformation about visual elements. There were some fine-tuning submissions, but these generally underperformed, primarily due to the use of smaller models constrained by limited resources. The most widely used models were Gemini-2.5 and Qwen2.5-VL, with the former consistently outperforming across all leaderboards, showing that the most recent advances in reasoning models can compete on graduate exams with complex visual elements.

8   The ToPicto Task

The second edition of the ToPicto task focuses on the automatic generation of pictogram translations from two input modalities: written text and speech. This challenge introduces a novel multimodal dataset to support the training of machine learning models in a cross-modal translation setting.

Compared to the first edition, the dataset has been significantly extended to include a wider variety of acoustic domains (from read to spontaneous speech) and a broader set of thematic domains, including both medical and everyday-life contexts. Participants were tasked with building models that can generalize effectively and perform robustly across these diverse conditions.

8.1   Task Setup

The ToPicto 2025 task consists of two sub-tasks: Text-to-Picto and Speech-to-Picto. Participants were allowed to submit to one or both sub-tasks, with a maximum of 10 submissions in total.

- **Subtask 1: From Text to Pictogram Sequence** – The Text-to-Picto sub-task focuses on the automatic generation of a corresponding sequence of pictogram terms from a French text.
- **Subtask 2: From Speech to Pictogram sequence** – The Speech-to-Picto sub-task focuses on two modalities: speech and pictograms, and aims to directly map a speech input to pictogram concepts.

8.2   Data Set

The benchmarking data are curated from three aligned multimodal corpora: Propicto-commonvoice, Propicto-orféo, and Propicto-eval [16–18]. Propicto-commonvoice is based on the French portion of CommonVoice v15 [4], containing 967 h of read speech from 17,911 speakers, with pictogram sequences generated using the method described in [16]. Propicto-orféo, built from the CEFC corpus [10], consists of 233 h of spontaneous speech from various domains and interaction types, with corresponding pictogram translations. Propicto-eval is a controlled evaluation set with multi-speaker read speech derived from children's stories, everyday scenarios, and medical texts, intended to assess model performance across different content domains.

8.3   Participating Groups and Submitted Runs

In 2024, a total of 16 teams participated in the ToPicto challenge, and four teams completed the Text-to-Picto task and submitted their results. In 2025, a total of 41 teams participated in the ToPicto challenge and registered for both tasks. Only three teams completed the Text-to-Picto task (with 4 runs), and one team completed the Speech-to-Picto task (with 2 runs). Two teams merged their contributions for final submissions, resulting in two working notes provided.

8.4   Results

Table 14 shows the different submission results. Note that majahj and indira collaborated on both tasks and submitted a single working note.

Table 14. Performance of participating teams in the ToPicto 2025 task. Scores for sacreBLEU, METEOR, and PictoER are reported and ordered by the highest sacreBLEU score.

Sub-Task	Team Name	SacreBLEU↑	METEOR↑	PictoER (%)↓	Rank
Text-to-Picto	majahj	76,98	88,66	13,48	1
	sudharshan07	69,01	85,09	18,56	2
	indira	52,41	74,50	29,23	3
	indira	37,72	64,61	42,70	4
Speech-to-Picto	majahj	62,87	73,41	29,49	1
	majahj	54,71	65,90	40,02	2

8.5   Lessons Learned and Next Steps

In 2025, we observed a significant increase in registrations; however, only two teams submitted final working notes. Both teams conducted their work seriously and presented interesting results on the fine-tuning of Large Language Models for the pictogram translation task. One team focused on analyzing the impact of model size and number of training epochs, while the other designed a lightweight architecture tailored to the source language. For the next steps, emphasizing the place of this task in the NLP domain and providing participants with more documentation on pictograms should attract a wider range of profiles and more diverse solutions for solving both proposed tasks.

9   Conclusion

This paper presents the overview of the ImageCLEF 2025 benchmarking campaign. We introduce the four main tasks, introducing interesting challenges in medicine (caption analysis, medical data generation and assessment, medical image segmentation for question answering, visual question answering), multimodal and multilingual question answering, pictogram to text and audio translation and image retrieval/generation for arguments.

The majority of the approaches by the participants were deep learning-based. In the ImageCLEF Medical-Caption task, the top-performing teams used Convolutional Neural Networks for the Concept detection sub-task, as well as Vision-Language Models, for the Caption prediction problem. For the GANs task, the participating teams used a wide array of machine learning models, with a Vision Transformer architecture achieving the best results on the first subtask. For the second subtask, multiple teams achieved a performance over 98% using deep learning approaches. For MEDIQA-MAGIC, the participants relied on Large Language Models. In the MedVQA task, the majority of teams used transformer-based architectures for the first subtask, while Stable Diffusion models were widely used in the second one. For the newly introduced Multimodal Reasoning task, most of the approaches used a combination of proprietary and open-source Vision Language Models, while employing zero-shot or few shot techniques. For ToPicto, the translation of text to pictograms was done using Large Language Models.

ImageCLEF continues to serve as a platform for innovation, learning, and advancement across a wide range of fields. Future editions will focus on enhancing existing tasks, expanding into new domains, attracting more participants, and fostering experimentation and learning in emerging areas. Addressing current limitations—such as clarifying task descriptions, allocating resources effectively, refining evaluation metrics, and exploring new assessment methods and collaboration opportunities—will also be a priority. Our goal remains to continually raise the quality of this benchmarking campaign and contribute meaningfully to progress in the field.

Acknowledgements. The work of Louise Bloch, Raphael Brüngel and Benjamin Bracke was partially funded by a PhD grant from the University of Applied Sciences and Arts Dortmund (FH Dortmund), Germany. The work of Ahmad Idrissi-Yaghir, Tabea M. G. Pakull, Hendrik Damm, Henning Schäfer, Bahadir Eryilmaz, and Helmut Becker was funded by a PhD grant from the DFG Research Training Group 2535 Knowledge- and data-based personalisation of medicine at the point of care (WisPerMed). The work of Dimitar Dimitrov and Ivan Koychev is partially funded by the EU NextGenerationEU, through the National Recovery and Resilience Plan of the Republic of Bulgaria, project SUMMIT, No BG-RRP-2.004-0008. The ToPicto task was funded by the Agence Nationale de la Recherche (ANR) through the project PANTAGRUEL (ANR-23-IAS1-0001). This work is also carried out as part of the AugmentIA Chair, led by Didier Schwab and hosted by the Grenoble INP Foundation, with sponsorship from the Artelia Group. The chair also receives support from the French government, managed by the National Research Agency (ANR), under the France 2030 program with reference ANR-23-IACL-0006 (MIAI Cluster). The pictographic symbols used are the property of the Government of Aragón and have been created by Sergio Palao for ARASAAC (http://www.arasaac.org), that distributes them under Creative Commons License BY-NC-SA. The work of Bogdan Ionescu, Alexandra Andrei, Dan-Cristian Stanciu is supported by a grant of the Ministry of Research, Innovation and Digitization, CCCDI - UEFISCDI, project number PN-IV-P6-6.3-SOL-2024-2-0320, within PNCDI IV. The work of Liviu-Daniel Stefan was supported by a grant of the Ministry of Research, Innovation and Digitization, CCCDI - UEFISCDI, project number PN-IV-P6-6.3-SOL-2024-0049, within PNCDI IV. The work of Mihai Gabriel Constantin was supported by a grant of the Ministry of Research, Innovation and Digitization, CCCDI - UEFISCDI, project number PN-IV-P6-6.3-SOL-2024-0060, within PNCDI IV. This work was partly supported by the project GRESEL-UNED PID2023-151280OB-C22 funded by MICIU/AEI/ AEI 501100011033.

References

1. Andrei, A.G., et al.: Overview of ImageCLEFMedical 2025 GANs task: training data analysis and fingerprint detection. In: CLEF2025 Working Notes. CEUR Workshop Proceedings, CEUR-WS.org, Madrid, Spain (2025)
2. Andrei, A., Radzhabov, A., Coman, I., Kovalev, V., Ionescu, B., Müller, H.: Overview of ImageCLEFmedical GANs 2023 task – identifying training data "fingerprints" in synthetic biomedical images generated by GANs for medical image security. In: CLEF2023 Working Notes. CEUR Workshop Proceedings, CEUR-WS.org, Thessaloniki, Greece (2023)
3. Andrei, A., et al.: Overview of 2024 ImageCLEFmedical GANs task – investigating generative models' impact on biomedical synthetic images. In: CLEF2024 Working Notes. CEUR Workshop Proceedings, CEUR-WS.org, Grenoble, France (2024)
4. Ardila, R., et al.: Common voice: a massively-multilingual speech corpus. In: Calzolari, N., et al. (eds.) Proceedings of the Twelfth Language Resources and Evaluation Conference, pp. 4218–4222. European Language Resources Association, Marseille, France (2020)
5. Carrillo de Albornoz, J., et al.: Experimental IR meets multilinguality, multimodality, and interaction. Proceedings of the sixteenth international conference of the CLEF association (CLEF 2025). In: Proceedings of the 15th International Conference of the CLEF Association, CLEF 2024. Lecture Notes in Computer Science, Madrid, Spain (2025)
6. Chatzipapadopoulou, A., et al.: AUEB NLP group at ImageCLEFmedical Caption 2025. In: CLEF2025 Working Notes. CEUR Workshop Proceedings, CEUR-WS.org, Madrid, Spain (2025)
7. Clough, P., Sanderson, M.: The CLEF 2003 cross language image retrieval task. In: Proceedings of the Cross Language Evaluation Forum (CLEF 2003) (2004)
8. Damm, H., et al.: Overview of ImageCLEFmedical 2025 – medical concept detection and interpretable caption generation. In: CLEF 2025 Working Notes. CEUR Workshop Proceedings, CEUR-WS.org, Madrid, Spain (2025)
9. Das, R., Hristov, S., Li, H., Dimitrov, D., Koychev, I., Nakov, P.: EXAMS-V: a multi-discipline multilingual multimodal exam benchmark for evaluating vision language models. In: Ku, L.W., Martins, A., Srikumar, V. (eds.) Proceedings of the 62nd Annual Meeting of the Association for Computational Linguistics (Volume 1:

Long Papers), pp. 7768–7791. Association for Computational Linguistics, Bangkok, Thailand (2024)

10. Debaisieux, J.M., Benzitoun, C., Deulofeu, H.J.: Le projet ORFEO: Un corpus d'études pour le français contemporain. Corpus **15**, 91–114 (2016). https://doi.org/10.4000/corpus.2936. https://hal.science/hal-01449600

11. Dimitrov, D., et al.: Overview of imageclef 2025 – multimodal reasoning. In: CLEF 2025 Working Notes. CEUR Workshop Proceedings, CEUR-WS.org, Madrid, Spain (2025)

12. Gautam, S., Halvorsen, P., Riegler, M.A., Thambawita, V., Hicks, S.A.: Overview of ImageCLEFmedical 2025 – medical visual question answering for gastrointestinal tract. In: CLEF2025 Working Notes. CEUR Workshop Proceedings, CEUR-WS.org, Madrid, Spain (2025)

13. Hicks, S.A., Storås, A., Halvorsen, P., Riegler, M.A., Thambawita, V.: Overview of ImageCLEFmedical 2024 – medical visual question answering for gastrointestinal tract. In: CLEF2024 Working Notes. CEUR Workshop Proceedings, CEUR-WS.org, Grenoble, France (2024)

14. Ionescu, B., et al.: ImageCLEF 2019: multimedia retrieval in medicine, lifelogging, security and nature. In: Crestani, F., et al. (eds.) CLEF 2019. LNCS, vol. 11696, pp. 358–386. Springer, Cham (2019). https://doi.org/10.1007/978-3-030-28577-7_28

15. Kiesel, J., et al.: Overview of Touché 2025: argumentation systems. In: de Albornoz, J.C., Gonzalo, J., et al. (eds.) Experimental IR Meets Multilinguality, Multimodality, and Interaction. 16th International Conference of the CLEF Association (CLEF 2025). Lecture Notes in Computer Science. Springer, Heidelberg (2025)

16. Macaire, C., et al.: A multimodal French corpus of aligned speech, text, and pictogram sequences for speech-to-pictogram machine translation. In: Calzolari, N., Kan, M.Y., Hoste, V., Lenci, A., Sakti, S., Xue, N. (eds.) Proceedings of the 2024 Joint International Conference on Computational Linguistics, Language Resources and Evaluation (LREC-COLING 2024), pp. 839–849. ELRA and ICCL, Torino, Italia (2024). https://aclanthology.org/2024.lrec-main.76/

17. Macaire, C., Dion, C., Schwab, D., Lecouteux, B., Esperança-Rodier, E.: Approches cascade et de bout-en-bout pour la traduction automatique de la parole en pictogrammes. In: Balaguer, M., Bendahman, N., Ho-dac, L.M., Mauclair, J., G Moreno, J., Pinquier, J. (eds.) Actes de la 31ème Conférence sur le Traitement Automatique des Langues Naturelles, volume 1: articles longs et prises de position, pp. 22–35. ATALA and AFPC, Toulouse, France (2024)

18. Macaire, C., Dion, C., Schwab, D., Lecouteux, B., Esperança-Rodier, E.: Towards speech-to-pictograms translation. In: Interspeech 2024, pp. 857–861 (2024)

19. Macaire, C., Fabre, D., Lecouteux, B., Schwab, D.: Overview of the 2025 ImageCLEFtoPicto task – investigating the generation of pictogram sequences from text and speech. In: CLEF2025 Working Notes. CEUR Workshop Proceedings, CEUR-WS.org, Madrid, Spain (2025)

20. Pan, R., Bernal Beltrán, T., García Díaz, J.A., Valencia-García, R.: UMUTeam at ImageCLEF 2025: Fine-tuning a vision-language model for medical image captioning and SapBERT-based reranking for concept detection. In: CLEF2025 Working Notes. CEUR Workshop Proceedings, CEUR-WS.org, Madrid, Spain (2025)

21. Popescu, A., Deshayes-Chossart, J., Schindler, H., Ionescu, B.: Overview of the imageclef 2022 aware task. In: Experimental IR Meets Multilinguality, Multimodality, and Interaction. Proceedings of the 13th International Conference of the CLEF Association (CLEF 2022), LNCS Lecture Notes in Computer Science, Bologna, Italy. Springer (2022)

22. Rückert, J., et al.: Overview of ImageCLEFmedical 2024 – caption prediction and concept detection. In: CLEF2024 Working Notes. CEUR Workshop Proceedings, CEUR-WS.org, Grenoble, France (2024)

23. Ștefan, L.D., Constantin, M.G., Dogariu, M., Ionescu, B.: Overview of imagecleffusion 2023 task - testing ensembling methods in diverse scenarios. In: Experimental IR Meets Multilinguality, Multimodality, and Interaction. CEUR Workshop Proceedings, CEUR-WS.org, Thessaloniki, Greece (2023)

24. Tsikrika, T., de Herrera, A.G.S., Müller, H.: Assessing the scholarly impact of ImageCLEF. In: Forner, P., Gonzalo, J., Kekäläinen, J., Lalmas, M., de Rijke, M. (eds.) CLEF 2011. LNCS, vol. 6941, pp. 95–106. Springer, Heidelberg (2011). https://doi.org/10.1007/978-3-642-23708-9_12

25. Tsikrika, T., Larsen, B., Müller, H., Endrullis, S., Rahm, E.: The scholarly impact of CLEF (2000–2009). In: Forner, P., Müller, H., Paredes, R., Rosso, P., Stein, B. (eds.) CLEF 2013. LNCS, vol. 8138, pp. 1–12. Springer, Heidelberg (2013). https://doi.org/10.1007/978-3-642-40802-1_1

26. Yim, W., Ben Abacha, A., Codella, N., Novoa, R.A., Malvehy, J.: Overview of the MEDIQA-MAGIC task at ImageCLEF 2025: multimodal and generative telemedicine in dermatology. In: CLEF 2025 Working Notes. CEUR Workshop Proceedings, CEUR-WS.org, Madrid, Spain (2025)

27. Yim, W.W., et al.: Overview of the MEDIQA-M3G 2024 shared task on multilingual multimodal medical answer generation. In: Naumann, T., Ben Abacha, A., Bethard, S., Roberts, K., Bitterman, D. (eds.) Proceedings of the 6th Clinical Natural Language Processing Workshop, pp. 581–589. Association for Computational Linguistics, Mexico City, Mexico (2024)

28. Yim, W., Ben Abacha, A., Fu, Y., Sun, Z., Yetisgen, M., Xia, F.: Overview of the MEDIQA-MAGIC task at ImageCLEF 2024: multimodal and generative telemedicine in dermatology. In: Conference and Labs of the Evaluation Forum (2024)

29. Yim, W., Ben Abacha, A., Snider, N., Adams, G., Yetisgen, M.: Overview of the MEDIQA-Sum task at imageclef 2023: summarization and classification of doctor-patient conversations. In: CLEF 2023 Working Notes. CEUR Workshop Proceedings, CEUR-WS.org, Thessaloniki, Greece (2023)

30. Yim, W., et al.: Dermavqa-das: dermatology assessment schema (das) and datasets for closed-ended question answering and segmentation in patient-generated dermatology images. CoRR (2025)

31. Yim, W., Fu, Y., Sun, Z., Ben Abacha, A., Yetisgen-Yildiz, M., Xia, F.: Dermavqa: a multilingual visual question answering dataset for dermatology. In: CLEF2024 Working Notes. CEUR Workshop Proceedings, CEUR-WS.org, Grenoble, France (2024)

Overview of the CLEF 2025 JOKER Lab: Humour in Machine

Liana Ermakova[1]([envelope]) [ID], Ricardo Campos[2,3] [ID], Anne-Gwenn Bosser[4] [ID], and Tristan Miller[5,6] [ID]

[1] Université de Bretagne Occidentale, HCTI, Brest, France
liana.ermakova@univ-brest.fr
[2] INESC TEC, Porto, Portugal
[3] University of Beira Interior, Covilhã, Portugal
[4] École Nationale d'Ingénieurs de Brest, Lab-STICC CNRS UMR 6285, Brest, France
[5] Department of Computer Science, University of Manitoba, Winnipeg, Canada
[6] Austrian Research Institute for Artificial Intelligence (OFAI), Vienna, Austria

Abstract. Humour poses a unique challenge for artificial intelligence, as it often relies on non-literal language, cultural references, and linguistic creativity. The JOKER Lab, now in its fourth year, aims to advance computational humour research through shared tasks on curated, multilingual datasets, with applications in education, computer-mediated communication and translation, and conversational AI. This paper provides an overview of the JOKER Lab held at CLEF 2025, detailing the setup and results of its three main tasks: (1) humour-aware information retrieval, which involves searching a document collection for humorous texts relevant to user queries in either English or Portuguese; (2) pun translation, focussed on humour-preserving translation of paronomastic jokes from English into French; and (3) onomastic wordplay translation, a task addressing the translation of name-based wordplay from English into French. The 2025 edition builds upon previous iterations by expanding datasets and emphasising nuanced, manual evaluation methods. The Task 1 results show a marked improvement this year, apparently due to participants' judicious combination of retrieval and filtering techniques. Tasks 2 and 3 remain challenging, not only in terms of system performance but also in terms of defining meaningful and reliable evaluation metrics.

Keywords: Wordplay · Puns · Machine translation · Humour-aware Information Retrieval · Computational Humour · Parallel corpora

1 Introduction

Humour plays a vital role in social interaction. Understanding it, however, can be challenging for humans, often requiring a good grasp of cultural references and double meanings. State-of-the-art artificial intelligence (AI), natural language processing (NLP), and information retrieval (IR) systems still struggle

J. Carrillo-de-Albornoz et al. (Eds.): CLEF 2025, LNCS 16089, pp. 315–337, 2026.
https://doi.org/10.1007/978-3-032-04354-2_18

with humour or other non-literal meanings aspects of texts [10]. This challenge is particularly evident in tasks like wordplay detection or humour analysis, which often rely on surface-level clues such as orthography or pronunciation, features not directly encoded in the deep semantic embeddings of modern AI models. Moreover, current pre-training approaches, typically based on next-word prediction objectives, favor the learning of literal and statistically probable language patterns. As a result, they often miss the nuanced, non-literal meanings crucial to understanding humour, irony, or sarcasm.

Despite these limitations, the ability to process and understand humour has growing practical relevance with humour-aware technologies benefiting a range of applications. For instance, Bell [4] has noted how second language learners struggle with humour understanding. In this context, humour is considered of pedagogical value [3,6,24] with IR tools working as a support to teachers in selecting engaging, level-appropriate content. Learners, too, could explore humour in other languages and cultures through such systems. Translators might also benefit, as humour, when identified in a source text, is often omitted in translations into a non-native language [35]. Another example of the importance of humour-aware systems can be found in the online sphere, where humour plays a central role in meme culture [40], helping propaganda and misinformation [33] spread widely across social networks. Humour-aware tools could thus aid in the detection and analysis of such communication campaigns, contributing to the design of effective countermeasures [46]. Finally, conversational agents such as chatbots or embodied conversational agents (ECAs) would be better equipped to engage users if they could identify or generate contextually appropriate humour [5,29]. A recent work of Kalloniatis & Adamidis [25], presented an extensive and in-depth review of literature on computational humour recognition.

With this in mind, the JOKER Lab[1], now in its fourth year, continues to bring together researchers from both the social sciences and computer science communities to foster work on automatic humour analysis. This paper presents an overview of the CLEF 2025 JOKER track, including its three shared tasks:

Task 1 Humour-aware Information Retrieval [14]
Task 2 Pun Translation [15]
Task 3 Onomastic Wordplay Translation [16]

More in-depth descriptions of the individual tasks can be found in the corresponding task overview papers [14–16] in the CEUR-WS Working Notes [23].

A total of 56 teams registered their intention to participate in JOKER 2025, with 20 of these teams going on to formally register for our individual tasks on Codabench [45], the open-source web-based platform for organising AI benchmarks. Of these 20 teams, 13 produced systems that contributed 136 distinct runs across the three shared tasks. All participants who submitted runs also submitted system description papers to the Working Notes volume [23]. Three teams from the same university (alecs, fhelms, and kamps) submitted a single joint report, as did teams cryptix and sarath_kumar, resulting in a total of

---

[1] https://www.joker-project.com/.

10 Working Notes from all participants. Table 1 provides a summary of these submissions. The team names correspond to user names at Codabench.

Table 1. Statistics on the runs submitted to JOKER 2025 per task

Team	Paper	Task 1 EN	Task 1 PT	Task 2	Task 3	Total
alecs	[26]			8		8
arampageos	[1]	3	9	12	15	39
cryptix	[30]	3		1		4
fhelms	[26]	4				4
igoranchik	[27]	3	2	13	2	20
kamps	[26]	4	4	2		10
mariapazr20	[2]				1	1
pjmathematician	[43]	7	4	4	3	18
rasion	[9]	2	2			4
rdtaylorjr	[41]			4		4
sarath_kumar	[30]	1		5	1	7
tanishc228	[8]	14				14
verbanex	[42]			3		3
baselines					1	1
Total		41	21	52	22	136

2   Task 1: Humour-aware Information Retrieval

Humour-aware Information Retrieval (IR) plays a vital role in several scenarios, where users search for content that is not only topically relevant but also humorous. It can be especially relevant to help comedians or writers quickly find humorous material related to a specific topic, assist teachers in making lessons more engaging through subject-related jokes (e.g., math jokes in a classroom), enable chatbots to respond with humorous content, or even support people seeking to learn a second language.

In this task, the aim is to retrieve short humorous texts from a document collection based on a given query. The retrieved texts should fulfill the dual criteria of being relevant to the query and being instances of wordplay. The typical use case would be searching for a joke on a specific topic – e.g., a query of "math" means that the goal is to find math jokes such as "Why don't mathematicians argue? Because they always try to find common denominators!"

2.1 Data

In the 2025 edition, the English data is an extension of that used in Task 1: Humour-aware Information Retrieval from JOKER 2024 [11, 13], which was constructed based on an English wordplay detection corpus [10, 20] and valid translations [12, 13]. We grouped the humorous texts into clusters of related topics and created queries based on these clusters. We added a significant number of topically relevant but non-humorous texts by extracting relevant passages from Wikipedia and by generating passages using Meta's Llama-2 (7B) models. Due to the number of queries, the corpus contains a large fraction of non-relevant content. In 2024, the total number of documents in the corpus was 61,268, with 4,492 humorous texts and 56,776 non-humorous ones. For 57 queries, 11,831 documents were considered topically relevant. The data was made available for training [11, 13]. For the 2025 edition, we expanded this data by new manually created jokes and texts generated by the LLMs Bard, Claude, ChatGPT, and Phi-3 Mini. The resulting corpus contains 77,658 texts in total, of which 5,198 are humorous. Detailed statistics on the English-language data sources for Task 1: Humour-aware Information Retrieval is given in Table 2. For 219 queries, 6,655 documents were judged humourous and topically relevant. As in 2024, we used 11 queries for the train and the rest for the test. The detailed statistics on the number of relevant humourous documents per query for the English dataset is given in Fig. 1.

Table 2. Number of humorous and non-humorous documents in the English Task 1 data by source

source	non-humorous	humorous	total
Bard	36	4	40
Claude	0	74	74
ChatGPT	149	381	530
JOKER	4,954	3,507	8,461
Llama-2	12,523	0	12,523
Phi-3 Mini	8,204	0	8,204
manual	2	247	249
translations	985	0	985
Wikipedia	46,592	0	46,592
total	72,460	5,198	77,658

To broaden the multilingual and multicultural scope of the task, this year we enriched the dataset with a substantial collection of humorous texts in Portuguese. This extension offers researchers a valuable opportunity to explore language-specific nuances in humour detection and contributes to the development of more robust cross-linguistic NLP models. To this regard, we added a

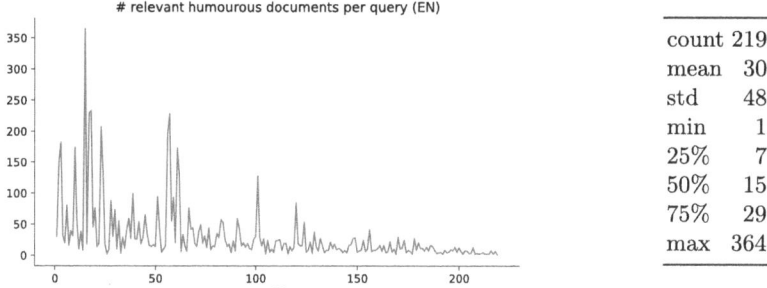

Fig. 1. Statistics on the number of relevant humourous documents per query in the English Task 1 data

substantial collection of 1,199 humorous texts and 43,927 non-humorous texts in European Portuguese. To compile the 1,199 humorous texts, we followed a three-stage process. First, 660 sentences from the English dataset were translated into Portuguese using DeepL. Second, 421 texts were gathered from diverse Portuguese-language websites. Lastly, 118 additional humorous texts were generated using ChatGPT (4o-mini model). Queries were derived from these humorous texts through a systematic process. Initially, puns were grouped by topic using GPT-3.5-turbo, creating clusters based on shared themes. For example, clusters like "grapes" and "oranges" were grouped under the broader query "fruit". Puns without a clear association to any query were labelled as irrelevant. A manual curation process refined these clusters into 98 general queries for a total number of 1,199 humorous texts. The detailed statistics on the number of relevant humourous documents per query for the Portuguese dataset is given in Fig. 2.

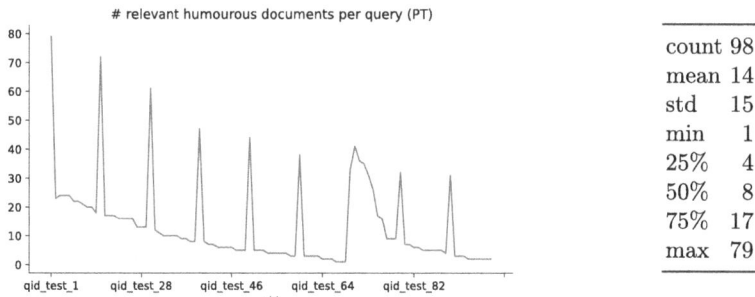

Fig. 2. Statistics on the number of relevant humourous documents per query in the Portuguese Task 1 data

To compile the 43,927 non-humorous sentences for the 98 queries, we employed a two-step approach. First, 41,028 sentences were retrieved using the Wikipedia API, following the same methodology as used for the English dataset.

Second, GPT-3.5-turbo was used to generate an additional 2,899 non-humorous sentences. To ensure consistency with the European Portuguese variant, all texts underwent a rigorous curation process. The PtVId model [39] was used to classify texts as either European or Brazilian Portuguese. Any texts identified as the latter were translated into the former using ChatGPT-4o-mini. Finally, a manual curation process verified the accuracy of the translations.

2.2 Evaluation

Performance was measured with standard information retrieval metrics as implemented in the `TrecTools`, an open-source Python library for information retrieval [31]. For each run we report the number of documents retrieved (#ret), the number of relevant documents retrieved (#rel), mean average precision (MAP; the mean of average precision scores across queries), geometric mean average precision (GMAP), precision at the number of relevant documents (P@R), mean reciprocal rank (MRR; the average of the reciprocal rank of the first relevant item across queries), precision (P@n; the proportion of relevant items retrieved at the top $n = 5, 10, 100, 1000$ positions), normalised discounted cumulative gain (NDCG; accounting for the relevance and position of documents in the ranking, normalised against the ideal ranking), and (for Portuguese only) the binary preference score (bpref).

2.3 Participants' Approaches

Team rasion [9] proposed a dual-screening architecture that separates humour-aware information retrieval into two distinct stages. The first employs a semantic similarity model that uses the paraphrase-multilingual-mpnet-base-v2 model to encode queries and documents into dense vector representations, and distance-based metrics and cosine similarity to quantify semantic alignment and filter query-relevant documents. This step is followed by a transformer-based classifier (xlm-roberta-base) that identifies humorous texts containing puns. The method, applied to both English and Portuguese datasets, aims to reduce task complexity through modularisation. Their system achieved strong performance in Portuguese, highlighting the effectiveness of separating relevance and humour detection subtasks.

Teams cryptix and sarath_kumar [30] employed a fine-tuned Sentence-BERT (SBERT) model to generate semantic embeddings of queries and documents. They trained the model using a cosine similarity loss on humour-labelled query–document pairs, aiming to capture implicit humour such as irony or exaggeration. The resulting vectors were indexed using the Facebook AI Similarity Search (FAISS) for efficient retrieval, and results were re-ranked using human-annotated humour intensity scores.

Team igoranchik [27] implemented a hybrid retrieval pipeline combining dense and lexical retrieval, followed by cross-encoder reranking. They fine-tuned the intfloat/multilingual-e5-small model using contrastive objectives – Multiple

Negative Ranking Loss (MNRL) and an Adaptive Margin Loss – on humour-annotated data, including synthetic queries generated with GPT-4o-mini. BM25 was used for lexical retrieval via Anserini, while dense vectors were stored in Qdrant. The top 1000 documents from both retrieval methods were merged using reciprocal rank fusion and re-ranked using the cross-encoder/ms-marco-MiniLM-L12-v2.

Team pjmathematician [43] implemented a two-stage pipeline using the Qwen family of large language models (LLMs). First, they applied large Qwen models (Qwen3-14B and Qwen3-32B) to analyse the entire document corpus, generating humour-related metadata such as a binary 'isJoke' flag and textual explanations for each document. These enriched representations were then used in a dense retrieval step, where smaller Qwen embedding models (Qwen3-4B and Qwen3-8B) indexed either the original text or the explanation-augmented versions. Retrieval was performed using both generic and humour-specific query prompts.

Team tanishc228 [8] proposed a multi-stage ensemble retrieval system combining traditional IR methods with neural rerankers (ColBERT and a BERT-based cross-encoder), complemented by handcrafted wordplay features. Their pipeline retrieves documents using both lexical and semantic methods, followed by contextual reranking and score fusion. The system aims to capture humorous content by incorporating features such as punctuation, repetition, and alliteration.

Teams kamps and fhelms [26] submitted baseline runs using Anserini BM25 or BM25+RM3 and zero-shot MSMARCO-trained neural cross-encoder rerankings of the top 100 results.

2.4   Results

Tables 3 and 4 report the Task 1 results for English and Portuguese, respectively.

For the English subtask, we received 39 distinct valid non-zero scored runs. Across all metrics, team pjmathematician [43], who applied two-stage Qwen LLM filter–explainer and dense retriever, obtained the best scores. In terms of MAP, their best approach (MAP = 0.3501) outperformed the next-best team by a factor of two. This latter team, from the University of Amsterdam [26], applied RM3 RoBERTa with drop 60 (MAP = 0.1672). The difference in NDCG@5 is even more significant, with 0.608 for pjmathematician compared to just 0.0152 for the University of Amsterdam. Close results were achieved by team rasion [9], who applied pre-trained models for a semantic matching network for relevance and a humor classification network for wordplay detection by RoBERTa. This approach had comparable NDCG@5 scores to the run pjmathematician_Q14-Q8-R. The application of cross-encoders by the University of Amsterdam [26] achieved significantly lower results than the RM3 and BM25 baselines in terms of MAP (0.0027 vs. 0.1237) and NDCG@5 (0.0038 vs. 0.14). This year's best run nearly tripled last year's top MAP =0.12 from the University of Amsterdam, who used RM3 with a T5 filter [11,13]. This year, the best results are comparable with those of topical relevance only in 2024 [11]. However, the baseline BM25 and RM3 runs by the University of Amsterdam this year and in 2024 [36] show comparable performance, with only slight improvements. This suggests that the core properties of the dataset have remained largely stable.

For the Portuguese subtask, we received 19 distinct valid non-zero scored runs. The best-scoring teams pjmathematician [43] and rasion [9] have very close results in terms of MAP (around 0.4) and NDCG@5 (around 0.5); they are followed by the University of Amsterdam [26] submitting the BM25 baseline with MAP = 0.08.

For the best runs, we observe weakly opposite trends for the English and Portuguese subtasks – namely, better results in terms of MAP for Portuguese but lower results in terms of P@5 and NDCG@5. This might be related to the higher average number of relevant humorous documents per query in English. This hypothesis is supported by the fact that for English, the best P@10 was 0.4 with a median of 15 relevant humorous documents per query, while for Portuguese P@10 reaches 0.34 with a median of 8.

3   Task 2: Pun Translation

This is the oldest JOKER task. As in previous iterations of JOKER, the goal here is to translate English punning jokes into French [12, 19, 22]. Translations should ideally be an instance of wordplay and preserve the meaning of the original as closely as possible. This task is difficult for modern machine translation models and humans translators alike, as literal translation seldom preserves both the form and meaning of figurative language, and may lead to nonsense and/or the loss of the discourse's pragmatic force. It is perhaps even more challenging for language models, which are trained on surface-level features and assume a level of linguistic predictability that may be subverted in punning jokes; their pattern-based training leads them to generate predictable, trope-filled content rather than original ideas [7, 28]. However, the incongruity theory of humour suggests that humour arises from mismatches between expectations and reality [37, 38].

3.1   Data

The training data for Task 2, which builds on previous editions [12, 20, 21], consists of 1,405 instances of wordplay in English, with a total of 5,838 French translations sourced from human professionals. For the 2025 edition, we collected 2,615 new manual translations of 1,682 distinct puns in English with manually annotated words or phrases with multiple meanings (pun locations) that we used for the test data. Some of the pun location annotations were collected for previous JOKER evaluation campaigns [10, 20]. We expanded this data with annotations of new references. Table 5 shows a histogram of the number of references and distinct locations per English pun in the test set. For 25% of English puns, we have multiple references and multiple locations. There are 1,382 English puns with a single location, while only 1,252 of them have a single reference.

Table 3. Results for Task 1 (English)

Run ID	#ret	#rel	MAP	GMAP	P@R	MRR	P@5	P@10	P@100	P@1000	NDCG@5
pjmathematician_Q14-Q8-Q32	207000	3007	35.01	24.65	38.68	79.04	54.88	40.63	9.05	1.45	60.80
pjmathematician_Q14-Q8-Q14	207000	2954	34.86	24.31	39.07	78.74	54.49	40.29	9.00	1.43	60.59
pjmathematician_Q32-Q8-Q14	207000	2932	34.38	23.98	38.98	80.78	54.49	41.01	8.82	1.42	60.94
pjmathematician_Q32-Q8-Q32	207000	3011	32.91	23.03	36.99	76.20	50.34	39.95	9.00	1.45	55.98
pjmathematician_Q14-Q8-R	207000	2835	23.88	16.03	27.10	64.27	36.62	28.70	7.84	1.37	41.23
UAms_RM3RoBERTa_drop60	82818	1448	16.72	7.04	23.05	54.46	30.82	23.09	5.98	0.70	1.52
Rasion_SenTransF+Roberta	4588	811	16.21	NaN	20.59	64.93	35.92	24.47	3.92	0.39	41.34
Rasion_SenTransF+Roberta	53552	1475	15.79	5.53	20.21	55.94	30.82	22.75	5.72	0.71	34.02
Cryptix_SBERT	207000	1914	15.07	5.52	19.44	56.94	28.70	20.97	4.75	0.92	33.46
UAms_RM3	207000	1864	15.02	7.22	19.53	40.87	24.35	20.00	6.22	0.90	25.66
UAms_RM3RoBERTa	186164	1798	14.94	7.16	19.56	42.47	25.51	19.61	6.07	0.87	26.76
CCC_Ensemble_ColBERT_RM3	103500	1967	14.15	7.44	16.29	33.69	16.71	17.73	6.70	0.95	18.17
CCC_Ensemble	206091	2050	14.03	7.12	17.01	40.33	20.48	18.65	5.87	0.99	22.55
UAms_RM3	207000	1872	12.16	5.77	15.36	33.30	18.55	18.07	5.72	0.90	19.57
UAms_en_bm25	41270	1884	11.91	5.64	12.23	26.28	12.95	12.71	5.94	0.91	14.00
CCC_TFIDF_Rerank	100185	1764	11.26	NaN	15.25	40.59	20.59	17.07	5.75	0.86	22.04
UAms_Anserini	207000	2134	10.76	5.35	10.56	25.03	11.88	12.22	6.14	1.03	12.37
UAms_en_rm3	207000	2134	10.76	5.35	10.56	25.03	11.88	12.22	6.14	1.03	12.37
CCC_ColBERT_Enhanced	207000	1879	9.93	5.17	12.47	33.35	15.75	14.15	5.56	0.91	16.53
CCC_XLM_R_Rerank	207000	2227	9.66	5.03	9.83	36.41	13.82	11.98	5.52	1.08	16.19
CCC_ColBERT_Enhanced	207000	1418	6.69	2.06	9.34	31.24	13.43	11.40	3.82	0.69	14.55
CCC_XLM_R_Rerank	10350	918	6.30	2.56	9.65	23.88	9.18	9.08	4.43	0.44	19.10
CCC_ColBERT_Enhanced	207000	1367	6.21	1.82	8.67	28.52	11.30	10.34	3.64	0.66	12.42
CCC_Advanced_Ensemble_LTR	165600	2122	6.20	3.65	6.38	11.54	2.90	5.89	5.67	1.03	2.67
CCC_TFIDF	207000	1321	5.79	1.56	8.41	25.29	9.47	9.03	3.79	0.64	10.50
CCC_Ensemble_ColBERT_RM3	103500	1904	5.44	2.24	6.52	21.03	8.12	6.96	3.79	0.92	8.77
CCC_TF-IDF_Ensemble_ColBERT_RM3	103500	1922	5.31	2.22	5.98	20.16	7.15	6.47	3.79	0.93	7.90
Skommarkhos_BM25_E5_MiniLM	207000	2182	5.02	2.98	3.03	6.47	0.87	3.24	4.44	1.05	0.65
UAms_en_bm25_CE1K	41270	1884	4.88	2.60	2.47	5.68	0.48	2.75	4.30	0.91	0.38
UAms_en_rm3_CE1K	207000	2134	4.78	2.67	2.32	5.48	0.39	2.51	4.32	1.03	0.27
UAms_Anserini	10350	849	4.76	1.41	3.92	7.67	0.87	3.67	4.10	0.41	0.75
cryptix_crossencoder	20700	999	3.78	1.55	2.43	5.64	0.48	2.51	4.83	0.48	0.34
CCC_Ensemble_RoBERTa_RM3	41400	1718	3.33	1.69	4.02	11.01	3.77	3.82	4.23	0.83	23.65
Skommarkhos_BM25_E5_MiniLM	20700	517	2.49	0.49	2.62	6.15	0.77	3.19	2.50	0.25	0.57
CCC_pipeline	5175	211	2.43	0.05	3.93	11.81	4.54	4.15	1.02	0.10	0.80
team_reranker_EN	20700	824	1.38	0.23	2.40	6.36	2.13	2.03	3.98	0.40	2.05
yourteam_xlm_roberta_large	20700	271	0.42	0.02	1.29	4.71	1.35	1.50	1.31	0.13	1.30
duth_xanthi_en	20700	62	0.04	0.00	0.21	0.75	0.10	0.10	0.30	0.03	0.08
cryptix_crossencoder	414000	336	0.02	0.00	0.01	0.08	0.00	0.00	0.01	0.05	0.00

3.2 Evaluation

As in the previous year's lab, we evaluated the runs with two traditional machine translation metrics, and this year we include a new pun location–based metric:

BLEU (BiLingual Evaluation Understudy) computes the translation's overlap in vocabulary overlap with a reference translation [32]. We used the sacreBLEU implementation [34] with the default tokeniser 13a. We report the BLEU score (harmonic mean) and the BLEU precisions for n-grams for $n = 1, 2, 3, 4$.

BERTScore computes tokenwise similarity scores between the candidate translation and a reference translation using contextual embeddings [47]. We used the Python implementation from the bert-score package.[2] We report mean values of BERTScore precision, recall, and F_1 over all references.

---

[2] https://pypi.org/project/bert-score/.

Table 4. Results for Task 1 (Portuguese)

Run ID	#ret	#rel	MAP	GMAP	P@R	MRR	P@5	P@10	P@100	P@1000	NDCG@5	bpref
pjmathematician_Q32-Q4-R	69000	932	42.21	30.78	42.01	69.07	43.77	34.35	8.80	1.35	42.14	58.40
pjmathematician_Q14-Q4-R	69000	938	42.17	30.81	41.65	68.98	43.77	34.49	8.83	1.36	51.69	58.65
Rasion_SenTransF+Roberta	69000	905	40.51	28.90	40.17	66.57	44.93	38.41	8.61	1.31	50.15	83.68
Rasion_SenTransF+Roberta	62576	904	40.51	28.90	40.17	66.57	44.93	38.41	8.61	1.31	50.12	83.62
UAms_pt_bm25	12856	229	7.89	0.19	5.96	9.83	5.22	6.09	3.03	0.33	5.13	11.52
Skommarkhos_BM25_E5_MiniLM	69000	503	7.42	1.65	5.74	11.91	6.38	6.23	2.87	0.73	6.44	7.60
pjmathematician_Q06-gist	69000	562	6.95	1.75	4.99	11.20	5.51	6.38	2.64	0.81	5.46	7.16
Skommarkhos_BM25_E5_MiniLM	6900	228	6.90	0.28	5.58	12.65	5.22	5.94	3.30	0.33	5.35	6.90
results_pt_pt_finetuned	6900	199	6.71	0.41	6.74	20.21	7.54	7.10	2.88	0.29	9.29	32.38
UAms_pt_rm3	67994	262	6.54	0.25	5.91	9.51	4.64	5.65	2.78	0.38	4.47	10.28
myteam_BERT	69000	496	6.13	1.26	6.38	19.54	8.12	6.38	2.54	0.72	8.78	7.35
duth_xanthi_pt	6900	225	5.95	0.37	6.76	15.65	7.54	8.41	3.26	0.33	7.03	15.13
pjmathematician_Q06-gist-exp32	69000	512	4.91	1.35	2.92	7.15	2.61	3.48	2.42	0.74	2.73	4.03
UAms_pt_rm3_CE1K	67994	262	4.16	0.19	2.47	5.20	1.45	3.19	2.41	0.38	1.34	4.80
UAms_pt_bm25_CE1K	12856	229	3.84	0.12	1.99	4.47	1.16	3.04	2.35	0.33	0.91	4.31
team_xlmr_PT	6900	133	2.96	0.11	5.33	12.03	4.64	5.94	1.93	0.19	4.57	17.66
results_pt_large_pt_finetuned	6900	65	0.31	0.01	0.02	0.73	0.00	0.00	0.94	0.09	1.72	4.00
yourteam_pt_zeroshot	6900	46	0.27	0.01	0.28	1.13	0.00	0.29	0.67	0.07	0.00	5.25
xlm-roberta-triplet-pt	6900	28	0.22	0.00	0.33	2.48	0.58	0.43	0.41	0.04	0.70	2.19

Table 5. Histogram of the number of references and locations per English pun in the Task 2 data

#	References	Locations
1	1,252	1,382
2	172	220
3	133	68
4	53	10
5	39	1
6	22	1
7	8	—
8	2	—
9	1	—

Pun location–based evaluation allows for a more fine-grained analysis of generated translations. We checked for words or phrases with multiple meanings (pun locations) from the reference texts by combining French reference translations with pun location annotations from the dataset used for JOKER 2023's Task 2: Pun Location and Interpretation [18,20]. We completed this data with pun location annotations of the new references.

3.3 Participants' Approaches

Team arampageos [1] combined neural machine translation systems with a hand-crafted translation dictionary of particularly challenging puns. The machine translation systems included Google Translate, Argos Translate, the Helsinki-NLP/opus-mt-en-fr models, Facebook's M2M100 (418M and 1.2B), MBART50, and NLLB (1.3B and distilled 600M). Their two-stage pipelines first checked whether the input pun matched the curated set, otherwise forwarding the input to the machine translation system.

Team verbanex [42] relied on extensive data preprocessing, including sentiment classification and phoneme conversion, to help the trained translation model capture emotional tone and pronunciation ambiguities. They used two different fine-tuning strategies – full parameter optimisation and parameter-efficient adaptation techniques – with the mBART-50 English-to-French translation model.

Team rdtaylorjr [41] relied on a three-stage approach. The first stage consisted of training multiple LLMs (provided by openAI, Google, Mistral, or DeepSeek) using a contrastive learning approach. In addition to the training set we provided, they used data from the JOKER 2023 shared task on pun location and interpretation, as well as a contrastive learning dataset constructed by neutralising puns of their French dataset. The second stage of the approach is based on chain-of-thought prompting making use of semantic and phonetic embeddings for the French language. Finally, evaluator agents were used to iterate over various properties of the proposed translations (conserving literal/contextual meaning, emotion level, and understandability in the target language).

Teams alecs and kamps [26] used a finetuned MarianMT sequence-to-sequence model, T5ForConditionalGeneration, T5-base, Meta AI NLLB-200-1.3B, and mBART-large-cc25.

Team pjmathematician [43] fine-tuned different Qwen models, including the Qwen2.5-14B, experimenting with different LoRA parameters on the provided corpus. They then used a simple prompting approach for requesting translations of puns.

Team igoranchik [27] used supervised fine-tuning with the aim of forcing a model to learn higher quality responses. They also used an Adaptive Rejection Preference Optimisation (ARPO) [44] implementation[3] in an attempt to enhance humour retention.

Teams cryptix and sarath_kumar [30] used back-translation for data augmentation. They fine-tuned the MarianMT model and used a loss function combining humour preservation metrics from a rule-based module evaluating humour preservation with standard BLEU metrics.

3.4   Results

Tables 6, 7, and 8 report, respectively, the results based on BLEU, BERTScore, and the pun location–based metric for Task 2.

The top three runs according to BERTScore and BLEU – namely, UvA_finetunedNLLB-1.3B [26], Skommarkhos_Lucie_SFT [27], and Skommarkhos_skommarkhos_lucie7binstructv1_1_sft_v4 [27] – are very close

---

[3] https://github.com/felixxu/ALMA.

to each other. Interestingly, the fourth-best run according to BERT Score, UvA_finetunedMarianMT [26], drops to 14th position according to BLEU, while the fourth-best run according to BLEU, Skommarkhos_Lucie_SFT_ARPO [27], drops to 12th position according to BERTScore. The top 14 results according to BLEU are shared by the teams Skommarkhos [27] and the University of Amsterdam [26]; for BERTScore these positions are also shared with the teams arampageos [1] and Cryptix [30].

According to the pun location–based metric, the best runs are dsgt_o4_mini_multi_agent_discriminator and dsgt_o4_mini_chain_of_thought_phonetic_embeddings [41], with respectively 156 and 132 translations with locations shared with references. It is followed by teamX_aug and pjmathematician_Q25-14 [30], despite these runs placing in the second half of all runs according to BLEU and BERTScore. They are followed by the run UvA_finetunedMarianMT [26], which is also in the fourth place according to BERTScore.

The top five runs shared only 7%–9% of locations with references. This number corresponds to the percentage of successful manually evaluated machine translations we reported previously [12,20]. As we discussed previously [12], the percentage of locations shared with references is similar to the percentage of successful wordplay translations. However, to prove this hypothesis, we will manually evaluate the generated translations and we will provide detailed results in the Task 2 overview paper [15].

4   Task 3: Onomastic Wordplay Translation

In the 2025 edition of JOKER, we ran a new task based on the extensions of the JOKER 2022 corpus [17,21]. The objective of the task is to translate onomastic (i.e., name-related) wordplay from English to French. Such wordplay is widely used as a rhetorical device in novels, poetry, and plays. It is widespread in classic literature, such as in Shakespeare's characters' names, but also in names found in modern-day works such as the Pokémon, Harry Potter, and Asterix series, as well as video games. Meaningful proper names in fictional universes are often neologisms. Neologisms – that is, newly coined words – are among the most common forms of linguistic creativity. Due to their highly idiosyncratic nature, neologisms – particularly humorous ones – are challenging for both humans and machines to translate.

4.1   Data

We constructed a parallel corpus of wordplay in named entities in English and French. We collected from English-language video games, advertising slogans, literature, and other sources named entities containing wordplay, along with their translations into French. We added short contexts for the names, which is often necessary to recognise, understand, and translate the wordplay. Note that certain examples of onomastic wordplay, along with their translations, are derived from

Table 6. Task 2 results in terms of BLEU score and BLEU n-gram precision

Run ID	Score	$n = 1$	$n = 2$	$n = 3$	$n = 4$
Skommarkhos_Lucie_SFT	43.33	65.05	46.98	37.59	30.67
Skommarkhos_skommarkhos_lucie7binstructv1_1_sft_v4	43.20	64.73	46.74	37.50	30.69
UvA_finetunedNLLB-1.3B	42.55	64.74	46.26	36.70	29.83
Skommarkhos_Lucie_SFT_ARPO	42.48	63.76	45.92	36.86	30.17
Skommarkhos_skommarkhos_lucie7binstructv1_1_sft_v8	42.26	64.44	46.13	36.47	29.42
Skommarkhos_skommarkhos-lucie7binstructv1-1-sft-arpo-a5	42.15	63.33	45.50	36.56	29.95
Skommarkhos_skommarkhos-lucie7binstructv1-1-sft-arpo-a1	42.14	63.38	45.53	36.54	29.90
Skommarkhos_skommarkhos-lucie7binstructv1-1-sft-arpo-a7	42.14	63.37	45.55	36.56	29.87
Skommarkhos_Lucie-7B-Instruct-v1.1	42.14	63.43	45.54	36.51	29.88
Skommarkhos_Lucie-7B-Instruct-v1.1	42.12	63.41	45.54	36.50	29.86
Skommarkhos_skommarkhos-lucie7binstructv1-1-sft-arpo-a11	42.11	63.33	45.46	36.51	29.91
Skommarkhos_skommarkhos_lucie7binstructv1_1_sft_arpo_a19	42.00	63.29	45.41	36.40	29.74
UvA_finetunedNLLB-1.3B&finetunedroBERTa	41.80	63.86	45.49	36.01	29.17
UvA_finetunedMarianMT	41.19	63.37	44.74	35.31	28.76
duth_hybrid_fusion	41.11	63.45	44.62	35.17	28.70
yourteamid_marianmt_pun_postedit	41.01	63.40	44.52	35.07	28.58
duth_xanthi_helsinki	41.01	63.40	44.52	35.07	28.58
Cryptix	41.01	63.40	44.52	35.07	28.58
Cryptix_marianmt	40.98	63.36	44.49	35.04	28.55
duth_xanthi_GoogleTranslate_fallback	40.94	62.75	44.21	35.12	28.84
UvA_finetunedMarianMT&finetunedroBERTa	40.85	62.90	44.38	35.03	28.49
Cryptix	40.75	62.54	43.98	34.95	28.69
duth_google_flant5_fallback	40.74	62.60	43.99	34.92	28.65
duth_xanthi_GoogleTranslate	40.73	62.59	43.98	34.91	28.64
duth_xanthi_GoogleTranslate_fallback	40.73	62.60	43.99	34.91	28.63
duth_xanthi_argos	40.49	63.21	44.13	34.73	28.24
pjmathematician_Q25-14	39.08	62.57	43.10	33.19	26.05
pjmathematician_Q25-14	38.49	61.88	42.37	32.62	25.66
pjmathematician_Q25-14	38.24	61.56	42.12	32.40	25.46
UvA_finetunedT5-base	36.77	60.29	40.58	30.94	24.15
duth_xanthi_m2m100_1_2B	36.46	61.22	41.01	30.91	23.90
Skommarkhos_Croissant_SFT_ARPO	36.35	60.12	40.30	30.47	23.63
Skommarkhos_Croissant_SFT	36.20	59.93	40.22	30.35	23.47
UvA_T5-base&finetunedroBERTa	36.14	59.75	39.92	30.30	23.61
teamX_aug	33.86	54.03	37.18	28.70	22.80
duth_xanthi_mbart50	32.73	57.21	36.32	26.81	20.62
duth_xanthi_t5_base_gpu	32.44	56.04	36.26	26.82	20.33
duth_combined_m2m100	30.09	56.97	34.81	24.41	17.66
teamX_final	29.11	53.62	32.49	23.39	17.63
teamX_aug	28.63	55.38	33.30	22.66	16.07
dsgt_o4_mini_multi_agent_discriminator	21.41	46.61	25.26	16.35	10.92
UvA_mBARTcc25&finetunedroBERTa	18.20	40.86	21.09	13.74	9.26
duth_xanthi_bloomz3b_local	16.68	41.17	19.57	12.15	7.91
UvA_finetuneddmBARTcc25	16.55	39.64	19.49	12.22	7.95
dsgt_o4_mini_chain_of_thought_phonetic_embeddings	16.52	39.85	19.78	12.12	7.79
dsgt_simple_mistral_medium	14.94	37.47	17.74	10.90	6.88
dsgt_simple_o4_mini	8.15	29.13	9.80	5.17	2.99
Cryptix_finetunedmarian	0.37	13.75	0.82	0.13	0.02
Cryptix_rulebased	0.00	0.04	0.01	0.01	0.00
yourteam_rulebased	0.00	0.04	0.01	0.01	0.00

Table 7. Task 2 results in terms of BERTScore precision, recall, and F_1

Run ID	P	R	F_1
UvA_finetunedNLLB-1.3B	87.85	87.04	87.42
Skommarkhos_Lucie_SFT	87.74	87.15	87.42
Skommarkhos_skommarkhos_lucie7binstructv1_1_sft_v4	87.61	87.01	87.28
UvA_finetunedMarianMT	87.72	86.82	87.24
UvA_finetunedNLLB-1.3B&finetunedroBERTa	87.55	86.96	87.23
UvA_finetunedMarianMT&finetunedroBERTa	87.50	86.78	87.11
Skommarkhos_skommarkhos_lucie7binstructv1_1_sft_v8	87.31	86.91	87.08
duth_xanthi_GoogleTranslate_fallback	87.20	86.77	86.96
duth_xanthi_GoogleTranslate	87.20	86.77	86.96
duth_google_flant5_fallback	87.17	86.74	86.93
Cryptix	87.10	86.63	86.84
Skommarkhos_Lucie_SFT_ARPO	86.79	86.59	86.66
Skommarkhos_Lucie-7B-Instruct-v1.1	86.68	86.46	86.54
Skommarkhos_skommarkhos-lucie7binstructv1-1-sft-arpo-a1	86.68	86.45	86.53
Skommarkhos_Lucie-7B-Instruct-v1.1	86.66	86.46	86.53
Skommarkhos_skommarkhos-lucie7binstructv1-1-sft-arpo-a7	86.67	86.45	86.53
Skommarkhos_skommarkhos_lucie7binstructv1_1_sft_arpo_a19	86.67	86.43	86.52
Skommarkhos_skommarkhos-lucie7binstructv1-1-sft-arpo-a5	86.64	86.40	86.49
Skommarkhos_skommarkhos-lucie7binstructv1-1-sft-arpo-a11	86.63	86.39	86.48
pjmathematician_Q25-14	87.00	85.95	86.45
UvA_finetunedT5-base	86.69	86.24	86.44
duth_hybrid_fusion	87.18	85.79	86.43
Cryptix_marianmt	87.17	85.77	86.42
duth_xanthi_argos	87.00	85.91	86.42
yourteamid_marianmt_pun_postedit	87.17	85.74	86.40
Cryptix	87.17	85.74	86.40
duth_xanthi_helsinki	87.17	85.74	86.40
pjmathematician_Q25-14	86.84	85.97	86.37
pjmathematician_Q25-14	86.71	85.89	86.27
UvA_T5-base&finetunedroBERTa	86.42	86.12	86.24
duth_xanthi_GoogleTranslate_fallback	86.34	85.91	86.10
Skommarkhos_Croissant_SFT	85.71	85.65	85.65
Skommarkhos_Croissant_SFT_ARPO	85.66	85.64	85.62
duth_xanthi_m2m100_1_2B	85.90	85.28	85.56
teamX_aug	85.69	85.46	85.46
duth_xanthi_mbart50	85.14	84.14	84.60
duth_combined_m2m100	84.76	84.05	84.37
teamX_aug	84.61	84.15	84.35
teamX_final	84.01	83.50	83.73
duth_xanthi_t5_base_gpu	83.91	83.60	83.71
dsgt_o4_mini_multi_agent_discriminator	79.84	81.57	80.66
UvA_mBARTcc25&finetunedroBERTa	80.07	80.00	80.00
UvA_finetunedmBARTcc25	79.48	79.79	79.59
duth_xanthi_bloomz3b_local	79.30	78.66	78.94
dsgt_o4_mini_chain_of_thought_phonetic_embeddings	77.80	79.15	78.42
dsgt_simple_mistral_medium	77.60	79.12	78.30
Cryptix_finetunedmarian	75.06	73.56	74.27
dsgt_simple_o4_mini	73.82	73.95	73.85
yourteam_rulebased	64.26	54.35	58.86
Cryptix_rulebased	64.26	54.35	58.86

Table 8. Task 2 results in terms of the pun location–based metric

Run ID	Count	Location	%
dsgt_o4_mini_multi_agent_discriminator	1682	156	9.27
dsgt_o4_mini_chain_of_thought_phonetic_embeddings	1682	132	7.85
teamX_aug	1682	118	7.02
pjmathematician_Q25-14	1682	118	7.02
UvA_finetunedMarianMT	1682	114	6.78
UvA_finetunedMarianMT&finetunedroBERTa	1682	114	6.78
Cryptix	1682	113	6.72
Cryptix_marianmt	1682	113	6.72
duth_xanthi_helsinki	1682	113	6.72
yourteamid_marianmt_pun_postedit	1682	113	6.72
duth_xanthi_GoogleTranslate_fallback	1682	112	6.66
duth_hybrid_fusion	1682	112	6.66
Cryptix	1682	112	6.66
duth_google_flant5_fallback	1682	112	6.66
Skommarkhos_skommarkhos_lucie7binstructv1_1_sft_v4	1682	111	6.60
pjmathematician_Q25-14	1682	111	6.60
duth_xanthi_GoogleTranslate	1682	111	6.60
duth_xanthi_GoogleTranslate_fallback	1682	111	6.60
Skommarkhos_skommarkhos_lucie7binstructv1_1_sft_v8	1682	111	6.60
UvA_finetunedNLLB-1.3B&finetunedroBERTa	1682	111	6.60
duth_xanthi_argos	1682	109	6.48
Skommarkhos_skommarkhos-lucie7binstructv1-1-sft-arpo-a7	1682	109	6.48
Skommarkhos_skommarkhos_lucie7binstructv1_1_sft_arpo_a19	1682	109	6.48
Skommarkhos_Lucie-7B-Instruct-v1.1	1682	109	6.48
Skommarkhos_skommarkhos-lucie7binstructv1-1-sft-arpo-a5	1682	108	6.42
Skommarkhos_skommarkhos-lucie7binstructv1-1-sft-arpo-a11	1682	108	6.42
Skommarkhos_skommarkhos-lucie7binstructv1-1-sft-arpo-a1	1682	108	6.42
Skommarkhos_Lucie-7B-Instruct-v1.1	1682	107	6.36
UvA_finetunedNLLB-1.3B	1682	107	6.36
UvA_T5-base&finetunedroBERTa	1682	107	6.36
pjmathematician_Q25-14	1682	107	6.36
UvA_finetunedT5-base	1682	106	6.30
Skommarkhos_Lucie_SFT_ARPO	1682	100	5.95
Skommarkhos_Croissant_SFT	1682	96	5.71
duth_xanthi_t5_base_gpu	1682	95	5.65
Skommarkhos_Croissant_SFT_ARPO	1682	92	5.47
duth_xanthi_m2m100_1_2B	1682	91	5.41
Skommarkhos_Lucie_SFT	1682	90	5.35
teamX_aug	1682	90	5.35
UvA_mBARTcc25&finetunedroBERTa	1682	89	5.29
teamX_final	1682	84	4.99
duth_xanthi_mbart50	1682	81	4.82
duth_combined_m2m100	1682	71	4.22
UvA_finetunedmBARTcc25	1682	64	3.80
dsgt_simple_mistral_medium	1682	60	3.57
duth_xanthi_bloomz3b_local	1682	45	2.68
Cryptix_finetunedmarian	1682	25	1.49
dsgt_simple_o4_mini	1682	18	1.07
yourteam_rulebased	1682	0	0.00
Cryptix_rulebased	1682	0	0.00

official translations obtained from sources such as Wikipedia or other resources used in training LLMs. These translations, which may already be familiar to LLMs, are included as part of the training data. For training purposes, we released 353 onomastic wordplay instances in English with corresponding French translations and descriptions.

For testing purposes, we compiled our own dataset of instances manually translated by trained professionals, as well as instances of official translations. We used 2,333 of onomastic wordplay instances in English with corresponding French translations as our test set.

4.2   Evaluation

Participants' translations were evaluated by automatically checking them for case-insensitive exact matches against the manual reference translations. We also performed a manual evaluation of 1,737 translations of 203 distinct source wordplay instances sampled from the participants' runs. Although we tried to remove translations matching the references and the English sources to reduce the annotators working load, some of them were maintained due to the slight format differences. We added reference translations to calculate the percentage of successful translations in runs resulting in 1,833 distinct lower-cased stripped translations.

4.3   Participants' Approaches

Team mariapazr20 [2] used chain-of-thought prompting techniques with several large language models, including additional constraints identified from recurring translation patterns for each literary work of the provided corpus (such as favouring puns in a given semantic field over meaning preservation for instance).

Team arampageos [1] started by manually or semi-automatically classifying the names in the training data into four categories (alliteration, wordplay, realistic names and unclassified). They then used the same strategy as for Task 2, with a two-stage approach where a record of manually defined translations preceded using large language models.

Team sarath_kumar [30] prepared a dataset containing named entity recognition annotations and used it to source translations from the T5-base model. They used a beam search to prioritise phonetic matches between source and target names, and then ranked the translations according to their creativity, phonetic fidelity, and cultural relevance.

Team pjmathematician [43] used zero-shot prompting with Qwen models. The prompt consists of about 50 lines and includes guidance on how to translate wordplay (when to not translate, when to use a literal translation, or suggesting creative constraints such as considering characters' traits or relying on the story universe vocabulary).

4.4   Results

Table 9 reports the percentage of matching translations for each run, according to the aforementioned manual and automatic metrics. For context, it also reports the percentage of instances in each run where the translated French wordplay is identical to the English original; this figure is 100% for the run labelled "copy", is a naïve baseline that "translates" by copying input to output verbatim.

The best run according to both manual evaluation (62.56%) and exact match to the references (39%) is VerbaNex_gpt4o [2]. Following this, with about half the exact-match score, are two runs by the team pjmathematician [43]. However, the differences are much less stark according to the manual evaluation, which allowed alternative translations (62.56% vs. 46,31%). In both teams' top-scored runs, VerbaNex_gpt4o and pjmathematician_task_3_Q332, we observe that 23% of generations were considered to be successful alternative translations. The top-scored VerbaNex_gpt4o has only 8.53% of translations identical to the English source wordplay.

About 12% of reference translations are identical to the English source wordplay. The manual evaluation shows only 2.55% of appropriate translations, as we tried to remove translations matching the references and the English sources in order to reduce the annotators' working load. These 2.55% correspond to some translations identical to the source that were not filtered out due to the minor differences in formatting or typography. The identity baseline remains a strong one, outperforming more than half of submitted runs in terms of matching to the references. Half the runs have more than 40% French translations identical to English while 30% keep half the onomastic wordplay instances untranslated.

Among 1,737 manually evaluated translations, 172 (10%) were considered successful ones. Among these, 17 were nearly identical to the reference translations, with the only differences manifesting in diacritics, capitalization, and/ or punctuation – for example, "Oreilles de Soie" (run) vs. "Oreilles-De-Soie" (reference) for "Ears of Silk" (source). Less than 10% of manually evaluated translations (155 instances) were genuinely alternative translations, suggesting that translating onomastic wordplay remains a challenge despite the impressive capacities of LLMs. Among recurrent errors, 226 generations were identical to the English source. In 102 cases, we found the suffix "-ix" as in Celtic names, which might be a result of overfitting on the training set containing the names from the *Asterix* comics. In 11 cases, the translations lack the character's surname. Twenty-nine generations were blank or consisted only of punctuation (e.g., "???"), and in 13 cases we found spurious overgeneration such as "l'aide de" or seemingly random translations such as "l'intention des autorités fédérales, il" for "Chimchar". There were 226 translations containing extraneous articles (*le, l', la, les*) as in "Le Munchlax" for "Munchlax" or "Le Shinx" for "Shinx". The Pokémon name "Pidove" was inexplicably translated as "pédophile" in one run.

Table 9. Results for Task 3, showing the percentage of matching translation instances according to the automatic and manual evaluations, as well as the percentage of translation instances identical to the source text

Run ID	automatic	manual	identical
VerbaNex_gpt4o	39.05	62.56	8.53
pjmathematician_task_3_Q332	22.85	46.31	21.82
pjmathematician_task_3_Q314	21.13	39.60	33.48
duth_task_3_Helsinki	14.83	18.88	77.67
duth_xanthi_task_3_Helsinki-NLP-opus-mt-tc-big-en-fr	14.66	18.88	77.45
Cryptix_task_3_flanT5	14.49	13.43	38.15
duth_task_3_Helsinki	11.83	2.55	100.00
duth_xanthi_task_3_facebook-nllb-200-distilled-600M	10.72	16.75	41.83
duth_xanthi_task_3_facebook-nllb-200-1.3B	10.72	16.75	41.83
duth_xanthi_task_3_MarianMT_BLOOM	10.42	13.86	45.95
duth_xanthi_task_3_MarianMT_BLOOM	10.29	13.86	45.78
duth_xanthi_task_3_Helsinki-NLP-opus-mt-en-fr	10.29	13.86	45.78
duth_xanthi_task_3_t5-base	8.57	7.03	50.32
duth_xanthi_task_3_t5-small	8.53	6.00	58.04
duth_xanthi_task_3_facebook-m2m100_1.2B	4.71	9.50	19.12
duth_xanthi_task_3_facebook-m2m100_418M	4.37	4.00	20.15
team1_task_3_gemma2b_v2	4.20	2.99	27.69
duth_task3_hybrid_v1	0.04	1.47	0.21
duth_xanthi_task_3_MarianMT_LLM_Prompting	0.00	0.00	0.00
Skommarkhos_task3_Lucie-7B-Instruct_SFT_Q8B_LoRA	0.00	0.00	0.00
copy	11.83	2.55	100.00

5   Discussion and Conclusions

In this paper, we have described the Overview of JOKER and how its tasks relate to and expand on the challenges of previous editions. In addition to the usual French and English corpora provided in years past, one of this year's tasks (Task 1: Humour-aware Information Retrieval) used a novel Portuguese corpus.

This year's English Task 1: Humour-aware Information Retrieval showed remarkable progress, with the best run by team pjmathematician achieving a MAP of 0.3501 – nearly triple last year's top score – and outperforming all competitors by a wide margin across the various metrics. In contrast, the University of Amsterdam's cross-encoder approaches performed substantially worse than their RM3 and BM25 baselines, confirming the effectiveness of simpler retrieval strategies for this dataset. For the Portuguese subtask, results were more balanced, with pjmathematician and raison achieving similar MAP and NDCG@5 scores around 0.4 to 0.5, far ahead of the BM25 baseline. Interestingly, while the Portuguese runs achieved higher MAP scores, they trailed the English runs in precision and NDCG@5, likely due to the smaller pool of relevant humorous documents per query. Overall, these findings suggest that while the dataset's core properties have remained stable, combining retrieval and filtering remains key to advancing performance.

For Task 2: Pun Translation, we expanded the parallel corpus of wordplay translation by collecting 2,615 new manual translations of 1,682 unique English puns, with manually annotated pun locations, for the test set. Introducing this location annotation for the test set allowed us to evaluate automatically the runs. However, the trends obtained by BLEU and BERTScore are quite opposite to the trends observed in the pun location-based evaluation. The top three runs shared only 7% - 9% of locations with the references – a figure matching the previously reported percentage of successful manually evaluated translations [12,20]. As discussed, the proportion of shared locations closely aligns with the success rate of wordplay translations. We will perform a partial manual evaluation to confirm the validity of the method that we will report in the overview of Task 2: Pun Translation [15].

For Task 3: Onomastic Wordplay Translation, we built a parallel corpus of English and French onomastic wordplay collected from video games, slogans, literature, and other sources, including 353 training instances with translations and context descriptions. For testing, we assembled 2,333 English wordplay instances paired with professional or official French translations. Participants' submissions were evaluated both automatically by exact matching and manually on 1,737 translations, with some near-identical outputs retained due to minor formatting or typographical differences, resulting in 1,833 distinct normalised translations for analysis. Despite notable advances, translating onomastic wordplay remains highly challenging, with fewer than 10% of manually evaluated translations judged as genuine alternatives. While VerbaNex_gpt4o achieved the highest performance overall, a significant portion of the outputs of runs still relied on identity to the English source or near-verbatim adaptations, illustrating the limitations of current models. Recurrent errors – such as overfitting to training data, omission of surnames, and occasional nonsensical generations – highlight that further progress is needed to produce creative, culturally adapted translations at scale.

Our shared tasks are intended to contribute to the advancement of humour generation for dialogue systems, machine(-assisted) translation of humour, and humour-aware information retrieval. Prospective participants are encouraged to visit the JOKER website at https://joker-project.com for further details on the Lab.

Acknowledgments. This work has received a government grant managed by the National Research Agency under the program Investissements d'avenir integrated into France 2030, with the Reference ANR-19-GURE-0001. It was also financed by National Funds through the Portuguese funding agency FCT through the project LA/P/0063/2020 (DOI 10.54499/LA/P/0063/2020). Ricardo Campos would also like to acknowledge project StorySense, with reference 2022.09312.PTDC (DOI 10.54499/2022.09312.PTDC).

We would like to thank all colleagues and Master's students in translation at the University of Brest who contributed to the construction of our dataset. Their expertise and dedication were invaluable to the project. We are grateful for their careful work in collecting, translating, and reviewing the material.

Disclosure of Interests. The authors have no competing interests to declare that are relevant to the content of this article.

References

1. Arampatzis, G., Arampatzis, A.: DUTH at CLEF JOKER 2025 Tasks 2 and 3: translating puns and proper names with neural approaches. In: Faggioli et al. [23]
2. Atencio, M.P.R., Jimenez, J.D., Gómez, D., Serrano, J.E., Puertas, E.: VerbaNexAI at CLEF 2025 JOKER Task 3: multi-model LLM approach for onomastic wordplay translation. In: Faggioli et al. [23]
3. Bell, N.D.: Exploring L2 language play as an aid to SLL: a case study of humour in NS-NNS interaction. Appl. Linguis. **26**(2), 192–218 (2005). https://doi.org/10.1093/applin/amh043
4. Bell, N.D.: Learning about and through humor in the second language classroom. Lang. Teach. Res. **13**(3), 241–258 (2009). https://doi.org/10.1177/1362168809104697
5. Blinov, V., Mishchenko, K., Bolotova, V., Braslavski, P.: A pinch of humor for short-text conversation: an information retrieval approach. In: Jones, G.J., et al. (eds.) Experimental IR Meets Multilinguality, Multimodality, and Interaction, pp. 3–15. Springer, Cham (2017)
6. Bushnell, C.: "Lego my keego!": an analysis of language play in a beginning Japanese as a foreign language classroom". Appl. Linguis. **30**(1), 49–69 (2009). https://doi.org/10.1093/applin/amn033
7. Chakrabarty, T., Padmakumar, V., Brahman, F., Muresan, S.: Creativity support in the age of large language models: an empirical study involving professional writers. In: Proceedings of the 16th Conference on Creativity & Cognition, pp. 132–155. C&C 2024, Association for Computing Machinery, New York, NY, USA (2024). https://doi.org/10.1145/3635636.3656201
8. Chaudhari, T., Vora, A., Hotha, S., Sonawane, S.: PICT at CLEF 2025 JOKER Task 1: BERT-enhanced ensemble methods. In: Faggioli et al. [23]
9. Chen, B., Zhong, C., Kong, L.: CLEF 2025 JOKER Track Enhancing Humor-Aware Information Retrieval with Relevance-Aware Classification. In: Faggioli et al. [23]
10. Ermakova, L., Bosser, A.G., Jatowt, A., Miller, T.: The JOKER Corpus: English–French parallel data for multilingual wordplay recognition. In: SIGIR 2023: Proceedings of the 46th International ACM SIGIR Conference on Research and Development in Information Retrieval, pp. 2796–2806. Association for Computing Machinery, New York, NY (2023). https://doi.org/10.1145/3539618.3591885
11. Ermakova, L., Bosser, A.G., Miller, T., Jatowt, A.: Overview of the CLEF 2024 JOKER task 1: humour-aware information retrieval. In: Faggioli, G., Ferro, N., Galuščáková, P., Seco de Herrera, A.G. (eds.) Working Notes of the Conference and Labs of the Evaluation Forum (CLEF 2024). CEUR Workshop Proceedings, vol. 3740, pp. 1775–1785 (2024)
12. Ermakova, L., Bosser, A.G., Miller, T., Jatowt, A.: Overview of the CLEF 2024 JOKER task 3: translate puns from English to French. In: Faggioli, G., Ferro, N., Galuščáková, P., Seco de Herrera, A.G. (eds.) Working Notes of the Conference and Labs of the Evaluation Forum (CLEF 2024). CEUR Workshop Proceedings, vol. 3740, pp. 1800–1810 (2024)
13. Ermakova, L., Bosser, A.G., Miller, T., Palma Preciado, V.M., Sidorov, G., Jatowt, A.: Overview of the CLEF 2024 JOKER track: automatic humour analysis. In: Goeuriot, L., Mulhem, P., Quénot, G., Schwab, D., Soulier, L., Nunzio, G.M.D., Galuščáková, P., de Herrera, A.G.S., Faggioli, G., Ferro, N. (eds.) Experimental IR Meets Multilinguality, Multimodality, and Interaction: Proceedings of the Fifteenth International Conference of the CLEF Association (CLEF 2024). Lecture Notes in

Computer Science, vol. 14959, pp. 165–182. Springer, Cham (2024). https://doi.org/10.1007/978-3-031-71908-0_8

14. Ermakova, L., Campos, R., Bosser, A.G., Miller, T.: Overview of the CLEF 2025 JOKER Task 1: humour-aware information retrieval. In: Faggioli et al. [23]

15. Ermakova, L., Campos, R., Bosser, A.G., Miller, T.: Overview of the CLEF 2025 JOKER Task 2: Wordplay Translation from English into French. In: Faggioli et al. [23]

16. Ermakova, L., Miller, y., Naud, Y., Bosser, A-G., Campos, R.: Overview of the CLEF 2025 JOKER Task 3: Onomastic Wordplay Translation. In: Faggioli et al. [23]

17. Ermakova, L., Miller, T., Boccou, J., Digue, A., Damoy, A., Campen, P.: Overview of the CLEF 2022 JOKER Task 2: translate wordplay in named entities. In: Faggioli, G., Ferro, N., Hanbury, A., Potthast, M. (eds.) Proceedings of the Working Notes of CLEF 2022 – Conference and Labs of the Evaluation Forum, Bologna, Italy, September 5th to 8th, 2022. CEUR Workshop Proceedings, vol. 3180, pp. 1666–1680 (2022)

18. Ermakova, L., Miller, T., Bosser, A.G., Palma Preciado, V.M., Sidorov, G., Jatowt, A.: Overview of JOKER 2023 automatic wordplay analysis task 2 – pun location and interpretation. In: Aliannejadi, M., Faggioli, G., Ferro, N., Vlachos, M. (eds.) Working Notes of CLEF 2023 – Conference and Labs of the Evaluation Forum. CEUR Workshop Proceedings, vol. 3497, pp. 1804–1817 (2023)

19. Ermakova, L., Miller, T., Bosser, A.G., Palma Preciado, V.M., Sidorov, G., Jatowt, A.: Overview of JOKER 2023 automatic wordplay analysis task 3 – pun translation. In: Aliannejadi, M., Faggioli, G., Ferro, N., Vlachos, M. (eds.) Working Notes of CLEF 2023 – Conference and Labs of the Evaluation Forum. CEUR Workshop Proceedings, vol. 3497, pp. 1818–1827 (2023)

20. Ermakova, L., Miller, T., Bosser, A.G., Palma Preciado, V.M., Sidorov, G., Jatowt, A.: Overview of JOKER – CLEF-2023 track on automatic wordplay analysis. In: Arampatzis, A., Kanoulas, E., Tsikrika, T., Vrochidis, S., Giachanou, A., Li, D., Aliannejadi, M., Vlachos, M., Faggioli, G., Ferro, N. (eds.) Experimental IR Meets Multilinguality, Multimodality, and Interaction, vol. 14163, pp. 397–415. Springer, Cham (2023). https://doi.org/10.1007/978-3-031-42448-9_26

21. Ermakova, L., et al.: Overview of JOKER@CLEF 2022: automatic wordplay and humour translation workshop. In: Barrón-Cedeño, A., Martino, G.D.S., Esposti, M.D., Sebastiani, F., Macdonald, C., Pasi, G., Hanbury, A., Potthast, M., Faggioli, G., Ferro, N. (eds.) Experimental IR Meets Multilinguality, Multimodality, and Interaction: Proceedings of the Thirteenth International Conference of the CLEF Association (CLEF 2022). Lecture Notes in Computer Science, vol. 13390, pp. 447–469. Springer, Cham (2022).https://doi.org/10.1007/978-3-031-13643-6_27

22. Ermakova, L., et al.: Overview of the CLEF 2022 JOKER Task 3: pun translation from English into French. In: Faggioli, G., Ferro, N., Hanbury, A., Potthast, M. (eds.) Proceedings of the Working Notes of CLEF 2022: Conference and Labs of the Evaluation Forum. CEUR Workshop Proceedings (2022)

23. Faggioli, G., Ferro, N., Rosso, P., Spina, D.: Working notes of CLEF 2025. In: Conference and Labs of the Evaluation Forum. CEUR Workshop Proceedings, CEUR-WS.org (2025)

24. Forman, R.: Humorous language play in a Thai EFL classroom. Appl. Linguis. **32**(5), 541–565 (2011). https://doi.org/10.1093/applin/amr022

25. Kalloniatis, A., Adamidis, P.: Computational humor recognition: a systematic literature review. Artif. Intell. Rev. **58**(43), 1481–1490 (2024). https://doi.org/10.1007/s10462-024-11043-3

26. Kreefft-Libiu, A., Helms, F., Selçuk, C., Bakker, J., Kamps, J.: University of Amsterdam at the CLEF 2025 JOKER Track. In: Faggioli et al. [23]
27. Kuzmin, I.: No pun left behind: Skommarkos at CLEF 2025 JOKER lab tasks 1 & 2 - humour retrieval and translation. In: Faggioli et al. [23]
28. Li, S., et al.: Defining a New NLP playground. In: Bouamor, H., Pino, J., Bali, K. (eds.) Findings of the Association for Computational Linguistics: EMNLP 2023, pp. 11932–11951. Association for Computational Linguistics, Singapore (2023). https://doi.org/10.18653/v1/2023.findings-emnlp.799
29. Nijholt, A., Niculescu, A., Valitutti, A., Banchs, R.E.: Humor in human–computer interaction: a short survey. In: Joshi, A., Balkrishan, D.K., Dalvi, G., Winckler, M. (eds.) Adjunct Proceedings: INTERACT 2017 Mumbai, pp. 199–220. Industrial Design Centre, Indian Institute of Technology Bombay (2017). https://www.interact2017.org/downloads/INTERACT_2017_Adjunct_v4_final_24jan.pdf
30. Sarath Kumar P, Beulah A, Sushmitha M, Thanalaxmi S.: REC_Cryptix at JOKER CLEF 2025: Teaching Machines to Laugh: Multilingual Humor Detection and Translation. In: Faggioli et al. [23]
31. Palotti, J.A., Scells, H., Zuccon, G.: TrecTools: an open-source Python library for information retrieval practitioners involved in TREC-like campaigns. In: Proceedings of the 42nd International ACM SIGIR Conference on Research and Development in Information Retrieval, pp. 1325–1328. Association for Computing Machinery, New York (2019). https://doi.org/10.1145/3331184.3331399
32. Papineni, K., Roukos, S., Ward, T., Zhu, W.J.: BLEU: a method for automatic evaluation of machine translation. In: Proceedings of the 40th Annual Meeting of the Association for Computational Linguistics, pp. 311–318 (2002). https://doi.org/10.3115/1073083.1073135, https://www.aclweb.org/anthology/P02-1040
33. Piskorski, J., et al.: Overview of the CLEF-2024 CheckThat! lab task 3 on persuasion techniques. In: Faggioli, G., Ferro, N., Galuščáková, P., Seco de Herrera, A.G. (eds.) Working Notes of the Conference and Labs of the Evaluation Forum (CLEF 2024). CEUR Workshop Proceedings, vol. 3740, pp. 299–310 (2024)
34. Post, M.: A call for clarity in reporting BLEU scores. In: Bojar, O., et al. (eds.) Proceedings of the Third Conference on Machine Translation: Research Papers, pp. 186–191 (2018). https://doi.org/10.18653/v1/W18-6319
35. Prinzl, M.G.: Death to neologisms: domestication in the English retranslations of Thomas Mann's *Der Tod in Venedig*. Int. J. Literary Linguist. **5**(3) (2016). https://doi.org/10.15462/ijll.v5i3.73
36. Schuurman, E., Cazemier, M., Buijs, L., Kamps, J.: University of Amsterdam at the CLEF 2024 JOKER track. In: Working Notes of the Conference and Labs of the Evaluation Forum (CLEF 2024). CEUR Workshop Proceedings, pp. 1909–1922 (2024)
37. Shani, C., Libov, A., Tolmach, S., Lewin-Eytan, L., Maarek, Y., Shahaf, D.: "Alexa, what do you do for fun?" Characterizing playful requests with virtual assistants (2021). http://arxiv.org/abs/2105.05571, arXiv:2105.05571 [cs]
38. Shaw, J.: Philosophy of humor. Philos. Compass **5**(2), 112–126 (2010). https://doi.org/10.1111/j.1747-9991.2009.00281.x
39. Sousa, H., Almeida, R., Silvano, P., Cantante, I., Campos, R., Jorge, A.: Enhancing Portuguese variety identification with cross-domain approaches. In: Proceedings of the 39th Annual AAAI Conference on Artificial Intelligence (AAAI 2025), vol. 39, pp. 25192–25200 (2025)
40. Taecharungroj, V., Nueangjamnong, P.: Humour 2.0: styles and types of humour and virality of memes on Facebook. J. Creative Commun. **10**(3), 288–302 (2015). https://doi.org/10.1177/0973258615614420

41. Taylor, R., Herbert, B., Sana, M.: Pun intended: multi-agent translation of wordplay with contrastive learning and phonetic-semantic embeddings for CLEF JOKER 2025 Task 2. In: Faggioli et al. [23]

42. Tobon, D.A.M., Jimenez, J.D., Serrano, J., Santos, J.C.M., Puertas, E.: UTBNLP at CLEF JOKER 2025 Task 2: mBART-50 fine-tuning with dictionary-guided forced decoding and phoneme-based techniques for English-French pun translation. In: Faggioli et al. [23]

43. Vachharajani, P.: pjmathematician at the CLEF 2025 JOKER Lab Tasks 1, 2 & 3: a unified approach to humour retrieval and translation using the Qwen LLM family. In: Faggioli et al. [23]

44. Xu, H., Murray, K., Koehn, P., Hoang, H., Eriguchi, A., Khayrallah, H.: X-alma: Plug & play modules and adaptive rejection for quality translation at scale (2025). https://arxiv.org/abs/2410.03115

45. Xu, Z., et al.: Codabench: flexible, easy-to-use, and reproducible meta-benchmark platform. Patterns **3**(7) (2022). https://doi.org/10.1016/j.patter.2022.100543

46. Yeo, S.K., McKasy, M.: Emotion and humor as misinformation antidotes. Proc. Natl. Acad. Sci. **118**(15), e2002484118 (2021). https://doi.org/10.1073/pnas.2002484118

47. Zhang, T., Kishore, V., Wu, F., Weinberger, K.Q., Artzi, Y.: BERTScore: evaluating text generation with BERT. In: Proceedings of the International Conference on Learning Representations (2020). https://openreview.net/forum?id=SkeHuCVFDr

Overview of LifeCLEF 2025: Challenges on Species Presence Prediction and Identification, and Individual Animal Identification

Lukáš Picek[1,2(✉)] , Stefan Kahl[3,4] , Hervé Goëau[1,5] , Lukáš Adam[6] ,
Théo Larcher[1] , Cesar Leblanc[1] , Maximilien Servajean[7] ,
Klára Janoušková[8] , Jiří Matas[8] , Vojtěch Čermák[8] ,
Kostas Papafitsoros[9] , Robert Planqué[10] , Willem-Pier Vellinga[10] ,
Holger Klinck[6] , Tom Denton[11] , Juan Sebastián Cañas[12] ,
Giulio Martellucci[13] , Fabrice Vinatier[13] , Pierre Bonnet[5] ,
and Alexis Joly[1]

[1] Inria, LIRMM, Univ Montpellier, CNRS, Montpellier, France
Lukaspicek@gmail.com
[2] Department of Cybernetics, FAV, University of West Bohemia in Pilsen,
Pilsen, Czechia
[3] K. Lisa Yang Center for Conservation Bioacoustics, Cornell University, Ithaca, USA
[4] Chemnitz University of Technology, Chemnitz, Germany
[5] CIRAD, UMR AMAP, Montpellier, Occitanie, France
[6] RICE, FEL, University of West Bohemia, Pilsen, Czechia
[7] LIRMM, AMIS, Univ Paul Valéry Montpellier, Univ Montpellier, CNRS,
Montpellier Cedex 5, France
[8] Czech Technical University in Prague, Prague, Czechia
[9] School of Mathematical Sciences, Queen Mary University of London, London, UK
[10] Xeno-canto Foundation, The Hague, The Netherlands
[11] Google Research, San Francisco, USA
[12] University College London, London, UK
[13] LISAH, Univ Montpellier, INRAE, IRD, Montpellier, France

Abstract. Biodiversity monitoring using AI-powered tools has become vital for tracking species distributions and assessing ecosystem health on a large scale. Automated image- and sound-based species recognition, in particular, continues to accelerate conservation efforts by enabling rapid, low-cost surveys of vulnerable populations. However, the ever-growing variety of algorithms and data sources underscores the need for standardized benchmarks to assess real-world performance. Since 2011, the LifeCLEF lab has filled this role by organizing annual evaluations that promote collaboration among AI experts, citizen science, and ecologists. In this overview, we report on the LifeCLEF 2025 edition, which featured five distinct, data-driven tasks: (i) AnimalCLEF, focusing on open-set individual animal re-identification; (ii) BirdCLEF+, about species recognition in complex acoustic soundscape recordings; (iii) FungiCLEF, addressing few-shot classification of rare fungi species; (iv) GeoLifeCLEF, combining environmental and high-resolution remote

J. Carrillo-de-Albornoz et al. (Eds.): CLEF 2025, LNCS 16089, pp. 338–362, 2026.
https://doi.org/10.1007/978-3-032-04354-2_19

sensing with occurrence records to predict plant species presence; and (v) PlantCLEF, aiming to identify multiple co-occurring plant species in vegetation-plot imagery. This paper provides an overview of the motivation, methodology, and main outcomes of the five challenges.

1 LifeCLEF Lab Overview

Accurate identification of organisms in their natural habitats supports ecological research by enabling detailed studies of species interactions, population trends, and ecosystem dynamics [47, 83]. When researchers can reliably label observations, they gain insights into biodiversity patterns, temporal changes, and conservation needs. Yet, distinguishing among hundreds of thousands of species, such as the more than 300,000 vascular plants [12, 52], often requires specialized taxonomic expertise. This shortage of trained experts, commonly referred to as the *taxonomic impediment*, has long been recognized as a significant barrier to achieving the goals of the Convention on Biological Diversity.

In 2004, Gaston and O'Neill [21] argued that building a widely used automated species identification system requires (i) assembling large, well-labeled training datasets, (ii) precisely measuring error rates, (iii) ensuring scalability to large datasets, and (iv) adding methods to detect new species. Over the past two decades, research in automated identification has grown quickly—for example, fine-grained classification methods have been developed across many taxonomic groups [15, 44, 63, 64, 80], and machine learning models now power leading citizen science platforms such as iNaturalist, eBird, and Pl@ntNet.

At the same time, new sensing methods (i.e., camera traps, acoustic recorders, drones, and satellites) have produced huge volumes of image, audio, and hyperspectral data [17, 81, 85]. By including artificial intelligence into biodiversity monitoring, researchers can process diverse data sources on a larger scale, which in turn helps, for example, to create high-resolution maps of species distributions, assess population health, and identify early signs of ecosystem change [5, 7].

To provide a long-lasting method for tracking progress in AI-driven biodiversity monitoring, the LifeCLEF virtual lab was established in 2014, which evolved from the plant identification track at ImageCLEF [24–26]. LifeCLEF has since evolved into a workshop and with its challenges now covers all the kingdoms, i.e., plants, animals, and fungi, and includes audio and remote-sensing data alongside images [33–41, 43].

In the 2025 edition [42], LifeCLEF includes five challenges: (i) AnimalCLEF, focusing on open-set individual animal re-identification; (ii) BirdCLEF+, about species recognition in complex acoustic soundscape recordings; (iii) FungiCLEF, on few-shot classification of rare fungi species; (iv) GeoLifeCLEF, about plant species presence prediction with environmental and remote-sensing data; and (v) PlantCLEF, aiming to identify multiple plant species in vegetation plots. The type and quantity of the provided data is listed in Table 1. In total, 2456 research teams participated (2000 only for the BirdCLEF+ challenge) by submitting runs to at least one of the five challenges. Only some of them managed to get good results (going beyond baseline solutions), and 33 of them wrote and submitted a *working note* describing their approach and results (for publication in CEUR-WS proceedings). In the following sections, we provide a synthesis of the methodology and main outcomes of each of the five challenges. For more details, refer to the extended overview papers for each challenge [2,10,32,51,61] and the individual working notes of the participants.

Table 1. LifeCLEF challenges data overview. The provided datasets vary in modality, size, and complexity as each challenge suits different use cases.

	Modality	Categories	Items	Task	Metric
BirdCLEF+	*audio*	206	38K	Multi-label classification	ROC-AUC
AnimalCLEF	*images*	11,567	147K	Open-set individual-Re-ID	ad-hoc metric
FungiCLEF	*images metadata*	2,427	12K	Open-set classification	Recall@k
PlantCLEF	*images (SD+HD)*	7,806	1.4M	Multi-label classification	Sample-averaged F1
GeoLifeCLEF	*sat. images time-series tabular*	10,358	6.6M	Multi-label classification	Sample-averaged F1

All challenges are hosted on the Kaggle and are jointly organized with FGVC 12, an annual workshop dedicated to Fine-Grained Visual Categorization, held in conjunction with the CVPR international conference on computer vision and pattern recognition.

2   AnimalCLEF Challenge: Multi-species Individual Animal Identification

Comprehensive details on the challenge and an extensive discussion of the results are available in the extended overview paper [2].

2.1   Objective

Animal re-identification plays a key role in wildlife research, supporting disease tracking [54], ecosystem analysis [67], monitoring invasive species [8], and assessing human impact on habitats [6]. Its goal is to recognize individual animals within a species by detecting unique features such as markings or patterns. Automating this task enables large-scale data collection on movement, habitat use, and behavior, informing studies on survival [68], behavior [69], health [56], tourism impact [55], and species distribution [29]. Despite advances in machine learning, many models still overfit to the background rather than the animal itself [62], limiting their accuracy in new environments. Improving model generalization remains critical for reliable use in conservation.

2.2   Dataset

The AnimalCLEF 2025 dataset extends the WildlifeReID-10k collection [1], which was assembled using the wildlife-datasets Python toolkit [79] by aggregating multiple curated sources. Each image is cropped to highlight the region relevant for individual identification (e.g., turtle heads, tiger flanks, or whale tails) depending on the species. In addition to the training set, we assembled a test dataset with images never seen and shared on the internet, including three different species, i.e., *Lynx lynx*, *Salamander salamander*, and *Caretta caretta*. The lynx data is part of the newly released dataset, CzechLynx [57]. The turtle data is from an unreleased part of the SeaTurtleID2022 dataset [3]. The salamander data has not yet been released. Some images are accompanied by metadata such as capture time and orientation. In Fig. 1 we provide sample images, and in Table 2 we summarize their key characteristics.

Fig. 1. AnimalCLEF 2025 dataset. The provided datasets cover three species, each with different data collection procedures: photographed underwater, captured using camera traps, and documented through handheld photography.

Table 2. AnimalCLEF 2025 dataset statistics. Training/test split is time-aware (e.g., before/after cutoff year), with unseen identities in the test set. The *known* and *unknown* indicate individuals present/absent in the database.

Subset	Number of Images	Number of Individuals		
		Total	*Known*	*Unknown*
Lynx lynx	3,903	99	99	–
↳ *Database*	2,957	77	77	–
↳ *Query*	946	41	19	22
Salamander salamander	2,077	646	646	–
↳ *Database*	1,388	587	587	–
↳ *Query*	689	108	59	49
Caretta caretta	9,229	488	488	–
↳ *Database*	8,729	438	438	–
↳ *Query*	500	100	50	50
WildlifeReID-10k (*Training*)	140,488	10,772	10,772	–

2.3 Evaluation Protocol

For each query image, participants have to decide whether the individual was known (i.e., present in the database). If so, they also need to predict the correct identity. This makes the task inherently open-set. The training set is optional, allowing participants to trade off between training time and accuracy.

Evaluation is based on accuracy for both known and unknown individuals. To ensure fairness, accuracy is balanced across datasets and individuals, minimizing the influence of overrepresented classes. Formally:

$$\text{BAKS} = \frac{1}{D} \sum_{d=1}^{D} \frac{1}{|C_d|} \sum_{c \in C_d} \frac{1}{|I^{d,c}|} \sum_{i \in I^{d,c}} \mathbf{1}(\hat{y}_i = y_i),$$

$$\text{BAUS} = \frac{1}{D} \sum_{d=1}^{D} \frac{1}{|C_d^*|} \sum_{c \in C_d^*} \frac{1}{|I^{d,c}|} \sum_{i \in I^{d,c}} \mathbf{1}(\hat{y}_i = \text{new}),$$

$$(1)$$

where $\mathbf{1}$ is the indicator function, $D = 3$ is the number of datasets, C_d and C_d^* are the known and unknown classes in dataset d, and $I^{d,c}$ are the image indices for individual c. The overall score is the geometric mean of the two:

$$\text{normalized accuracy} = \sqrt{\text{BAKS} \cdot \text{BAUS}}. \qquad (2)$$

2.4 Participants and Results

To enable an easier start for all participants, we provided two baselines based on the pre-trained models. The first one is the MegaDescriptor-L-384 [79], which is a Swin [49] model trained on a dataset similar to the training set of the

AnimalCLEF 2025 challenge. The second one is WildFusion [11], which combines MegaDescriptor with the local-feature extractor ALIKED [84] by a calibration process. Neither of these models was fine-tuned on the database.

The first edition of AnimalCLEF drew strong interest, with 3,289 successful runs submitted by 251 participating teams. Out of these, 172 teams provided at least one solution that was different from the baseline. A subset of these teams also submitted working notes documenting their approaches. Among the participants, 136 teams outperformed the *MegaDescriptor-L-384* baseline (normalized accuracy 0.309), and 88 teams exceeded the performance of the stronger *MegaDescriptor+ALIKED* baseline (normalized accuracy 0.443). Details of the top-performing approaches are summarized in the extended overview paper [2], with further insights provided in the individual working notes.

In Fig. 2, we show the top 20 teams on the private leaderboard. Dash-dotted lines indicate per-dataset normalized accuracy, while the solid blue line shows the average. Performance on turtles (orange) was relatively consistent, while lynxes (red) and salamanders (black) had the most influence on ranking. The 2nd team (*webmaking*) [71] led on salamanders but placed only 25th on lynxes. In contrast, the winning team (*DataBoom*) [46] achieved balanced results across all datasets. A combined prediction from both teams would raise the accuracy to 0.734, compared to 0.713 for *DataBoom* and 0.675 for *webmaking*.

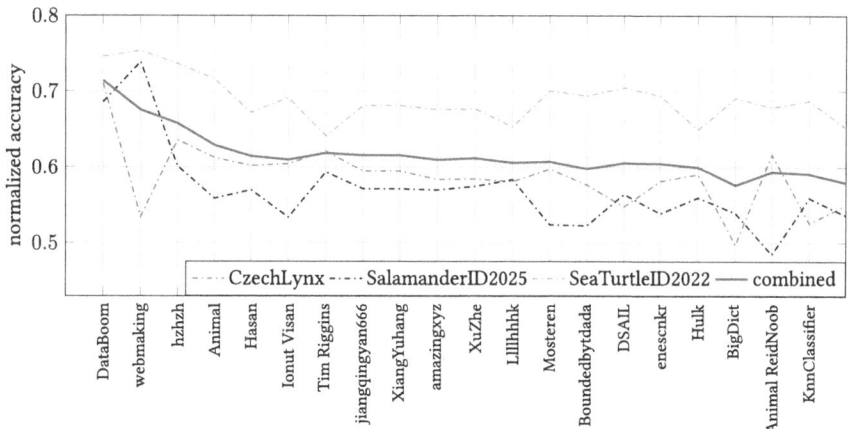

Fig. 2. Private Leaderboard of the AnimalCLEF 2025 competition. We report the performance of the top 50 (out of 230) teams. The track metrics are the normalized accuracy split on individual datasets (top subfigure) and BAKS, BAUS and precision combined on all datasets (bottom subfigure).

3 BirdCLEF+ Challenge: Multi-taxonomic Sound Identification

A detailed description of the challenge and a more complete discussion of the results can be found in the extended overview paper [10].

3.1 Objective

Mobile and habitat-diverse species serve as valuable indicators of biodiversity change, as shifts in their assemblages and population dynamics can signal the success or failure of ecological restoration efforts. These species often respond rapidly to environmental changes, making them particularly useful for detecting early signs of ecological improvement or degradation. However, traditional observer-based biodiversity surveys across large areas are both costly and logistically demanding, often requiring extensive fieldwork, expertise, and repeated visits to remote locations, challenges that limit the frequency and scale of monitoring. In contrast, passive acoustic monitoring (PAM), combined with AI, offers a scalable and non-invasive solution that enables conservationists to collect and analyze vast amounts of ecological data with minimal human presence. PAM systems can operate continuously over extended periods and in challenging environments, capturing the vocal activity of a wide range of taxa, including birds, amphibians, and insects. When paired with automated species identification, it enables researchers to monitor biodiversity across broad spatial and temporal scales, allowing more timely and data-driven reviews of restoration outcomes.

Humid tropical rainforests, the Earth's most biodiverse and ancient ecosystems, are vital for climate regulation and protection of water resources. However, rainforests face severe threats. In Colombia, a megadiverse country, the lowlands of the Magdalena Valley are a biodiversity hotspot, home to numerous endangered species. More than 70% of the lowland rainforests in the Magdalena Valley are replaced by vast pastures for cattle ranching, and illegal logging is common in the remnants of forest fragments. The protection of the last forest remnants and wetlands is an urgent need.

A significant part of the reserve, previously used for extensive livestock farming, is under an ecological restoration project. Through the Kaggle competition, we aim to automate detecting and classifying different taxonomic groups of soundscapes from El Silencio Natural Reserve, intending to provide a better understanding of the ecological process of the restoration projects.

This competition aimed to advance automated species identification in soundscape data from the Middle Magdalena Valley of Colombia, including the El Silencio Natural Reserve. Key objectives include detecting species across diverse taxonomic groups, developing machine learning models capable of recognizing rare and endangered species from limited training data, and leveraging unlabeled data to improve detection and classification performance.

3.2   Dataset

Building on lessons from previous editions, we refined the task to encourage participants to design models tailored to the unique challenges of the competition. Training and test data were carefully selected to reflect a range of bird and non-bird taxa[1], supporting this goal. As in past years, Xeno-canto remained the main source of training data, complemented by expertly annotated soundscape recordings for testing. This year, we expanded the dataset to include contributions from iNaturalist and the Colombian Sound Archive (CSA) of the Humboldt Institute, with a focus on underrepresented species, those ecologically important but difficult to detect due to rarity or elusive behavior. Common species were also included to support the development of robust models. Test data sources were chosen to reflect diverse acoustic conditions, including varying call densities, background noise, and recording qualities (mono vs. stereo). To further support innovation, we again provided unlabeled training data similar to the test set, allowing participants to explore semi- and self-supervised learning approaches.

3.3   Evaluation Protocol

The challenge was hosted on Kaggle, following a similar evaluation setup as in previous years, with hidden test data and a code competition format. We used a variant of macro-averaged ROC-AUC as the evaluation metric, excluding classes with no true positive labels, allowing us to assess model performance without relying on confidence threshold tuning and emphasizing species-level rather than segment-level accuracy. Participants were asked to identify species in short, 5-second audio clips extracted from labeled soundscape recordings, a length chosen to balance signal clarity with adequate context. The dataset was kept under 50 GB to ensure accessibility and ease of use. To further support participants, we provided starter code and documentation to help newcomers get started quickly.

3.4   Participants and Results

The BirdCLEF+ 2025 competition attracted 2,757 participants organized in 2,161 teams, who collectively submitted more than 75,000 runs. Figure 3 presents the performance of the top 25 submissions. The final ranking was determined by the private leaderboard score, which was revealed only after the submission deadline to prevent overfitting to hidden test data. During the competition, participants had access to a public leaderboard based on 35% of the test set, providing feedback while preserving the integrity of the final evaluation.

The baseline score for this year's competition was 0.5, reflecting the expected performance of a model assigning random confidence scores across all species and segments under the selected evaluation metric. The highest-scoring submission achieved a private leaderboard score of 0.930 (public: 0.933), with the top

---

[1] We therefore renamed the competition from BirdCLEF to BirdCLEF+.

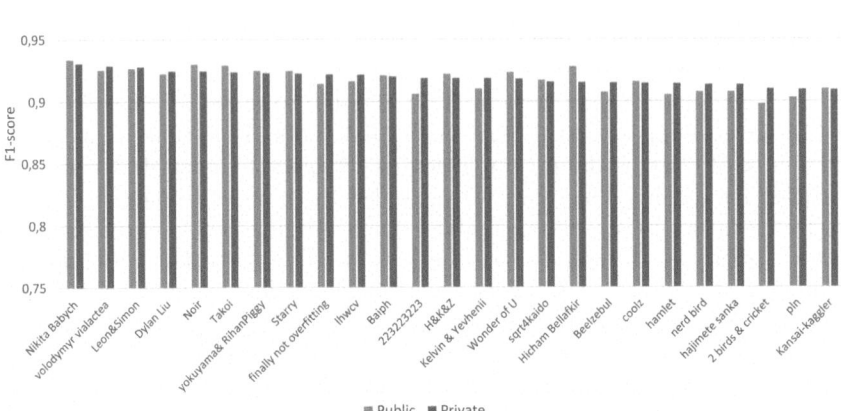

Fig. 3. BirdCLEF+ 2025 results. The top 25 teams, ranked by their private leaderboard scores. The top 10 submissions were separated by just 1.2%, and the close alignment between public and private leaderboard scores suggests that participants' models generalized well and did not overfit to the public test set.

10 teams separated by only 1.2%, highlighting a high level of overall performance and model generalization. Across submissions, several common strategies emerged. Data augmentation played a central role, with techniques such as Mixup, Cutmix, Sumix, Frequency and Time Masking, Gain adjustments, Resampling, and FilterAugment widely used. Some teams also introduced external noise, including human speech, to improve model robustness. Undersampled species were typically addressed through upsampling, while pseudo-labeling and self-training on the unlabeled soundscape data proved key for boosting accuracy. These strategies often involved generating pseudo-labels from preliminary models, applying transformations (e.g., power scaling, filtering low-confidence predictions), and iteratively refining the labels. Weighting more confident pseudo-labeled examples more heavily during training also contributed to improved outcomes.

For inference, teams commonly employed Test-Time Augmentation (TTA) by processing overlapping audio segments and smoothing predictions over time, sometimes with delta shifts. Post-processing steps - such as adjusting prediction confidence, applying power-based scaling, or calibrating outputs - were used to further refine model predictions. Ensemble methods, including blending models from different training folds or checkpoints, were instrumental in boosting final scores. To meet runtime constraints, many participants optimized inference speed using tools like ONNX, OpenVINO, and multiprocessing.

The dominant modeling approach was Sound Event Detection (SED), often enhanced with dedicated SED heads. CNN-based models were also widely used, sometimes in hybrid combinations with SED components. EfficientNet backbones were especially popular, though alternatives like RegNet and NFNet also saw successful implementations. Some teams trained separate models for taxonomic subgroups (e.g., Amphibia, Insecta), incorporating additional external datasets to improve representation. Input features were typically log-transformed Mel spectrograms, with variation in the number of mel bins, hop sizes, and frequency ranges. A variety of loss functions were explored, including Cross Entropy, BCE With Logits Loss, and Focal Loss variants, with some evidence suggesting Focal or Cross Entropy loss could offer marginal improvements with appropriate tuning. Pretraining model backbones on large external datasets such as Xeno-Canto prior to fine-tuning on the competition data significantly boosted early performance.

Further details on methodologies and a comprehensive analysis of results are available in the extended overview paper [10] and in participants' working notes.

4   GeoLifeCLEF Challenge: Plant Species Presence Prediction with Environmental and High-Resolution Remote Sensing Data

Comprehensive details on the challenge and an extensive discussion of the results are available in the extended overview paper [61].

4.1   Objective

Predicting plant species presence at a given location remains a fundamental challenge in biodiversity monitoring and ecological modeling. Accurate predictions are critical for tasks such as conservation prioritization, ecological forecasting, land management, and invasive species control. Yet, this task is complicated by spatial biases in observation data, complex ecological dependencies, and varying resolution across data sources. GeoLifeCLEF 2025 addresses these difficulties at scale, involving thousands of species and multimodal data sources with spatial resolutions up to 10 m. The competition combines Presence-Only (PO) and Presence-Absence (PA) data with satellite imagery, climatic time series, and rasterized environmental data to evaluate species presence prediction methods.

4.2   Dataset

The dataset for GeoLifeCLEF 2025 builds directly upon the GeoPlant dataset [59], i.e., the dataset used in the 2024 edition [58]. The training occurrence data remains the same and includes ~5M PO observations from GBIF and related repositories and ~100K standardized PA surveys from EVA, covering roughly 10K species. The dataset continues to provide multimodal inputs, including: (i) Sentinel-2 image patches (RGB+NIR, 64×64, 10 m resolution), (ii) Landsat-based satellite time series (6 spectral bands, spanning 84 seasons from 2000–2020), (iii) Monthly climatic time series (CHELSA, 2000–2019), and (iv) Raster-derived scalar and spatial predictors including elevation, land cover, soil, bioclimatic variables, human footprint, etc. However, several notable updates and improvements have been introduced:

– A test set was enriched with more than 9,000 surveys from different geographical origins (i.e., eastern and northern Europe), allowing for testing geospatial generalization.
– A new set of significantly more detailed human footprint rasters was added, now at a 30-meter resolution (compared to 200 m in previous editions). Derived from high-resolution OpenStreetMap data, these layers capture a wide range of anthropogenic features, including roads, buildings, infrastructure density, and land use intensity.
– The SoilGrids data, which was incorrectly exported in the 2024 dataset, was corrected and re-extracted.
– The Sentinel-2 satellite data underwent a major upgrade in format and processing. Instead of the previously used compressed JPEG images, this year's edition provides raw multi-band TIFF files, significantly improving radiometric fidelity and spatial integrity for geospatial modeling. These TIFFs include all four bands (RGB + NIR) at 10 m resolution. Updated preprocessing and normalization techniques were provided in official tutorial notebooks, enabling more accurate and flexible use of the remote sensing inputs.

4.3   Evaluation Protocol

As in the previous edition [58], the evaluation metric was selected as the sample-averaged F1 score (F_1). The F_1-score serves as a metric to measure the degree of agreement between the predicted and actual species composition observed within a specific geographical area and timeframe. In the context of ecological surveys, such as those conducted in protected areas, each survey instance i is associated with a ground-truth set of labels Y_i, representing the plant species found by experts within a defined grid. Given this setup, and a list of predicted labels $\widehat{Y}_{i,1}, \widehat{Y}_{i,2}, \ldots, \widehat{Y}_{i,R_i}$, the micro F_1-score can be computed as follows:

$$F_1 = \frac{1}{N} \sum_{i=1}^{N} \frac{2 \cdot TP_i}{2 \cdot TP_i + FP_i + FN_i}, \quad \text{where} \begin{cases} TP_i - \text{correctly predicted} \\ FP_i - \text{predicted but not observed} \\ FN_i - 4\text{not predicted but present} \end{cases}$$

(3)

This formulation encapsulates the precision and recall elements crucial for assessing the accuracy of predictive models in ecological studies.

4.4   Participants and Results

To support easy onboarding and early experimentation, we provided notebooks for downloading, exploring, and processing datasets, along with three single-modality baseline models using Landsat cubes, Biclimatic cubes, and Sentinel-2 patches. These baselines served as simple, interpretable entry points into the competition, each showcasing one data modality separately. Unlike in 2024, no official multimodal fusion or ensemble baseline was released this year. However, participants could use all the baselines from the previous edition, i.e., also those that included fusion models and additional CNN-based architectures. These were commonly reused or extended in participant workflows.

GeoLifeCLEF 2025 drew 41 participating teams, submitting a total of 750 entries. The final leaderboard, computed on approximately 77% of the test set, revealed a substantial drop in absolute performance compared to the last year. The top-performing team, *webmaking* [70], achieved an F_1 score of 0.2302, followed by *PredComX* [76] (0.2215) and *Miss Qiu* [48] (0.2169). The overall performance of the top 25 teams is visualized in Fig. 4.

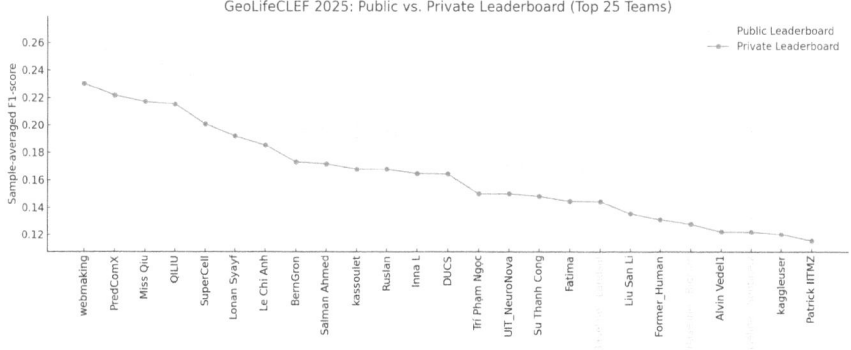

Fig. 4. GeoLifeCLEF 2025 results; top 25 teams on public and private leaderboards. Most teams show a drop in performance from public to private scores, highlighting the pure generalization. The performance gap varies by team, suggesting differing levels of overfitting. Baseline models highlighted in green. (Color figure online)

In comparison, the best-performing team in 2024 (also *webmaking*) achieved a much higher F_1 score of 0.4089, with over 20 teams surpassing the 0.30 mark on the final leaderboard. In 2025, however, no team exceeded an F_1 of 0.24, reflecting a considerably more challenging evaluation scenario. Several factors likely contributed to the overall lower scores, but we attribute the largest impact to the expanded and more ecologically diverse test set, which significantly increased the need for model generalization. Unlike the 2024 edition, where the test samples were geographically closer to the training data and many teams relied primarily on PA data, this year's setup required effective use of PO data to succeed, a task that remains difficult due to its inherent biases and lack of negative labels. Overall, while the absolute performance dropped, the technical quality and competitiveness remained high. The challenge successfully pushed participants to develop more generalizable, scalable, and multimodal solutions. Further technical details are available in participant working notes [48, 66, 70, 74, 76] and the extended overview paper [61].

5   FungiCLEF Challenge: Few-Shot Classification with Rare Fungi Species

Comprehensive details on the challenge and an extensive discussion of the results are available in the extended overview paper [32].

5.1   Objective

Recognizing fungal species accurately is vital for biodiversity monitoring, ecological research, and the safe identification of toxic or invasive species. This is a challenging task due to the vast diversity of fungi, their visual similarity, and variability in environmental conditions and image quality. FungiCLEF 2025 addresses these challenges through a few-shot learning setup, where most species have only 1–4 labeled examples. This mirrors real-world conditions of biodiversity datasets, where many species are rare and collecting additional data and annotations is often not possible.

5.2   Dataset

The FungiCLEF 2025 dataset is derived from the *FungiTastic* benchmark [60], which includes over 350,000 expert-verified observations and 630,000 photographs submitted to the Atlas of Danish Fungi before the end of 2023. Each observation contains one or more photos and rich contextual data, including GPS coordinates, timestamps, habitat and substrate information, elevation, land cover, biogeographical zone, toxicity status etc. Automatically generated image captions (Molmo-7B [16]) are also provided.

In this competition, we use the FungiTastic–FS subset, consisting of species with fewer than five observations. This subset contains 6,391 observations, 12,015 images, and 2,427 species (see more in Table 3). Most training classes have only 1–4 examples, encouraging models to generalize from limited data. The dataset is temporally split to simulate real-world generalization: training data includes observations up to 2021, while validation and test sets correspond to 2022 and 2023, respectively.

Table 3. FungiCLEF 2025 dataset statistics.

Subset	Observations	Images	Species
Training	4,293	7,819	2,427
Validation	1,099	2,285	570
Test	999	1,911	567

5.3 Evaluation Protocol

The official evaluation metric for FungiCLEF 2025 was Recall@5, which measures the proportion of test observations for which the true species label appears among the top five predictions. Formally, for a test set of size N, we define:

$$\text{Recall@}k \; = \; \frac{1}{N} \sum_{i=1}^{N} \mathbf{1}\big(y_i \in \hat{Y}_i^k\big),$$

where, y_i is the true label for the i-th test observation, $\hat{Y}_i^k = \{\hat{y}_{i,1}, \hat{y}_{i,2}, \ldots, \hat{y}_{i,k}\}$ is the set of the top-k predicted labels (ranked by confidence) for that observation, and $\mathbf{1}(\cdot)$ is the indicator function (1 if true, 0 otherwise).

5.4 Participants and Results

Common across most solutions was the use of pretrained vision transformers (especially DINOv2 [53] and BioCLIP [73]), few-shot metric learning via prototypical networks or contrastive loss, and techniques to mitigate class imbalance. Several teams enriched visual features with multimodal inputs, using metadata (e.g., habitat, substrate, collection date), automatically extracted textual descriptions, and ecological context (e.g., biogeographic region and taxonomic hierarchy). This multimodal fusion helped disambiguate visually similar species, especially in rare or underrepresented classes.

The winning solution by *Jack Etheredge* [19] combined embeddings from DINOv2, BEIT, and SAM [45] models into an ensemble of prototypical networks. Concatenated embeddings were processed through a learned projection network to further enhance class discriminability, with cosine similarity driving final predictions. This solution heavily benefited from both training- and test-time data augmentation. Besides, ensemble averaging across independently trained pipelines added stability and robustness to the final rankings.

Other top-performing teams adopted similarly strong few-shot learning strategies. *Yang Tuán Anh* [82] used a multimodal pipeline combining visual features (e.g., BioCLIP, SigLIP [78], DINOv2), textual descriptions, and metadata, trained in two stages: supervised pretraining followed by prototypical fine-tuning. Team *Parabellum* focused on improving prototype quality using BioCLIP embeddings and distance-based classification with refined prototype averaging. Team *I2C-UHU-Pegasus* developed a multimodal system that fine-tunes BioCLIP by integrating metadata and hierarchical taxonomic prompts to enhance the recognition of rare species. *DS@GT* [75] explored class-balanced sampling, Mixup augmentation applied directly on embedding space, and multi-objective learning combining category, genus, species, and poisonous annotations. Finally, *hard_work* [50] applied supervised contrastive learning over DINOv2 features using a custom Transformer encoder with an enhanced contrastive loss design to improve fine-grained feature disentanglement.

For a better comparison, refer to Fig. 5, which shows the top 5 team performance evaluated across multiple top-k metrics.

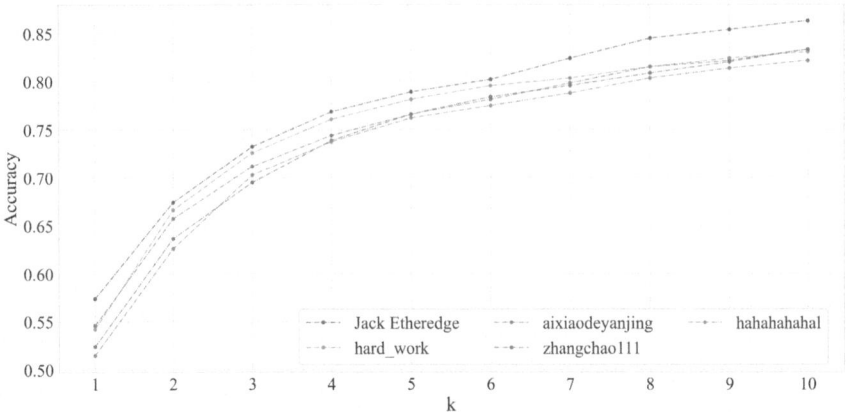

Fig. 5. FungiCLEF 2025 results. Evolution of the top-k accuracy metric as k increased for the top 5 teams. Results reported on the private leaderboard.

6 PlantCLEF Challenge: Multi-species Plant Identification in Vegetation Quadrat Images

A detailed description of the challenge and a more complete discussion of the results can be found in the extended overview paper [51] and participants' working notes [9,18,20,28,30,31,65,72,77].

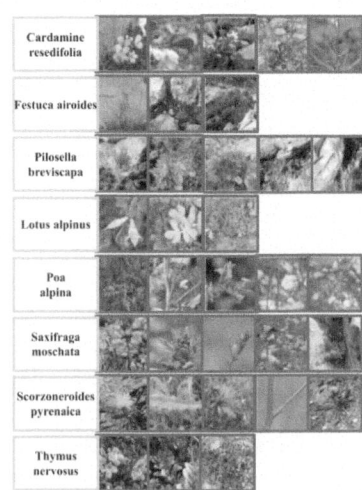

(a) Vegetative quadrat (test)   (b) Single plant images (training)

Fig. 6. PlantCLEF 2025 dataset. The test set (a) includes only top-down view photographs with multiple plant species, and the training set (b), based on images of individual plants and their organs (e.g., flowers, fruits, leaves, stems).

6.1 Objective

Vegetation quadrat inventories are crucial for ecological studies, as they enable standardized sampling, biodiversity assessment, long-term monitoring, and large-scale remote surveys. They provide valuable data on ecosystems, biodiversity conservation, and evidence-based environmental decision-making. Quadrat images, typically 0.5×0.5 m, are processed by botanists who identify all species present. They also quantify species abundance using indicators like biomass, qualification factors, and areas occupied in photographs. AI could significantly improve the efficiency of surveys (with, for example, the participation of non-specialists), thereby increasing the frequency and coverage of ecological studies.

While it is now possible to access very large volumes of images of individual plants and to train large classification models [22,23], a multi-label declination on high-resolution quadrat images would require complete annotation of all visible species to consider supervised learning of classification models. Unfortunately, such data no longer exists, and it would require considerable effort to be produced. The PlantCLEF 2025 challenge aims to evaluate approaches using classical observations of individual plants as training data, despite the discrepancies between the training and test data, as shown in Fig. 6. Specifically, the challenge is a weakly supervised multi-label classification task aimed at predicting all plant species visible in high-resolution quadrat images, using single-label plant images as training data. One of the main difficulties lies in the domain shift between the high-resolution test images of vegetation quadrats, which may contain many species, and the training data, primarily consisting of close-up images of individual plants collected through the collaborative platform Pl@ntNet [4].

Furthermore, different weather conditions and shooting angles, along with varying phenological stages, can increase data disparity. Collaborative data might be overrepresented by opportunistic views of flowers, which facilitate identification. In contrast, vegetation quadrats are typically observed multiple times over one or several years without prior assumptions about the plants' phenological stages (some may be flowering, others fruiting, some in seedling stage, and others senescent or affected by disease).

6.2 Dataset

The training set is composed of observations of individual plants, similar to those used in previous editions of PlantCLEF. More precisely, it is a subset of the Pl@ntNet training data focusing on south-western Europe and covering 7,806 plant species. It contains about 1.4 million images, extended with some images with trusted labels aggregated from the GBIF platform to complete the less illustrated species. Links to original images are provided in the 'url' column of the metadata csv file. The images have a relatively high resolution (the minimum side is 800 pixels) to allow the use of classification models that can handle relatively large resolution inputs and may reduce the difficulty of predicting small plants in large vegetative quadrat images. Images are pre-organized into subfolders by class (i.e., by species) and split into a predefined train-validation-test sets to facilitate the training of individual plant classification models.

In addition to the labeled data, an unlabeled dataset of high-resolution pseudo-quadrat images is included to facilitate better model adaptation to the target domain. This dataset is derived from the *LUCAS Cover Photos 2006– 2018 collection* [13] and contains 212,782 ground vegetation images that resemble quadrat photographs in framing and perspective, although they do not systematically cover a 50×50 cm area. It is hypothesized that this type of data can help reduce the domain shift between the labeled single-species training images and the more complex multi-species quadrat images used for testing. Although direct supervision is not possible, such data could be exploited through self-supervised learning techniques to pre-train models on the visual characteristics of natural vegetation scenes prior to fine-tuning.

The test set is a compilation of several image datasets of quadrats in different floristic contexts, including Pyrenean, Mediterranean and central French woodland floras. These datasets are all produced by experts and consist of a total of 2,105 high-resolution images. The shooting protocol can vary significantly from one context to another: the use of wooden frames or measuring tape to delimit the quadrat or not, angles of view more or less perpendicular to the ground. Additionally, the quality of the images may vary depending on the weather, which can result in more or less pronounced shadows, blurry areas, etc.

For participants with limited computational resources who are unable to train a plant image identification model on such a large volume of data, or to enable direct work with a pre-trained backbone, two pre-trained models are shared through Zenodo [27]. Both are based on a vision transformer architecture initially pretrained with the DINO2 self-supervised learning approach [14,53] and fine-tuned on PlantCLEF 2025 training data (with a classical softmax and cross-entropy).

6.3   Evaluation Protocol

The aim of the challenge is to exhaustively list the presence of every plant species on each high-resolution vegetation quadrat image, from among more than 7,800 species, bearing in mind that quadrats are generally 50×50 cm in size, and that it is rare to observe dozens of species simultaneously. Each quadrat represents a sample of a specific area within a selected site (e.g. a 5 m×1 m transect in the Pyrenees), following a common ecological protocol designed to assess biodiversity across a broader section of the landscape.

The evaluation metric used to rank participant submissions is the F1 score, which balances recall and precision—encouraging methods that avoid both over-prediction and under-detection of species. Among the available F1 variants, the sample-averaged version is selected as the primary metric (i.e. averaging the F1 scores computed individually for each quadrat), ensuring equal weighting of each image regardless of species richness. To further reduce bias from spatial sampling heterogeneity, F1 scores are first macro-averaged across transects before computing the final score.

The use of the metadata (image names, EXIF data, licenses) is authorized provided that, for each run using metadata, an equivalent run using only the visual information without metadata in submitted in order to assess the raw contribution of a purely visual analysis. The use of additional data is permitted provided that an equivalent run with only the data provided is submitted to enable more accurate and fair comparisons.

6.4   Participants and Results

The challenge is hosted on the kaggle platform, providing an opportunity for researchers and enthusiasts to contribute to the development of plant recognition in such new context. Among the 540 initial entrants who expressed interest in participating, a total of 55 participants grouped into 38 teams ultimately submitted 659 runs. Details of the best methods and systems used are synthesized in the overview working notes paper of the task [51]. In Fig. 7, we report the best run for each of the 24 teams that achieved a score strictly greater than 0 and different from the baselines. The main outcomes we can derive from those results are the following:

– The best score remains below 0.37 in Sample-Averaged F1 but shows clear progress compared to last year's result (0.29). Considering that there are about 8 species per quadrat on average, this score typically reflects 4–5 correct predictions per image, with relatively few false positives, indicating a good trade-off between precision and recall under challenging conditions.
– The image resizing method has a non-negligible impact on the performance of the provided pre-trained DINOv2 model. Variations in interpolation strategy can lead to measurable differences in final accuracy, suggesting a lack of robustness to preprocessing artifacts and highlighting the potential value of incorporating diverse resizing algorithms during training.
– Most participants likely explored various tiling strategies, experimenting with different tile sizes, overlap settings, fusion or decision mechanisms, among other configurations. While each team identified an optimal configuration tailored to their own method, there is no clear consensus yet on a universally best tiling approach. The close proximity of the top-performing scores suggests that multiple strategies can yield comparable performance under the current benchmark conditions.

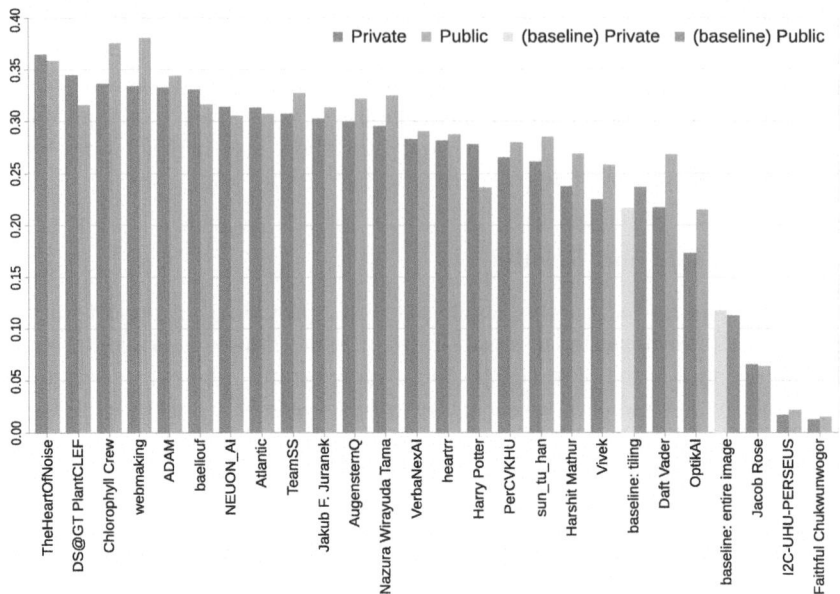

Fig. 7. PlantCLEF 2025 results. Best run for each team; mean sample-averaged F1 per transect (each including multiple quadrats from the same site).

7 Conclusions and Perspectives

The LifeCLEF 2025 edition reaffirms the importance of benchmark-driven research for advancing AI applications in biodiversity monitoring. By addressing a diverse set of tasks, ranging from open-set individual animal re-identification to few-shot classification of rare fungi and from large-scale species presence prediction to weakly supervised multi-species plant recognition, this edition pushes the boundaries of machine learning in ecological contexts.

A key insight from this year is the persistent and substantial impact of domain shift across all tasks, especially where training and test conditions diverge significantly in terms of geography, sensor modality, or species composition. While baseline models provided strong starting points, overcoming domain gaps required innovations in the form of pretraining, self-supervised learning, and multimodal data integration. In particular, the use of unlabeled data, either through semi-supervised or contrastive learning, proved promising in BirdCLEF+.

The performance gap observed in GeoLifeCLEF 2025, despite the use of high-resolution and multimodal data, highlights both the increased difficulty of the updated benchmark and the need for more robust generalization strategies. Similarly, FungiCLEF and PlantCLEF demonstrated that few-shot and weakly supervised scenarios remain challenging, although several teams achieved meaningful improvements by utilizing vision transformers, prototype-based classifiers, and metadata-aware pipelines.

Multimodality emerged as a critical driver of success across tasks. Models that effectively fused heterogeneous sources (e.g., images, soundscapes, text, and environment) demonstrated superior adaptability and resilience. In contrast, the search for lightweight, efficient architectures in some tasks (e.g., BirdCLEF+ and GeoLifeCLEF) stressed the practical need for scalable, deployable solutions.

Finally, the LifeCLEF community continues to grow in both scale and diversity, with thousands of participants contributing to a shared goal of ecological insight through AI. The increasing openness in sharing code, data pipelines, and pre-trained models accelerates collective progress and encourages reproducible science. Looking ahead, expanding the role of pretext tasks, enhancing interoperability between modalities, and exploring lifelong learning frameworks may further advance the field.

Acknowledgments. The research described in this paper was partly funded by the European Commission via the GUARDEN and MAMBO projects, which have received funding from the European Union's Horizon Europe research and innovation program under grant agreements 101060693 and 101060639. The opinions expressed in this work are those of the authors and are not necessarily those of the GUARDEN or MAMBO partners or the European Commission.

References

1. Adam, L., Cermak, V., Papafitsoros, K., Picek, L.: Wildlifereid-10k: wildlife re-identification dataset with 10k individual animals. In: Proceedings of the Computer Vision and Pattern Recognition Conference (CVPR) Workshops, pp. 2099–2109 (2025)
2. Adam, L., Papafitsoros, K., Kovář, R., Čermák, V., Picek, L.: Overview of AnimalCLEF 2025: recognizing individual animals in images. In: Working Notes of CLEF 2025 - Conference and Labs of the Evaluation Forum (2025)
3. Adam, L., Čermák, V., Papafitsoros, K., Picek, L.: Seaturtleid2022: a long-span dataset for reliable sea turtle re-identification. In: Proceedings of the IEEE/CVF Winter Conference on Applications of Computer Vision (WACV), pp. 7146–7156 (2024)
4. Affouard, A., Goeau, H., Bonnet, P., Lombardo, J.C., Joly, A.: Pl@ntnet app in the era of deep learning. In: 5th International Conference on Learning Representations (ICLR 2017), April 24-26 2017, Toulon, France (2017)
5. Besson, M., et al.: Towards the fully automated monitoring of ecological communities. Ecol. Lett. **25**(12), 2753–2775 (2022)
6. Blount, J.D., Chynoweth, M.W., Green, A.M., Şekercioğlu, Ç.H.: COVID-19 highlights the importance of camera traps for wildlife conservation research and management. Biol. Cons. **256**, 108984 (2021)
7. Buchelt, A., et al.: Exploring artificial intelligence for applications of drones in forest ecology and management. For. Ecol. Manage. **551**, 121530 (2024)
8. Caravaggi, A., et al.: An invasive-native mammalian species replacement process captured by camera trap survey random encounter models. Remote Sens. Ecol. Conserv. **2**(1), 45–58 (2016)

9. Carrillo, J.T., Araujo, D.P., Álvarez, V.P., Vázquez, J.M.: I2c in plantclef 2025: multi-label identification and classification of plant species in images using machine learning techniques. In: Working Notes of CLEF 2025 - Conference and Labs of the Evaluation Forum (2025)
10. Cañas, J.S., et al.: Overview of BirdCLEF+ 2025: multi-taxonomic sound identification in the middle Magdalena valley, Colombia. In: Working Notes of CLEF 2025 - Conference and Labs of the Evaluation Forum (2025)
11. Cermak, V., Picek, L., Adam, L., Neumann, L., Matas, J.: WildFusion: individual animal identification with calibrated similarity fusion. In: European Conference on Computer Vision, pp. 18–36. Springer, Cham (2025). https://doi.org/10.1007/978-3-031-92387-6_2
12. Chapman, A.D., et al.: Numbers of living species in Australia and the world (2009)
13. d'Andrimont, R., et al.: LUCAS cover photos 2006–2018 over the EU: 874 646 spatially distributed geo-tagged close-up photos with land cover and plant species label. Earth Syst. Sci. Data **14**(10), 4463–4472 (2022)
14. Darcet, T., Oquab, M., Mairal, J., Bojanowski, P.: Vision transformers need registers (2024)
15. Das, N., Mondal, A., Chaki, J., Padhy, N., Dey, N.: Machine learning models for bird species recognition based on vocalization: a succinct review. Inf. Technol. Intell. Transp. Syst., 117–124 (2020)
16. Deitke, M., et al.: Molmo and pixMo: open weights and open data for state-of-the-art vision-language models. In: Proceedings of the Computer Vision and Pattern Recognition Conference, pp. 91–104 (2025)
17. Dyrmann, M., Mortensen, A.K., Linneberg, L., Høye, T.T., Bjerge, K.: Camera assisted roadside monitoring for invasive alien plant species using deep learning. Sensors **21**(18), 6126 (2021)
18. Espitalier, V.: Preprocessing is all you need. In: Working Notes of CLEF 2025 - Conference and Labs of the Evaluation Forum (2025)
19. Etheredge, J.: Few-shot fungi classification with prototypical networks using multiple pretrained embedding models. In: Working Notes of CLEF 2025 - Conference and Labs of the Evaluation Forum (2025)
20. Filho, L.A.D., da Silva Neto, A.M., David, R.P., Calumby, R.T.: Zero-shot segmentation through prototype-guidance for multi-label plant species identification. In: Working Notes of CLEF 2025 - Conference and Labs of the Evaluation Forum (2025)
21. Gaston, K.J., O'Neill, M.A.: Automated species identification: why not? Philos. Trans. Royal Soc. London B: Biol. Sci. **359**(1444), 655–667 (2004)
22. Goëau, H., Bonnet, P., Joly, A.: Overview of PlantCLEF 2022: image-based plant identification at global scale. In: Working Notes of CLEF 2022 - Conference and Labs of the Evaluation Forum (2022)
23. Goëau, H., Bonnet, P., Joly, A.: Overview of PlantCLEF 2023: image-based plant identification at global scale. In: Working Notes of CLEF 2023 - Conference and Labs of the Evaluation Forum (2023)
24. Goëau, H., et al.: The ImageCLEF 2013 plant identification task. In: CLEF task overview 2013, CLEF: Conference and Labs of the Evaluation Forum, Sep. 2013, Valencia, Spain. Valencia (2013)
25. Goëau, H., et al.: The ImageCLEF 2011 plant images classification task. In: CLEF task overview 2011, CLEF: Conference and Labs of the Evaluation Forum, Sep. 2011, Amsterdam, Netherlands. (2011)

26. Goëau, H., et al.: ImageCLEF2012 plant images identification task. In: CLEF task overview 2012, CLEF: Conference and Labs of the Evaluation Forum, Sep. 2012, Rome, Italy. Rome (2012)

27. Goëau, H., Lombardo, J.C., Affouard, A., Espitalier, V., Bonnet, P., Joly, A.: PlantCLEF 2024 pretrained models on the flora of the south western Europe based on a subset of Pl@ntNet collaborative images and a ViT base patch 14 dinoV2 (2024). https://doi.org/10.5281/zenodo.10848263

28. Gustineli, M., Miyaguchi, A., Cheung, A., Khattak, D.: Multi-label plant species classification using vision transformers at PlantCLEF 2025. In: Working Notes of CLEF 2025 - Conference and Labs of the Evaluation Forum (2025)

29. Hanna, M.E., Chandler, E.M., Semmens, B.X., Eguchi, T., Lemons, G.E., Seminoff, J.A.: Citizen-sourced sightings and underwater photography reveal novel insights about green sea turtle distribution and ecology in southern California. Front. Marine Sci. **8**, 671061 (2021). https://doi.org/10.3389/fmars.2021.671061

30. Herasimchyk, H., Labryga, R., Prusina, T.: Multi-label plant species prediction with metadata-enhanced multi-head vision transformers. In: Working Notes of CLEF 2025 - Conference and Labs of the Evaluation Forum (2025)

31. Ishrat, H.A., Lee, S.H., Chang, Y.L., Chin, C.K.: Post-hoc aggregation as a competitive alternative to model-centric pipelines: neuon submission to PlantCLEF 2025. In: Working Notes of CLEF 2025 - Conference and Labs of the Evaluation Forum (2025)

32. Janouskova, K., Matas, J., Picek, L.: Overview of FungiCLEF 2025: few-shot classification with rare fungi species. In: Working Notes of CLEF 2025 - Conference and Labs of the Evaluation Forum (2025)

33. Joly, A., et al.: Overview of LifeCLEF 2023: evaluation of AI models for the identification and prediction of birds, plants, snakes and fungi. In: International Conference of the Cross-Language Evaluation Forum for European Languages, pp. 416–439. Springer, Cham (2023). https://doi.org/10.1007/978-3-031-42448-9_27

34. Joly, A., et al.: Overview of LifeCLEF 2018: a large-scale evaluation of species identification and recommendation algorithms in the era of AI. In: Jones, G.J., et al. (eds.) CLEF: Cross-Language Evaluation Forum for European Languages. Experimental IR Meets Multilinguality, Multimodality, and Interaction, vol. LNCS. Springer, Avigon, France (2018)

35. Joly, A., et al.: LifeCLEF 2016: multimedia life species identification challenges. In: Fuhr, N., et al. (eds.) CLEF 2016. LNCS, vol. 9822, pp. 286–310. Springer, Cham (2016). https://doi.org/10.1007/978-3-319-44564-9_26

36. Joly, A., et al.: LifeCLEF 2017 lab overview: multimedia species identification challenges. In: Jones, G.J.F., et al. (eds.) CLEF 2017. LNCS, vol. 10456, pp. 255–274. Springer, Cham (2017). https://doi.org/10.1007/978-3-319-65813-1_24

37. Joly, A., et al.: LifeCLEF 2014: Multimedia Life Species Identification Challenges. In: Kanoulas, E., et al. (eds.) CLEF 2014. LNCS, vol. 8685, pp. 229–249. Springer, Cham (2014). https://doi.org/10.1007/978-3-319-11382-1_20

38. Joly, A., et al.: Lifeclef 2015: multimedia life species identification challenges. In: Experimental IR Meets Multilinguality, Multimodality, and Interaction, pp. 462–483. Springer, Cham (2015). https://doi.org/10.1007/978-3-319-24027-5_46

39. Joly, A., et al.: Overview of LifeCLEF 2020: a system-oriented evaluation of automated species identification and species distribution prediction. In: International Conference of the Cross-Language Evaluation Forum for European Languages, pp. 342–363. Springer, Cham (2020). https://doi.org/10.1007/978-3-030-58219-7_23

40. Joly, A., et al.: Overview of LifeCLEF 2022: an evaluation of machine-learning based species identification and species distribution prediction. In: International Conference of the Cross-Language Evaluation Forum for European Languages. Springer, Cham (2022). https://doi.org/10.1007/978-3-031-13643-6_19

41. Joly, A., et al.: Overview of LifeCLEF 2021: an evaluation of machine-learning based species identification and species distribution prediction. In: International Conference of the Cross-Language Evaluation Forum for European Languages, pp. 371–393. Springer (2021)

42. Joly, A., et al.: LifeCLEF 2025 teaser: challenges on species presence prediction and identification, and individual animal identification. In: European Conference on Information Retrieval, pp. 373–381. Springer, Cham (2025). https://doi.org/10.1007/978-3-031-88720-8_57

43. Joly, A., et al.: Overview of LifeCLEF 2024: challenges on species distribution prediction and identification. In: International Conference of the Cross-Language Evaluation Forum for European Languages, pp. 183–207. Springer (2024)

44. Júnior, T.D.C., Rieder, R.: Automatic identification of insects from digital images: a survey. Comput. Electron. Agric. **178**, 105784 (2020)

45. Kirillov, A., et al.: Segment anything. In: Proceedings of the IEEE/CVF International Conference on Computer Vision, pp. 4015–4026 (2023)

46. Lanskikh, S., Demidov, G., Dinmuhametov, D., Khlopotnukh, A., Bogdan, K., Pakhomov, R.: Matchers with weights. In: Working Notes of CLEF 2025 - Conference and Labs of the Evaluation Forum (2025)

47. Lindenmayer, D.B., Likens, G.E.: The science and application of ecological monitoring. Biol. Cons. **143**(6), 1317–1328 (2010)

48. Liu, H., Wang, Y., Shi, C., Xu, T., Xing, H.: Tighnari v2: Mitigating label noise and distribution shift in multimodal plant distribution prediction via mixture of experts. In: Working Notes of CLEF 2025 - Conference and Labs of the Evaluation Forum (2025)

49. Liu, Z., et al.: Swin transformer v2: scaling up capacity and resolution. In: Proceedings of the IEEE/CVF Conference on Computer Vision and Pattern Recognition, pp. 12009–12019 (2022)

50. Lu, L., Yang, H., Li, S., Liu, F., Chen, P., Ma, W.: Few-shot fine-grained classification of fungi species using contrastive representation learning. In: Working Notes of CLEF 2025 - Conference and Labs of the Evaluation Forum (2025)

51. Martellucci, G., Goëau, H., Bonnet, P., Vinatier, F., Joly, A.: Overview of PlantCLEF 2025: multi-species plant identification in vegetation quadrat images. In: Working Notes of CLEF 2025 - Conference and Labs of the Evaluation Forum (2025)

52. Mora, C., Tittensor, D.P., Adl, S., Simpson, A.G., Worm, B.: How many species are there on earth and in the ocean? PLoS Biol. **9**(8), e1001127 (2011)

53. Oquab, M., et al.: DINOv2: learning robust visual features without supervision. arXiv preprint arXiv:2304.07193 (2023)

54. Palencia, P., Vada, R., Zanet, S., Calvini, M., De Giovanni, A., Gola, G., Ferroglio, E.: Not just pictures: utility of camera trapping in the context of African swine fever and wild boar management. Transboundary Emerging Diseases, 1–9 (2023)

55. Papafitsoros, K., Adam, L., Schofield, G.: A social media-based framework for quantifying temporal changes to wildlife viewing intensity. Ecol. Model. **476**, 110223 (2023)

56. Papafitsoros, K., Panagopoulou, A., Schofield, G.: Social media reveals consistently disproportionate tourism pressure on a threatened marine vertebrate. Anim. Conserv. **24**(4), 568–579 (2021)

57. Picek, L., et al.: CzechLynx: a dataset for individual identification and pose estimation of the eurasian lynx. arXiv preprint arXiv:1804.07177 (2025)
58. Picek, L., et al.: Overview of GeoLifeCLEF 2024: species presence prediction based on occurrence data and high-resolution remote sensing images. In: Working Notes of CLEF 2024 - Conference and Labs of the Evaluation Forum (2024)
59. Picek, L., et al.: GeoPlant: spatial plant species prediction dataset. In: Globerson, A., et al. (eds.) Advances in Neural Information Processing Systems, vol. 37, pp. 126653–126676. Curran Associates, Inc. (2024)
60. Picek, L., Janouskova, K., Cermak, V., Matas, J.: FungiTastic: a multi-modal dataset and benchmark for image categorization. In: Proceedings of the Computer Vision and Pattern Recognition Conference (CVPR) Workshops, pp. 2046–2056 (2025)
61. Picek, L., Leblanc, C., Larcher, T., Servajean, M., Bonnet, P., Joly, A.: Overview of GeoLifeCLEF 2025: plant species presence prediction with environmental and high-resolution remote sensing data. In: Working Notes of CLEF 2025 - Conference and Labs of the Evaluation Forum (2025)
62. Picek, L., Neumann, L., Matas, J.: Animal identification with independent foreground and background modeling. In: DAGM German Conference on Pattern Recognition, pp. 241–257. Springer (2024)
63. Picek, L., Šulc, M., Matas, J., Heilmann-Clausen, J., Jeppesen, T.S., Lind, E.: Automatic fungi recognition: deep learning meets mycology. Sensors **22**(2), 633 (2022)
64. Picek, L., Šulc, M., Patel, Y., Matas, J.: Plant recognition by AI: deep neural nets, transformers, and KNN in deep embeddings. Front. Plant Sci. **13**, 787527 (2022)
65. R, J., Mirunalini, P., K, K.M., J, H.: Multi-label plant species classification using tiling-based inference. In: Working Notes of CLEF 2025 - Conference and Labs of the Evaluation Forum (2025)
66. Rawlings, D., Chopard, T.: Enhancing presence-absence identification models using presence-only data. In: Working Notes of CLEF 2025 - Conference and Labs of the Evaluation Forum (2025)
67. Rowcliffe, J.M., Field, J., Turvey, S.T., Carbone, C.: Estimating animal density using camera traps without the need for individual recognition. J. Appl. Ecol., 1228–1236 (2008)
68. Schofield, G., Klaassen, M., Papafitsoros, K., Lilley, M., Katselidis, K.A., Hays, G.C.: Long-term photo-id and satellite tracking reveal sex-biased survival linked to movements in an endangered species. Ecology **11**(7), e03027 (2020)
69. Schofield, G., et al.: More aggressive sea turtles win fights over foraging resources independent of body size and years of presence. Anim. Behav. **190**, 209–219 (2022)
70. Semenova, N.: Addressing class imbalance and spatial shift in GeoLifeCLEF 2025. In: Working Notes of CLEF 2025 - Conference and Labs of the Evaluation Forum (2025)
71. Semenova, N.: Dual-stream meta-algorithm for animal re-id: WildFusion global-local matching and XGBoost on megadescriptor-miew embeddings. In: Working Notes of CLEF 2025 - Conference and Labs of the Evaluation Forum (2025)
72. Semenova, N.: Optimizing a vision transformer with ecological context for multi-label plant species identification. In: Working Notes of CLEF 2025 - Conference and Labs of the Evaluation Forum (2025)
73. Stevens, S., et al.: BioCLIP: a vision foundation model for the tree of life. In: Proceedings of the IEEE/CVF Conference on Computer Vision and Pattern Recognition, pp. 19412–19424 (2024)

74. Syayfetdinov, A.: Swin-t based multimodal networks for GeoLifeCLEF 2025. In: Working Notes of CLEF 2025 - Conference and Labs of the Evaluation Forum (2025)
75. Tam, J.K., Gustineli, M., Miyaguchi, A.: Transfer learning and mixup for fine-grained few-shot fungi classification. In: Working Notes of CLEF 2025 - Conference and Labs of the Evaluation Forum (2025)
76. Tikhonov, G., Tikhonov, D.: Synthesizing joint and deep species distribution modeling to enhance spatial prediction of plant communities at continental scale. In: Working Notes of CLEF 2025 - Conference and Labs of the Evaluation Forum (2025)
77. Tovar, A.M., Serrano, J., Martinez-Santos, J.C., Puertas, E.: Patch-based segmentation with vision transformers for multi-species classification in vegetation images. In: Working Notes of CLEF 2025 - Conference and Labs of the Evaluation Forum (2025)
78. Tschannen, M., et al.: Siglip 2: Multilingual vision-language encoders with improved semantic understanding, localization, and dense features. arXiv preprint arXiv:2502.14786 (2025)
79. Čermák, V., Picek, L., Adam, L., Papafitsoros, K.: WildlifeDatasets: an opensource toolkit for animal re-identification. In: Proceedings of the IEEE/CVF Winter Conference on Applications of Computer Vision (WACV), pp. 5953–5963 (2024)
80. Wäldchen, J., Rzanny, M., Seeland, M., Mäder, P.: Automated plant species identification–trends and future directions. PLoS Comput. Biol. **14**(4), e1005993 (2018)
81. Wan, F., Wan, H., Zhang, Z., Gao, J., Sun, C., Wang, Y.: The application potential of unmanned aerial vehicle surveys in grassland plant diversity. Biodivers. Sci. **32**(3), 23381 (2024)
82. Yang, T.A., Nguyen, M.Q.: Mushroom for improvement: prototypical few-shot learning with multimodal fungal features. In: Working Notes of CLEF 2025 - Conference and Labs of the Evaluation Forum (2025)
83. Yoccoz, N.G., Nichols, J.D., Boulinier, T.: Monitoring of biological diversity in space and time. Trends Ecol. Evol. **16**(8), 446–453 (2001)
84. Zhao, X., Wu, X., Chen, W., Chen, P.C.Y., Xu, Q., Li, Z.: ALIKED: a lighter keypoint and descriptor extraction network via deformable transformation. IEEE Trans. Instrum. Measur. **72**, 1–16 (2023)
85. Zwerts, J.A., et al.: Methods for wildlife monitoring in tropical forests: Comparing human observations, camera traps, and passive acoustic sensors. Conserv. Sci. Pract. **3**(12) (2021)

LongEval at CLEF 2025: Longitudinal Evaluation of IR Systems on Web and Scientific Data

Matteo Cancellieri[1] , Alaa El-Ebshihy[2,3] , Tobias Fink[2,3] , Maik Fröbe[4] ,
Petra Galuščáková[5] , Gabriela Gonzalez-Saez[6] , Lorraine Goeuriot[6] ,
David Iommi[2] , Jüri Keller[7(✉)] , Petr Knoth[1] , Philippe Mulhem[6] ,
Florina Piroi[2,3] , David Pride[1] , and Philipp Schaer[7]

[1] The Open University, Milton Keynes, UK
[2] Research Studios Austria, Data Science Studio, Vienna, Austria
[3] TU Wien, Vienna, Austria
[4] Friedrich-Schiller-Universität Jena, Jena, Germany
[5] University of Stavanger, Stavanger, Norway
[6] Univ. Grenoble Alpes, CNRS, Grenoble INP (Institute of Engineering Univ.
Grenoble Alpes), LIG, Grenoble, France
[7] TH Köln - University of Applied Sciences, Cologne, Germany
jueri.keller@th-koein.de

Abstract. The LongEval lab focuses on the evaluation of information retrieval systems over time. Two datasets are provided that capture evolving search scenarios with changing documents, queries, and relevance assessments. Systems are assessed from a temporal perspective—that is, evaluating retrieval effectiveness as the data they operate on changes. In its third edition, LongEval featured two retrieval tasks: one in the area of ad-hoc web retrieval, and another focusing on scientific article retrieval. We present an overview of this year's tasks and datasets, as well as the participating systems. A total of 19 teams submitted their approaches, which we evaluated using nDCG and a variety of measures that quantify changes in retrieval effectiveness over time.

Keywords: Longitudinal Evaluation · Temporal Persistence · Temporal Generalisability · Temporal Change · Information Retrieval

1 Introduction

Information Retrieval (IR) systems are constantly challenged by the evolving search setting [17]. Document collections evolve, user information needs shift, and relevance judgments may vary over time [1, 35, 39]. These temporal dynamics have strong implications to the long-term effectiveness of retrieval models. It is known that search is sensitive to temporal shifts [25, 30] and that incorporating historical signals can enhance retrieval robustness [3, 27]. Additionally, in modern IR, the systems are updated or retrained often, making them a dynamic component themselves in the evolving search setting.

While these temporal factors strongly influence retrieval effectiveness, they are often overlooked in standard IR evaluation protocols, which typically assume a static test collection. The advantages of such datasets are that they are easily used to evaluate and test systems. Some data sets, like CORD19, contain documents collected at different points in time, showing differences in the set of documents from one collection time to another. We have shown previously that the ranking of systems varies over time and that the most effective system is not necessarily also the system that performs the most consistently [4,5,26]. This shows how the experimental setup strongly influences the measured effectiveness.

With the aim of tackling this challenge of making models have persistent quality over time, the objective of the LongEval lab is twofold: (i) to systematically assess how performance of retrieval systems changes over time as test collections evolve, and (ii) to propose improved methods that mitigate performance drop by making models more robust over time.

The third edition of the LongEval lab [13] was part of the Conference and Labs of the Evaluation Forum (CLEF) 2025. In this edition it consisted of two retrieval tasks: Task 1 - WebRetrieval, which is the classical web case, and Task 2 - SciRetrieval, which is for scientific search. Task 1 evaluates retrieval robustness over time using an evolving web search collection, while Task 2 follows a similar setup but uses scholarly publications as the underlying document corpus.

2   Tasks Description

In contrast to traditional IR evaluation that rely on one static datasets, LongEval 2025 explores the effect of changing datasets on the retrieval systems and measured effectiveness. Similarly to the LongEval 2023 and 2024 Retrieval Tasks [4,5], we focus on a setup in which the datasets are evolving. In concrete terms, differences between datasets can be the addition, removal, or the change of documents and queries. Each evolved state of a dataset is captured as a new snapshot, which forms the basis for new experiments. We evaluate systems by computing efficiency metrics on the experimental approaches (or runs) submitted by the participants who designed those systems. The two main scenarios considered in our evaluation focus on single systems and multiple systems:

A single-system in evolving setup: Each system is trained once, on a collection snapshot at a given timestamp, and evaluated on later collection snapshots. This setup assesses how well a system maintains retrieval effectiveness over time without retraining.

Multi-system comparison: Each system is compared to the other systems, across several snapshots. This evaluation setup is to provide more information about systems' stability and robustness compared to each-other.

2.1   Task 1: WebRetrieval

The WebRetrieval task investigates how IR systems cope with evolving web search collections. The dataset consists of a series of monthly snapshots between

June 2022 to August 2023, extracted from the French search engine Qwant[1], each containing documents and corresponding user queries.

2.2 Task 2: SciRetrieval

The SciRetrieval task extends the LongEval Lab to the domain of academic search. It aims to evaluate the temporal robustness of IR systems when retrieving scholarly publications from an evolving corpus. The dataset is derived from the CORE collection[2], which aggregates open access research outputs from repositories and journals worldwide. As of this edition, CORE is one of the largest platforms of openly available full-text scholarly documents.

3 Datasets

This section describes the datasets used in the two retrieval tasks, including data sources, snapshot construction, document and query statistics, and relevance judgments methods. All datasets are available from the TU Wien Research Data Repository [18–20] or through the LongEval IR_datasets integration [28,31][3].

3.1 WebRetrieval Dataset

The dataset for this task was provided by the French search engine Qwant. It consist of the queries issued by the users of this search engine, cleaned Web documents, which were 1) selected to correspond to the queries, and 2) to add additional noise, and relevance judgments, which were created using a click model. The dataset as of 2022 is fully described in [24]. Later additions to the LongEval WebRetrieval collection have followed the same collection procedure.

The 2025 dataset includes all data from the 2023 and 2024 editions, along with newly added, previously unreleased months. The training dataset consists of 19 million French documents (June 2022 - February 2023) and 119,341 queries with computed relevance assessments based on a simplified Dynamic Bayesian Network (sDBN) Click Model [15,16], acquired from real users of the French Qwant search engine. Compared to the previous datasets, the data were processed differently to combine similar queries and unify IDs. Therefore, direct comparisons are not possible, although all datasets have overlapping snapshots.

The test collection spans 7 months of data (March 2023 - August 2023) and consists of 14 million documents and 63,416 queries. Each month is captured in one snapshot. They are similar in structure as the training snapshot, except that they do not contain any relevance assessments. Participants submitted their runs for each snapshot, using the same system trained only on the training dataset. The total data for this task consists of 33 million documents and 182,757 queries, provided by Qwant.

---

[1] Qwant search engine: https://www.qwant.com/.

[2] CORE search engine: https://core.ac.uk/.

[3] GitHub: https://github.com/clef-longeval/ir-datasets-longeval.

Figure 1 shows the document overlap between each pair of monthly snapshots in the WebRetrieval dataset. The values reflect the proportion of shared documents across snapshots. As shown, overlap is highest between consecutive months and decreases over time. More specifically, earlier snapshots (e.g., mid-2022) share substantially fewer documents with later ones (e.g., mid to late 2023), demonstrating the evolving nature of the dataset. This highlights the challenge posed to the retrieval models, which should maintain effectiveness despite the changes of the collection over time.

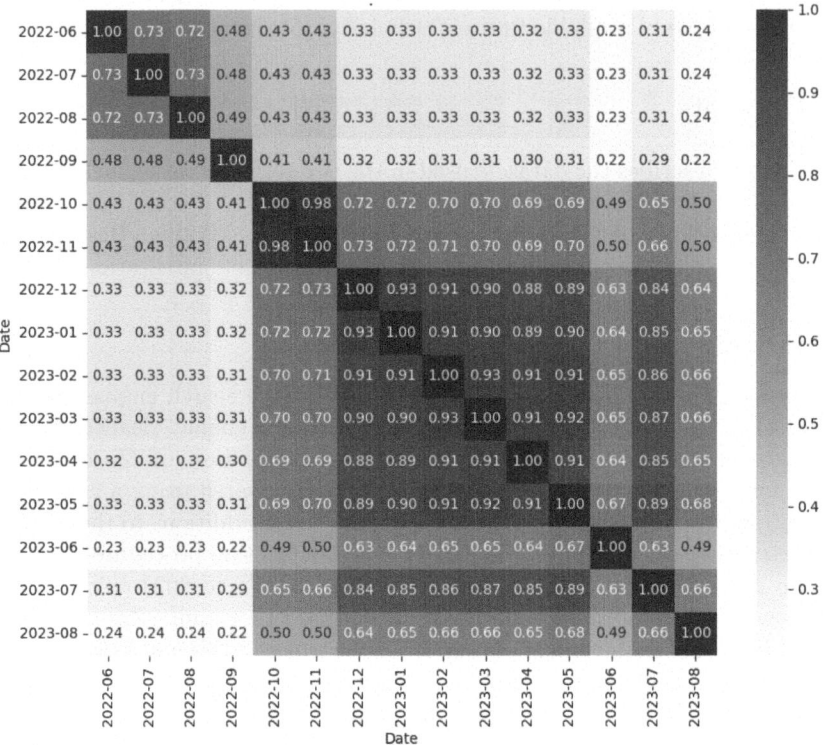

Fig. 1. Document overlap between monthly snapshots of the WebRetrieval dataset. Lighter colors indicate lower overlap between collections. (Color figure online)

3.2 SciRetrieval Dataset

The data for this task were acquired from the CORE[4] collection of scholarly documents. The dataset includes cleaned user queries and documents returned as results to these queries, along with additional negative, randomly selected

---

[4] CORE (COnnecting REpositories) https://core.ac.uk/.

Table 1. Teams participating to the LongEval 2025 lab.

Team	Team size (# persons)	Task	Location	Notebook
3DS2A	6	Web	Padua, Italy	[12]
BASETTE	4	Web	Padua, Italy	[8]
CIR	19	Web	Cologne, Germany	[9]
CIR_cluster	3	Web	Cologne, Germany	[34]
DataHunter	5	Web	Padua, Italy	[33]
DS@GT	3	Web	Georgia, USA	[32]
EAIiIB	4	Sci	Cracow, Poland	[38]
OpenWebSearch	7	Sci	Nijmegen, The Netherlands; Jena and Kassel, Germany	[2]
RACOON	5	Web	Padua, Italy	[23]
RAND	7	Web	Padua, Italy	[6]
RISE	7	Web	Padua, Italy	[22]
SARD	5	Web	Padua, Italy	[14]

scientific articles. Similar to the WebRetrieval task, the relevance judgments were created using a click model [15,16], based on click signals from user logs (e.g., PDF downloads).

As this is the first iteration of the SciRetrieval task, the dataset includes only two snapshots: (1) The first snapshot spans mid-November 2024 to mid-December 2024, and (2) The second snapshot was collected in January 2025.

The training dataset (from the first snapshot) includes approximately 2 million documents, 393 queries, and 4,262 relevance assessments. The test dataset includes over 1 million documents, 492 new queries (from the second snapshot), and 99 held-out queries selected from the first snapshot. Like in the WebRetrieval task, the test collections mirror the structure of the training data but do not contain relevance assessments. Participants submitted their runs for both test sets using a system trained only on the training dataset. The total data for this task consists of over 3 million documents and 984 queries.

4   Submissions

We received 45 runs to the WebRetrieval task and 23 runs to the SciRetrieval task from 19 participating teams. Of these 19 teams 12 submitted a notebook paper. The submissions were done through the TIRA platform [21] where participants could upload either runs, software, or both. 36 further teams registered in tira but never submitted any run. We collected the descriptions and metadata of each run in the ir_metadata specification [11]. Table 2 gives an overview of the ir_metadata of the submitted approaches.

In the following, we give a brief summary of the submissions based on the information provided by the participants.

3DS2A [12]: Besides a traditional BM25 search pipeline, the system by Team 3DS2A utilizes more sophisticated techniques like chunk-based search, where documents are divided into semantically coherent text segments - called chunks - before indexing. The effectiveness of chunk-based approaches has been highlighted in both neural and classical IR frameworks. Additionally, pseudo-relevance feedback and reranking based on sentence embeddings were used to enhance the retrieval capability of the queries. The system was based on Lucene.

BASETTE [8]: Team BASETTE's main priority was optimizing performance on limited consumer hardware, deliberately avoiding the use of GPUs or other specialized computational resources. The system relies on classical IR techniques and is designed to run both indexing and retrieval in a multithreaded fashion to ensure high execution speed. One example of this was to process the indexing in-memory and never write it to disk. Parameters of the search system were optimised using Optuna. The team chose to discard a bunch of approaches due to limited hardware resources or integration complexity, like WordNet term expansion, the integration of the Duckling framework to extract temporal expressions from the documents, or neural ranking with CamemBERT.

CIR [9]: This multi-team submission summarized five different teams: CIR_SchaeredRetrieval, CIR_SuperTeam123, CIR_Sauerkraut, CIR_JMFT, and CIR_fair_schaer. The groups had five different approaches and motivations to test in the LongEval setting: (1) Finding time-dependent queries with the help of LLMs and to treat these queries differently by boosting their retrieval scores based on the categorization; (2) Finding time-dependent queries and scoring them on a scale from 0 to 1 and to use that score to influence the final ranking; (3) Using relevance information from older sub-collections and to use relevance feedback on the current sub-collection by using query expansion using tf-idf; (4) Boosting known relevant documents-query pairs from older sub-collections but comparing the similarity of old and recent documents; Finally, (5) a neural relevance re-ranking based on a topcial semantic clustering. All systems used PyTerrier's BM25 as the foundational retrieval system.

CIR_cluster [34]: The submission by Team CIR_cluster aims to leverage historical information, such as past relevance judgments. In a previous submission, this information was used in two different ways: Query Boost and Relevance Feedback. Both methods are limited when no prior information is available, which is the case when queries can not be mapped due to slight variations. It was observed that many similar queries are captured in the test collections. Often, they even differ only on a lexical level, such as spelling or word order. Therefore, Team CIR_cluster aimed to identify query variants – queries that relate to the same information need but express it in different ways, as documents relevant to a given query might also be relevant to its semantic variants. Based on this assumption, they cluster queries and link previously unseen queries to the history of their query variants.

DataHunter [33]: Team DataHunter compared a traditional BM25 query-based searcher with a neural reranking system. The latter used an inverse square rank fusion based on BM25 with RM3 expansion, SPLADE (sparse transformer-based retrieval), and a cross-encoder CamemBERT re-ranker (top-100 reranking). Next

Table 2. Overview of the collected ir_metadata according to (a) frequently used software libraries, (b) implemented retrieval paradigms, and (c) available git repositories.

(a) Most popular libraries

Library	Teams
python-terrier	7
numpy	4
SQLAlchemy	4
scikit-learn	4
pandas	4
transformers	3
torch	3
ir-datasets-longeval	3
Lucene	3
Yaml	2

(b) Retrieval paradigms

Paradigm	**Used**	
	Yes	No
Lexical	84	23
Deep Neural	60	47
Sparse Neural	5	102
Dense Neural	59	48
Single Stage	67	40

(c) Repositories

Hoster	**Public**	**Private**
Bitbucket	0	8
Github	10	4

to the retrieval system, the team also conducted a statistical analysis using ANOVA and Tukey HSD tests to determine whether performance differences are significant over time, providing insights into the robustness and generalizability of IR models in dynamic environments.

DS@GT [32]: Team DS@GT developed an IR pipeline to observe temporal variance using topic modeling and a two-phase retrieval system involving query expansion based on the Gemini LLM. The workflow begins with a Parquet ingestion pipeline for the LDA topic modeling and sentence transformer processing. The transformed documents are stored in a shared directory where the Pyserini system was used to perform BM25 retrieval. When an input query is received, the system references a precomputed mapping index of Qwant search queries, each mapped to expanded queries generated using Gemini.

EAIiIB [38]: The EAIiIB team participated in the Sci task and compared classical lexical retrieval, dense vector-based retrieval, and hybrid approaches, incorporating reranking via cross-encoders. They experimented with a reduced and the full data set and used the following models to build their baseline: BM25, MiniLM-L6-v2, E5-large-v2. Additionally, some hybrid pipelines were introduced, combining different approaches. In their setting, dense and cross-encoder reranking outperforms all lexical and hybrid configurations.

OpenWebSearch [2]: Team OpenWebSearch used different web crawls of the CORE search engine to build their submissions on. They crawled the top-25 documents from CORE for quries based on different fields (title, abstract, and full text). They had two underlying assumptions: First, a practical search engine should make only incremental improvements to existing rankings, and second, that most relevant documents are already among the top results shown by the CORE engine, especially since LongEval defines relevance based on user clicks.

Building on these initial rankings, different techniques to enhance retrieval effectiveness were tested: (1) qrel-boosting based on past relevance, RM3 keyquery expansion, cluster-based boosting, monoT5 re-ranking of top results, and user intent prediction. Each of these methods is designed to re-rank or selectively boost documents within the CORE results to increase the likelihood of retrieving relevant content.

RACOON [23]: The search system developed by Team RACOON incorporates a French lemmatizer within the analyzer, a LLM-based Query Expansion (QE) module, a Relevance Feedback (RF) mechanism based on query assessment over the preceding months, and an LLM-based Elo reranking strategy. The custom lexicon lemmatizer is based on the Lexique des formes fléchies du français (LEFFF), a large-scale morphological and syntactic lexicon for French. The QE pipeline was using LLaMA 3 70B and LLaMA 4 Scout 17B and instructed the models to (1) generate 20 semantically related expressions to a query, (2) provide a passage that answers the initial query, (3) listing synonyms, and (4) infer the user's intent explicitly. The RF was based on worked presented in 2024 [27]. The reranker was finally implemented using either an SBERT embedding cosine similarity score, an ms-marco-MiniLM-L6-v2 setting, or an ELO-pseudo ranking method, where two documents were randomly compared at a time and an LLM decided on their appropriateness to the query.

RAND [6]: Team RAND built upon a classic Lucene pipeline and tested various configurations, including different tokenizers, filters, stoplists, stemmers, etc. They also used the pre-trained sentence embedding model multi-qa-MiniLM-L6-cos-v1 and Llama-3.2-1B-Instruct to include a semantic scoring for re-ranking the initial Lucene results. Additionally, they included the Lucene ICUFoldingFilter, which applies search term folding to Unicode text, including accent removal and case folding, among other features. This proved to be useful, as the French language contains many words with diacritical marks (accents), which can impact search and term matching.

RISE [22]: The system of Team RISE incorporates a modular architecture, including a parser, an analyzer, an indexer, and a searcher. It also includes query translation and expansion using the Gemini LLM, and a non-neural reranking component to enhance retrieval quality. The emphasis was placed on optimizing indexing and search speed through multithreading, improving relevance by crafting a title for each document, and improving the content of the URL-based document based on the alignment between user queries and the document's URL. The system extracted titles from documents and query terms from the URL. Both was not marked explicitly in the dataset to give the extracted titles a higher weight in the final ranking.

SARD [14]: The SARD team focused on the development of custom language-specific analyzers for the preprocessing of documents and queries, and also the correction of user queries using GPT-4 Turbo. The use of GPT-4 Turbo aims to rewrite and clarify ill-formed or noisy queries without altering their intent, resulting in more precise and effective matching with the indexed content. The

Basesystem is implemented on Lucene, and a language-detection process was introduced to differentiate the two main languages of the corpus. Additionally, two distinct query processes were employed: BooleanQuery and PhraseQuery. An LLM-based QE approach was discarded due to its high costs and processing time, and was replaced with a dictionary-based QE (WordNet for English and WOLF for French).

5   Results

The submitted retrieval approaches are evaluated along different dimensions and measures, and different snapshots. For the SciRetrieval task two test snapshots were available to compare the system effectiveness. For the WebRetrieval task, six test snapshots were available. In this paper we detail the evaluation for the WebRetrieval task for the following two cases: (a) short-term effectiveness change evaluation, comparing scores for the snapshots 2023-03 and 2023-05, and (b) long-term effectiveness change evaluation for scores on snapshots 2023-03 and 2023-08.

5.1   Effectiveness

The retrieval effectiveness is assessed by nDCG and nDCG@10. The results per snapshot and approach are presented in Tables 3 for the Web task and Table 4 for the Sci task. The effectiveness measured by nDCG@10 for the test snapshots is also visualized in Fig. 2 for the Web task and in Fig. 4a for the Sci task. Especially for the Web task, nDCG@10 is considered to match the retrieval setting of Web search well, where users often focus on the first retrieval results. Some participants submitted multiple versions of the same approach that do not or only slightly differ. In these cases, only the best or newest approach was considered. For the Sci task, one team submitted only for the 2025-01 snapshot, and one team submitted runs with really few documents ranked.

The top three approaches for the Web task appear to be relatively stable over time. For the later snapshots, the approach `clef25-seupd2425-rise` from team RISE outperforms the baseline `qrel-boost`. These two approaches are followed by the `run1` of team RACOON, `rr-ps` of team 3DS2A, and `query-variants-qrel-boost-kmeans` of team CIR-CLUSTER as measured by nDCG@10. For nDCG@1000, this order slightly varies. Especially in 2023-03, the approaches `run1` and `run2` from team RACOON perform better. The gradient of both result tables recommends a relatively stable system ranking for both measures.

The top three approaches for the Sci task vary by the snapshot. For the first snapshot 2024-11 (within time), the approaches `ows-cluster-boosting` and `ows-bm25` are on par, followed by `rm3-on-qrel-boost`. All approaches were submitted by the Open Web Search team. Regarding the second snapshot 2025-01 (short term), the three approaches `BM25_k1_1p0_b_0p7_stop_stem_fullText`, `BM25_k1_0p2_b_0p8_stop_stem_fullText`, and `BM25_k1_0p95_b_0p75_stop_stem_fullText` by the team Academy Retrievals dominate. As depicted by the

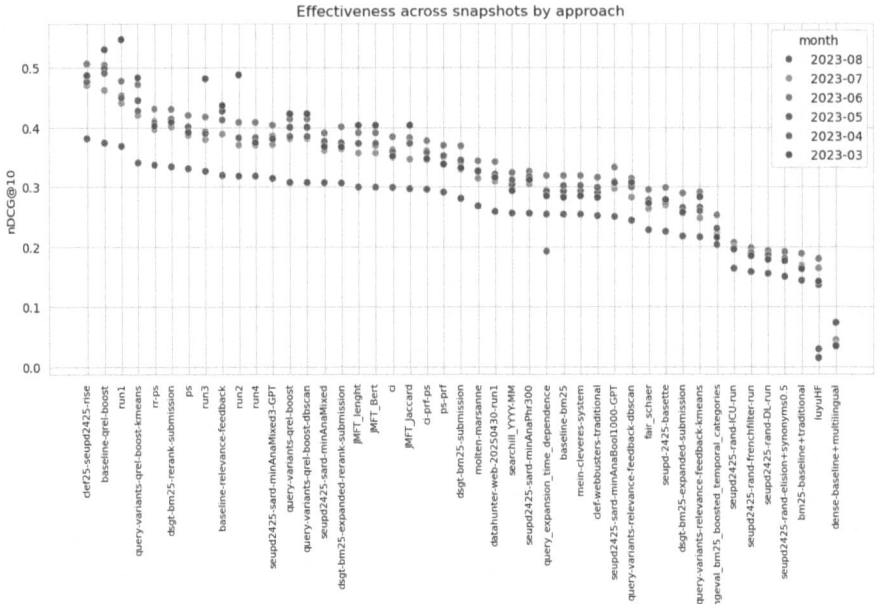

Fig. 2. nDCG@10 for all approaches and test snapshots in the WebRetrieval task.

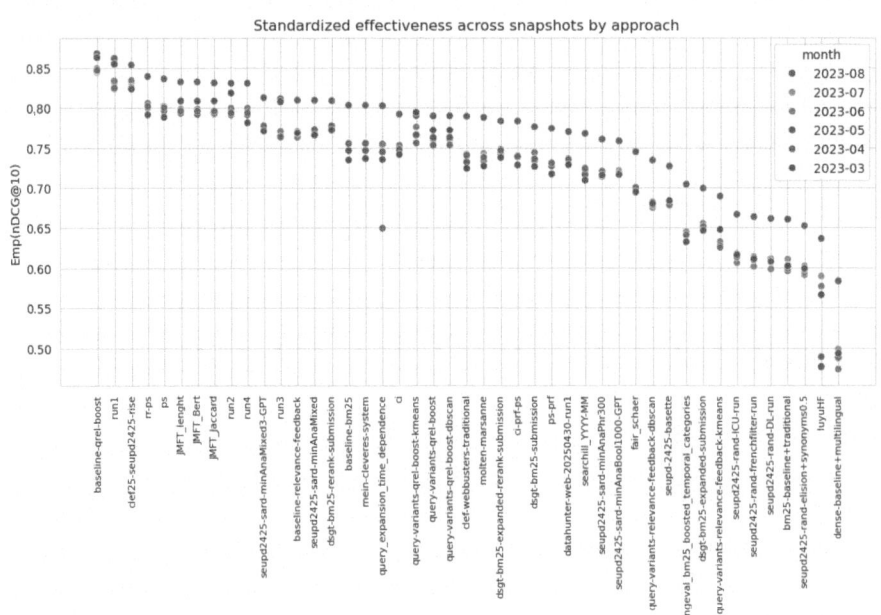

Fig. 3. Standardized nDCG@10 for all approaches and test snapshots in the WebRetrieval task.

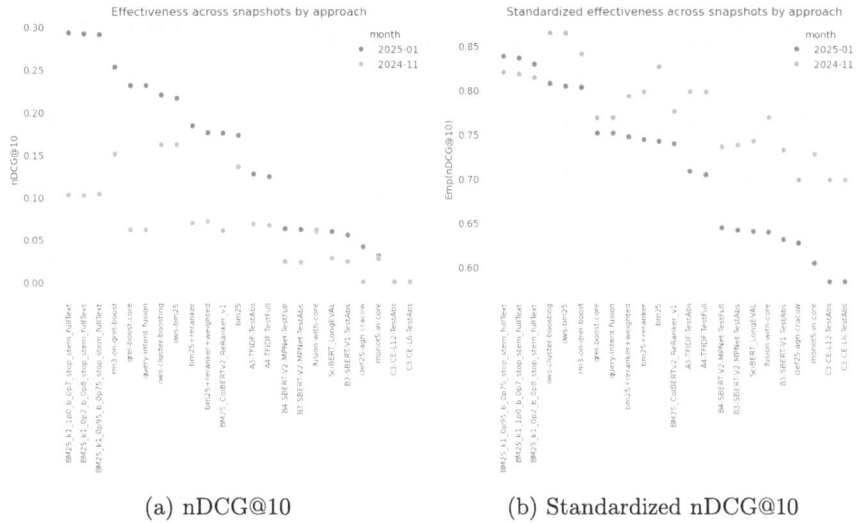

(a) nDCG@10   (b) Standardized nDCG@10

Fig. 4. Effectiveness measured by nDCG@10 across all approaches and test snapshots of the SciRetrieval task.

gradient in Table 4, the ranking of systems appears to be more stable across measures than time.

To compare the systems' effectiveness across snapshots, we also present standardized nDCG@10 scores using empirical standardization, as in [40] in Fig. 3 for the Web task and in Fig. 4b for the Sci task. Each query of the snapshot is standardized according to the performance of all participating systems. Finally, a standardized nDCG@10 mean is computed.

When considering the standardized performance, the best systems in the Web task remain the same but in a different order (Fig. 3): the baseline `qrel-boost` achieves the highest standardized performance in snapshot 2023-08, with `run1` from team RACOON in second place. The baseline `qrel-boost` also stands out for its stability. Since it is built using historical data, we can hypothesize that this contributes to its temporal stability, as its performance variations reflect the differences between snapshots. We observed similar results for the Sci task

(Fig. 4b): the best systems in snapshot 2025-01 remain the same and have the most stable standardized performance. This stability suggests that their performance varies in line with the general trend of the ensemble of participating systems, as represented by the empirical standardization function.

5.2   Ranking of Systems

We inspect system rankings between two snapshots to better understand how the approaches evolve relative to each other over time. This evaluation should provide insights into how robust the system ranking is. As discussed before, the gradients of Table 3 indicate a similar ranking of systems for the WebRetrieval task. Table 4 shows a less similar system ranking for the SciRetrieval case.

To further investigate this, we compare the system order between two snapshots based on the systems' ranks with Kendall's Tau and based on the measured effectiveness scores by Pearson's correlation. All correlations are based on the nDCG@10 scores and rankings.

For the Web task, the short-term change (2023-03 to 2023-05) and long-term change (2023-03 to 2023-08) are compared. The Pearson correlation for the short-term change is 0.954, indicating an extremely strong positive linear relationship. The long-term change (2023-03 to 2023-08) shows an even slightly stronger correlation with 0.965. The Kendall also indicates a high correlation. For the short-term change, 0.930 was measured, and for the long-term, 0.893. All p-values indicate significance.

For the Sci task, only two snapshots are available for comparison. The Person's correlation is much weaker compared to the Web correlations but is with 0.708 still strong. The Kendall's tau of 0.500 indicates a moderate-to-strong positive. Like before, the p-values are significant.

The ranking of systems shows a stronger correlation when standardized performances are compared. Table 5 presents the results for the scientific task, with a Pearson correlation of 0.848 and a Kendall's tau of 0.587.

Table 3. Evaluation results for the short-term and long-term changes of the WebRetrieval task. The results are sorted by nDCG@10 for the 2023-08 snapshot.

Approach	nDCG@10			nDCG@1000		
	2023-03	2023-05	2023-08	2023-03	2023-05	2023-08
clef25-seupd2425-rise [22]	0.487	0.484	0.381	0.541	0.543	0.422
baseline-qrel-boost [34]	0.529	0.498	0.374	0.561	0.540	0.407
run1 [23]	0.546	0.452	0.368	0.588	0.508	0.405
query-variants-qrel-boost-kmeans [34]	0.483	0.445	0.340	0.517	0.490	0.376
rr-ps [12]	0.402	0.410	0.337	0.462	0.475	0.378
dsgt-bm25-rerank-submission [32]	0.409	0.412	0.334	0.447	0.456	0.363
ps [12]	0.391	0.401	0.330	0.453	0.468	0.374
run3 [23]	0.481	0.392	0.326	0.530	0.455	0.367
baseline-relevance-feedback [34]	0.436	0.427	0.319	0.494	0.490	0.366
run4 [23]	0.374	0.383	0.318	0.441	0.454	0.365
run2 [23]	0.487	0.383	0.318	0.543	0.454	0.365
seupd2425-sard-minAnaMixed3-GPT [14]	0.380	0.385	0.314	0.444	0.452	0.359
query-variants-qrel-boost-dbscan [34]	0.422	0.400	0.307	0.474	0.460	0.354
query-variants-qrel-boost [34]	0.422	0.400	0.307	0.474	0.460	0.354
seupd2425-sard-minAnaMixed [14]	0.367	0.374	0.307	0.428	0.438	0.351
dsgt-bm25-expanded-rerank-submission [32]	0.367	0.374	0.306	0.398	0.410	0.331
JMFT_lenght [9]	0.403	0.373	0.299	0.456	0.436	0.346
JMFT_Bert [9]	0.403	0.370	0.299	0.456	0.433	0.346
ci [12]	0.351	0.362	0.299	0.418	0.433	0.346
JMFT_Jaccard [9]	0.403	0.373	0.297	0.456	0.436	0.343
ci-prf-ps [12]	0.346	0.360	0.296	0.414	0.431	0.342
ps-prf [12]	0.337	0.350	0.291	0.405	0.424	0.339
dsgt-bm25-submission [32]	0.331	0.341	0.280	0.385	0.399	0.320
molten-marsanne [9]	0.326	0.325	0.268	0.398	0.401	0.321
datahunter-web-20250430-run1 [33]	0.315	0.316	0.259	0.376	0.384	0.305
searchill_YYYY-MM [searchill]	0.293	0.304	0.256	0.363	0.379	0.308
seupd2425-sard-minAnaPhr300 [14]	0.311	0.314	0.256	0.340	0.342	0.272
query_expansion_time_dependence [9]	0.285	0.293	0.254	0.355	0.367	0.307
baseline-bm25 [34]	0.282	0.293	0.254	0.352	0.367	0.307
clef-webbusters-traditional [web-busters]	0.282	0.290	0.251	0.352	0.366	0.305
seupd2425-sard-minAnaBool1000-GPT [14]	0.307	0.308	0.250	0.376	0.382	0.301
query-variants-relevance-feedback-dbscan [34]	0.307	0.305	0.244	0.377	0.381	0.299
fair_schaer [9]	0.272	0.272	0.228	0.340	0.344	0.282
seupd-2425-basette [8]	0.278	0.276	0.225	0.297	0.295	0.238
dsgt-bm25-expanded-submission [32]	0.257	0.263	0.217	0.313	0.323	0.262
query-variants-relevance-feedback-kmeans [34]	0.283	0.265	0.216	0.354	0.344	0.274
bm25_boosted_temporal_categories [9]	0.216	0.229	0.203	0.294	0.311	0.263
seupd2425-rand-ICU-run [6]	0.196	0.199	0.164	0.260	0.267	0.215
seupd2425-rand-frenchfilter-run [6]	0.185	0.190	0.158	0.248	0.257	0.208
seupd2425-rand-DL-run [6]	0.179	0.185	0.155	0.220	0.229	0.189
seupd2425-rand-elision+synonyms0.5 [6]	0.176	0.180	0.150	0.238	0.246	0.201
bm25-baseline+traditional [air5]	0.163	0.163	0.144	0.179	0.183	0.159
luyuHF [air5]	0.142	0.014	0.136	0.163	0.014	0.136
dense-baseline+multilingual [air5]	0.034	0.033	0.073	0.040	0.040	0.084

5.3   Changes in Effectiveness

Especially interesting to this lab is to investigate how the retrieval effectiveness changes over time and which approaches are more resilient against an evolving search setting. Like in previous iterations of this lab, the Relative Improvement (RI) in effectiveness is measured [4,5]. Additionally, we employ the Delta Relative Improvement (DRI) and Effect Ratio (ER) to further describe the changes in the effectiveness with a focus on the system effect [10,26,27]. Both measures first relate the measured scores to a pivot system, BM25 [36] in our case, and then compare the emerging deltas across time. Since the experimental system of interest and the pivot system are exposed to the same changes, the changes introduced by the evolving search setting are dampened, and the results emphasize the effect of the experimental system. This is especially of interest when the experimental system is updated, retrained, or directly interacts with the previous snapshots. In these cases, a robust system effect is desirable. The change and persistence-based measures are complemented with the Average Retrieval Performance (ARP) across all topics at the evolved snapshot based on the instantiation measure nDCG@10 and the Mean Average Retrieval Performance (MARP) that calculates the mean between both compared snapshots.

For the Web task, short-term and long-term changes are compared and the results are displayed in the Tables 8 and 7. The reference system BM25 did not rank any results for up to 73 topics. To calculate the per-topic differences in the experimental systems, these topics are removed. Therefore, the ARP may vary slightly between the Tables 3, 8, and 7. The RI and DRI measure how the effectiveness changes relative to the first snapshot 2023-03. Due to their definition, values below zero indicate improving effectiveness. Almost half of the systems show an improved effectiveness over the short timespan. In contrast, only the three approaches ci-prf-ps, longeval_bm25_boosted_temporal_categories, and luyuHF indicate an improving system effect. This disagreement is also reflected in the top three approaches with the lowest change. RI rank the approaches seupd2425-rand-ICU-run, and seupd2425-rand-elision+synonyms0.5 followed by dsgt-bm25-rerank-submission and seupd2425-sard-minAnaBool1000-GPT with an RI of 0.003. The systems ci-prf-ps, ps-prf, and searchill_YYYY-MM indicate the lowest system effect change according to DRI. These systems also show a low RI. Since the effectiveness is generally not changing much, all differences are small. Regarding the long-term changes, only the dense-baseline+multilingual improves in effectiveness measured by RI. The approaches luyuHF, longeval_bm25_boosted_temporal_categories and query_expansion_time_dependence show the lowest changes. Regarding the system effects measured by DRI, almost all approaches impair again. query_expansion_time_dependence, mein-cleveres-system, and clef-webbusters-traditional and seem to be most robust.

The ER describes the extent to which the system effect is recovered. If the same system effect is achieved in the evolved setting, the ER is 1, which describes a perfectly persistent system. The approaches ci, dsgt-bm25-submission, and ps show the highest recovery of the system effect over the short term. For the

Table 4. Evaluation results for the SciRetrieval task. The results are sorted by nDCG@10 for the 2025-01 snapshot.

Approach	nDCG@10 2024-11	nDCG@10 2025-01	nDCG@1000 2024-11	nDCG@1000 2025-01
BM25_k1_1p0_b_0p7_stop_stem_fullText [academy-retrievals]	0.102	0.292	0.156	0.393
BM25_k1_0p2_b_0p8_stop_stem_fullText [academy-retrievals]	0.102	0.291	0.154	0.391
BM25_k1_0p95_b_0p75_stop_stem_fullText [academy-retrievals]	0.103	0.290	0.156	0.391
rm3-on-qrel-boost [2]	0.150	0.252	0.331	0.384
qrel-boost-core [2]	0.061	0.231	0.274	0.393
query-intent-fusion [2]	0.061	0.231	0.274	0.393
ows-cluster-boosting [2]	0.161	0.220	0.344	0.350
ows-bm25 [2]	0.161	0.216	0.344	0.348
bm25+reranker [tf-idk]	0.069	0.183	0.121	0.253
bm25+reranker+weighted [tf-idk]	0.071	0.176	0.126	0.254
BM25_ColBERTv2_ReRanker_v1 [academy-retrievals]	0.060	0.175	0.125	0.283
bm25 [tf-idk]	0.135	0.172	0.242	0.245
A3-TFIDF-TestAbs [sambs]	0.068	0.126	0.119	0.204
A4-TFIDF-TestFull [sambs]	0.066	0.123	0.117	0.201
B4-SBERT-V2-MPNet-TestFull [sambs]	0.024	0.062	0.043	0.097
B3-SBERT-V2-MPNet-TestAbs [sambs]	0.023	0.061	0.041	0.095
fusion-with-core [2]	0.061	0.059	0.274	0.263
SciBERT_LongEVAL [long-eval-sci-group-5]	0.028	0.059	0.059	0.117
B3-SBERT-V1-TestAbs [sambs]	0.024	0.055	0.046	0.092
clef25-agh-cracow [38]	0.000	0.041	0.000	0.045
monot5-in-core [2]	0.027	0.031	0.229	0.218
C3-CE-L6-TestAbs [sambs]	0.000	0.000	0.000	0.000
C3-CE-L12-TestAbs [sambs]	0.000	0.000	0.000	0.000
A7-BM25-TestAbs [sambs]	0.000	0.000	0.000	0.000
A8-BM25-TestFull [sambs]	0.000	0.000	0.000	0.000
B4-SBERT-V1-TestFull [sambs]	0.000	0.000	0.000	0.000
C4-CE-L6-TestFull [sambs]	0.000	0.000	0.000	0.000
C4-CE-L12-TestFull [sambs]	0.000	0.000	0.000	0.000

long-term change setting, the scores diverge strongly. `seupd2425-rand-DL-run`, `seupd2425-rand-elision+synonyms0.5`, and `bm25-baseline+traditional` have an ER closest to 1.

The results for the Sci task are presented in Table 6. The systems that ranked too few documents and the BM25 reference approach were excluded from this evaluation. The change measured by RI and DRI does not necessarily agree with each other. The approaches with the lowest change according to the RI are `fusion-with-core`, `monot5-in-core`, and `ows-bm25`. The DRI states `monot5-in-core`, `monot5-in-core`, `ows-bm25`, and `ows-cluster-boosting`. Besides the `fusion-with-core`, all approaches improve in effectiveness. This approach and additionally `monot5-in-core` do not show an improved system effect while the difference for `monot5-in-core` is only marginal.

Regarding the ER, the systems `B4-SBERT-V2-MPNet-TestFull`, `B3-SBERT-V2-MPNet-TestAbs`, and `BM25_ColBERTv2_ReRanker_v1` have an ER closest to 1, indicating the highest persistence. The measure seems not to agree with the change-based measures described before.

Table 5. Standardized results for the SciRetrieval task. The results are sorted by nDCG@10 for the 2025-01 snapshot, as in Table 4.

Approach	Emp(nDCG@10) 2024-11	2025-01	Emp(nDCG@1000) 2024-11	2025-01
BM25_k1_1p0_b_0p7_stop_stem_fullText [academy-retrievals]	0.818	0.836	0.654	0.761
BM25_k1_0p2_b_0p8_stop_stem_fullText [academy-retrievals]	0.814	0.829	0.637	0.741
BM25_k1_0p95_b_0p75_stop_stem_fullText [academy-retrievals]	0.820	0.838	0.653	0.755
rm3-on-qrel-boost [2]	0.841	0.803	0.813	0.776
qrel-boost-core [2]	0.769	0.751	0.819	0.767
query-intent-fusion [2]	0.769	0.751	0.819	0.767
ows-cluster-boosting [2]	0.864	0.807	0.863	0.777
ows-bm25 [2]	0.864	0.804	0.863	0.772
bm25+reranker [tf-idk]	0.798	0.744	0.609	0.611
bm25+reranker+weighted [tf-idk]	0.793	0.747	0.605	0.613
BM25_ColBERTv2_ReRanker_v1 [academy-retrievals]	0.776	0.739	0.608	0.646
bm25 [tf-idk]	0.826	0.742	0.696	0.599
A3-TFIDF-TestAbs [sambs]	0.798	0.708	0.610	0.545
A4-TFIDF-TestFull [sambs]	0.798	0.704	0.609	0.540
B4-SBERT-V2-MPNet-TestFull [sambs]	0.736	0.644	0.464	0.431
B3-SBERT-V2-MPNet-TestAbs [sambs]	0.738	0.641	0.458	0.426
fusion-with-core [2]	0.769	0.639	0.819	0.666
SciBERT_LongEVAL [long-eval-sci-group-5]	0.742	0.640	0.502	0.460
B3-SBERT-V1-TestAbs [sambs]	0.732	0.630	0.466	0.421
clef25-agh-cracow [38]	0.698	0.627	0.403	0.372
monot5-in-core [2]	0.727	0.604	0.656	0.529
C3-CE-L6-TestAbs [sambs]	0.698	0.583	0.403	0.316
C3-CE-L12-TestAbs [sambs]	0.698	0.583	0.403	0.316
A7-BM25-TestAbs [sambs]	0.000	0.000	0.000	0.000
A8-BM25-TestFull [sambs]	0.000	0.000	0.000	0.000
B4-SBERT-V1-TestFull [sambs]	0.000	0.000	0.000	0.000
C4-CE-L6-TestFull [sambs]	0.000	0.000	0.000	0.000
C4-CE-L12-TestFull [sambs]	0.000	0.000	0.000	0.000

Table 6. Change results for the SciRetrieval task in comparison to the snapshot 2024-11. The results are sorted by RI. Darker colors indicate less change or higher effectiveness.

Approach	ARP	AARP	RI	DRI	ER
fusion-with-core [2]	0.059	0.060	0.029	0.108	1.523
monot5-in-core [2]	0.031	0.029	-0.120	0.025	1.312
ows-bm25 [2]	0.218	0.191	-0.339	-0.058	1.673
ows-cluster-boosting [2]	0.222	0.193	-0.363	-0.080	1.816
rm3-on-qrel-boost [2]	0.253	0.201	-0.680	-0.351	5.257
A3-TFIDF-TestAbs [sambs]	0.137	0.102	-1.028	-0.294	0.585
A4-TFIDF-TestFull [sambs]	0.134	0.100	-1.028	-0.287	0.619
SciBERT_LongEVAL [long-eval-sci-group-5]	0.059	0.043	-1.132	-0.137	1.057
bm25+reranker+weighted [tf-idk]	0.176	0.123	-1.482	-0.494	-0.048
B3-SBERT-V1-TestAbs [sambs]	0.060	0.042	-1.517	-0.170	1.051
bm25+reranker [tf-idk]	0.183	0.126	-1.663	-0.553	-0.165
BM25_k1_0p95_b_0p75_stop_stem_fullText [academy-retrievals]	0.290	0.197	-1.817	-0.920	-3.696
BM25_k1_0p2_b_0p8_stop_stem_fullText [academy-retrievals]	0.291	0.196	-1.862	-0.935	-3.569
BM25_k1_1p0_b_0p7_stop_stem_fullText [academy-retrievals]	0.292	0.197	-1.865	-0.939	-3.628
B4-SBERT-V2-MPNet-TestFull [sambs]	0.068	0.046	-1.867	-0.218	0.978
B3-SBERT-V2-MPNet-TestAbs [sambs]	0.067	0.045	-1.892	-0.215	0.983
BM25_ColBERTv2_ReRanker_v1 [academy-retrievals]	0.177	0.119	-1.916	-0.575	-0.031
qrel-boost-core [2]	0.231	0.146	-2.789	-0.888	-0.800
query-intent-fusion [2]	0.231	0.146	-2.789	-0.888	-0.800

Table 7. Short-term change results for the WebRetrieval task in comparison to the snapshot 2023-03. The results are sorted by RI for the 2023-05 snapshot. Darker colors indicate less change or higher effectiveness.

Approach	ARP	MARP	RI	DRI	ER
run2 [23]	0.385	0.436	0.210	0.400	0.448
run3 [23]	0.394	0.438	0.182	0.351	0.507
run1 [23]	0.455	0.501	0.168	0.372	0.613
JMFT_Bert [9]	0.372	0.390	0.086	0.164	0.637
query-variants-qrel-boost-kmeans [34]	0.447	0.467	0.083	0.191	0.755
JMFT_Jaccard [9]	0.375	0.391	0.079	0.154	0.661
JMFT_lenght [9]	0.375	0.391	0.079	0.154	0.661
query-variants-relevance-feedback-kmeans [34]	0.265	0.275	0.064	0.093	18.201
baseline-qrel-boost [34]	0.501	0.518	0.063	0.173	0.828
query-variants-qrel-boost-dbscan [34]	0.402	0.414	0.057	0.130	0.762
query-variants-qrel-boost [34]	0.402	0.414	0.057	0.130	0.762
baseline-relevance-feedback [34]	0.426	0.430	0.021	0.078	0.878
dense-baseline+multilingual [air5]	0.034	0.034	0.019	0.006	1.039
seupd-2425-basette [8]	0.279	0.281	0.013	0.043	5.459
query-variants-relevance-feedback-dbscan [34]	0.305	0.306	0.008	0.042	0.470
bm25-baseline+traditional [air5]	0.170	0.170	0.007	0.023	1.105
clef25-seupd2425-rise [22]	0.485	0.487	0.007	0.064	0.939
datahunter-web-20250430-run1 [33]	0.326	0.327	0.006	0.043	0.709
fair_schaer [9]	0.276	0.277	0.006	0.036	2.326
molten-marsanne [9]	0.327	0.328	0.004	0.040	0.750
seupd2425-rand-ICU-run [6]	0.220	0.220	-0.001	0.023	1.145
seupd2425-rand-elision+synonyms0.5 [6]	0.198	0.198	-0.002	0.020	1.106
dsgt-bm25-rerank-submission [32]	0.416	0.415	-0.003	0.042	0.936
seupd2425-sard-minAnaBool1000-GPT [14]	0.311	0.310	-0.003	0.031	0.643
seupd2425-sard-minAnaPhr300 [14]	0.358	0.357	-0.005	0.033	0.891
seupd2425-sard-minAnaMixed3-GPT [14]	0.387	0.385	-0.010	0.029	0.943
seupd2425-rand-frenchfilter-run [6]	0.210	0.209	-0.010	0.015	1.097
seupd2425-sard-minAnaMixed [14]	0.378	0.375	-0.011	0.026	0.939
rr-ps [12]	0.415	0.412	-0.014	0.025	0.970
run4 [23]	0.385	0.381	-0.018	0.018	0.975
ps [12]	0.406	0.402	-0.019	0.018	0.985
seupd2425-rand-DL-run [6]	0.205	0.203	-0.020	0.008	1.068
query_expansion_time_dependence [9]	0.294	0.291	-0.023	0.009	-0.035
ci [12]	0.366	0.362	-0.024	0.010	0.989
mein-cleveres-system [9]	0.294	0.291	-0.024	0.008	0.000
dsgt-bm25-submission [32]	0.345	0.340	-0.025	0.007	0.986
clef-webbusters-traditional [web-busters]	0.292	0.288	-0.026	0.006	4.277
searchill_YYYY-MM [searchill]	0.306	0.302	-0.030	0.002	0.973
ps-prf [12]	0.354	0.349	-0.031	0.001	1.025
ci-prf-ps [12]	0.364	0.358	-0.033	-0.001	1.034
bm25_boosted_temporal_categories [9]	0.232	0.226	-0.054	-0.016	0.955
luyuHF [air5]	0.165	0.157	-0.108	-0.038	1.049

Table 8. Long-term change results for the WebRetrieval task in comparison to the snapshot 2023-03. The results are sorted by RI for the 2023-08 snapshot. Darker colors indicate less change or higher effectiveness.

Approach	ARP	MARP	RI	DRI	ER
run2 [23]	0.320	0.403	0.344	0.456	0.318
run1 [23]	0.371	0.459	0.322	0.466	0.440
run3 [23]	0.328	0.405	0.320	0.407	0.369
query-variants-qrel-boost-kmeans [34]	0.342	0.415	0.298	0.370	0.429
baseline-qrel-boost [34]	0.376	0.456	0.297	0.402	0.484
query-variants-qrel-boost [34]	0.309	0.368	0.276	0.286	0.379
query-variants-qrel-boost-dbscan [34]	0.309	0.368	0.276	0.286	0.379
JMFT_Jaccard [9]	0.299	0.353	0.267	0.259	0.353
baseline-relevance-feedback [34]	0.319	0.377	0.266	0.275	0.425
JMFT_Bert [9]	0.301	0.354	0.260	0.249	0.374
JMFT_lenght [9]	0.301	0.354	0.260	0.249	0.374
query-variants-relevance-feedback-kmeans [34]	0.217	0.250	0.235	0.145	24.220
clef25-seupd2425-rise [22]	0.383	0.436	0.217	0.215	0.626
query-variants-relevance-feedback-dbscan [34]	0.244	0.276	0.205	0.121	-0.522
seupd-2425-basette [8]	0.228	0.255	0.196	0.101	9.776
dsgt-bm25-rerank-submission [32]	0.337	0.376	0.188	0.137	0.628
datahunter-web-20250430-run1 [33]	0.267	0.298	0.188	0.107	0.217
seupd2425-sard-minAnaBool1000-GPT [14]	0.252	0.281	0.187	0.100	-0.130
seupd2425-rand-ICU-run [6]	0.180	0.200	0.182	0.067	1.177
molten-marsanne [9]	0.270	0.299	0.179	0.096	0.325
seupd2425-sard-minAnaMixed3-GPT [14]	0.315	0.349	0.177	0.109	0.611
seupd2425-sard-minAnaMixed [14]	0.310	0.341	0.171	0.097	0.614
rr-ps [12]	0.340	0.375	0.169	0.103	0.683
fair_schaer [9]	0.231	0.255	0.169	0.070	3.130
seupd2425-rand-elision+synonyms0.5 [6]	0.165	0.181	0.166	0.048	1.055
seupd2425-rand-frenchfilter-run [6]	0.174	0.191	0.163	0.048	1.074
ps [12]	0.334	0.366	0.161	0.089	0.695
dsgt-bm25-submission [32]	0.283	0.309	0.159	0.071	0.540
seupd2425-sard-minAnaPhr300 [14]	0.300	0.328	0.157	0.074	0.579
ci [12]	0.302	0.330	0.155	0.071	0.645
run4 [23]	0.320	0.349	0.154	0.074	0.693
ci-prf-ps [12]	0.299	0.326	0.153	0.067	0.642
seupd2425-rand-DL-run [6]	0.171	0.186	0.149	0.035	1.023
ps-prf [12]	0.294	0.319	0.144	0.054	0.660
searchill_YYYY-MM [searchill]	0.258	0.277	0.132	0.033	0.166
bm25-baseline+traditional [air5]	0.149	0.160	0.127	0.015	0.934
clef-webbusters-traditional [web-busters]	0.253	0.269	0.112	0.008	4.657
mein-cleveres-system [9]	0.255	0.271	0.111	0.008	0.000
query_expansion_time_dependence [9]	0.256	0.272	0.111	0.007	0.147
bm25_boosted_temporal_categories [9]	0.206	0.213	0.064	-0.035	0.760
luyuHF [air5]	0.141	0.145	0.053	-0.030	0.842
dense-baseline+multilingual [air5]	0.073	0.054	-1.129	-0.165	0.727

Table 9. Change results for the SciRetrieval task in comparison to the snapshot 2024-11. The results are sorted by RI. Darker colors indicate less change or higher effectiveness.

Approach	ARP	AARP	RI	DRI	ER
fusion-with-core [2]	0.059	0.060	0.029	0.108	1.523
monot5-in-core [2]	0.031	0.029	-0.120	0.025	1.312
ows-bm25 [2]	0.218	0.191	-0.339	-0.058	1.673
ows-cluster-boosting [2]	0.222	0.193	-0.363	-0.080	1.816
rm3-on-qrel-boost [2]	0.253	0.201	-0.680	-0.351	5.257
A3-TFIDF-TestAbs [sambs]	0.137	0.102	-1.028	-0.294	0.585
A4-TFIDF-TestFull [sambs]	0.134	0.100	-1.028	-0.287	0.619
SciBERT_LongEVAL [long-eval-sci-group-5]	0.059	0.043	-1.132	-0.137	1.057
bm25+reranker+weighted [tf-idk]	0.176	0.123	-1.482	-0.494	-0.048
B3-SBERT-V1-TestAbs [sambs]	0.060	0.042	-1.517	-0.170	1.051
bm25+reranker [tf-idk]	0.183	0.126	-1.663	-0.553	-0.165
BM25_k1_0p95_b_0p75_stop_stem_fullText [academy-retrievals]	0.290	0.197	-1.817	-0.920	-3.696
BM25_k1_0p2_b_0p8_stop_stem_fullText [academy-retrievals]	0.291	0.196	-1.862	-0.935	-3.569
BM25_k1_1p0_b_0p7_stop_stem_fullText [academy-retrievals]	0.292	0.197	-1.865	-0.939	-3.628
B4-SBERT-V2-MPNet-TestFull [sambs]	0.068	0.046	-1.867	-0.218	0.978
B3-SBERT-V2-MPNet-TestAbs [sambs]	0.067	0.045	-1.892	-0.215	0.983
BM25_ColBERTv2_ReRanker_v1 [academy-retrievals]	0.177	0.119	-1.916	-0.575	-0.031
qrel-boost-core [2]	0.231	0.146	-2.789	-0.888	-0.800
query-intent-fusion [2]	0.231	0.146	-2.789	-0.888	-0.800

6 Discussion

This year, more teams than ever registered for the LongEval Lab. In total, 56 teams registered at CLEF, and 55 teams also registered at TIRA to submit an approach. Of those, 19 teams submitted at least one approach, which is the highest number of teams to date. 68 approaches were submitted, almost twice as many as last year, but still six submissions fewer compared to the first iteration in 2023. Twelve notebook papers were submitted, while 14 were received in 2023 and nine in 2024. For the first time, LongEval incorporated a new retrieval task: SciRetrieval. As expected, this task received fewer submissions compared to the established WebRetrieval task. Besides being new, this might also be because fewer snapshots are available so far, an important factor for temporal systems. Nevertheless, the task received many interesting approaches and was generally well received.

The new task is a big step, as longitudinal evaluation is not primarily a web search-related task but is imminent in very different retrieval settings. Therefore, a task designed for a new and disjoint dataset from a different retrieval domain is a very fruitful addition to LongEval. SciRetrieval originates from the CORE platform, one of the largest scientific retrieval systems worldwide, with currently 441 million publicly available research papers indexed. This dataset is especially interesting because it covers the unique dynamics of a very different retrieval setting that can be described as academic search. While some other labs also used scientific papers to create a dataset or retrieval task, like the TREC domain-specific cross-language IR task [29] or the LiLAS lab at CLEF [37], none of them included a developing test collection with different time frames and snapshots. The only academic search dataset that allows for studying longitudinal effects to some extent is TREC-COVID [41].

Academic search and our new collaboration with CORE enables the community to create a test collection and shared task that includes new interaction paradigms, like faceted search or queries using classic patterns like block strategies or specific scientific search stratagems such as "Footnote Chasing", "Citation Searching", 'Keyword Searching", "Author Searching" and "Journal Run" as described in the seminal paper of Marcia Bates in 1990 [7]. These stratagems enable going beyond simple ad-hoc search settings and are a step toward evaluating interactive retrieval settings and search sessions.

Both tasks follow the same two main objectives of the lab: To explore how the retrieval effectiveness drops over time and to develop systems resilient to that. Regarding the drop in effectiveness, this year the focus shifted towards measuring changes. In the first iteration of the lab in 2023, the effectiveness improved on average, while in 2024 it actually dropped [4,5]. Overall, this year the effectiveness is only slightly decreasing, although last years datasets were also based on the snapshots 2023-06 and 2023-08. These observations highlight the importance of the reference point to which the results are compared.

Specifically, as shown in Table 3, a clear decline in nDCG@10 can be observed across systems from snapshot 2023-03 to 2023-08 in most approaches. This aligns with the document overlap matrix in Fig. 1, which illustrates that later snapshots contain increasingly different content compared to earlier ones (i.e. starting from 2023-06). This decreasing overlap suggests a shifting document distribution over time, which presents a major challenge for static systems. This highlights again the importance of designing retrieval models that remain effective across temporally distant test collections.

It was observed again how the most effective approaches do not necessarily align with the approaches with the fewest change. This motivates us to continue this objective and research more complete definitions of temporal robustness and persistence. Investigations in this regard were also adopted from some teams as complementary evaluations in their notebook papers. In contrast regarding the second objective, improving the retrieval effectiveness remained the main motivation for all teams and no team specifically focused on developing a system with an especially stable effectiveness.

The submitted approaches increasingly rely on past snapshots and approaches that directly use previous qrels show to be highly effective especially in the WebRetrieval task where a long history of snapshots is available. While many interesting relevance signals are developed from this data, no team submitted a retrieval model that was directly trained or fine-tuned on prior snapshots. This remains as an interesting direction for future work.

7 Conclusion

We have given an overview of the LongEval 2025 Lab, describing briefly the data provided for the two tasks, the participant submissions, and the carried out evaluation. Compared to the previous years, participants have designed a variety of solutions to handle the temporal aspects of the given document collections. The approaches adopted by the participants to LongEval 2025 highlight the fast-paced evolution of IR systems. While established retrieval methods such as BM25 and Lucene remain main building blocks, we notice a shift toward integrating advanced methods like chunk-based indexing, neural reranking, and query expansion powered by large language models (LLMs). Some teams prioritized computational efficiency and broad accessibility, optimizing classical IR frameworks for use on standard hardware with multithreaded processing. Others combined neural solutions, incorporating statistical analyses and topic modeling to more effectively address the temporal and semantic challenges presented by contemporary queries. Retrieval effectiveness was measured using nDCG and nDCG@10, looking at how well systems ranked relevant documents in the top results for the WebRetrieval task. Only the best or latest versions of similar submissions were considered in our analysis. The best performing approaches to the WebRetrieval task remained stable over time, with Team RISE leading in later snapshots. For the SciRetrieval task, leading approaches varied by snapshot, but system rankings were generally consistent across evaluation measures.

We also analysed the changes in retrieval effectiveness changes over time to identify which systems remain robust as the collections to search in evolve. We used Relative Improvement (RI), Delta Relative Improvement (DRI), and Effect Ratio (ER) metrics, which we compared to the BM25 baseline. We claim that these measures help isolate the effect of system changes from general shifts in the search environment. Results show that while many systems improved in the short-term setting, only a few maintained or even enhanced their effectiveness long-term, with most changes being very small. For the Sciretrieval task, the persistence and improvement of system effectiveness varied depending on the metric used, with some systems showing high robustness (ER close to 1), though the different measures did not always agree on which systems performed best over time.

Acknowledgments. This work is supported by the ANR Kodicare bi-lateral project, grant ANR-19-CE23-0029 of the French Agence Nationale de la Recherche, the Austrian Science Fund (FWF, grant I4471-N), the UKRI/EPSRC Turing AI Fellowship to Maria Liakata (grant no. EP/V030302/1), the Ministry of Education, Youth and Sports of the Czech Republic, Project No. LM2023062 LINDAT/CLARIAH-CZ, and the German Research Foundation (DFG) through project grant No. 407518790. This work has been using services provided by the LINDAT/CLARIAH-CZ Research Infrastructure (https://lindat.cz), supported by the Ministry of Education, Youth and Sports of the Czech Republic (Project No. LM2023062).

Disclosure of Interests. The authors have no competing interests to declare that are relevant to the content of this article.

References

1. Adar, E., Teevan, J., Dumais, S.T., Elsas, J.L.: The web changes everything: understanding the dynamics of web content. In: WSDM, pp. 282–291. ACM (2009)
2. Alexander, D., et al.: Team OpenWebSearch at CLEF 2025: LongEval. In: Faggioli, G., Ferro, N., Rosso, P., Spina, D. (eds.) Working Notes of CLEF 2025 – Conference and Labs of the Evaluation Forum, CEUR Workshop Proceedings (2025)
3. Alexander, D., et al.: Team openwebsearch at CLEF 2024: Longeval. In: CLEF (Working Notes). CEUR Workshop Proceedings, vol. 3740, pp. 2304–2313. CEUR-WS.org (2024)
4. Alkhalifa, R., et al.: Overview of the clef-2023 longeval lab on longitudinal evaluation of model performance. In: Experimental IR Meets Multilinguality, Multimodality, and Interaction. Proceedings of the Fourteenth International Conference of the CLEF Association (CLEF 2023). Lecture Notes in Computer Science (LNCS), Springer, Thessaliniki, Greece (2023)
5. Alkhalifa, R., et al.: Overview of the CLEF 2024 LongEval Lab on Longitudinal Evaluation of Model Performance. In: Goeuriot, L., et al. (eds.) Experimental IR Meets Multilinguality, Multimodality, and Interaction. Proceedings of the Fifteenth International Conference of the CLEF Association (CLEF 2024). Lecture Notes in Computer Science (LNCS), Springer, Heidelberg, Germany (2024)
6. Amato, G., et al.: Team RAND at LongEval 2025: composable information retrieval with semantic and language-aware components. In: Faggioli, G., Ferro, N., Rosso, P., Spina, D. (eds.) Working Notes of CLEF 2025 – Conference and Labs of the Evaluation Forum, CEUR Workshop Proceedings (2025)
7. Bates, M.J.: Where should the person stop and the information search interface start? Inf. Process. Manag. **26**(5), 575–591 (1990). https://doi.org/10.1016/0306-4573(90)90103-9
8. Bottari, A., Croce, L., Abadi, F.M.H., Ferro, N.: SEUPD@CLEF: team BASETTE on an IR system for basic hardware. In: Faggioli, G., Ferro, N., Rosso, P., Spina, D. (eds.) Working Notes of CLEF 2025 – Conference and Labs of the Evaluation Forum, CEUR Workshop Proceedings (2025)
9. Braun, F., et al.: CIR at LongEval 2025: exploring temporal sensitivity in web retrieval. In: Faggioli, G., Ferro, N., Rosso, P., Spina, D. (eds.) Working Notes of CLEF 2025 – Conference and Labs of the Evaluation Forum, CEUR Workshop Proceedings (2025)
10. Breuer, T., Ferro, N., Fuhr, N., Maistro, M., Sakai, T., Schaer, P., Soboroff, I.: How to measure the reproducibility of system-oriented IR experiments. In: Huang, J.X., et al. (eds.) Proceedings of the 43rd International ACM SIGIR conference on research and development in Information Retrieval, SIGIR 2020, Virtual Event, China, July 25-30, 2020, pp. 349–358. ACM (2020). https://doi.org/10.1145/3397271.3401036
11. Breuer, T., Keller, J., Schaer, P.: ir_metadata: an extensible metadata schema for IR experiments. In: Amigó, E., Castells, P., Gonzalo, J., Carterette, B., Culpepper, J.S., Kazai, G. (eds.) SIGIR '22: The 45th International ACM SIGIR Conference on Research and Development in Information Retrieval, Madrid, Spain, July 11 - 15, 2022, pp. 3078–3089. ACM (2022). https://doi.org/10.1145/3477495.3531738
12. Bruttomesso, A., Cavazza, D., Corrò, A., Peraro, S., Seghetto, D., Ferro, N.: SEUPD@CLEF: team 3DS2A on performance evaluation over time of IR systems with proximity search and reranking components. In: Faggioli, G., Ferro, N., Rosso, P., Spina, D. (eds.) Working Notes of CLEF 2025 – Conference and Labs of the Evaluation Forum, CEUR Workshop Proceedings (2025)

13. Cancellieri, M., et al.: Longeval at clef 2025: longitudinal evaluation of IR model performance. In: Hauff, C., et al. (eds.) Advances in Information Retrieval, pp. 382–388. Springer Nature Switzerland, Cham (2025)

14. Caon, D., Maschio, R.D., Disarò, A., Maule, S., Ferro, N.: SARD at LongEval 2025: on longitudinal evaluation of IR systems by using query rewriting and hybrid queries. In: Faggioli, G., Ferro, N., Rosso, P., Spina, D. (eds.) Working Notes of CLEF 2025 – Conference and Labs of the Evaluation Forum, CEUR Workshop Proceedings (2025)

15. Chapelle, O., Zhang, Y.: A dynamic Bayesian network click model for web search ranking. In: Proceedings of the 18th International Conference on World Wide Web, pp. 1–10. WWW '09, Association for Computing Machinery, New York, NY, USA (2009). https://doi.org/10.1145/1526709.1526711

16. Chuklin, A., Markov, I., Rijke, M.D.: Click models for web search. Synth. Lect. Inf. Concepts, Retrieval, Serv. **7**(3), 1–115 (2015). https://doi.org/10.2200/S00654ED1V01Y201507ICR043

17. Dumais, S.T.: Putting searchers into search. In: SIGIR, pp. 1–2. ACM (2014)

18. Fink, T., et al.: Longeval 2025 web retrieval collection (2025). https://doi.org/10.48436/d987t-2qf34

19. Fink, T., et al.: Longeval 2025 core retrieval test collection (2025). https://doi.org/10.48436/kd962-hym06

20. Fink, T., et al.: Longeval 2025 core retrieval train collection (2025). https://doi.org/10.48436/st7d6-wbe25

21. Fröbe, M., et al.: Continuous integration for reproducible shared tasks with TIRA.io. In: Kamps, J., et al. (eds.) Advances in Information Retrieval. 45th European Conference on IR Research (ECIR 2023), pp. 236–241. Lecture Notes in Computer Science, Springer, Berlin Heidelberg New York (2023). https://doi.org/10.1007/978-3-031-28241-6_20

22. Furlan, D., et al.: SEUPD@CLEF: team RISE on improving search by crafting titles and matching URLs. In: Faggioli, G., Ferro, N., Rosso, P., Spina, D. (eds.) Working Notes of CLEF 2025 – Conference and Labs of the Evaluation Forum, CEUR Workshop Proceedings (2025)

23. Gaio, G., Mazzarotto, F., Meneghin, M., Saro, E., Visonà, F.: SEUPD2425-RACOON at LongEval 2025: a novel approach to Information Retrieval with LLM-based query expansion and temporal relevance feedback techniques. In: Faggioli, G., Ferro, N., Rosso, P., Spina, D. (eds.) Working Notes of CLEF 2025 – Conference and Labs of the Evaluation Forum, CEUR Workshop Proceedings (2025)

24. Galuščáková, P., et al.: Longeval-retrieval: French-English dynamic test collection for continuous web search evaluation (2023)

25. Kanhabua, N., Blanco, R., Nørvåg, K.: Temporal information retrieval. Found. Trends Inf. Retr. **9**(2), 91–208 (2015)

26. Keller, J., Breuer, T., Schaer, P.: Evaluation of temporal change in IR test collections. In: Oosterhuis, H., Bast, H., Xiong, C. (eds.) Proceedings of the 2024 ACM SIGIR International Conference on Theory of Information Retrieval, ICTIR 2024, Washington, DC, USA, 13 July 2024, pp. 3–13. ACM (2024). https://doi.org/10.1145/3664190.3672530

27. Keller, J., Breuer, T., Schaer, P.: Leveraging prior relevance signals in web search. In: CLEF (Working Notes). CEUR Workshop Proceedings, vol. 3740, pp. 2396–2406. CEUR-WS.org (2024)

28. Keller, J., Fröbe, M., Hendriksen, G., Alexander, D., Potthast, M., Schaer, P.: Simplified longitudinal retrieval experiments: a case study on query expansion and document boosting. In: Experimental IR Meets Multilinguality, Multimodality, and Interaction - 16th International Conference of the CLEF Association, CLEF 2024, Madrid, Spain, September 9-12, 2025, Proceedings, Part I. Lecture Notes in Computer Science, Springer (2025)

29. Kluck, M., Gey, F.C.: The domain-specific task of CLEF - specific evaluation strategies in cross-language information retrieval. In: Peters, C. (ed.) Cross-Language Information Retrieval and Evaluation, Workshop of Cross-Language Evaluation Forum, CLEF 2000, Lisbon, Portugal, September 21-22, 2000, Revised Papers. Lecture Notes in Computer Science, vol. 2069, pp. 48–56. Springer, Berlin, Heidelberg (2000). https://doi.org/10.1007/3-540-44645-1_5

30. Liu, Y.A., Zhang, R., Guo, J., de Rijke, M., Fan, Y., Cheng, X.: Robust neural information retrieval: an adversarial and out-of-distribution perspective (2024). https://arxiv.org/abs/2407.06992

31. MacAvaney, S., Yates, A., Feldman, S., Downey, D., Cohan, A., Goharian, N.: Simplified data wrangling with ir_datasets. In: SIGIR, pp. 2429–2436. ACM (2021)

32. Miyaguchi, A., Afrulbasha, I., Pramov, A.: DS@GT at LongEval: evaluating temporal performance in web search systems and topics with two-stage retrieval. In: Faggioli, G., Ferro, N., Rosso, P., Spina, D. (eds.) Working Notes of CLEF 2025 – Conference and Labs of the Evaluation Forum, CEUR Workshop Proceedings (2025)

33. Mukhtar, A., Leonardo, P., Zaccarin, F., Shen, Z., Ferro, N.: SEUPD@CLEF: Team [DataHunter] on temporal stability analysis of Boolean and CamemBERT-based retrieval systems. In: Faggioli, G., Ferro, N., Rosso, P., Spina, D. (eds.) Working Notes of CLEF 2025 – Conference and Labs of the Evaluation Forum, CEUR Workshop Proceedings (2025)

34. Ndiema, A.M., Keller, J., Schaer, P.: LongEval: CIR_cluster at LongEval 2025: clustering query variants for temporal generalization. In: Faggioli, G., Ferro, N., Rosso, P., Spina, D. (eds.) Working Notes of CLEF 2025 – Conference and Labs of the Evaluation Forum, CEUR Workshop Proceedings (2025)

35. Roberts, K., et al.: Searching for scientific evidence in a pandemic: an overview of TREC-COVID. J. Biomed. Inform. **121**, 103865 (2021)

36. Robertson, S., Walker, S., Jones, S., Hancock-Beaulieu, M., Gatford, M.: Okapi at TREC-3 (1994)

37. Schaer, P., Breuer, T., Castro, L.J., Wolff, B., Schaible, J., Tavakolpoursaleh, N.: Overview of lilas 2021 - living labs for academic search. In: Candan, K.S., et al. (eds.) Experimental IR Meets Multilinguality, Multimodality, and Interaction - 12th International Conference of the CLEF Association, CLEF 2021, Virtual Event, September 21-24, 2021, Proceedings. Lecture Notes in Computer Science, vol. 12880, pp. 394–418. Springer, Cham (2021). https://doi.org/10.1007/978-3-030-85251-1_25

38. Stryszewski, J., Prosowicz, W., Kawiak, T., Jaśkowiec, A.: Improving scientific information retrieval with dense representations and cross-encoder re-ranking. In: Faggioli, G., Ferro, N., Rosso, P., Spina, D. (eds.) Working Notes of CLEF 2025 – Conference and Labs of the Evaluation Forum, CEUR Workshop Proceedings (2025)

39. Tikhonov, A., Bogatyy, I., Burangulov, P., Ostroumova, L., Koshelev, V., Gusev, G.: Studying page life patterns in dynamical web. In: SIGIR, pp. 905–908. ACM (2013)

40. Urbano, J., Lima, H., Hanjalic, A.: A new perspective on score standardization. In: Proceedings of the 42nd International ACM SIGIR Conference on Research and Development in Information Retrieval, pp. 1061–1064 (2019)
41. Voorhees, E.M., et al.: TREC-COVID: constructing a pandemic information retrieval test collection. SIGIR Forum **54**(1), 1:1–1:12 (2020). https://doi.org/10.1145/3451964.3451965

Overview of PAN 2025: Voight-Kampff Generative AI Detection, Multilingual Text Detoxification, Multi-author Writing Style Analysis, and Generative Plagiarism Detection

Janek Bevendorff[1,13](✉), Daryna Dementieva[2], Maik Fröbe[3], Bela Gipp[4], André Greiner-Petter[4], Jussi Karlgren[5], Maximilian Mayerl[6], Preslav Nakov[7], Alexander Panchenko[8], Martin Potthast[9,10,11], Artem Shelmanov[7], Efstathios Stamatatos[12], Benno Stein[13], Yuxia Wang[7], Matti Wiegmann[13], and Eva Zangerle[14]

[1] Leipzig University, Leipzig, Germany
[2] Technical University of Munich, Munich, Germany
[3] Friedrich Schiller University Jena, Jena, Germany
[4] Georg-August-Universität, Göttingen, Germany
[5] University of Helsinki, Helsinki, Finland
[6] University of Applied Sciences BFI Vienna, Vienna, Austria
[7] Mohamed bin Zayed University of Artificial Intelligence, Abu Dhabi, UAE
[8] Skoltech & AIRI, Moscow, Russia
[9] University of Kassel, Kassel, Germany
[10] hessian.ai, Darmstadt, Germany
[11] ScaDS.AI, Leipzig, Germany
[12] University of the Aegean, Samos, Greece
[13] Bauhaus-Universität Weimar, Weimar, Germany
[14] University of Innsbruck, Innsbruck, Austria
pan@webis.de, https://pan.webis.de/

Abstract. The goal of the PAN lab is to advance the state of the art in text forensics and stylometry through an objective evaluation of new and established methods on new benchmark datasets. In 2025, we organized four shared tasks: (1) generative AI detection, particularly in mixed and obfuscated authorship scenarios, (2) multilingual text detoxification, a continued task that aims re-formulate text in a non-toxic way for multiple languages, and (3) multi-author writing style analysis, a continued task that aims to find positions of authorship change, and (4) generative plagiarism detection, a new task that targets source retrieval and text alignment between generated text and source documents. PAN 2025 concluded successfully with 56 notebook papers.

1 Introduction

PAN is a workshop series and a networking initiative for stylometry and digital text forensics. PAN hosts computational shared tasks on authorship analy-

J. Carrillo-de-Albornoz et al. (Eds.): CLEF 2025, LNCS 16089, pp. 388–411, 2026.
https://doi.org/10.1007/978-3-032-04354-2_21

sis, computational ethics, and the originality of writing. Since the workshop's inception in 2007, we organized 77 shared tasks[1] and assembled 60 evaluation datasets[2] plus nine datasets contributed by the community. In 2025, we organized four tasks that concluded in 57 notebook papers.

First, the *Voight-Kampff Generative AI Detection* task asks to distinguish between human and machine-written text, with a focus on detector sensitivity in the presence of obfuscation and mixed human-machine authorship. The subtask 1 continues the research from 2024 in collaboration with the ELOQUENT lab and frames AI detection as an authorship verification task, tested across a large number of domains and obfuscation techniques to test detector robustness. The subtask 2 asks to distinguish between 6 different forms of human-AI collaboration in a given document, ranging from fully human-written to text with deep AI intervention. The Voight-Kampff Generative AI Detection task resulted in 30 notebook submissions. The task details are described in Sect. 2.

Second, the continuation of the *Multilingual Text Detoxification* task asks to, given a toxic piece of text, re-write it in a non-toxic way while saving the main content as much as possible. The task was extended to include texts from 15 languages—adding to 2024 edition Italian, French, Hebrew, Hinglish, Japanese, and Tatar—and had cross-lingual and multilingual as well as supervised and unsupervised challenges. The Multilingual Text Detoxification task resulted in 12 notebook submissions. The task details are described in Sect. 3.

Third, the continuation of the *Multi-Author Writing Style Analysis* task asks to, given a document, determine at which positions the author changes. This task was revamped for 2023 with a new dataset and structured around topical heterogeneity as an indicator of difficulty. While the previous iterations asked to separate authors at a paragraph level, we increased the difficulty for this year and asked participants to separate at the sentence level. The Multi-Author Writing Style Analysis task resulted in 11 notebook submissions. The task details are described in Sect. 4.

Fourth, the new *Generated Plagiarism Detection* task asks to, given a source and an LLM-obfuscated, suspicious document, determine the positions where the suspicious document reuses text from the source. The task resulted in 3 notebook submissions. The task details are described in Sect. 5.

PAN is committed to reproducible research in IR and NLP, hence all participants are asked to submit their software (instead of just their predictions) through the submission software TIRA. With the recent updates to the TIRA platform [30], a majority of the submissions to PAN are publicly available as docker containers. In the following sections, we briefly outline the 2025 tasks and their results.

---

[1] Find PAN's past shared tasks at https://pan.webis.de/shared-tasks.html.
[2] Find PAN's datasets at https://pan.webis.de/data.html.

Input / Task	Possible Assignment Patterns

Input / Task

1. { $\boxed{?}$, $\boxed{?}$ }

2. { $\boxed{?}$, $\boxed{?}$ }
3. { $\boxed{?}$, $\boxed{?}$ } $\longrightarrow$
4. { $\boxed{?}$, $\boxed{?}$ }
5. { $\boxed{?}$, $\boxed{?}$ }
6. { $\boxed{?}$, $\boxed{?}$ }

7. $\boxed{?}$

Possible Assignment Patterns

1. { $\boxed{A}$, $\boxed{M}$ }

2. { $\boxed{A}$, $\boxed{M}$ }, { $\boxed{A}$, $\boxed{A}$ }
3. { $\boxed{A}$, $\boxed{M}$ }, { $\boxed{M}$, $\boxed{M}$ }
4. { $\boxed{A}$, $\boxed{M}$ }, { $\boxed{A}$, $\boxed{A}$ }, { $\boxed{M}$, $\boxed{M}$ }
5. { $\boxed{A}$, $\boxed{M}$ }, { $\boxed{A}$, $\boxed{A}$ }, { $\boxed{A}$, $\boxed{B}$ }
6. { $\boxed{A}$, $\boxed{M}$ }, { $\boxed{A}$, $\boxed{A}$ }, { $\boxed{A}$, $\boxed{B}$ }, { $\boxed{M}$, $\boxed{M}$ }

7. $\boxed{A}$, $\boxed{M}$

Fig. 1. Hierarchy of authorship verification problems from "easiest" (1) to "hardest" (7), involving LLM-generated text. Ignoring mixed human and machine authorship, the difficulty arises from the pairing constraints imposed by the possible assignment patterns. $\boxed{M}$ denotes LLM-generated text, while $\boxed{A}$ and $\boxed{B}$ denote human-authored text (same letter meaning same human author).

2   Voight-Kampff Generative AI Detection

Authorship verification is a fundamental task in author identification. PAN has continuously been organizing authorship verification tasks for years [8–11] and with generative AI / LLM detection being fundamentally also an authorship verification task [15], decided to "delve" into that realm. So, in 2024 we offered, for the first time, the *"Voight-Kampff" Generative AI Authorship Verification* task [3,14], which attracted a large number of submissions.

For the 2024 installment, we formalized different task variants and ordered them from easiest to hardest (Fig. 1). To establish a baseline, we decided to start with the easiest variant, in which participants were given a pair of texts of which exactly one was of human and the other of machine origin. This year, we move on to the harder variant, in which participants are given only one text. This variant reflects a more realistic scenario of authorship verification "in the wild," aligning with the settings commonly addressed in other LLM detection shared tasks.

Moreover, we extend the task to two distinct subtasks: (1) The classic binary *"Voight-Kampff" AI Detection Sensitivity* task, and (2) a multi-class *Human-AI Collaborative Text Classification* task. The subtask 1 is organized in collaboration with the ELOQUENT Lab in a builder-breaker style similar to the previous year: PAN participants build systems to identify machine authorship, while ELOQUENT participants supply datasets to try to break the systems.

A more detailed description and analysis of the submissions and the results can be found in the joint PAN and ELOQUENT task overview paper [13].

2.1   Subtask 1: Voight-Kampff AI Detection Sensitivity

The subtask 1 is in essence the classic binary detection task known also from other LLM detection shared tasks. However, we are testing the limits of the detectors by crafting a test set with text "obfuscations" that try to evade detection. Apart from drastic text length restrictions, the obfuscations we tested or received from ELOQUENT participants in the previous year turned out to be mostly ineffective. So this year, we tested what happens when the human writers obfuscate their style and whether machines can replicate this.

Dataset. We created the task datasets from a selection of 19th-century English fiction from Project Gutenberg, as well as the extended Brennan-Greenstadt [19] and Riddell-Juola [93] corpora. The latter two were constructed by collecting existing essays and then asking the authors to write another text describing their neighborhood but, in doing so, try to conceal their identity. No further instructions were given on how to achieve that. To generate LLM versions for all texts, we used the same summarize-then-expand technique as last year by prompting GPT-4o to generate bullet-point summaries of the input texts. The model was instructed to extract the main topic, a list of key points, the narrative point of view, the grammatical tense, and certain apparent style or obfuscation markers. We then used 13 LLMs to replicate both the original essays and the obfuscations from the summaries and style instructions. In addition to the neighborhood prompts, we asked the LLMs to also generate texts in the style of a 7-year-old, in subject-object-verb "Yoda" grammar, or with alliterations. Further, we added random words to the prompts which we asked the model to ignore, and we increased the temperature to the highest value that still produced sensible text.

Participants were given a training and a validation split of the dataset, which included only the original human fiction and essay texts and plain LLM versions of them. The obfuscated texts (both human and LLM) were held back for the test set. Participants were allowed to use external training and validation data, including last year's training set. The test set included both obfuscated and unobfuscated texts, as well as a small subsample of human and LLM U.S. news articles from last year's test dataset (which we never published).

Baselines. We provided implementations of the following three baseline systems: As zero-shot baselines, we provided (1) Binoculars [36] (using Llama 3.1) and (2) a simple PPMd-based compression model using the compression-based cosine measure [35,77]. The operating points for both were tuned on the validation set that was handed out to participants. As a supervised baseline, (3) we trained a linear SVM on the top-1000 TF-IDF 1–4-grams from the validation set. The TF-IDF detector and Binoculars can be considered state of the art, the compression model marks a more conservative lower baseline.

Evaluation. All systems were submitted and evaluated on Tira [30]. At test time, the participants had to calculate a score between 0 and 1 for each text, indicating the likelihood that the text was LLM-generated. A score of exactly 0.5 could be given to signal a non-decision.

For each participant, we computed a confusion table and the following scores, which we used in previous authorship verification shared tasks as well:

Table 1. Arithmetic mean of all evaluation measures per submission for subtask 1.

Team	Score	System
Macko [59]	0.899	LoRA-tuned Qwen3 and data augmentation [60]
Seeliger [78]	0.880	Document-word correlations
Zaidi [99]	0.879	Fine-tuned BERT and data augmentation
Yang [98]	0.877	RoBERTa with contrastive learning
Teja [85]	0.874	Ensemble: Mixture of experts with PLMs
Marchitan [61]	0.872	Ensemble: LightGBM, XGBoost, Log. Regression, SVM with Qwen3 embeddings
Liu [57]	0.871	Ensemble: Fine-tuned PLM with contrastive loss
Valdez-V. [89]	0.869	Syntactic graphs and embeddings with GNNs
Voznyuk [92]	0.863	DeBERTa-v3 with multi-task learning (task, genre, model family classification)
TF-IDF SVM	0.856	*Baseline TF-IDF SVM*
Pudasaini [72]	0.852	Ensemble: SVM bagging of fine-tuned PLMs
Ostrower [67]	0.851	XGBoost with binoculars + stylometric features
Ochab [66]	0.844	LightGBM classifier with stylometric features
Völpel [90]	0.843	MLP with syntax n-gram features
Jimeno-G. [42]	0.838	Stacking ensemble with stylometric and word features
Sun [83]	0.835	Bi-CE [34] loss function + 25 stylometric features
Basani [6]	0.831	XGBoost classifier with token surprisal features
Titze [86]	0.827	Logistic regression on surprisal scores, entropy and JSD from two LLMs
Binoculars	0.818	*Baseline Binoculars Llama3.1 [36]*
Larson [50]	0.814	SVM with word and punctuation frequency features
Huang [38]	0.807	Fine-tuned RoBERTa + training data augmentation
Kumar [47]	0.788	Fine-tuned DistillBERT + stylometric features
PPMd CBC	0.758	*Baseline PPMd Compression-based Cosine [35, 77]*
Liang [53]	0.753	ModernBERT fine-tuning + loss-weighting based on example difficulty

- ROC-AUC: The area under the Receiver Operating Characteristic curve.
- BRIER: The complement of the Brier score (mean squared loss)
- C@1: A modified accuracy score that assigns non-answers (score = 0.5) the average accuracy of the remaining cases [68].
- F_1: The harmonic mean of precision and recall.
- $F_{0.5u}$: A modified $F_{0.5}$ measure (precision-weighted F measure) that treats non-answers (score = 0.5) as false negatives [12].
- MEAN: The arithmetic mean of all previous measures

Submitted Systems. We received 20 submissions of which 7 beat the strongest baseline (TF-IDF SVM) and 9 more beat the second-strongest baseline (Binoculars). Overall, most systems had quite high mean scores above 0.9 with the best approach being almost perfect at 0.991. Table 1 shows the ranking of all participating teams ordered by their systems' MEAN scores on the test set (excluding ELOQUENT submissions). If teams submitted multiple systems, only the highest score is shown. A more detailed break-down of how systems respond to individual obfuscations is described in the extended task overview paper [13].

Table 2. Subtask 2 training, development and test set distribution across six categories.

Label	Text Category	Train	Dev	Test
0	Fully human-written	75,270	12,330	34,509
1	Human-written, then machine-polished	95,398	12,289	43,154
2	Machine-written, then machine-humanized	91,232	10,137	25,234
3	Human-initiated, then machine-continued	10,740	37,170	22,802
4	Deeply-mixed text	14,910	225	12,500
5	Machine-written, then human-edited	1,368	510	2,557
Total		288,918	72,661	140,756

In total, this subtask attracted 20 teams to submit systems in addition to the baseline systems we provided. Table 1 shows the best-performing system of each team that submitted notebook papers and a brief description of their approach.

2.2 Subtask 2: Human-AI Collaboration

The integration of AI technologies into the writing process has significantly altered traditional notions of authorship. The line between human and AI contributions has become increasingly ambiguous. AI involvement increasingly rises from *none* to *complete* [39]. From the perspective of ethical and intellectual accountability, we identify the role of humans and AIs for six types of text. Given a document collaboratively authored by humans and AIs, the subtask 2 is to classify it into one of the following six categories:

 i. fully human-written;
 ii. human-written, then machine-polished;
iii. machine-written, then machine-humanized (obfuscated);
iv. human-initiated, then machine-continued;
 v. deeply mixed text; where some parts are written by a human and some are generated by a machine;
vi. machine-written, then human-edited.

Dataset. The training and validation sets were constructed from existing datasets for fine-grained machine-generated text detection, comprising 288,918 examples for training and 72,661 for validation. For constructing the test set, we collected student essays, research papers, and peer reviews. We also incorporated several newly released datasets to comprehensively evaluate the generalization of detection systems across unseen generators and domains. The result test set consists of 140,756 instances. Detailed data distribution across six categories is shown in Table 2.

Participants were given the training and development sets. Although they were not allowed to use external training and validation data, data augmentation strategies such as back-translation, synonym replacement, random word deletion, and replacement were allowed.

Table 3. Subtask 2 evaluation results of 22 submissions, ranking by macro-recall, along with macro-F1 and accuracy, with one delayed submission.

Rank	Team	Recall	F1	Acc	System Description
1	mdok [59]	64.46	65.06	74.09	QLoRa PEFT fine-tuned Qwen3-4B-Base.
2	Bohan Li [51]	61.72	61.73	69.28	Under-sample high-frequency classes and adopt data augmentation for underrepresented classes, along with R-Drop regularization for DeBERTa-v3-base fine-tuning.
3	Advacheck [92]	60.16	60.85	69.04	Shared Transformer Encoder between several classification heads trained to distinguish the domains.
4	StarBERT [108]	57.46	56.31	66.81	Combine the deep language understanding of DeBERTa-v3-large and the high-dimensional mapping ability of StarBlock2d.
5	Atu [96]	56.87	56.45	66.30	DeBERTa enhanced by contextual and geometric attention
6	TaoLi [52]	56.74	55.39	66.27	Use DeBERTa-v3-Large
7	ReText.Ai [40]	56.11	55.25	64.79	Fine-tune Gemma-2 2B for sequence classification with multiple classification heads.
8	DetectTeam [82]	54.49	54.40	62.89	Fine-tune DeBERTa-V3-Large and combining multi-scale features.
9	WeiDongWu [95]	54.09	53.57	63.01	Combine the contextual strength of BERT with the sequence modeling capabilities of Transformer layers.
10	zhangzhiliang [107]	54.06	52.81	61.65	Fine-tune DeBERTa-V3-Large and combine it with BiLSTM and attention mechanism.
11	CNLP-NITS-PP [85]	54.05	53.49	62.23	Soft and Hard Mixture of Experts (MoE) architectures with DeBERTa-V3-Large
12	a.dusuki	52.83	51.44	60.45	–
13	Steely [78]	52.14	51.81	59.88	Cumulative sum of token-Level correlation signals
14	a.elnenaey	49.56	50.10	58.96	–
	Baseline	48.32	47.82	57.09	Fine-tune RoBerTa
15	VerbaNex AI [32]	47.15	47.15	56.24	Fine-tune Roberta with class balancing, data augmentation, and calculation of specific weights for each unbalanced class.
16	Unibuc-NLP [61]	44.33	42.76	51.42	Combine features at different layers extracted using Transformers with layer-wise projection and attentive pooling.
	Nexus Interrogators [99]	33.86	31.86	35.45	Fine-tune transformer models with data augmentation strategies on underrepresented classes.
17	johanjthomas	33.71	31.63	37.85	–
18	lza	32.90	31.98	33.20	–
19	NanMu	32.87	31.79	34.52	–
20	hkkk	32.79	31.95	34.21	–
21	YoussefAhmed21	16.48	14.98	21.22	–

Baseline. To establish a baseline, we fine-tuned a pre-trained transformer-based model RoBERTa on the training set. Fine-tuning was performed using the Hugging Face `Trainer` API with the following configuration: learning rate of 2×10^{-5}, batch size of 16 for both training and evaluation, weight decay of 0.1, and a total of 3 training epochs. Checkpoints were evaluated at the end of each epoch, and the best-performing model on the development set was retained for subsequent testing. The baseline achieved a macro-recall of 68.67% on the development set, with corresponding macro-F1 and accuracy scores of 61.26% and 56.71%, respectively.

Evaluation. Predictions of all systems were submitted and evaluated in CodaLab. At test time, participants assigned the predicted label among [0, 1, 2, 3, 4, 5] for each text, indicating its category. Participants in the leaderboard were ranked by macro-recall. Macro-recall is selected as the primary evaluation metric for two reasons: *(i.)* it gives equal importance to each class, preventing performance for majority classes from dominating the overall score on an unbalanced test set; and *(ii.)* macro-recall provides a more focused view on the model's ability to capture all positive instances for every class, compared with macro-F1 balancing precision and recall for each class. As additional evaluation metrics, we computed accuracy and macro-F1.

Submitted Systems. 22 teams submitted their predictions to CodaLab, of which 16 submitted notebook papers [31,32,40,51,52,59,61,78,82,85,92,95,96, 99,107,108]. The performance of 14 teams is above the baseline, and 8 teams are below fine-tuned RoBERTa-base, as shown in Table 3. Many teams fine-tuned DeBERTa-v3-large and achieved better results than RoBERTa. Larger language models such as Qwen-3 4B and Gemma-2 2B were superior to DeBERTa and RoBERTa. The performance drop observed on the test set compared to the development set highlights the need for further improvement in fine-grained human-AI collaborative text detection.

3 Multilingual Text Detoxification

Text detoxification is a subtask of style transfer, aiming to transform toxic text into a neutral version while preserving its original meaning. With the rapid advancement of language models, concerns have intensified around their potential to generate harmful or biased content with many works developing toxicity mitigation in LLMs approaches [94]. A key challenge in this space is designing detoxification techniques that generalize effectively across languages. Building on our 2024 release of a multilingual parallel detoxification corpus covering 9 languages [27] (English, Spanish, German, Chinese, Arabic, Hindi, Ukrainian, Russian, Amharic), we now extend the task to explore both multilingual and cross-lingual generalization. This year's shared task introduces 6 additional languages—Italian, French, Hebrew, Hinglish, Japanese, and Tatar— offering new challenges for scalable and inclusive detoxification methods.

Dataset. We provided several datasets for participants to train their models and enhance their approaches:

- **Multilingual ParaDetox**: Train part of parallel toxic-neutral 400 pairs per 9 languages from 2024 edition;
- **Multilingual Toxic Lexicon**: Collection from open corpora of toxic keywords for all 15 languages;

Table 4. Results of the final evaluation of the TextDetox test phase. Scores are sorted by the average **J**oint scores: with parallel (**P**) and without parallel (**NP**) training data. Baselines are highlighted with gray , Human References are highlighted with green .

Team	AvgP	AvgNP	System
Human References	0.854	0.847	Human paraphrases from our Multilingual ParaDetox
ducanhhbtt [23]	0.685	0.643	LoRA fine-tuning and advanced prompting with Gemma3-12B
MetaDetox [18]	0.685	0.609	CoT prompting of DeepSeek with outputs re-ranking
sky.Duan [97]	0.676	0.501	Combination of our mT0-detox baseline with Qwen3
Pratham [79]	0.676	0.575	Fine-tuned mT0 with lexical refining
jellyproll	0.675	0.605	mT0 baseline with improved vocab
mT0	0.675	0.572	Fine-tuned mT0 on 9 languages train ParaDetox
Jiaozipi [58]	0.656	0.607	Ensemble of LLMs with RISE framework
SVATS [44]	0.656	0.599	Combination of fine-tuned Qwen2 and Gemma2
nikita.sushko [91]	0.628	0.512	Additionally tuned mT0 with our and synthetic data
ylmmcl [48]	0.612	0.471	Combination of BART, mT0, and LLaMa3.1 for outputs ranking
Gopal [45]	0.611	0.595	Replacement of toxic spans with GPT4o-mini
d1n910 [69]	0.604	0.575	CoT with DeepSeek-R1
GPT-o3	0.562	0.484	Few-shot Prompting of GPT-o3mini
GPT-o4	0.560	0.535	Few-shot Prompting of GPT-o4
Something Awful	0.549	0.511	Llama3.1 with Reasoning with top5 selection
Delete	0.536	0.510	Elimination of toxic keywords
Backtr.	0.481	0.342	Translation of data to English+BART-detox
Duplicate	0.475	0.482	Simple duplication of toxic input

- **Multilingual Toxic Spans**: Toxic collocations extracted with GPT-4 from 9 languages from the train Multilingual ParaDetox dataset [26];
- **Multilingual Toxicity Classification Data**: Collection of binary toxicity classification corpora for all 15 languages.

Then, we extended our test set to 6 new languages for which no parallel training data were provided: Italian, French, Hebrew, Hinglish, Japanese, and Tatar. The language stakeholders utilized various opensourced toxicity or hate speech classification datasets then rewriting the texts into neutral version with native speakers. We provided the same annotation instructions as for 2024 edition [27]. The goal of annotation was to obtain detoxification pairs for 600 unique toxic original instances per each language to form the test set.

Phases and Tracks. We structured our shared task into two phases: (i) **Development phase**: Participants were provided with the Multilingual ParaDetox parallel training data for 9 languages, alongside 600 test toxic instances for each of these languages and an additional 100 toxic instances for each of 6 new languages. (ii) **Test phase**: Participants received the full 600 toxic test instances for all 15 languages, including the newly introduced ones.

To emphasize both multilingual and cross-lingual generalization, we reported results across two evaluation tracks in each phase:

– **AvgP**: The average performance across the 9 languages with available *Parallel* training data according (hence the name). This track focuses on building *multilingual* detoxification models that generalize well across multiple high-resource settings.

– **AvgNP**: The average performance on the 6 new languages for which *No Parallel* training data was released—only test sets were provided. This track presents a *cross-lingual* challenge, encouraging participants to develop approaches that transfer knowledge from the training languages or leverage other external resources to perform well in low-resource settings.

Evaluation. For both phases, we provided the leaderboard based on an automatic evaluation setup. We evaluate the outputs based on three parameters—style of text, content preservation, and conformity to human references—combining them into the final Joint score:

– **Style Transfer Accuracy (STA)** ensures that the generated text is indeed more non-toxic. It was estimated with XLM-R [22] `large` instance fine-tuned for the binary toxicity classification task for our target languages. We compared the non-toxicity scores of models outputs with human references.

– **Content Similarity (SIM)** is the cosine similarity between LaBSE embeddings [29] of both the toxic source and human references and the generated texts.

– **Fluency** is used to estimate the proximity of the detoxified texts to human references and their fluency estimated with xCOMET [49].

Final Joint Score (J) was the aggregation of the three above metrics:

$$\mathbf{J} = \frac{1}{n}\sum_{i=1}^{n}\mathbf{STA}(x^{ref},y_i)\cdot(0.4*\mathbf{SIM}(x_i,y_i)+0.6*\mathbf{SIM}(x_i^{ref},y_i))\cdot\mathbf{FL}(x_i,x^{ref},y_i)$$

We calculated all the metrics separately per each language. In the end, we calculated the **Average** score of **J**oint scores per all languages in the track.

Baselines. We provided several both unsupervised and more modern baselines. For the easy start, we provided:

i. Duplicate: a simple duplication of the toxic input.
ii. Delete: elimination of a toxic keywords based on a predefined dictionary for each language.
iii. Backtranslation: translation of any input to English and detoxification with BART-detox model and translation back.
iv. LLMs prompting: GPT-4o and GPT-o3-mini zero-shot prompting.

For supervised approaches, we provided mBART [26] and mT0 [75] models fine-tuned on 9 languages training ParaDetox.

Submitted Systems. Per both *development* and *test* phases, we got 31 systems submitted that resulted in 12 notebooks submissions [18,23,28,44,45,48, 58,69,79,87,91,97]. While there is indeed a very big tendency of LLMs prompting solutions, still, many submissions were based on various improvements over *seq2seq* generative models or LLMs. Thus, many participants tried chain-of-thoughts or other advanced prompting techniques over recent powerful LLMs like DeepSeek [25], LLaMa3 [1], Qwen [4], and Gemma [24], as well as special fine-tuning and cross-lingual inference with mT0 [65].

Results. The results of the most interesting submissions are presented in Table 4. First, only five submissions outperformed our strongest baseline, mT0, and even these remained well below human reference performance. Additionally, many systems showed imbalanced results between languages with and without training data. Nevertheless, several creative approaches demonstrated that effective cross-lingual text detoxification is feasible with modern language models.

4   Multi-author Writing Style Analysis

Writing style analysis serves as the cornerstone for authorship identification. The multi-author writing style analysis task within PAN@CLEF has continuously advanced this essential research domain by developing challenges. The task has undergone substantial transformation across multiple iterations: beginning with the identification and clustering of individual authors [74], progressing to distinguishing between single-author and multi-author documents [43,88,106], advancing to determining the precise number of contributing authors [105], and paragraph-level detection of style changes within documents [100–103].

In the 2025 edition of the PAN multi-author writing style analysis task, we asked participants to identify positions of writing style changes within a set of documents. Building on previous editions that focused on the detection of paragraph-level style changes, this year's task advances to detecting style changes at the sentence level, making the setting more realistic.

The dataset provided to participants consists of three datasets varying in the difficulty of detection style changes: *Easy:* Each document covers a variety of topics, allowing participants to leverage topic information as a cue for detecting changes in writing style. Furthermore, the stylistic similarity between sentences in the document is rather low. *Medium:* The topics within a document are more homogeneous, requiring approaches to rely more heavily on stylistic features rather than topic differences to identify style changes. The stylistic similarity between sentences is moderate. *Hard:* All sentences within a document are of a single topic and stylistically similar.

We control for topical diversity across the datasets to ensure that the focus is on stylistic changes. In particular, the hard dataset eliminates topical differences as a proxy signal for authorship, requiring the use of writing style analysis to detect changes.

Data Set and Evaluation

We leverage data from the Reddit platform[3] for the multi-author writing analysis task. In particular, we select user posts from topic-specific subreddits, including *r/worldnews*, *r/politics*, *r/askhistorians*, and *r/legaladvice*. This diverse selection of sources allows for curating documents with varying levels of topical coherence. To construct individual documents, we extract posts from these subreddits, apply preprocessing steps (such as removing quotes, whitespace, emojis, and hyperlinks), and then split the posts into individual sentences.

Based on this data, we construct documents by extracting sentences from a single Reddit post, authored by two to four users. For each sentence, we compute semantic and stylistic feature vectors, enabling the computation of topical (semantic) and stylistic similarity between individual sentences. Based on these similarities, we apply a mixing approach for all sentences of the individual authors of the given Reddit post. We then concatenate sentences based on their topical and stylistic similarity, allowing us to control for the difficulty of the style detection task. For the three datasets, we configure the similarity threshold for consecutive sentences to be (1) relatively high for the *easy* dataset, (2) moderate for the *medium* dataset, and (3) small for the *hard* dataset. Each of the easy, medium, and hard datasets contains 6,000 documents. We provided participants with training, validation, and test splits for all three datasets. The training sets contain 70% of the documents in each dataset, while the validation and test sets contain 15% each. The test sets were withheld for the evaluation phase of the competition.

The submitted approaches are evaluated on each dataset using the macro-averaged F1-score calculated across all documents.

Results

The task received twelve valid software submissions and working notes papers. The F1-scores for each task achieved by the participants are shown in Table 5. The best average F1-score across the three datasets was achieved by team wqd, reaching a score of 0.870. For the easy dataset, Team stylospies achieved a marginally better result, while scoring the fifth and third best results for the medium and hard datasets, respectively. For the medium dataset, xxsu-team achieved a marginally higher score. Generally, we observe that the individual approaches perform quite differently on the three datasets. For instance, teams cornell-1 and better-call-claude perform better on the medium dataset than on the easy and the hard datasets. Most submissions were able to outperform the two simple baselines: one baseline that predicted a style change for each pair of sentences, and one that predicted no style change for each pair of sentences. Further details on the approaches taken can be found in the overview paper [104].

---

[3] https://www.reddit.com/.

Table 5. Overall results for the multi-author writing style analysis task, ranked by average F_1 performance across all three datasets. Best results are marked in bold.

Team	Easy F_1	Medium F_1	Hard F_1
wqd [55]	0.958	0.823	**0.830**
xxsu-team [54]	0.955	**0.825**	0.829
stylospies [17]	**0.959**	0.786	0.791
team-tmu [37]	0.950	0.792	0.792
better-call-claude [76]	0.929	0.815	0.731
openfact [46]	0.919	0.771	0.752
cornell-1 [16]	0.909	0.793	0.698
batatavada-pict [73]	0.823	0.766	0.667
hhu [62]	0.761	0.666	0.642
ksu [2]	0.507	0.747	0.467
hellojie [20]	0.461	0.583	0.484
team-of-bf [56]	0.486	0.443	0.473
Baseline Predict 1	0.178	0.177	0.147
Baseline Predict 0	0.439	0.440	0.453

5 Generative Plagiarism Detection

Plagiarism detection has a long-standing tradition in PAN, with main tasks running from 2009 [71] to 2015 [80]. Over time, the focus gradually shifted toward more specialized intrinsic tasks, such as the still active authorship analysis challenges. However, the recent breakthrough of generative AI has dramatically transformed the landscape of plagiarism detection. For the first time in history, LLMs can serve as so-called automatic plagiarists [5]. This shift inspired us to revive a classic plagiarism detection task for 2025, this time centered on automatically generated plagiarism using LLMs.

For the 2025 edition, we adhered to the well-established foundations of the 2015 plagiarism detection task, particularly in evaluation methodology and dataset formatting [5]. Participants received an annotated synthetic dataset of pairs of documents (S, P), where S is a source document and P is the plagiarism document in which the paragraphs p were replaced with paraphrased versions of paragraphs s in S using LLMs without citation. This setup closely mirrors the 2015 PAN text alignment task[4], allowing us to evaluate how well past approaches have aged.

---

[4] http://www.uni-weimar.de/medien/webis/events/pan-15/pan15-web/plagiarism-detection.html.

5.1   Dataset

The synthetic dataset was constructed by first identifying the most semantically similar document pairs on arXiv, using embeddings from the SPECTER model [21] applied to the 2025 release of ar5iv[5]. We then sampled a subset of 100,000 documents with an even distribution across all arXiv categories (also known as archives), to ensure a wide variety of topics. For each remaining document pair (S, P), we aligned the most semantically similar paragraphs s and p from S and P, respectively, based on three criteria. The alignment score was computed as a weighted aggregate: 50% semantic similarity via SciBERT sentence embeddings [7], 40% lexical similarity using TF-IDF vector similarity, and 10% section title similarity using SciBERT embeddings. The inclusion of similarity in the title of the section helped discourage the alignment of paragraphs from unrelated sections of the documents.

For each pair (S, P), we selected one of three LLMs: LLaMA-3 [1] (3.3 70B Instruct), DeepSeek-R1 [25] (Distill-Qwen-32B) or Mistral [63] (7B Instruct v0.3), and replaced all p in each aligned paragraph (s, p) with LLM-paraphrased versions s' derived from paragraphs s in S. To support a more detailed analysis of system performance, we established several categories of document pairs. First, 5% of the 100,000 pairs remained unchanged, i.e., both S and P are original arXiv documents. An additional 20% of pairs do not contain any plagiarism, but some paragraphs in P have been paraphrased by an LLM independently of S. These examples are useful for evaluating systems that aim to detect LLM-generated content rather than plagiarism specifically. The remaining 75% of document pairs were constructed as described above.

We further classified the severity of plagiarism in P into three levels: low, medium, and high. These refer to the proportion of paragraphs in P that were replaced with paraphrased versions from S. In 30% of the document pairs, the severity was *low*, with 20% to 40% of paragraphs replaced. In 40% of the pairs, severity was *medium*, with 40% to 60% replaced. The remaining 30% had *high* severity, where 70% to 100% of paragraphs in P were substituted.

Paraphrasing Prompts. Each LLM used three types of prompts to generate paraphrased plagiarism. These were distributed across document pairs as follows. 60% of the pairs used a *simple prompt*:

```
Paraphrase the given paragraph for a professional audience.
```

30% used a *medium prompt*:

```
Reformulate the given paragraph in a sophisticated manner
while preserving its meaning. Modify sentence structure,
reword phrases, and incorporate elements of general knowledge
to ensure coherence. The less token overlap, the better.
```

---

[5] https://ar5iv.labs.arxiv.org/.

Table 6. Plagiarism alignment dataset and LLM splits.

Splits/LLMs	Llama-3		DeepSeek-R1		Mistral		Altered	Original	Total
Train	18,423	79.80%	18,452	79.46%	6,265	79.65%	15,101	3,918	62,159
Validation	2,353	10.19%	2,383	10.26%	802	10.20%	1,919	518	7,975
Test	2,310	10.01%	2,386	10.28%	799	10.16%	1,919	490	7,904
Total	23,086	42.62%	23,221	42.86%	7,866	14.52%	18,939	4,926	78,038

The final 10% used a *hard prompt* that incorporated immediate context to help the generated paragraph blend into its surrounding text. The prompt took the following form:

```
Completely rephrase the given paragraph in your own words.
Feel free to incorporate elements from general knowledge to
ensure coherence, flow, and better understanding.

{context_before}
```

All prompts included additional instructions to output only the paraphrased content, avoiding any explanatory text. Special tokens were used to suppress verbose output, tailored to each LLM. For DeepSeek-R1, a custom `<thinking>...</thinking>` block was used to suppress the model's internal reasoning steps, which would otherwise significantly slow down the generation. It is worth noting that Mistral performed poorly in following prompt instructions. It often produced explanatory content, hallucinated facts, or entered repetitive output loops, an issue reminiscent of neural network architectures before the attention mechanism era. In total, the final dataset consisted of 78,038 document pairs, divided into training, validation, and test subsets. The training and validation sets were provided to participants, while the test set was kept private for the evaluation phase. The data splits and sizes is given in Table 6.

5.2   Evaluation

All systems were submitted and evaluated on the TIRA platform. The participants were tasked with identifying all the paragraphs s' in P and aligning each with the corresponding paragraph s in S. The training and validation sets contained all alignments (s, s') for each pair of documents (S, P), together with the full text of both documents. The evaluation was carried out using the original scripts from the 2015 PAN plagiarism detection task. The metrics included micro and macro F1 scores as well as the established `plagdet` metric [70].

Four teams participated in the task by submitting software. Table 7 shows the aggregated evaluation results for all submissions that we also compared to the PAN baseline from 2012. We report the arithmetic mean of all evaluation measures (micro precision, macro precision, micro recall, macro recall, micro

Table 7. Arithmetic mean of all evaluation measures per submission for the plagiarism detection alignment task.

Team	Score	System
chi-zi-zhi-xin-dui [81]	0.440	Sentence-BERT, MPNet, TF-IDF
jrluo [41]	0.263	E5 and MiniLM-L6
foshan-university [84]	0.400	TF-IDF and BERT classifier
yukino [64]	**0.471**	Glove embeddings
Baseline PAN-12	0.233	Lexical near-duplicate detection
Baseline Llama-3.3 [1]	0.269	Llama-3.3 70B embeddings
Baseline Qwen2 [4]	0.375	Qwen2 7b Instruct embeddings

plagdet, and macro pladget) as main evaluation score. All submissions substantially improve upon the PAN-12 baseline that used lexical near-duplicate detection. All submissions used some form of semantic similarity embeddings. Therefore, we added two additional baselines relying upon two typical embedding models: Llama-3.3 70B and Qwen2 7B instruct. Team Yukino achieving the highest score relying on Glove embeddings closely followed by Team Su, which used an ensemble of multiple semantic embeddings combined with lexical TF-IDF similarity. An extended evaluation will be available in the task overview [33].

Acknowledgments. The work of Janek Bevendorff, Matti Wiegmann, Maik Fröbe, Martin Potthast, and Benno Stein has been funded as part of the OpenWebSearch project by the European Commission (OpenWebSearch.eu, GA 101070014). The work of Andre Greiner-Petter has been funded by the Deutsche Forschungsgemeinschaft (DFG, German Research Foundation) – 554559555.

References

1. AI@Meta: Llama 3 Model Card (2024). https://github.com/meta-llama/llama3/blob/main/MODEL_CARD.md. Accessed 14 Dec 2024
2. Alsheddi, A., El Bachir Menai, M.: Style change detection in multi-authored english texts based on graph convolutional networks. In: Working Notes of CLEF 2025 - Conference and Labs of the Evaluation Forum, CEUR-WS.org (2025)
3. Ayele, A.A., et al.: Overview of PAN 2024: multi-author writing style analysis, multilingual text detoxification, oppositional thinking analysis, and generative AI authorship verification. In: Experimental IR Meets Multilinguality, Multimodality, and Interaction. 15th International Conference of the CLEF Association (CLEF 2024). Lecture Notes in Computer Science, vol. 14959, pp. 231–259. Springer, Heidelberg (2024). https://doi.org/10.1007/978-3-031-71908-0_11
4. Bai, J., et al.: Qwen technical report. arXiv preprint arXiv:2309.16609 (2023)
5. Barrón-Cedeño, A., Potthast, M., Rosso, P., Stein, B.: Corpus and evaluation measures for automatic plagiarism detection. In: Proceedings of the International

Conference on Language Resources and Evaluation, LREC 2010, 17–23 May 2010, Valletta, Malta, European Language Resources Association (2010). http://www.lrec-conf.org/proceedings/lrec2010/summaries/35.html

6. Basani, A.R., Chen, P.: DivEye at PAN 2025: diversity boosts AI-generated text detection. In: Working Notes of CLEF 2025 - Conference and Labs of the Evaluation Forum, CEUR-WS.org (2025)

7. Beltagy, I., Lo, K., Cohan, A.: Scibert: a pretrained language model for scientific text. In: Proceedings of the 2019 Conference on Empirical Methods in Natural Language Processing and the 9th International Joint Conference on Natural Language Processing, EMNLP-IJCNLP 2019, pp. 3613–3618. ACL (2019). https://doi.org/10.18653/V1/D19-1371

8. Bevendorff, J., et al.: Overview of PAN 2023: authorship verification, multi-author writing style analysis, profiling cryptocurrency influencers, and trigger detection. In: Experimental IR Meets Multilinguality, Multimodality, and Interaction. 14th International Conference of the CLEF Association (CLEF 2023), Lecture Notes in Computer Science, vol. 14163, pp. 459–481. Springer, Heidelberg (2023). https://doi.org/10.1007/978-3-031-42448-9_29

9. Bevendorff, J., et al.: Overview of PAN 2022: authorship verification, profiling irony and stereotype spreaders, and style change detection. In: Experimental IR Meets Multilinguality, Multimodality, and Interaction. 13th International Conference of the CLEF Association (CLEF 2022), Lecture Notes in Computer Science, vol. 13186. Springer, Heidelberg (2022). https://doi.org/10.1007/978-3-031-13643-6

10. Bevendorff, J., et al.: Overview of PAN 2021: authorship verification, profiling hate speech spreaders on twitter, and style change detection. In: Candan, K.S., et al. (eds.) CLEF 2021. LNCS, vol. 12880, pp. 419–431. Springer, Cham (2021). https://doi.org/10.1007/978-3-030-85251-1_26

11. Bevendorff, J., et al.: Overview of PAN 2020: authorship verification, celebrity profiling, profiling fake news spreaders on twitter, and style change detection. In: Experimental IR Meets Multilinguality, Multimodality, and Interaction. 11th International Conference of the CLEF Initiative (CLEF 2020), Lecture Notes in Computer Science, vol. 12260, pp. 372–383. Springer, Heidelberg (2020). https://doi.org/10.1007/978-3-030-58219-7_25

12. Bevendorff, J., Stein, B., Hagen, M., Potthast, M.: Generalizing unmasking for short texts. In: 14th Conference of the North American Chapter of the Association for Computational Linguistics: Human Language Technologies (NAACL 2019), pp. 654–659. Association for Computational Linguistics (2019). https://aclanthology.org/N19-1068/

13. Bevendorff, J., et al.: Overview of the "Voight-Kampff" generative AI authorship verification task at PAN and ELOQUENT 2025. In: Working Notes of CLEF 2025 – Conference and Labs of the Evaluation Forum, CEUR Workshop Proceedings. CEUR-WS.org (2025)

14. Bevendorff, J., et al.: Overview of the "Voight-Kampff" generative AI authorship verification task at PAN and ELOQUENT 2024. In: Working Notes of CLEF 2024 – Conference and Labs of the Evaluation Forum, pp. 2486–2506, CEUR Workshop Proceedings, CEUR-WS.org (2024). http://ceur-ws.org/Vol-3740/paper-225.pdf

15. Bevendorff, J., Wiegmann, M., Richter, E., Potthast, M., Stein, B.: The two paradigms of LLM detection: authorship attribution vs. authorship verification. In: The 63rd Annual Meeting of the Association for Computational Linguistics (ACL 2025) (Findings). Association for Computational Linguistics (2025)

16. Boloni-Turgut, D., Verma, D., Cardie, C.: Team cornell-1 at PAN: ensembling fine-tuned transformer models for writing style analysis. In: Working Notes of CLEF 2025 - Conference and Labs of the Evaluation Forum. CEUR-WS.org (2025)

17. Boriceanu, I., Băltoiu, A.: Style change detection using graph and structural-linguistic features for multi-author writing analysis. In: Working Notes of CLEF 2025 - Conference and Labs of the Evaluation Forum. CEUR-WS.org (2025)

18. Bourbour, S., Kelishami, A.S., Gheysari, M., Rahimzadeh, F.: Cross-lingual detoxification with few-chain prompting: a competitive system for TextDetox 2025. In: Working Notes of CLEF 2025 - Conference and Labs of the Evaluation Forum. CEUR-WS.org (2025)

19. Brennan, M., Afroz, S., Greenstadt, R.: Adversarial stylometry: circumventing authorship recognition to preserve privacy and anonymity. ACM Trans. Inf. Syst. Secur. **15**(3) (2012). https://doi.org/10.1145/2382448.2382450

20. Chen, D., Li, J., Qi, H.: Llama-3 with 4-bit quantization and IA^3 tuning for multi-author writing style analysis. In: Working Notes of CLEF 2025 - Conference and Labs of the Evaluation Forum. CEUR-WS.org (2025)

21. Cohan, A., Feldman, S., Beltagy, I., Downey, D., Weld, D.S.: SPECTER: document-level representation learning using citation-informed transformers. In: Proceedings of the 58th Annual Meeting of the Association for Computational Linguistics, ACL 2020, Online, 5–10 July 2020, pp. 2270–2282. Association for Computational Linguistics (2020). https://doi.org/10.18653/V1/2020.ACL-MAIN.207

22. Conneau, A., et al.: Unsupervised cross-lingual representation learning at scale. In: Proceedings of the 58th Annual Meeting of the Association for Computational Linguistics, ACL 2020, Online, 5–10 July 2020, pp. 8440–8451. Association for Computational Linguistics (2020). https://doi.org/10.18653/V1/2020.ACL-MAIN.747

23. Dang, T.D.A., D'Elia, F.P.: GemDetox: enhancing a massively multilingual model for text detoxification on low-resource languages. In: Working Notes of CLEF 2025 - Conference and Labs of the Evaluation Forum. CEUR-WS.org (2025)

24. DeepMind: Gemma Model Card (2024). https://github.com/google-deepmind/gemma. Accessed 09 June 2025

25. DeepSeek-AI: Deepseek-v3 technical report (2024). https://arxiv.org/abs/2412.19437

26. Dementieva, D., et al.: Multilingual and explainable text detoxification with parallel corpora. In: Proceedings of the 31st International Conference on Computational Linguistics, pp. 7998–8025. Association for Computational Linguistics, Abu Dhabi, UAE (2025). https://aclanthology.org/2025.coling-main.535/

27. Dementieva, D., et al.: Overview of the multilingual text detoxification task at pan 2024 (2024)

28. Farid, H., Ahmad, Z., Mahmood, A., Ameer, I.: HF_Detox at PAN 2025 TextDetox: prompt-driven multilingual detoxification. In: Working Notes of CLEF 2025 - Conference and Labs of the Evaluation Forum. CEUR-WS.org (2025)

29. Feng, F., Yang, Y., Cer, D., Arivazhagan, N., Wang, W.: Language-agnostic BERT sentence embedding. In: Proceedings of the 60th Annual Meeting of the Association for Computational Linguistics (Volume 1: Long Papers), ACL 2022, Dublin, Ireland, 22–27 May 2022, pp. 878–891. Association for Computational Linguistics (2022). https://doi.org/10.18653/V1/2022.ACL-LONG.62

30. Fröbe, M., et al.: Continuous integration for reproducible shared tasks with TIRA.io. In: Advances in Information Retrieval. 45th European Conference on

IR Research (ECIR 2023). Lecture Notes in Computer Science, pp. 236–241. Springer, Heidelberg (2023)

31. Fuchuan, Y., Cao, H., Zhongyuan, H.: Sentence-level AI-generated text detection with fine-tuned BERT. In: Working Notes of CLEF 2025 - Conference and Labs of the Evaluation Forum. CEUR-WS.org (2025)

32. Gómez Sánchez, D., Jimenez, J., Ramírez, M., Martinez, J.: RoBERT-IA: human-AI collaborative text classification. In: Working Notes of CLEF 2025 - Conference and Labs of the Evaluation Forum. CEUR-WS.org (2025)

33. Greiner-Petter, A., et al.: Overview of the generative plagiarism detection task at PAN 2025. In: CLEF 2025 Working Notes. CEUR-WS.org (2025)

34. Guo, H., et al.: Biscope: AI-generated text detection by checking memorization of preceding tokens. Adv. Neural. Inf. Process. Syst. **37**, 104065–104090 (2024)

35. Halvani, O., Winter, C., Graner, L.: On the usefulness of compression models for authorship verification. In: Proceedings of the 12th International Conference on Availability, Reliability and Security, vol. Part F1305. ACM, New York (2017). https://doi.org/10.1145/3098954.3104050

36. Hans, A., et al.: Spotting LLMs with binoculars: zero-shot detection of machine-generated text. In: International Conference on Machine Learning abs/2401. 12070, pp. 17519–17537 (2024). https://doi.org/10.48550/arXiv.2401.12070

37. Hosseinbeigi, S.B., Mehrani, A.: Team TMU at PAN 2025: an ensemble of fine-tuned LaBSE and siamese neural network for multi-author writing style analysis. In: Working Notes of CLEF 2025 - Conference and Labs of the Evaluation Forum. CEUR-WS.org (2025)

38. Huang, J., Cao, H., Lin, X., Han, Z.: Application and analysis of roberta-base model fine tuning based on data enhancement in AI text detection. In: Working Notes of CLEF 2025 - Conference and Labs of the Evaluation Forum. CEUR-WS.org (2025)

39. Hutson, J.: Human-AI collaboration in writing: a multidimensional framework for creative and intellectual authorship. Int. J. Changes Educ. (2025)

40. Ignatenko, D., Zaitsev, K., Shkriaba, O.: ReText.Ai team at PAN 2025: applying a multiple classification heads to a transformer model for human-AI collaborative text classification. In: Working Notes of CLEF 2025 - Conference and Labs of the Evaluation Forum. CEUR-WS.org (2025)

41. Jieren, L., Mancheng, H., Biao, L., Zhongyuan, H.: Two-stage generative plagiarism detection: from TF-IDF/Jaccard filtering to transformer classification. In: Working Notes of CLEF 2025 - Conference and Labs of the Evaluation Forum. CEUR-WS.org (2025)

42. Jimeno-Gonzalez, M., Martínez-Cámara, E., Noelia Fernandez, P.G., na López, L.A.U.: Team SINAI-INTA at PAN 2025: uncovering machine generated text with linguistic features. In: Working Notes of CLEF 2025 - Conference and Labs of the Evaluation Forum. CEUR-WS.org (2025)

43. Kestemont, M., et al.: Overview of the author identification task at PAN 2018: cross-domain authorship attribution and style change detection. In: Working Notes of CLEF 2018 - Conference and Labs of the Evaluation Forum. CEUR-WS.org (2018)

44. Kozlovskiy, V., Ploskin, A., Tantry, S., Matveeva, T., Savelyeva, S.: Can small models outperform large ones in text detoxification? In: Working Notes of CLEF 2025 - Conference and Labs of the Evaluation Forum. CEUR-WS.org (2025)

45. Krishna, N., Sai Teja, L., Mishra, A.: Team detox at PAN: multilingual text detoxification using LLM. In: Working Notes of CLEF 2025 - Conference and Labs of the Evaluation Forum. CEUR-WS.org (2025)

46. Księżniak, E., Węcel, K., Sawiński, M.: OpenFact at PAN 2025: punctuation-guided pretraining for sentence-level style change detection. In: Working Notes of CLEF 2025 - Conference and Labs of the Evaluation Forum. CEUR-WS.org (2025)

47. Kumar, R., Trivedi, A., Varshney, O.: Voight-Kampff AI detection sensitivity: IIITS@CLEF'25. In: Working Notes of CLEF 2025 - Conference and Labs of the Evaluation Forum. CEUR-WS.org (2025)

48. Lai-Lopez, N., Yuan, S., Wang, L., Zhang, L.: Lexicon-guided detoxification and classifier-gated rewriting: a PAN 2025 submission. In: Working Notes of CLEF 2025 - Conference and Labs of the Evaluation Forum. CEUR-WS.org (2025)

49. Larionov, D., Seleznyov, M., Viskov, V., Panchenko, A., Eger, S.: xCOMET-lite: bridging the gap between efficiency and quality in learned MT evaluation metrics. In: Proceedings of the 2024 Conference on Empirical Methods in Natural Language Processing, pp. 21934–21949. Association for Computational Linguistics, Miami, Florida, USA (2024). https://aclanthology.org/2024.emnlp-main.1223

50. Larson, J.: Generative AI detection using simple Feature Selection and SVM. In: Working Notes of CLEF 2025 - Conference and Labs of the Evaluation Forum. CEUR-WS.org (2025)

51. Li, B., Qi, H., Yan, K.: Team Bohan Li at PAN: DeBERTa-v3 with R-drop regularization for human-AI collaborative text classification. In: Working Notes of CLEF 2025 - Conference and Labs of the Evaluation Forum. CEUR-WS.org (2025)

52. Li, T.: Fine-grained human-AI collaborative text classification using DeBERTa. In: Working Notes of CLEF 2025 - Conference and Labs of the Evaluation Forum. CEUR-WS.org (2025)

53. Liang, Z., Sun, K., Cao, H., Luo, J., Han, Z.: Research on text author classification based on ModernBERT and gradient loss function. In: Working Notes of CLEF 2025 - Conference and Labs of the Evaluation Forum. CEUR-WS.org (2025)

54. Lin, K., Liu, C., Ye, F., Han, Z.: SCL-DeBERTa: multi-author writing style change detection enhanced by supervised contrastive learning. In: Working Notes of CLEF 2025 - Conference and Labs of the Evaluation Forum. CEUR-WS.org (2025)

55. Lin, X., Han, Z., Liu, C., Duan, X.: Style change detection in multi-author writing: a deep learning approach based on DeBERTa. In: Working Notes of CLEF 2025 - Conference and Labs of the Evaluation Forum. CEUR-WS.org (2025)

56. Liu, B., Yang, L., Qi, H.: Integrating adversarial-contrastive learning and large language model for multi-author writing style analysis. In: Working Notes of CLEF 2025 - Conference and Labs of the Evaluation Forum. CEUR-WS.org (2025)

57. Liu, J., Kong, L., Peng, Z., Chen, F.: Generative AI authorship verification based on contrastive-enhanced dual-model decision system. In: Working Notes of CLEF 2025 - Conference and Labs of the Evaluation Forum. CEUR-WS.org (2025)

58. Liu, X., et al.: Jiaozipi at CLEF 2025: a multilingual text detoxification method based on large language model-based ensemble learning. In: Working Notes of CLEF 2025 - Conference and Labs of the Evaluation Forum. CEUR-WS.org (2025)

59. Macko, D.: mdok of KInIT: robustly fine-tuned LLM for binary and multiclass AI-generated text detection. In: Working Notes of CLEF 2025 - Conference and Labs of the Evaluation Forum. CEUR-WS.org (2025)

60. Macko, D., Moro, R., Srba, I.: Increasing the robustness of the fine-tuned multilingual machine-generated text detectors. arXiv preprint arXiv:2503.15128 (2025)

61. Marchitan, T., Creanga, C., Dinu, L.: Unibuc - NLP at "Voight-Kampff" generative AI detection PAN 2025. In: Working Notes of CLEF 2025 - Conference and Labs of the Evaluation Forum. CEUR-WS.org (2025)

62. Meier, P., Boland, K., Kallmeyer, L., Dietze, S.: Team HHU - an ensemble-based approach to multi-author writing style analysis combining experts for different difficulty levels. In: Working Notes of CLEF 2025 - Conference and Labs of the Evaluation Forum. CEUR-WS.org (2025)

63. MistralAI: Mistral 7b instruct v0.3 Model Card (2024). https://huggingface.co/mistralai/Mistral-7B-Instruct-v0.3. Accessed 14 Feb 2025

64. Mo, D., Zhang, H., Zhang, X., Kong, L.: Using GloVe for fragment feature matching and overlap ratio optimized generated plagiarism detection method. In: Working Notes of CLEF 2025 - Conference and Labs of the Evaluation Forum. CEUR-WS.org (2025)

65. Muennighoff, N., et al.: Crosslingual generalization through multitask finetuning. In: Proceedings of the 61st Annual Meeting of the Association for Computational Linguistics (Volume 1: Long Papers), ACL 2023, Toronto, Canada, 9–14 July 2023, pp. 15991–16111. Association for Computational Linguistics (2023). https://doi.org/10.18653/V1/2023.ACL-LONG.891

66. Ochab, J., Matias, M., Boba, T., Walkowiak, T.: StylOch at PAN: gradient-boosted trees with frequency-based stylometric features. In: Working Notes of CLEF 2025 - Conference and Labs of the Evaluation Forum. CEUR-WS.org (2025)

67. Ostrower, B., Doongare, P., Unnikrishnan, M.: Binoculars, BART, and adversaries: multi-faceted AI text detection for PAN 2025. In: Working Notes of CLEF 2025 - Conference and Labs of the Evaluation Forum. CEUR-WS.org (2025)

68. Peñas, A., Rodrigo, Á.: A simple measure to assess non-response. In: Proceedings of the 49th Annual Meeting of the Association for Computational Linguistics: Human Language Technologies, pp. 1415–1424 (2011). https://aclanthology.org/P11-1142.pdf

69. Peng, J., Kaiyin, S., Kaichuan, L., Zhankeng, L., Zhongyuan, H.: A multilingual text detoxification method based on chain-of-thoughts prompting approach. In: Working Notes of CLEF 2025 - Conference and Labs of the Evaluation Forum. CEUR-WS.org (2025)

70. Potthast, M., Stein, B., Barrón-Cedeño, A., Rosso, P.: An evaluation framework for plagiarism detection. In: COLING 2010, 23rd International Conference on Computational Linguistics, Posters Volume, 23–27 August 2010, Beijing, China, pp. 997–1005. Chinese Information Processing Society of China (2010). https://aclanthology.org/C10-2115/

71. Potthast, M., Stein, B., Eiselt, A., Barrón-Cedeño, A., Rosso, P.: PAN plagiarism corpus 2009 (PAN-PC-09) (version 1) (2009). https://doi.org/10.5281/zenodo.3250083

72. Pudasaini, S., Miralles-Pechuán, L., Lillis, D., Salvador, M.L.: Enhancing AI text detection with frozen pretrained encoders and ensemble learning. In: Working Notes of CLEF 2025 - Conference and Labs of the Evaluation Forum. CEUR-WS.org (2025)

73. Rohra, H., Shah, N., Sonawane, S.: Team BatataVada at PAN: sentence-level style change detection with RoBERTa for multi-author writing style analysis. In: Working Notes of CLEF 2025 - Conference and Labs of the Evaluation Forum. CEUR-WS.org (2025)

74. Rosso, P., Rangel, F., Potthast, M., Stamatatos, E., Tschuggnall, M., Stein, B.: Overview of PAN'16—new challenges for authorship analysis: cross-genre profiling, clustering, diarization, and obfuscation. In: Experimental IR Meets Multilinguality, Multimodality, and Interaction. 7th International Conference of the CLEF Initiative (CLEF 16) (2016)

75. Rykov, E., Zaytsev, K., Anisimov, I., Voronin, A.: Smurfcat at PAN 2024 textdetox: alignment of multilingual transformers for text detoxification. In: Working Notes of the Conference and Labs of the Evaluation Forum (CLEF 2024), Grenoble, France, 9–12 September 2024. CEUR Workshop Proceedings, vol. 3740, pp. 2866–2871. CEUR-WS.org (2024). https://ceur-ws.org/Vol-3740/paper-276.pdf

76. Schmidt, G., Römisch, J., Halchynska, M., Gorovaia, S., Yamshchikov, I.: better_call_claude: sequential style shift model for fine-grained multi-author style change detection. In: Working Notes of CLEF 2025 - Conference and Labs of the Evaluation Forum. CEUR-WS.org (2025)

77. Sculley, D., Brodley, C.E.: Compression and machine learning: a new perspective on feature space vectors. In: Data Compression Conference (DCC'06), pp. 332–341. IEEE (2006). ISBN 9780769525457, ISSN 1068-0314,2375-0359. https://doi.org/10.1109/dcc.2006.13

78. Seeliger, M., Styll, P., Staudinger, M., Hanbury, A.: Human or not? Light-weight and interpretable detection of AI-generated text. In: Working Notes of CLEF 2025 - Conference and Labs of the Evaluation Forum. CEUR-WS.org (2025)

79. Shah, P., Shah, V., Kale, S.: Multilingual text detoxification via prompted MT0-XL and lexical filtering. In: Working Notes of CLEF 2025 - Conference and Labs of the Evaluation Forum. CEUR-WS.org (2025)

80. Stamatatos, E., Potthast, M., Pardo, F.M.R., Rosso, P., Stein, B.: Overview of the PAN/CLEF 2015 evaluation lab. In: Experimental IR Meets Multilinguality, Multimodality, and Interaction - 6th International Conference of the CLEF Association, CLEF 2015, Toulouse, France, 8–11 September 2015, Proceedings, Lecture Notes in Computer Science, vol. 9283, pp. 518–538. Springer (2015). https://doi.org/10.1007/978-3-319-24027-5_49

81. Su, Z., Han, Y., Jia, Y., Kong, L.: Hierarchical generative plagiarism detection method. In: Working Notes of CLEF 2025 - Conference and Labs of the Evaluation Forum. CEUR-WS.org (2025)

82. Sun, Q., et al.: DeBERTa-FPN: fusion feature pyramid network for human-AI collaborative text classification. In: Working Notes of CLEF 2025 - Conference and Labs of the Evaluation Forum. CEUR-WS.org (2025)

83. Sun, Y., Afanaseva, S., Stowe, K., Patil, K.: Bi-directional cross-entropy loss and stylometric feature combined classifier. In: Working Notes of CLEF 2025 - Conference and Labs of the Evaluation Forum. CEUR-WS.org (2025)

84. Tang, J., Hu, Q., Han, Z.: Efficient plagiarism detection via sentence embeddings and FAISS-based retrieval. In: Working Notes of CLEF 2025 - Conference and Labs of the Evaluation Forum. CEUR-WS.org (2025)

85. Teja, L.S., Yadagiri, A., Pakray, P.: Team CNLP-NITS-PP at PAN: advancing generative AI detection: mixture of experts with transformer models. In: Working Notes of CLEF 2025 - Conference and Labs of the Evaluation Forum. CEUR-WS.org (2025)

86. Titze, S., Halvani, O.: LOG-AID: logit-based statistical features for AI text detection. In: Working Notes of CLEF 2025 - Conference and Labs of the Evaluation Forum. CEUR-WS.org (2025)

87. Totok, A., Ermolaev, A., Izyumova, A., Finogeev, E.: The evolution of methods for text detoxification: the role of language in method selection. In: Working Notes of CLEF 2025 - Conference and Labs of the Evaluation Forum. CEUR-WS.org (2025)

88. Tschuggnall, M., et al.: Overview of the author identification task at PAN 2017: style breach detection and author clustering. In: CLEF 2017 Labs and Workshops, Notebook Papers (2017)

89. Valdez-Valenzuela, A., Gómez-Adorno, H.: AI-generated text detection using ISGraphs and graph neural networks. In: Working Notes of CLEF 2025 - Conference and Labs of the Evaluation Forum. CEUR-WS.org (2025)

90. Völpel, F., Halvani, O.: Adept: AI-generated text detection based on phrasal category N-grams. In: Working Notes of CLEF 2025 - Conference and Labs of the Evaluation Forum. CEUR-WS.org (2025)

91. Voronin, A., Moskovsky, D., Sushko, N.: PAN 2025 Textdetox: exploring a sage-T5-like approach for text detoxification. In: Working Notes of CLEF 2025 - Conference and Labs of the Evaluation Forum. CEUR-WS.org (2025)

92. Voznyuk, A., Gritsai, G., Grabovoy, A.: Team advacheck at PAN: multitasking does all the magic. In: Working Notes of CLEF 2025 - Conference and Labs of the Evaluation Forum. CEUR-WS.org (2025)

93. Wang, H., Juola, P., Riddell, A.: Reproduction and replication of an adversarial stylometry experiment. arXiv [cs.CL] (2022). http://arxiv.org/abs/2208.07395

94. Wang, M., et al.: Detoxifying large language models via knowledge editing. In: Proceedings of the 62nd Annual Meeting of the Association for Computational Linguistics (Volume 1: Long Papers), pp. 3093–3118, Association for Computational Linguistics, Bangkok, Thailand (2024). https://doi.org/10.18653/v1/2024.acl-long.171

95. Wu, W., et al.: Bert_T for human-AI collaborative text classification. In: Working Notes of CLEF 2025 - Conference and Labs of the Evaluation Forum. CEUR-WS.org (2025)

96. Xian, T., et al.: DBG: human-AI collaborative text classification with DeBERTa-enhanced contextual and geometric attention. In: Working Notes of CLEF 2025 - Conference and Labs of the Evaluation Forum. CEUR-WS.org (2025)

97. Xianbing, D., Zhongyuan, H., Jiangao, P., Kaiyin, S.: Multilingual text detoxification system based on parallel architecture: an intelligent approach integrating local models and large language models. In: Working Notes of CLEF 2025 - Conference and Labs of the Evaluation Forum. CEUR-WS.org (2025)

98. Yang, J., Yan, K.: Genre-aware contrastive learning for AI text detection: a RoBERTa-based approach. In: Working Notes of CLEF 2025 - Conference and Labs of the Evaluation Forum. CEUR-WS.org (2025)

99. Zaidi, S., Ahmed, H., Akbar, S., Shakeel, Z., Alvi, F., Samad, A.: Team nexus interrogators at PAN: Voight-Kampff generative AI detection. In: Working Notes of CLEF 2025 - Conference and Labs of the Evaluation Forum. CEUR-WS.org (2025)

100. Zangerle, E., Mayerl, M., , Potthast, M., Stein, B.: Overview of the style change detection task at PAN 2021. In: CLEF 2021 Labs and Workshops, Notebook Papers. CEUR-WS.org (2021)

101. Zangerle, E., Mayerl, M., , Potthast, M., Stein, B.: Overview of the style change detection task at PAN 2022. In: CLEF 2022 Labs and Workshops, Notebook Papers. CEUR-WS.org (2022)

102. Zangerle, E., Mayerl, M., Potthast, M., Stein, B.: Overview of the style change detection task at PAN 2023. In: CLEF 2023 Labs and Workshops, Notebook Papers. CEUR-WS.org (2023)

103. Zangerle, E., Mayerl, M., Potthast, M., Stein, B.: Overview of the multi-author writing style analysis task at PAN 2024. In: Working Notes of CLEF 2024 - Conference and Labs of the Evaluation Forum. CEUR-WS.org (2024)

104. Zangerle, E., Mayerl, M., Potthast, M., Stein, B.: Overview of the multi-author writing style analysis task at PAN 2025. In: CLEF 2025 Working Notes. CEUR-WS.org (2025)

105. Zangerle, E., Mayerl, M., Specht, G., Potthast, M., Stein, B.: Overview of the style change detection task at PAN 2020. In: CLEF 2020 Labs and Workshops, Notebook Papers (2020)

106. Zangerle, E., Tschuggnall, M., Specht, G., Stein, B., Potthast, M.: Overview of the style change detection task at PAN 2019. In: CLEF 2019 Labs and Workshops, Notebook Papers (2019)

107. Zhang, Z., et al.: DBA: a hybrid neural network model for generative human-AI collaborative text classification. In: Working Notes of CLEF 2025 - Conference and Labs of the Evaluation Forum. CEUR-WS.org (2025)

108. Zheng, M., et al.: StarBERT: a hybrid neural network model for human-AI collaborative text classification. In: Working Notes of CLEF 2025 - Conference and Labs of the Evaluation Forum. CEUR-WS.org (2025)

Overview of QuantumCLEF 2025: The Second Quantum Computing Challenge for Information Retrieval and Recommender Systems at CLEF

Andrea Pasin[1]([⊠]), Maurizio Ferrari Dacrema[2], Washington Cunha[3], Marcos André Gonçalves[3], Paolo Cremonesi[2], and Nicola Ferro[1]

[1] University of Padua, Padua, Italy
`andrea.pasin.1@phd.unipd.it, nicola.ferro@unipd.it`
[2] Politecnico di Milano, Milan, Italy
`{maurizio.ferrari,paolo.cremonesi}@polimi.it`
[3] Universidade Federal de Minas Gerais, Belo Horizonte, Brazil
`{washingtoncunha,mgoncalv}@dcc.ufmg.br`

Abstract. Quantum Computing (QC) is an emerging research field that is attracting significant interest from the scientific community due to its potential to solve complex problems more efficiently than traditional computers by leveraging the principles of quantum physics. Even though real quantum computers exist, at the moment we are still in the early stages of development of these innovative technologies, and many of their capabilities and limitations are yet to be discovered.

In this work, we present an overview of the second edition of QuantumCLEF, a lab that focuses on the application of Quantum Annealing (QA), a specific QC paradigm, for different tasks related to IR and RS. The main objective of the QuantumCLEF lab is to investigate QC, raise awareness, and develop and evaluate new QC algorithms for different applications. This lab represents a great chance for researchers and industry practitioners to understand more about this new field by having access to real quantum computers, which are still not easily accessible nowadays.

This edition consisted of three different tasks: Feature Selection for IR and RS systems, Instance Selection for IR systems, and Clustering for IR systems. There have been a total of 44 teams that registered for this lab, and eventually, 5 teams managed to successfully submit their runs following the lab guidelines. Participants have been provided with examples, tutorials, and comprehensive materials due to the novelty of the QC field, allowing them to understand how QA works and how to program quantum annealers.

1 Introduction

Quantum Computing (QC) is a new computational paradigm that focuses on leveraging the quantum mechanical principles such as superposition, entanglement, and tunnelling to perform calculations. Quantum computers have the potential to revolutionize the way we solve tasks in several fields, especially when dealing with complex combinatorial problems, due to their capabilities in exploring large search spaces very efficiently [20].

Nowadays, there already exist real quantum computers. However, we are still in their early stages of development, and researchers are studying the capabilities of these machines while, at the same time, trying to overcome some of their current limitations. In fact, quantum computers are really delicate to work with, since noise (e.g., thermal fluctuations, electromagnetic interferences) can easily break computation. Furthermore, existing quantum computers have a limited number of qubits, which limits the size of problems that can be targeted. Nevertheless, it is already possible to have access to these technologies to learn how they can be applied and what benefits they offer.

In 2024, the QuantumCLEF lab [31–34] was started with the goal of studying how QC technologies could be used in the fields of IR and RS. It represented the first evaluation challenge, giving participants access to real quantum computers to develop and evaluate algorithms for different practical tasks. This year, a second edition of the lab was conducted to further shed light on the potential of QC for IR and RS, with the following goals:

– develop and evaluate new QC algorithms for IR and RS, comparing the results (efficiency and effectiveness) with traditional approaches;
– gather all resources and data for future researchers to compare their results with the ones achieved during the lab;
– allow participants to learn more about QC through comprehensive materials and to use real quantum computers, which are still not easily accessible to the public;
– raise the awareness of the potential of QC and form a new research community around this new field.

In this paper, we present the overview of the second edition of QuantumCLEF held in 2025 [30]. Similarly to the previous edition, this edition has focused on the usage of Quantum Annealing (QA), a specific QC paradigm that can be used to tackle optimization problems. We have granted participants access to the state-of-the-art QA devices (quantum annealers) produced by D-Wave, one of the leading companies in this sector. The QA paradigm is easier to understand with respect to the Universal Gate-Based paradigm. Furthermore, D-Wave provides several tools and libraries to program quantum annealers without requiring a particularly deep knowledge of the quantum physics governing these devices.

QuantumCLEF 2025 was composed of three main tasks:

– **Task 1**: Feature Selection for IR and RS;
– **Task 2**: Instance Selection [11, 12, 29] for IR;
– **Task 3**: Clustering for IR.

Participants were invited to design and implement their own algorithms to address the proposed tasks using both QA and Simulated Annealing (SA). SA is a well-established optimization technique that shares some similarities with QA, but it operates entirely on classical hardware and does not exploit or simulate any quantum phenomena. Given the novelty and technical complexity, participants were provided with extensive support materials, including instructional videos, slides, tutorials, and code examples, to facilitate their understanding of QA and to help them solve the tasks using quantum annealers.

To ensure easy accessibility, the KIMERA infrastructure [35] has been used. This platform not only allowed participants to run their algorithms on actual quantum annealers but also helped improve reproducibility, comparability, and simplification of the workflow management.

A total of 44 teams registered for the lab, out of which 6 teams actively participated. Finally, 5 teams submitted at least one run. Due to an unforeseen circumstance that prevented access to quantum computers in the last 5 days of the challenge, many of the teams' submissions involve the usage of SA only. Despite this issue, the submitted solutions featured innovative and well-structured techniques that remain highly relevant to the proposed problems. Many of these approaches are readily adaptable to QA, underlining their potential for future quantum implementations. Overall, the experimental results show similar trends to the previous 2024 edition: QA or Hybrid (H) approaches usually perform on par with SA and other classical techniques, with the added advantage of generally higher efficiency in terms of annealing time.

The paper is organized as follows: Sect. 2 discusses related works; Sect. 3 presents the tasks of the QuantumCLEF 2024 lab while Sect. 4 introduces the lab's setup; Sect. 5 shows and discusses the results achieved by the participants; finally, Sect. 6 draws some conclusions and outlooks some future work.

2   Related Works

In this section, we provide an introduction to QA and SA. Furthermore, we summarize the tasks and the main outcomes of the previous QuantumCLEF edition.

2.1   Background on Quantum and Simulated Annealing

Quantum Annealing. QA is a QC paradigm based on special-purpose devices, known as quantum annealers, designed to solve optimization problems that are properly formulated using specific formats. The fundamental idea behind a quantum annealer is to encode the problem into the energy configuration of a physical system. Quantum-mechanical phenomena such as superposition, entanglement, and tunneling are then exploited to guide the system toward its lowest energy state, which corresponds to the optimal solution of the original problem.

To use quantum annealers for a given optimization problem, first, it must be formulated as a minimization problem using the Quadratic Unconstrained Binary Optimization (QUBO) formulation [19], a well-established mathematical framework. The QUBO problem is defined as:

$$\min \quad y = x^T Q x \tag{1}$$

where x is a vector of binary decision variables, and Q is a matrix of constant values encoding the problem to be solved.

An additional step, known as *minor embedding*, is required to map this mathematical formulation onto the physical hardware of the quantum annealer, taking into account the limited number of available qubits and their connectivity. In fact, each Quantum Processing Unit (QPU) has a specific hardware topology that can be represented as a graph: vertices correspond to qubits, and edges correspond to couplers (interactions) between qubits. Minor embedding involves selecting physical qubits to represent logical decision variables. If a variable requires more connections than physically available on the QPU, a chain of qubits is used to represent it, and its connections are distributed across these qubits.

As a consequence, the number of physical qubits required to represent a problem can be significantly larger than the number of logical variables. Minor embedding is, in itself, an *NP*-hard problem, typically solved using heuristic algorithms [9].

If the problem is too large to fit directly onto the QPU, D-Wave offers a H approach, which decomposes the problem into smaller sub-problems, and solves them using a combination of classical and quantum methods.

Constraints can be incorporated into the problem formulation using penalty terms $P(x)$ [45], which assign a cost to infeasible solutions. These represent *soft constraints*, meaning they push quantum annealers to favor feasible solutions but do not strictly enforce compliance. In other words, solutions violating these constraints are penalized within the optimization process and thus unlikely to be returned by quantum annealers. These penalties are added to the original objective function to obtain the final objective function:

$$\min \quad C(x) = y + P(x) \tag{2}$$

The relative influence of these penalties can be tuned via hyperparameters, depending on how *strict* the constraints need to be.

In summary, the process of solving a problem using a quantum annealer involves the following stages [45]:

1. **Formulation**: express the problem as a QUBO.
2. **Embedding**: generate a minor embedding of the QUBO onto the specific QPU architecture.
3. **Data Transfer**: submit the problem and its embedding to the quantum annealer via the global network.
4. **Annealing**: execute the quantum annealing process, which involves programming the QPU, sampling multiple solutions, and reading the results. Due to the stochastic nature of the process, it is typically repeated hundreds of times to obtain a distribution of candidate solutions. These are then filtered and evaluated to select the most optimal and feasible one.

Once the QUBO has been successfully embedded and transferred to the quantum annealer, solving the problem typically takes only a few milliseconds.

Simulated Annealing. SA is a consolidated meta-heuristic that can be run on traditional hardware [6,7,44]. It is a probabilistic algorithm that can be used to find the global minimum of a given cost function, even in the presence of many local minima. It is based on an iterative process that starts from an initial solution and tries to improve it by randomly perturbing it. The cost function is represented by the QUBO problem formulation, similar to what would be used for QA. In SA, there is no minor embedding phase since the problem is directly solved on a traditional machine.

We underline that SA is an optimization algorithm different from QA; it is not a simulation of QA on a traditional machine, and, therefore these two algorithms are not equivalent. However, SA can be used for benchmarking purposes to compare the performance of QA with respect to a traditional hardware counterpart.

Moreover, access to quantum annealers in QuantumCLEF is limited to ensure a fair distribution of resources. Therefore, SA can also be used to perform initial experiments to assess a QUBO formulation feasibility without affecting the available quota in the quantum environment.

2.2 QuantumCLEF 2024

The QuantumCLEF 2024 lab [32,34], presented at CLEF 2024, explored the application of QA for IR and RS. The lab consisted in two tasks:

- **Feature Selection:** this task focused on selecting optimal feature subsets for training IR and RS systems using QA.
- **Clustering:** Leveraging document embeddings, this task involved grouping similar documents via QA to speed up dense retrieval.

Participants were given access to D-Wave hardware through CINECA and used the KIMERA [35] infrastructure for enhanced resource access, comparability, and reproducibility.

Out of 26 registered teams, 7 submitted official runs [1,2,17,26,28,36,41]. The results demonstrated the feasibility of using real quantum annealers in IR and RS, encouraging interdisciplinary research and offering benchmarks for further exploration of QC in real-world applications.

3 Tasks

QuantumCLEF 2025, addresses three different tasks involving computationally intensive problems: Feature Selection, Instance Selection, and Clustering. The main goals for each task are:

- finding some possible QUBO formulations of the considered problem;
- map the QUBO formulation to the physical QPU;
- evaluating the QA approach compared to corresponding traditional approaches to assess both efficiency and effectiveness.

For each task, we provided Jupyter Notebooks as entry points for partici-pants, helping them to learn how to program quantum annealers and to success-fully complete the tasks following the submission guidelines. Additionally, we supplied the tutorial slides presented at ECIR and SIGIR [15,16], which intro-duce the core concepts of QC and QA. A video tutorial[1] was also made available, demonstrating the usage of our KIMERA infrastructure and the provided note-books.

For each tasks, participants are asked to submit their runs using both QA and SA. It was recommended that participants use SA for testing purposes during the development phase of their algorithms, due to the limited availability of QA resources.

3.1 Task 1 Quantum Feature Selection

This task focuses on formulating the well-known *NP-Hard* feature selection prob-lem to make it solvable through a quantum annealer, similarly to what has already been done in previous works [14,27].

Objectives. Feature Selection is a widespread problem for both IR and RS. It consists of identifying a subset of the available features (e.g., the most infor-mative, less noisy, less redundant, etc.) to train a learning model for improved efficiency and, in some cases, effectiveness. This problem is very impacting since many IR and RS systems involve the optimization of complex learning models, and reducing the dimensionality of the input data can improve their perfor-mance. Therefore, in this task, we aim to understand if QA can be applied to solve this problem more efficiently and effectively, exploiting its capability of exploring large problem spaces in a short amount of time.

Feature selection fits very well the QUBO formulation, in which there is one variable x per feature and its value indicates whether it should be selected or not. The challenge lies in designing the objective function, i.e., the matrix Q (see Eq. 1).

Sub-Tasks. Task 1 is divided into two sub-tasks:

- **Task 1A**: Feature Selection for IR. This task involves selecting the optimal subset of features using QA and SA that will be used to train a LambdaMART [8] model according to a Learning-To-Rank framework;
- **Task 1B**: Feature Selection for RS. This task involves selecting the optimal subset of features using QA and SA that will be used to train a kNN recom-mendation system model. The item-item similarity is computed with cosine on the feature vectors, a shrinkage of 5 is added to the denominator and the number of selected neighbors for each item is 100.

---

[1] https://www.youtube.com/watch?v=fKrnaJn40Kk/ (accessed June 17, 2025).

Datasets. For Task 1A, we decided to employ the well-known MQ2007 [38] and the Istella S-LETOR [22] datasets. MQ2007 represents an easier challenge since it has 46 features, allowing direct embedding of the problem formulations inside the QPU of quantum annealers. Istella instead has 220 features, making it impossible to embed problem formulations directly, thus requiring some further processing steps for the participants to fit the problem into the current physical QPU hardware.

For Task 1B instead, we decided to employ a custom dataset of music recommendations containing 1.9 thousand users and 18 thousand items. The dataset contains both collaborative data, with 92 thousand implicit user-item interactions, as well as two different sets of item features that are derived from item descriptions and user-provided tags, called Item Content Matrix (ICM). The small set, ICM_100, includes 100 features and can be embedded directly on the QPU with small adjustments, the large set, ICM_400, has 400 features and requires significant pruning to fit in the QPU or the use of Hybrid methods. Both sets of features contain noisy and redundant features.

Evaluation Measures. The official evaluation measure for both Task 1A and Task 1B is nDCG@10.

Baseline. For sub-task 1A the baseline is a Feature Selection model that uses a Recursive Feature Elimination approach paired with a Linear Regression model to select the most relevant subset of features.

For sub-task 1B the baseline is a kNN recommendation system model that uses all the available features. The hyperparameters are the same used for the model computed on the selected features, i.e., the item-item similarity is computed with cosine, adding a shrink term of 5 to the denominator, and the number of neighbors is 100.

Runs Format. Participants in both tasks 1A and 1B can submit a maximum of 5 runs per dataset using QA or Hybrid methods and a maximum of 5 runs using SA. Each run that uses QA or Hybrid methods should correspond to a run that employs SA to make a fair comparison between them.

The results of the run must be a text file which lists the features that were selected, one per line. The discarded features are not reported in the run file. Furthermore, the last line must report the list of IDs associated with the problems solved using QA, SA, or Hybrid to obtain the final subset of features by the considered approach.

Each run file must be left in each team's workspace in a specific directory called /config/workspace/submissions, which is already available.

The submission file name should comply with the format
[Task]_[Dataset]_[Method]_[Groupname]_[SubmissionID].txt, where:

- **[Task]**: it should be either *1A* or *1B* based on the task the submission refers to;

- **[Dataset]**: it should be either *MQ2007*, *Istella*, *100_ICM* or *400_ICM* based on the dataset used;
- **[Method]**: it should be either *QA* or *SA* based on the method used;
- **[Groupname]**: the team name;
- **[SubmissionID]**: a custom submission ID that must be the same for the submissions using the same algorithm but performed with different methods (e.g., QA or SA).

3.2   Task 2 Quantum Instance Selection

This task focuses on formulating the Instance Selection problem and solving it with quantum annealers. Instance Selection focuses on identifying and selecting the most representative samples in a dataset to train a learning model. In this way, the training phase will be more efficient and the trained model will achieve a comparable level of effectiveness.

Objectives. Instance Selection is an important task in the fields of IR and RS, particularly when dealing with large-scale datasets. By selecting a representative subset of instances from a larger dataset, Instance Selection aims to reduce the time required to train a learning model and maintain (or even improve) its effectiveness [11–13]. It has already been shown that using QA for this task is possible and it allowed to reduce the training dataset size by up to 28% without compromising model performance [29].

In this task, the participants should identify subsets of the considered datasets that are then used to train a Llama3.1 7B model [42] for Text Classification and Sentiment Analysis. A good Instance Selection approach should improve the training efficiency while retaining the model's effectiveness.

Datasets. For this task, we considered two different datasets:

- Vader NYT: A sentiment-labeled dataset from New York Times articles;
- Yelp Reviews: A collection of customer reviews labeled with sentiment scores.

Both datasets have been split into training and test sets. The training sets have also been split into 5 folds following a 5-fold cross-validation approach.

Evaluation Measures. The performance will be evaluated on the test sets from the 5-fold cross-validation splits using the Macro-F1 score, ensuring a fair and comprehensive assessment of effectiveness. Furthermore, the evaluation will also consider the time required to fine-tune the model and the dataset reduction rate.

Baseline. For this task, the baseline is the Llama3.1 7B model trained on the whole initial datasets.

Runs Format. Participants can submit a maximum of 5 runs per dataset using QA or Hybrid methods and a maximum of 5 runs using SA. Each run that uses QA or Hybrid methods should correspond to a run that employs SA. In this way, it is possible to make a fair comparison between them.

The results of the run must be a text file which lists the documents that were selected, one per line. The discarded documents are not reported in the run file. Furthermore, the last line must report the list of IDs associated with the problems solved using QA, SA, or Hybrid to obtain the final subset of features by the considered approach.

Each run file must be left in each team's workspace in a specific directory called */config/workspace/submissions*, which is already available.

The submission file name should comply with the format *[Dataset]_[FoldNumber]_[Method]_[Groupname]_[SubmissionID].txt*, where:

- **[Dataset]**: it should be either *Vader* or *Yelp* based on the dataset used;
- **[FoldNumber]**: it should be a number in $[0, 4]$, representing the considered fold;
- **[Method]**: it should be either *QA* or *SA* based on the method used;
- **[Groupname]**: the team name;
- **[SubmissionID]**: a custom submission ID that must be the same for the submissions using the same algorithm but performed with different methods (e.g., QA or SA).

3.3   Task 3 Quantum Clustering

This task focuses on the formulation of the Clustering problem in such a way that it can be solved with a quantum annealer. It involves grouping the items according to their characteristics. Thus, "similar" items fall into the same group while different items belong to distinct groups.

Objectives. Clustering is important for IR and RS as it helps organize large collections, assists users in exploring content, and provides similar search results for a query.

This task focuses on IR in a document retrieval setting where documents are represented as embeddings from a Transformer model. Each document is a vector, and clusters are based on distances between vectors, representing dissimilarity. Participants should use QA and SA methods to cluster documents into 10, 25, and 50 clusters, reporting the centroids and their associated documents.

Clustering enables faster search by matching the query to the nearest centroid and retrieving documents only from that cluster instead of the entire collection.

Clustering fits the QUBO formulation, and various methods have already been proposed [3,4,43]. Most of these methods involve the usage of one variable per document, thus making it very hard to consider large datasets due to the limited number of physical qubits and interconnections between them. There are ways to overcome this issue, such as by applying a coarsening or a hierarchical approach.

Datasets. For this task, we considered a custom split of the ANTIQUE [21] dataset containing 6486 documents, 200 queries, and manual relevance judgments. Each document and each query have been transformed into a corresponding embedding with the pre-trained **all-mpnet-base-v2** model[2]. The queries are divided into 50 for the Training Dataset and 150 for the Test Dataset.

Evaluation Measures. The official evaluation measures for Task 2 are:

– the Davies-Bouldin Index to measure the overall cluster quality without considering the document retrieval phase;
– nDCG@10 to measure the retrieval effectiveness based on the clusters found.

Baseline. For this task, the baseline is a traditional k-Medoids approach using the cosine distance as a distance function.

Runs Format. Participants can submit a maximum of 5 runs for each number of clusters (i.e., 10, 25, 50) using QA or Hybrid methods and a maximum of 5 runs using SA. Each run that uses QA or Hybrid methods should correspond to a run that employs SA. In this way, it is possible to make a fair comparison between them.

The run file must be a text file (JSON formatted) with a list of 10, 25, and 50 vectors that represent the final centroids achieved through their clustering algorithm. Each centroid should also be followed by the list of documents that belong to the given cluster. Furthermore, the last line must report the list of IDs associated with the problems solved using QA, SA, or Hybrid to obtain the final clusters by the considered approach.

Each run file must be left in each team's workspace in a specific directory called */config/workspace/submissions*, which is already available.

The submission file name should comply with the format
[Centroids]_ [Method]_[Groupname]_ [SubmissionID].txt, where:

– **[Centroids]**: it should be either 10, 25, or 50 based on the number of centroids;
– **[Method]**: it should be either *QA* or *SA* based on the method used;
– **[Groupname]**: the team name;
– **[SubmissionID]**: a custom submission ID that must be the same for the submissions using the same algorithm but performed with different methods (e.g., QA or SA).

4 Lab Setup

In this section, we detail the infrastructure used during this lab, and we present the guidelines the participants had to comply with to submit their runs.

---

[2] https://huggingface.co/sentence-transformers/all-mpnet-base-v2.

Table 1. The hardware resources corresponding to the machine where KIMERA was deployed and to the participants' workspaces.

Hardware resources.			
-	**CPU**	**RAM**	**Hard Drive**
Infrastructure	32 cores	128 GB RAM	1 TB HDD
Workspace	1200 millicores	12 GB RAM	20 GB HDD

Table 2. The monthly quotas to use quantum resources according to the tasks.

Monthly quotas for the tasks.				
Task	February	March	April	May
Task 1: Feature Selection	30 s	30 s	30 s	30 s*
Task 2: Instance Selection	120 s	120 s	120 s	120 s*
Task 3: Clustering	120 s	120 s	120 s	120 s*

* From 05/05/2025 to 10/05/2025 (submissions deadline), quantum annealers were unavailable

4.1 Infrastructure

Access to quantum annealers is limited. D-Wave enforces it through monthly time quotas and the use of API keys to monitor and control usage. Since sharing our API key with participants is not possible, we decided to use KIMERA, which enables participants to access quantum annealers without requiring their own API keys or agreements with D-Wave. It also ensures fairness and reproducibility, since all participants are provided with identical computational workspaces, each equipped with the same CPU and RAM resources. This uniform environment facilitates consistent efficiency measurements. Moreover, it also allows participants to monitor quotas and develop, test, and run code directly from their browsers, eliminating the need for local experiments and for owning computational resources. Finally, all participants' submissions are stored in a database to track quota usage and collect data for analysis and reporting.

The final infrastructure was deployed on a machine hosted at the Department of Information Engineering at the University of Padua. Table 1 reports the specifications of the hardware resources corresponding to the machine and to each team's workspace. All participants were given the same monthly quota to use quantum resources. Table 2 reports the monthly quotas according to the three tasks.

4.2 General Guidelines

Each team has access to its personal area inside our infrastructure with the credentials that have been provided to them. All runs must be executed by using the workspaces that have been created for each one of the participating teams, thus ensuring a fair comparison and easy reproducibility.

Table 3. The teams that participated and submitted to QuantumCLEF 2025.

Team	Affiliation	Country
DS@GT qClef [37]	Georgia Institute of Technology	United States
FAST-NU [40]	National University of Computer and Emerging Sciences	Pakistan
GPLSI [10]	Language Processing and Information Systems (GPLSI), University of Alicante.	Spain
Malto [18]	Politecnico di Torino	Italy
SINAI-UJA [24]	Universidad de Jaén	Spain

All participants cannot exceed their given quotas (see Table 2) to execute problems on quantum devices. The quotas can be monitored by each participating team through a dashboard that is constantly being automatically updated, reporting usages of the different methods (i.e., QA, H, and SA) and some general statistics.

All participants' runs must follow the file formats that are detailed in Section 3.1,3.2, and 3.3 for an easy automated evaluation with the prepared scripts.

5   Results

In this section, we present the results achieved by the participants and we discuss their approaches. Out of the 44 registered teams, 5 teams managed to upload some final runs. In total, the number of runs is 69, considering both SA, QA, and H(H was introduced in Sect. 2.1). Table 3 reports the 5 teams that correctly participated and submitted some final runs.

In total, throughout the entire lab, participants have submitted 7183 problems (vs 976 in QuantumCLEF 2024). Specifically, 6333 of them were solved with SA, while 848 were solved using QA and 2 with the H method. The total execution time of SA has been more than 4 h, while the total QA and H execution time has been roughly 1 min.

The QA execution time in this whole Section refers to the *Annealing* phase as described in Sect. 2.1, therefore it includes the time required to program the QPU, sampling, and reading the result. The embedding time and network latencies are not taken into account and are left to be considered for possible future editions of the QuantumCLEF lab.

5.1   Task 1A: Quantum Feature Selection for IR

Here we present the results achieved by the teams participating in task 1A, considering each dataset separately.

MQ2007 Dataset. As it is possible to see in Table 4, teams considered different numbers of features in their submissions. In general, we can observe that most of the submissions achieve similar nDCG@10 values, especially when considering a number of features $n_f \geq 20$.

Table 4. The results for Task 1A on the MQ2007 dataset. Rows marked in grey ( ) represent the results achieved with QA/H, rows marked in yellow ( ) refer to the baselines' results, and the remaining refer to results achieved with SA.

Group	Submission id	nDCG@10	Annealing time (ms)	Type	N° features
DS@GT qClef	1A_MQ2007_SA_DS@GT qClef_pfi-k-25-cmi	0.4500	2219	SA	25
DS@GT qClef	1A_MQ2007_SA_DS@GT qClef_pfi-k-20-cpfi	0.4318	2185	SA	20
DS@GT qClef	1A_MQ2007_SA_DS@GT qClef_mi-k-25	0.4510	2214	SA	25
DS@GT qClef	1A_MQ2007_SA_DS@GT qClef_mi-k-15	0.4485	2136	SA	15
DS@GT qClef	1A_MQ2007_SA_DS@GT qClef_pfi-k-30-cmi	0.4523	2157	SA	30
DS@GT qClef	1A_MQ2007_QA_DS@GT qClef_mi-1	0.4436	183	QA	15
DS@GT qClef	1A_MQ2007_QA_DS@GT qClef_mi-2	0.4552	160	QA	13
FAST-NU	MQ2007_SA_FAST-NU_SA-2918	0.4212	4073	SA	15
FAST-NU	1A_MQ2007_SA_FAST-NU_SA-2915	0.3358	4164	SA	15
FAST-NU	1A_MQ2007_QA_FAST-NU_ae194be3-5267-45dd-aa0e-36a58579d719	0.4311	339	QA	15
FAST-NU	1A_MQ2007_QA_FAST-NU_26065450-e42a-4d92-bfb9-ff367d132142	0.4409	287	QA	15
FAST-NU	1A_MQ2007_QA_FAST-NU_1bba5207-9919-4048-b4a0-80f89b03f603	0.4375	275	QA	15
SINAI-UJA	response_k21_nr3000	0.4530	3448	SA	21
SINAI-UJA	response_k23_nr3000	0.4478	6632	SA	23
SINAI-UJA	response_k25_nr3000	0.4510	2998	SA	25
SINAI-UJA	response_k27_nr3000	0.4438	6637	SA	27
SINAI-UJA	response_k29_nr3000	0.4491	6614	SA	29
SINAI-UJA	response_k21_nr100	0.4580	34	QA	21
SINAI-UJA	response_k23_nr100	0.4437	37	QA	23
SINAI-UJA	response_k25_nr100	0.4550	31	QA	25
SINAI-UJA	response_k27_nr100	0.4425	34	QA	27
SINAI-UJA	response_k29_nr100	0.4528	34	QA	29
BASELINE	ALL_FEATURES	0.4473	-	-	46
BASELINE	RFE HALF_FEATURES	0.4450	-	-	23

Figure 1 shows the nDCG@10 values and Annealing timings of the runs that used QA and SA. From this figure it is possible to see that, in terms of efficiency (i.e., Annealing time), runs using QA required a substantially shorter amount of time with respect to SA. On average, QA required ≈ 26.83 times less compared to SA, thus representing a more efficient alternative.

Considering effectiveness, QA seems to be performing more consistently. In fact, on average it performs ≈ 1.02 times better compared to SA. Moreover, SA led to 2 outliers that underperformed with respect to the overall results.

Teams adopted different approaches to address this task:

- Team **DS@GT qClef**, inspired by a previous work [25], employed QUBO formulations that focused on different combinations of importance measures and redundancy measures for feature selection [37];
- Team **FAST-NU** adopted a Mutual Information-based approach, selecting features that maximize the information associated with relevance [40];
- Team **SINAI-UJA** also leveraged a Mutual Information-based approach, but also focused on post-processing the returned samples through normalization and projection techniques, aiming to find better solutions [24].

Istella Dataset. As it is possible to see in Table 5, also in this case, different numbers of features were considered among the submissions. It is interesting to

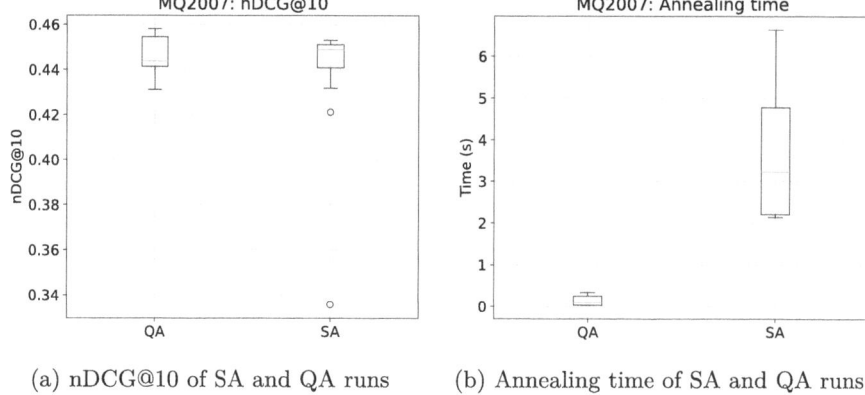

(a) nDCG@10 of SA and QA runs   (b) Annealing time of SA and QA runs

Fig. 1. The box plots of the nDCG@10 values and Annealing timings associated with the runs using QA and SA on the MQ2007 dataset.

Table 5. The results for Task 1A on the Istella dataset. Rows marked in grey ( ) represent the results achieved with QA/H, rows marked in yellow ( ) refer to the baselines' results, and the remaining refer to results achieved with SA.

Group	Submission id	nDCG@10	Annealing time (ms)	Type	N° features
DS@GT qClef	1A_Istella_SA_DS@GT qClef_mi_25	0.6025	126631	SA	25
DS@GT qClef	1A_Istella_SA_DS@GT qClef_mi_30	0.5104	133814	SA	30
DS@GT qClef	1A_Istella_SA_DS@GT qClef_mi_50	0.6524	159964	SA	50
DS@GT qClef	1A_Istella_SA_DS@GT qClef_mi_60	0.6682	173222	SA	60
DS@GT qClef	1A_Istella_SA_DS@GT qClef_mi_70	0.6523	184047	SA	70
DS@GT qClef	1A_Istella_QA_DS@GT qClef_mi_50	0.5586	9987	H	50
BASELINE	ALL_FEATURES	0.7146	-	-	220
BASELINE	RFE HALF_FEATURES	0.5560	-	-	110

see that the baseline method employing Recursive Feature Elimination for the extraction of 110 features performed worse than most of the participants' runs that considered a much lower number of features. Furthermore, running Recursive Feature Elimination to keep the top 110 features required a considerable amount of time (almost 2 h of computation) and a considerable amount of RAM (24 GB), which is much higher than the teams' workspace specifications.

Overall, it is possible to see how a different choice of features can lead to very different outcomes. In particular, the baseline *RFE_HALF_FEATURES*, performed poorly compared to other submissions that considered a lower number of kept features. This can probably be due to a poor feature choice from the baseline. Team **DS@GT qClef** leveraged QUBO formulations including importance measures and redundancy measures for feature selection [37].

Table 6. The results for Task 1B. Rows marked in grey ( ) represent the results achieved with QA/H, rows marked in yellow( ) refer to the baselines' results, and the remaining refer to results achieved with SA.

Dataset	Group	Submission id	nDCG@10	Annealing time (ms)	Type	N° features
ICM_100	Malto	1B_100_ICM_SA_MALTO_1B - 100_ICM submission	0.0207	6149	SA	51
	BASELINE	ALL_FEATURES	0.0226	-	-	100
ICM_400	Malto	1B_400_ICM_SA_MALTO_1B - 400_ICM submission - 200	0.0294	80781	SA	200
	Malto	1B_400_ICM_SA_MALTO_1B - 400_ICM submission	0.0182	70269	SA	53
	BASELINE	ALL_FEATURES	0.0328	-	-	400

In general, we can also see that the H approach required a much lower Annealing time than the SA counterparts, making use of a combination of QA and traditional hardware computations.

5.2   Task 1B: Quantum Feature Selection for RS

Here we present the results achieved in task 1B. Results are divided according to the two feature sets.

Table 6 presents the performance of different submissions for Task 1B, focusing on the impact of feature selection on the recommendation quality, as measured by nDCG@10, as well as on the computational cost, measured via annealing time.

For the ICM_100 dataset, the SA-based submission from the Malto group using 51 features achieved an nDCG@10 of 0.0207, which is slightly below the baseline score of 0.0226 obtained using all 100 features. This suggests a marginal performance degradation due to feature reduction, but with a 49% reduction in the number of features used could represent a good trade-off for improving efficiency at run-time.

In contrast, for the ICM_400 dataset, results are more varied. The best SA-based configuration (with 200 features) achieved an nDCG@10 of 0.0294, closer to the baseline performance of 0.0328, though still lower. Another SA configuration using only 53 features underperformed (nDCG@10 of 0.0182), indicating that excessive feature reduction can harm recommendation quality. The annealing times were considerably higher for this dataset (around 70–80 s), due to the increased search space associated with the larger feature set.

Team **Malto** addressed Task 1B by computing feature importance using a Random Forest classifier trained on the full feature set. They constructed a QUBO objective function that incorporated these importance scores along with pairwise Pearson correlations to penalize redundant features [18].

Overall, these results highlight that SA can effectively reduce the number of features while maintaining competitive recommendation quality, especially when a moderate number of features are retained. However, aggressive feature selection may lead to substantial performance degradation, and the method incurs non-trivial computational costs, particularly with larger datasets. These observations underscore the importance of balancing performance, computational efficiency,

Table 7. The results for Task 2 on the Yelp dataset averaged over 5 folds. Rows marked in grey ( ) represent the results achieved with QA/H, rows marked in yellow( ) refer to the baselines' results, and the remaining refer to results achieved with SA.

Group	Submission id	Avg Macro F1	Avg Reduction	Avg Fine-Tuning time (s)	Avg Annealing time (ms)	Type
DS@GT qClef	Yelp_SA_qclef_bcos_075	99.5(0.2)	0.25	1548.5(2.8)	25997	SA
DS@GT qClef	Yelp_SA_qclef_it_del_075	99.3(0.3)	0.25	1549.2(1.5)	25784	SA
DS@GT qClef	Yelp_SA_qclef_svc_075	99.3(0.4)	0.25	1550.5(2.6)	25917	SA
DS@GT qClef	Yelp_QA_qclef_bcos	99.4(0.2)	0.274	1500(54.7)	1767	QA
GPLSI	Yelp_SA_gplsi_2-SentimentPairs(docs****Remove***)=just-final...	90.8(5.7)	0.963	170.8(3.8)	35810	SA
GPLSI	Yelp_SA_gplsi_2-SentimentPairs(docs****Remove***)=pair-relatesl...	99.2(0.3)	0.627	822.2(395)	35810	SA
GPLSI	Yelp_SA_gplsi_2-LocalSets	99.4(0.2)	0.512	1045.5(5.3)	28789	SA
GPLSI	Yelp_SA_gplsi_2-SentimentKmeansCard	98.5(1.1)	0.875	338.8(21)	17823	SA
GPLSI	Yelp_SA_gplsi_2-emoconflictCard	98.6(0.5)	0.728	628.2(65.9)	34024	SA
GPLSI	Yelp_QA_gplsi_2-SentimentKmeansCard	98.7(0.2)	0.869	351(25.1)	553	QA
GPLSI	Yelp_QA_gplsi_2-emoconflictCard	98.8(0.6)	0.702	678.8(80.9)	549	QA
Malto	Yelp_SA_MALTO_2-vader_nyt_2L_0	99.2(0.2)	0.751	582(2)	142949	SA
BASELINE	BASELINE_ALL	99.4(0.1)	-	2027.1(1.1)	-	-

and the level of feature reduction when applying feature selection techniques in real-world recommender systems.

5.3 Task 2: Quantum Instance Selection for IR

Here we present the results achieved by the teams participating in Task 2, divided by dataset.

Yelp Dataset. Table 7 reports the results achieved by the different teams on the Yelp dataset for Task 2. As it is possible to see, the teams approached the task by trying different reduction rates (from $\approx 25\%$ to $\approx 96\%$). In particular, the submission *Yelp_SA_qclef_bcos_075* managed to improve the effectiveness of the Llama3.1 7b model with respect to the baseline. This could be due to the removal of noisy documents in that dataset, which lowered the performance of the model if used during the fine-tuning.

Notably, the submission *Yelp_QA_gplsi_2-SentimentKmeansCard* shows how QA was able to produce a reduction rate of $\approx 87\%$ and a high level of effectiveness (i.e., 98.7 vs 99.4 of the full dataset). Generally, it is possible to observe that QA requires less Annealing time than SA, while its performance is overall on par with SA. We list here some of the approaches considered by the teams:

– Team **GPLSI** [10] considered different approaches, such as prioritizing diversity by selecting pairs with very high or low similarity and minimizing semantic overlap (Sentiment Pairs) or selecting training instances using local set geometry through the combination of noise filtering and clustering with Euclidean distance (Local Sets);
– Team **DS@GT qClef** extended a previous approach [29] for selecting document embeddings. The team used cosine similarity for off-diagonal Q-matrix entries, but introduced two new strategies for diagonal terms: weighting instances by their distance to an Support Vector Machine (SVM)decision

Table 8. The results for Task 2 on the Vader dataset averaged over 5 folds. Rows marked in grey ( ) represent the results achieved with QA/H, rows marked in yellow( ) refer to the baselines' results, and the remaining refer to results achieved with SA.

Group	Submission id	Avg Macro F1	Avg Reduction	Avg Fine-Tuning time (s)	Avg Annealing time (ms)	Type
DS@GT qClef	Vader_SA_qclef_svc_075	65.4(7.1)	0.25	1529(2.4)	25530	SA
DS@GT qClef	Vader_SA_qclef_combined_075	65.9(4.7)	0.25	1529.4(3)	25300	SA
DS@GT qClef	Vader_SA_qclef_it_del_075	65.6(3)	0.25	1529.5(2.3)	25348	SA
DS@GT qClef	Vader_SA_qclef_bcos_075	62.5(10.4)	0.25	1528.6(2.2)	25735	SA
DS@GT qClef	Vader_QA_qclef_bcos	62.6(7.5)	0.283	1493.3(83)	1874	QA
GPLSI	Vader_SA_gplsi_2-LocalSets	63.3(4.9)	0.505	1048.3(6.7)	29110	SA
GPLSI	Vader_SA_gplsi_2-SentimentPairs-docs=just-final...	47.4(5.4)	0.962	172.8(5.7)	42408	SA
GPLSI	Vader_SA_gplsi_2-SentimentPairs-docs=pair-related...	62.2(4.1)	0.7	671.8(352.8)	42408	SA
GPLSI	Vader_QA_gplsi_2-SentimentPairs-docs=just-final...	50(64)*	0.835*	172.9(26.9)*	545*	QA
GPLSI	Vader_QA_gplsi_2-SentimentPairs(docs***Remove**)=pair-related...	62.1(1.8)*	0.658*	750.7(2653.2)*	545*	QA
Malto	Vader_SA_MALTO_2 - vader_nyt_2L	63.1(2.5)	0.751	574.5(1.7)	126087	SA
BASELINE	BASELINE_ALL	88.9(0.8)	-	1997.3(5.7)	-	-

* The submission did not include all 5 folds

boundary and measuring their influence using a logistic regression leave-one-out approach. The team tested each method individually and in combination, using batching to handle the datasets [37].

Overall, the results clearly highlight the critical importance of Instance Selection, particularly in the current context where computational efficiency and sustainability have a constantly increasing impact. By carefully selecting representative subsets of data, it is possible to significantly reduce the computational cost associated with fine-tuning large language models. Specifically, the fine-tuning time for the Llama3.1 7B model was reduced by approximately up to a factor of 9, with a negligible performance trade-off (i.e., losing less than 1 absolute point in macro-F1 score).

This demonstrates that substantial gains in efficiency can be achieved without severely compromising model effectiveness, making Instance Selection a key technique for training or fine-tuning models, especially when they are computationally expensive.

Vader Dataset. Table 8 reports the results achieved by the different teams on the Vader dataset for Task 2. Also in this case, the teams considered different reduction rates (from ≈ 25% to ≈ 96%). However, differently from the previous results, all the subsets produced by the different approaches lead to a higher loss in terms of effectiveness of the fine-tuned Llama3.1 7b model.

It is possible to observe that submission *Vader_SA_MALTO_2 - vader nyt_2L* produced a subset of ≈ 25% of the original dataset size while allowing to achieve a Average Macro F1 score which is higher than the subset produced by *Vader_SA_qclef_bcos_075* subset that instead was ≈ 75% of the original dataset size, showing how different datasets can yield to potentially very different results.

For this task, the participating teams adopted similar approaches to the ones used for the other Yelp dataset. Also in this case, the QA approaches required

Table 9. The results for Task 3. Rows marked in grey ( ) represent the results achieved with QA/H, rows marked in yellow( ) refer to the baselines' results, and the remaining refer to results achieved with SA.

N° centroids	Team	Submission id	nDCG@10	DBI	Annealing Time (ms)	Type
	GPLSI	10_SA_gplsi_3-FPS-Medoids	0.5783	7.5147	15375	SA
	GPLSI	10_SA_gplsi_3-SubMedoidsQUBO	0.5579	6.8779	15305	SA
	GPLSI	10_SA_gplsi_CLARA-CLARANS	0.5444	6.6710	15395	SA
10	GPLSI	10_SA_gplsi_MBK-Medoids	0.5600	6.4258	15510	SA
	DS@GT qClef	10_SA_DS@GT qClef_1	0.5800	7.4776	83	SA
	DS@GT qClef	10_SA_DS@GT qClef_2 *	0.0172	4.4706	83	SA
	BASELINE	BASELINE_10	0.5509	7.9892	-	-
	GPLSI	25_SA_gplsi_3-FPS-Medoids	0.5475	5.5577	20875	SA
	GPLSI	25_SA_gplsi_3-SubMedoidsQUBO	0.5298	5.6255	40687	SA
25	GPLSI	25_SA_gplsi_CLARA-CLARANS	0.5310	5.6507	20532	SA
	GPLSI	25_SA_gplsi_MBK-Medoids	0.5193	5.3755	20758	SA
	BASELINE	BASELINE_25	0.5284	6.1201	-	-
	GPLSI	50_SA_gplsi_3-FPS-Medoids	0.5592	4.4531	9869	SA
	GPLSI	50_SA_gplsi_3-SubMedoidsQUBO	0.5148	4.9325	23719	SA
50	GPLSI	50_SA_gplsi_CLARA-CLARANS	0.5017	5.1703	9976	SA
	GPLSI	50_SA_gplsi_MBK-Medoids	0.5383	4.5025	24004	SA
	DS@GT qClef	50_SA_DS@GT qClef_3 *	0.0064	3.4217	228	SA
	BASELINE	BASELINE_50	0.4656	5.3679	-	-

*Dimensionality reduction was applied

considerably less time with respect to the SA approaches while achieving similar trends in terms of effectiveness.

5.4   Task 3: Quantum Clustering for IR

Here we present the results achieved by the teams participating in Task 3. Table 9 reports the results achieved in this task.

It is possible to see that in this task, teams focused only on the usage of SA to solve the clustering problem. From the achieved results, we can notice how both GPLSI and DS@GT qCLEF teams managed to provide some submissions that performed better with respect to a traditional k-medoids baseline in terms of nDCG@10 and Davies-Bouldin Index. This suggests that their proposed approaches managed to successfully identify representative clusters of vectors that could help efficiently and effectively retrieve documents corresponding to queries.

We briefly detail some of the approaches considered by the teams:

– Team **GPLSI** [10] developed a method that filters the dataset embeddings to 150 pivots, optimizes centroid selection through annealing, and assigns documents to improve retrieval efficiency. The pivot selection was carried out in different ways, using heuristic approaches (FPS and CLARA–CLARANS), k-Means, and a technique inspired by the qIIMAS team from the previous QuantumCLEF edition [36] (SubMedoids). These techniques try to choose pivots that provide good coverage of the whole dataset;

– Team **DS@GT qClef** used a two-step approach. First, they applied classical clustering methods (k-Medoids, HDBSCAN [23], GMM, and GMM-HDBSCAN), optionally with dimensionality reduction (e.g., UMAP [5], PaCMAP), to select a manageable subset of instances. In the second step, they applied the QUBO-based k-medoids formulation on this reduced subset [37].

It is really interesting to see how the submission with id *50_SA_gplsi_3-FPS-Medoids* managed to achieve a higher level of nDCG@10 with respect to *BASELINE_10*. This improvement is especially impressive given that the method utilized a substantially larger number of clusters, 50 in total, versus the 10 used by the baseline. Despite the increase in cluster count, which could potentially lead to over-segmentation and reduced retrieval performance, the clusters generated by the GPLSI's approach proved to be more effective.

Submissions *10_SA_DS@GT_qClef_2* and *50_SA_DS@GT_qClef_3* marked with an asterisk (*), involved the usage of UMAP [5] dimensionality reduction technique. The dimensionality of the document vector embeddings was reduced to only 2 dimensions, potentially causing a big loss of information. Due to this aggressive reduction, the effectiveness of these runs was negatively impacted, leading to low nDCG@10 values.

6   Conclusions and Future Work

In this paper, we have presented the overview of the second edition of the QuantumCLEF lab that was held in 2025. QuantumCLEF represents the first lab at CLEF focusing on the study, development, and evaluation of QC algorithms using real quantum computers. This lab was composed of three tasks concerning the problems of Feature Selection, Instance Selection, and Clustering, focusing on computationally complex problems faced by IR and RS systems. Participants used the KIMERA [35] infrastructure for a smooth workflow. The infrastructure has granted participants access to both computational resources and cutting-edge quantum annealers provided by D-Wave, thus giving the possibility of experimenting with real quantum computers.

A total of 44 teams registered for the lab, and 5 of them successfully managed to submit their runs. The results have shown that QA and H managed to achieve comparable results in terms of effectiveness with respect to SA while achieving a higher level of efficiency in terms of Annealing time. This shows that QC is starting to become a powerful technology that could help in the resolution of complex problems, especially in the future once it has matured enough. Furthermore, the QA results are competitive with respect to traditional baselines, showing that QA solutions are able to achieve good levels of effectiveness.

This second edition of the QuantumCLEF lab represented a great opportunity not only to develop and evaluate QC algorithms on **real quantum computers** (quantum technologies are still not easily accessible to the general public) but also to raise awareness of the potential of QC, which is likely to become a powerful technology in the future. In fact, participants were provided with comprehensive materials such as videos, slides, and examples that allowed them to learn how QC and QA work. Moreover, we opted for maximum transparency, allowing participants to work with the actual D-Wave libraries. In this way, participants familiarized themselves with them and, thus, are now able to program quantum annealers even outside our infrastructure to solve other problems in their research field.

In the future, we plan to organize a third edition of QuantumCLEF with different tasks and more challenges. We would like to invest in a more powerful infrastructure that will grant access to more participants and that will provide more resources (in terms of CPU and RAM) to each workspace. If possible, we would also like to extend the infrastructure to include gate-based quantum computers [39], in addition to the already available quantum annealers.

Acknowledgments. We acknowledge the CINECA award under the ISCRA initiative, for the availability of high-performance computing resources and support.

References

1. Almeida, T.M., Matos, S.: Towards a hyperparameter-free QUBO formulation for feature selection in IR. In: Faggioli, G., Ferro, N., Galuscáková, P., de Herrera, A.G.S. (eds.) Working Notes of the Conference and Labs of the Evaluation Forum (CLEF 2024), Grenoble, France, 9-12 September, 2024, CEUR Workshop Proceedings, vol. 3740, pp. 3054–3063, CEUR-WS.org (2024). https://ceur-ws.org/Vol-3740/paper-298.pdf
2. Alvarez-Giron, W., Tellezz-Torres, J., Tovar-Cortes, J., Gómez-Adorno, H.: Team qiimas on task 2 - clustering. In: Faggioli, G., Ferro, N., Galuscáková, P., de Herrera, A.G.S. (eds.) Working Notes of the Conference and Labs of the Evaluation Forum (CLEF 2024), Grenoble, France, 9-12 September, 2024, CEUR Workshop Proceedings, vol. 3740, pp. 3064–3074, CEUR-WS.org (2024). https://ceur-ws.org/Vol-3740/paper-299.pdf
3. Arthur, D., Date, P.: Balanced k-means clustering on an adiabatic quantum computer. Quantum Inf. Process. **20**(9), 1–30 (2021). https://doi.org/10.1007/s11128-021-03240-8
4. Bauckhage, C., Piatkowski, N., Sifa, R., Hecker, D., Wrobel, S.: A QUBO formulation of the k-medoids problem. In: Jäschke, R., Weidlich, M. (eds.) Proceedings of the Conference on "Lernen, Wissen, Daten, Analysen", Berlin, Germany, September 30 - October 2, 2019, CEUR Workshop Proceedings, vol. 2454, pp. 54–63, CEUR-WS.org (2019). https://ceur-ws.org/Vol-2454/paper_39.pdf
5. Becht, E., McInnes, L., Healy, J., Dutertre, C.A., Kwok, I.W., Ng, L.G., Ginhoux, F., Newell, E.W.: Dimensionality reduction for visualizing single-cell data using umap. Nat. Biotechnol. **37**(1), 38–44 (2019)
6. Bertsimas, D., Nohadani, O.: Robust optimization with simulated annealing. J. Glob. Optim. **48**(2), 323–334 (2010). https://doi.org/10.1007/S10898-009-9496-X
7. Bertsimas, D., Tsitsiklis, J.: Simulated annealing. Stat. Sci. **8**(1), 10–15 (1993)
8. Burges, C.J.C.: From RankNet to LambdaRank to LambdaMART: An Overview. Technical Report, Microsoft Research, MSR-TR-2010-82 (2010)

9. Cai, J., Macready, W.G., Roy, A.: A practical heuristic for finding graph minors. CoRR **abs/1406.2741** (2014). http://arxiv.org/abs/1406.2741

10. Consuegra-Ayala, J.P., Morote-Martínez, A., Valero-Abellón, F., Lloret, E., Moreda, P., Palomar, M.: Team gplsi at qclef 2025: Quantum-inspired instance selection and clustering. In: Faggioli, G., Ferro, N., Rosso, P., Spina, D. (eds.) Working Notes of CLEF 2025 - Conference and Labs of the Evaluation Forum, CEUR Workshop Proceedings (2025)

11. Cunha, W., Fernández, A.M., Esuli, A., Sebastiani, F., Rocha, L., Gonçalves, M.A.: A noise-oriented and redundancy-aware instance selection framework. ACM Trans. Inf. Syst. **43**(2), 45:1–45:33 (2025). https://doi.org/10.1145/3705000

12. Cunha, W., França, C., Fonseca, G., Rocha, L., Gonçalves, M.A.: An effective, efficient, and scalable confidence-based instance selection framework for transformer-based text classification. In: Chen, H., Duh, W.E., Huang, H., Kato, M.P., Mothe, J., Poblete, B. (eds.) Proceedings of the 46th International ACM SIGIR Conference on Research and Development in Information Retrieval, SIGIR 2023, Taipei, Taiwan, July 23-27, 2023, pp. 665–674, ACM (2023). https://doi.org/10.1145/3539618.3591638

13. Cunha, W., Viegas, F., França, C., Rosa, T., Rocha, L., Gonçalves, M.A.: A comparative survey of instance selection methods applied to non-neural and transformer-based text classification. ACM Comput. Surv. **55**(13s), 265:1–265:52 (2023). https://doi.org/10.1145/3582000

14. Ferrari Dacrema, M., Moroni, F., Nembrini, R., Ferro, N., Faggioli, G., Cremonesi, P.: Towards feature selection for ranking and classification exploiting quantum annealers. In: Amigó, E., Castells, P., Gonzalo, J., Carterette, B., Culpepper, J.S., Kazai, G. (eds.) SIGIR '22: The 45th International ACM SIGIR Conference on Research and Development in Information Retrieval, Madrid, Spain, July 11 - 15, 2022, pp. 2814–2824, ACM (2022). https://doi.org/10.1145/3477495.3531755

15. Ferrari Dacrema, M., Pasin, A., Cremonesi, P., Ferro, N.: Quantum computing for information retrieval and recommender systems. In: Goharian, N., et al. (eds.) Advances in Information Retrieval - 46th European Conference on Information Retrieval, ECIR 2024, Glasgow, UK, March 24-28, 2024, Proceedings, Part V, LNCS, vol. 14612, pp. 358–362, Springer (2024). https://doi.org/10.1007/978-3-031-56069-9_47

16. Ferrari Dacrema, M., Pasin, A., Cremonesi, P., Ferro, N.: Using and evaluating quantum computing for information retrieval and recommender systems. In: Yang, G.H., Wang, H., Han, S., Hauff, C., Zuccon, G., Zhang, Y. (eds.) Proceedings of the 47th International ACM SIGIR Conference on Research and Development in Information Retrieval, SIGIR 2024, Washington DC, USA, July 14-18, 2024, pp. 3017–3020, ACM (2024). https://doi.org/10.1145/3626772.3661378

17. Fröbe, M., Alexander, D., Hendriksen, G., Schlatt, F., Hagen, M., Potthast, M.: Team openwebsearch at CLEF 2024: quantumclef. In: Faggioli, G., Ferro, N., Galuscáková, P., de Herrera, A.G.S. (eds.) Working Notes of the Conference and Labs of the Evaluation Forum (CLEF 2024), Grenoble, France, 9-12 September, 2024, CEUR Workshop Proceedings, vol. 3740, pp. 3075–3081, CEUR-WS.org (2024). https://ceur-ws.org/Vol-3740/paper-300.pdf

18. Giobergia, F., Savelli, C., Koudounas, A., Baralis, E.: Quantum feature selection from interpretable models using QUBO formulation. In: Faggioli, G., Ferro, N., Rosso, P., Spina, D. (eds.) Working Notes of CLEF 2025 - Conference and Labs of the Evaluation Forum, CEUR Workshop Proceedings (2025)

19. Glover, F., Kochenberger, G., Hennig, R., Du, Y.: Quantum bridge analytics I: a tutorial on formulating and using QUBO models. Ann. Oper. Res. **314**, 141–183 (2022)
20. Gyongyosi, L., Imre, S.: A survey on quantum computing technology. Comput. Sci. Rev. **31**, 51–71 (2019)
21. Hashemi, H., Aliannejadi, M., Zamani, H., Croft, W.B.: ANTIQUE: a non-factoid question answering benchmark. In: Jose, J.M., et al. (eds.) Advances in Information Retrieval - 42nd European Conference on IR Research, ECIR 2020, Lisbon, Portugal, April 14-17, 2020, Proceedings, Part II, LNCS, vol. 12036, pp. 166–173, Springer (2020). https://doi.org/10.1007/978-3-030-45442-5_21
22. Lucchese, C., Nardini, F.M., Orlando, S., Perego, R., Silvestri, F., Trani, S.: Post-learning optimization of tree ensembles for efficient ranking. In: Perego, R., Sebastiani, F., Aslam, J.A., Ruthven, I., Zobel, J. (eds.) Proceedings of the 39th International ACM SIGIR conference on Research and Development in Information Retrieval, SIGIR 2016, Pisa, Italy, July 17-21, 2016, pp. 949–952, ACM (2016). https://doi.org/10.1145/2911451.2914763
23. McInnes, L., Healy, J., Astels, S., et al.: hdbscan: Hierarchical density based clustering. J. Open Source Softw. **2**(11), 205 (2017)
24. Molino-Piñar, L., Collado-Montañez, J., Montejo-Ráez, A.: Sinai team at quantumclef 2025: Quantum feature selection based on energy with d-wave. In: Faggioli, G., Ferro, N., Rosso, P., Spina, D. (eds.) Working Notes of CLEF 2025 - Conference and Labs of the Evaluation Forum, CEUR Workshop Proceedings (2025)
25. Mücke, S., Heese, R., Müller, S., Wolter, M., Piatkowski, N.: Feature selection on quantum computers. Quantum Mach. Intell. **5**(1), 11 (2023)
26. Naebzadeh, A., Eetemadi, S.: NICA at quantum computing CLEF tasks 2024. In: Faggioli, G., Ferro, N., Galuscáková, P., de Herrera, A.G.S. (eds.) Working Notes of the Conference and Labs of the Evaluation Forum (CLEF 2024), Grenoble, France, 9-12 September, 2024, CEUR Workshop Proceedings, vol. 3740, pp. 3087–3095, CEUR-WS.org (2024). https://ceur-ws.org/Vol-3740/paper-302.pdf
27. Nembrini, R., Ferrari Dacrema, M., Cremonesi, P.: Feature selection for recommender systems with quantum computing. Entropy **23**(8), 970 (2021). https://doi.org/10.3390/E23080970
28. Niu, J., Li, J., Deng, K., Ren, Y.: CRUISE on quantum computing for feature selection in recommender systems. In: Faggioli, G., Ferro, N., Galuscáková, P., de Herrera, A.G.S. (eds.) Working Notes of the Conference and Labs of the Evaluation Forum (CLEF 2024), Grenoble, France, 9-12 September, 2024, CEUR Workshop Proceedings, vol. 3740, pp. 3096–3104, CEUR-WS.org (2024). https://ceur-ws.org/Vol-3740/paper-303.pdf
29. Pasin, A., Cunha, W., Gonçalves, M.A., Ferro, N.: A quantum annealing instance selection approach for efficient and effective transformer fine-tuning. In: Oosterhuis, H., Bast, H., Xiong, C. (eds.) Proceedings of the 2024 ACM SIGIR International Conference on Theory of Information Retrieval, ICTIR 2024, Washington, DC, USA, 13 July 2024, pp. 205–214, ACM (2024). https://doi.org/10.1145/3664190.3672515
30. Pasin, A., Ferrari Dacrema, M., Cremonesi, P., Cunha, W., Gonçalves, M.A., Ferro, N.: Quantumclef 2025 - the second edition of the quantum computing lab at CLEF. In: Hauff, C., Macdonald, C., Jannach, D., Kazai, G., Nardini, F.M., Pinelli, F., Silvestri, F., Tonellotto, N. (eds.) Advances in Information Retrieval - 47th European Conference on Information Retrieval, ECIR 2025, Lucca, Italy, April 6-10, 2025, Proceedings, Part V, LNCS, vol. 15576, pp. 450–458, Springer (2025). https://doi.org/10.1007/978-3-031-88720-8_66

434   A. Pasin et al.

31. Pasin, A., Ferrari Dacrema, M., Cremonesi, P., Ferro, N.: qclef: A proposal to evaluate quantum annealing for information retrieval and recommender systems. In: Arampatzis, A., et al. (eds.) Experimental IR Meets Multilinguality, Multimodality, and Interaction - 14th International Conference of the CLEF Association, CLEF 2023, Thessaloniki, Greece, September 18-21, 2023, Proceedings, LNCS, vol. 14163, pp. 97–108, Springer (2023). https://doi.org/10.1007/978-3-031-42448-9_9

32. Pasin, A., Ferrari Dacrema, M., Cremonesi, P., Ferro, N.: Overview of quantumclef 2024: the quantum computing challenge for information retrieval and recommender systems at CLEF. In: Goeuriot, L., et al. (eds.) Experimental IR Meets Multilinguality, Multimodality, and Interaction - 15th International Conference of the CLEF Association, CLEF 2024, Grenoble, France, September 9-12, 2024, Proceedings, Part II, Lecture Notes in Computer Science, vol. 14959, pp. 260–282, Springer (2024). https://doi.org/10.1007/978-3-031-71908-0_12

33. Pasin, A., Ferrari Dacrema, M., Cremonesi, P., Ferro, N.: Quantumclef - quantum computing at CLEF. In: Goharian, N., et al. (eds.) Advances in Information Retrieval - 46th European Conference on Information Retrieval, ECIR 2024, Glasgow, UK, March 24-28, 2024, Proceedings, Part V, Lecture Notes in Computer Science, vol. 14612, pp. 482–489, Springer (2024). https://doi.org/10.1007/978-3-031-56069-9_66

34. Pasin, A., Ferrari Dacrema, M., Cremonesi, P., Ferro, N.: Quantumclef 2024: overview of the quantum computing challenge for information retrieval and recommender systems at CLEF. In: Faggioli, G., Ferro, N., Galuscáková, P., de Herrera, A.G.S. (eds.) Working Notes of the Conference and Labs of the Evaluation Forum (CLEF 2024), Grenoble, France, 9-12 September, 2024, CEUR Workshop Proceedings, vol. 3740, pp. 3032–3053, CEUR-WS.org (2024). https://ceur-ws.org/Vol-3740/paper-297.pdf

35. Pasin, A., Ferro, N.: Kimera: From evaluation-as-a-service to evaluation-in-the-cloud. In: Proceedings of the 48th International ACM SIGIR Conference on Research and Development in Information Retrieval, SIGIR 2025, Padova, Italy, July 13-18, 2025, ACM (2025). https://doi.org/10.1145/3726302.3730298

36. Payares, E., Puertas, E., Santos, J.C.M.: Team QTB on feature selection via quantum annealing and hybrid models. In: Faggioli, G., Ferro, N., Galuscáková, P., de Herrera, A.G.S. (eds.) Working Notes of the Conference and Labs of the Evaluation Forum (CLEF 2024), Grenoble, France, 9-12 September, 2024, CEUR Workshop Proceedings, vol. 3740, pp. 3105–3114, CEUR-WS.org (2024). https://ceur-ws.org/Vol-3740/paper-304.pdf

37. Pomeroy, C., Pramov, A., Thakrar, K., Yendapalli, L.: Quantum annealing for machine learning: Applications in feature selection, instance selection, and clustering. In: Faggioli, G., Ferro, N., Rosso, P., Spina, D. (eds.) Working Notes of CLEF 2025 - Conference and Labs of the Evaluation Forum, CEUR Workshop Proceedings (2025)

38. Qin, T., Liu, T.: Introducing LETOR 4.0 datasets. CoRR **abs/1306.2597** (2013). http://arxiv.org/abs/1306.2597

39. Rieffel, E., Polak, W.: An introduction to quantum computing for non-physicists. ACM Comput. Surv. (CSUR) **32**(3), 300–335 (2000)

40. Shaikh, M.T., Hamza, M., Ali, S.B., Rafi, M., Zahid, S.: Feature selection using quantum annealing: a mutual information based QUBO approach. In: Faggioli, G., Ferro, N., Rosso, P., Spina, D. (eds.) Working Notes of CLEF 2025 - Conference and Labs of the Evaluation Forum, CEUR Workshop Proceedings (2025)

41. Shimi, G., C, J.M., Thenmozhi, D.: Quantum feature selection. In: Faggioli, G., Ferro, N., Galuscáková, P., de Herrera, A.G.S. (eds.) Working Notes of the Conference and Labs of the Evaluation Forum (CLEF 2024), Grenoble, France, 9-12 September, 2024, CEUR Workshop Proceedings, vol. 3740, pp. 3082–3086, CEUR-WS.org (2024). https://ceur-ws.org/Vol-3740/paper-301.pdf
42. Touvron, H., et al.: Llama: Open and efficient foundation language models. CoRR **abs/2302.13971** (2023). https://doi.org/10.48550/ARXIV.2302.13971
43. Ushijima-Mwesigwa, H., Negre, C.F.A., Mniszewski, S.M.: Graph partitioning using quantum annealing on the d-wave system. CoRR **abs/1705.03082** (2017). http://arxiv.org/abs/1705.03082
44. Van Laarhoven, P.J., Aarts, E.H., van Laarhoven, P.J., Aarts, E.H.: Simulated annealing. Springer (1987)
45. Yarkoni, S., Raponi, E., Bäck, T., Schmitt, S.: Quantum annealing for industry applications: introduction and review. Reports on Progress in Physics **85**(10), 104001:1–104001:27 (2022)

Overview of the CLEF 2025 SimpleText Track
Simplify Scientific Text (and Nothing More)

Liana Ermakova[1], Hosein Azarbonyad[2], Jan Bakker[3],
Benjamin Vendeville[1], and Jaap Kamps[3(✉)]

[1] Université de Bretagne Occidentale, HCTI, Brest, France
{liana.ermakova,benjamin.vendeville}@univ-brest.fr
[2] Elsevier, Amsterdam, The Netherlands
h.azarbonyad@elsevier.com
[3] University of Amsterdam, Amsterdam, The Netherlands
{j.bakker,kamps}@uva.nl

Abstract. Building on the success and insights from previous years, the CLEF 2025 SimpleText Track continues to advance the mission of making scientific information more accessible to a broader audience. In 2025, we introduced a new biomedical corpus, based on aligned Cochrane abstracts and plain language summaries, for the main scientific text simplification task. In addition, we devote particular attention to remaining issues of current generative models, focusing on indentifying and classifying overgeneration and other information distortion in the predictions, as well as promoting grounded text generation approaches. This paper presents an overview of the CLEF 2025 SimpleText Track. Task 1 focuses on Text Simplification, aiming to simplify complex scientific texts. Task 2 addresses Controlled Creativity, emphasizing the detection, classification, and avoidance of hallucinations in generated content. Task 3 revisits selected tasks from SimpleText 2024 by popular demand. We discuss the data and benchmarks provided for these tasks, along with preliminary insights and anticipated challenges.

Keywords: Scientific text simplification · Biomedical AI · Generative AI · Information access · Natural language processing

1 Introduction

Becoming science-literate is more important than ever before. Objective scientific information helps any user navigate a world where misinformation, disinformation, or generated and unfounded information is only a single mouse click away. Everyone acknowledges the importance of objective scientific information, but the general public seldom consults scientific sources. The question is: how can we improve accessibility scientific text for everyone?

The value of objective scientific information cannot be overstated. Biomedical research can directly impact people's decisions about health. However, the

J. Carrillo-de-Albornoz et al. (Eds.): CLEF 2025, LNCS 16089, pp. 436–463, 2026.
https://doi.org/10.1007/978-3-032-04354-2_23

most reliable and up-to-date sources in biomedicine contain complex language and assume a high degree of background knowledge, making them difficult for the general public to understand. Rather than an afterthought, our track is motivated by a concrete use case of societal importance. This use case is one of the main differences with earlier work on text simplification, which looked only at lexical and grammatical complexity in isolation.

Scientific text simplification approaches can be applied to make these sources more accessible. In recent years, the evolution of large language models has empowered progress in text generation tasks such as scientific text simplification. While their performance is impressive, there are still issues with scientific jargon and information distortion. Training modern neural models to simplify scientific documents is a complex task that requires high-quality training data. Evaluating the performance of these models requires carefully crafted evaluation data, including detailed human evaluation and systematic analysis of the remaining limitations of generative summarization and simplification models.

The SimpleText track aims to lead in relevant NLP/IR state-of-the-art opportunities and challenges. In 2021, SimpleText was run as a workshop focusing on text simplification in light of the emerging generative models for NLP/IR tasks and the potential of AI/NLP models to improve the prototypical IR task of scientific information access [12]. From 2022 until 2024, we ran the track as a pipeline of three interrelated tasks of content selection, complexity spotting, and text simplification [16–18]. In 2025, after running the current track setup for three consecutive years (2022–2024), we made significant changes in the track to address some of today's main NLP/IR challenges.

To address these challenges, the CLEF 2025 Simple Track has three aims. First, we push the research frontier in text simplification by further expanding the scientific text simplification corpora, focusing on true paragraph-level and document-level simplification with greater variation, and considering the complex discourse structure. This setup fits current models, such as LLMs, that operate on a long input. This new biomedical corpus is constructed from aligned Cochrane abstracts and plain language summaries [3]. Second, we exploit the text simplification setup with aligned sources, references, and the output of generative models to detect, quantify, and avoid spurious information introduced gratuitously by the generative model. This is what is informally referred to as "hallucinations," addressing the remaining limitations of large generative models is crucial for the scientific use case, as current evaluation measures are "blind" and don't punish the unwarranted generation of additional content. This task addresses one of the main challenges in the Track, CLEF, and the fields of NLP and IR in general. Third, by popular demand, we revisited and reran some earlier tasks to ensure that the transition to the new track setup retains the active track participants of earlier years.

The CLEF 2025 SimpleText track is based on three interrelated tasks:

- **Task 1: Text Simplification** simplify scientific text.
- **Task 2: Controlled Creativity** identify and avoid hallucination.
- **Task 3: SimpleText 2024 Revisited** selected tasks by popular request.

L. Ermakova et al.

Table 1. CLEF 2025 Simpletext official run submission statistics

Team	Task 1		Task 2			Task 1	Task 2	Total runs
	1.1	1.2	2.1	2.2	2.3			
AIIRLab [25]	4	2	5	5		6	10	16
ASM [9]		10				10		10
DSGT [26]	2	1	6	6	3	3	15	18
DUTH [2]	3		2	2		3	4	7
EngKh (no paper)	2					2		2
Fujitsu [1]	19					19		19
LIA [21]		9				9		9
Mtest (no paper)	1	1	1	1		2	2	4
PICT [38]	1	1				2		2
RECAIDS [19]	1	1	1	1		2	2	4
Scalar [10]	10	1			1	11	1	12
SINAI [5]	2	2	15	15		4	30	34
THM [22]	22					22		22
UBO [36]	5	7	1	1		12	2	14
UM-FHS [24]	4	5				9		9
UvA [29]	5	9				14		14
Unknown (no paper)	2					2		2
Total	83	49	31	31	4	132	66	198

A total of 74 teams registered for our SimpleText track at CLEF 2025. A total of 18 teams submitted 198 runs in total for Tasks 1 and 2. The statistics for these runs submitted are presented in Table 1.[1] However, some runs had problems that we could not resolve. We do not detail them in the rest of the paper and leave out the 0-scoring runs. In addition, Task 3 asked for additional experiments on the CLEF 2024 SimpleText test collections. For Task 3, one participant, LIA [23], conducted extensive experiments on CLEF 2024 data. More details about individual runs and experiments can be found in the participants' papers, also shown in Table 1.

This paper gives an overview of the CLEF 2025 SimpleText Track. Further detail per track is provided in the Track overview papers published in the CEUR CLEF Working Notes, specifically for Task 1 on *Text Simplification* [4] and Task 2 on *Controlled Creativity* [34]. For Task 3 on *SimpleText 2024 Revisited*, we refer to the respective participants' papers for further details.

In the rest of this paper, we will provide a detailed description of each task of the CLEF 2025 edition in three self-contained sections: In the rest of this paper, we will provide a detailed description of each task of the CLEF 2025

---

[1] The table includes submissions in the Tasks 1 and 2 CodaBench evaluation platform, where we were privileged to have 29 (Task 1) and 13 (Task 2) participants.

edition in three self-contained sections: Task 1: Text Simplification in Sect. 2, Task 2: Controlled Creativity in Sect. 3, and Task 3: SimpleText 2024 Revisited in Sect. 4. We end with a discussion and conclusions in Sect. 5.

2   Task 1: Simplify Scientific Text

This section details *Task 1: Text Simplification* on simplify scientific text.

2.1   Description

The *Text Simplification* task aims to *simplify scientific text*. We created a new CLEF 2025 SimpleText corpus based on biomedical literature abstracts and lay summaries from Cochrane systematic reviews, called Cochrane-auto [3]. An example is shown in Fig. 1. This corpus was created by closely following the construction methods of the existing Wiki-auto and Newsela-auto datasets. It introduces a new domain of scientific text. The Cochrane-auto corpus is publicly available data, which is highly relevant and interesting to general and specialized users. It contains authentic parallel data produced by the same author. Cochrane-auto represents true document-level text simplification, incorporating greater variation, such as sentence merging and order rearrangements, while considering the discourse structure. Paragraph-level and sentence-level data are carefully realigned and restricted to matching paragraphs and sentences.

Complex paragraph

Fifteen heterogeneous trials, involving 1022 adults with dorsally displaced and potentially or evidently unstable distal radial fractures, were included. While all trials compared external fixation versus plaster cast immobilisation, there was considerable variation especially in terms of patient characteristics and interventions. Methodological weaknesses among these trials included lack of allocation concealment and inadequate outcome assessment.

Simple paragraph

Fifteen trials, involving 1022 adults with potentially or evidently unstable fractures, were included. While all trials compared external fixation versus plaster cast immobilisation, there was considerable variation in their characteristics especially in terms of patient characteristics and the method of external fixation.

Fig. 1. A complex-simple paragraph pair from Cochrane-auto (reproduced from [3])

Table 2 compares the new scientific text simplification corpus to the main existing text simplification corpora. The aligned references also free our translation students and professionals from the task of manually simplifying texts. This allows them to concentrate on analyzing and annotating samples of the evaluation data for various types of information distortion, providing ground truth for Task 2.

Table 2. Statistics for the automatically aligned Cochrane-auto, Newsela-auto, and Wiki-auto datasets, versus the manual simplifications of the track in 2022–2024.

	Cochrane-auto	Newsela-auto	Wiki-auto	SimpleText
Domain	Biomedical	News	General	Science
# Document Pairs	5,585	18,820	138,095	278
# Sentence Pairs	35,800	813,972	685,769	1,536

Task Description. This is the core NLP task of the track, and we continue with both sentence-level (*Task 1.1*) and document-level (*Task 1.2*) scientific text simplification. The main innovation is the very large new corpus we constructed in 2024, and the shift to the biomedical domain.

Data. As discussed above, we constructed a large scientific text simplification corpus, based on realigning abstracts and lay summaries at scale at the sentence, paragraph, and document levels. In 2025, we used this Cochrane-auto corpus as the training data.

Train data The specific train data for Task 1 consists of 1,085 documents, 4,171 paragraphs, and 14,719 sentences, with paired content from the abstract and the plain language summary. While the track distinguishes only between sentence and document level text simplification, the paragraph level of the sentence input is included, allowing also for paragraph level text simplification submission to Task 1.2.

Test data The primary test data consists of 217 new Cochrane abstracts with paired plain English summaries, composed of 4,293 source sentences.

These are new systematic reviews published by Cochrane over the last year. We process these paired abstracts and plain language summaries in two different ways.

- We process these as Cochrane-auto [3] to ensure a high-quality sentence and paragraph alignment. This results in a subset of 37 abstracts and 587 sentences, paired with 37 plain language summaries with 388 sentences. The processing is identical to Cochrane-auto and other text simplification data sets.
- For document-level text simplification, we can also use the original pairs of abstracts and plain language summaries, using only the *results and conclusions* sections, similar to [7]. This results in 217 abstracts with 4,293 source sentences, paired with 217 plain language summaries with 3,641 sentences.

We use the aligned subset as the main evaluation and also report the scores over the whole subset.

Analysis data For further analysis, we extended the test data with the Cochrane-auto validation and test splits (part of the train data), Medline abstracts for which TREC PLABA references exist,[2] and SimpleText 2024

---

[2] https://bionlp.nlm.nih.gov/plaba2024/.

abstracts for which we have references. The combined test file, including additional data sources, contains 666 documents with 9,160 sentences.

Evaluation. In 2025, we emphasize large-scale automatic evaluation measures (SARI, BLEU, compression, readability) that provide a reusable test collection. For further details on these evaluation measures for scientific text simplification, see [6]. This automatic evaluation will be supplemented with a detailed human evaluation of other aspects, essential for deeper analysis.

Almost all participants used generative models for text simplification, yet existing evaluation measures are blind to potential hallucinations with extra or distorted content. In 2025, we continue to provide further analysis of ways to detect and quantify spurious content in the output, potentially corresponding to what is informally called "hallucinations."

2.2   Participant's Approaches

A total of 18 teams submitted 132 runs in total. In the detailed results, we only include runs without errors, which got a non-zero score.

AIIRLab Largey et al. [25] submitted 6 runs in total for Task 1. They submitted 4 runs for Task 1.1 and 2 runs for Task 1.2.

ASM Djoudi et al. [9] submitted 10 runs in total for Task 1. They submitted 0 runs for Task 1.1 and 10 runs for Task 1.2.

DSGT Marturi and Elwazzan [26] submitted 3 runs in total for Task 1. They submitted 2 runs for Task 1.1 and 1 runs for Task 1.2.

DUTH Arampatzis and Arampatzis [2] submitted 3 runs in total for Task 1. They submitted 3 runs for Task 1.1 and 0 runs for Task 1.2.

EngKh (no paper) submitted 2 runs in total for Task 1. They submitted 2 runs for Task 1.1 and 0 runs for Task 1.2.

Fujitsu Aguero-Torales et al. [1] submitted 19 runs in total for Task 1. They submitted 19 runs for Task 1.1 and 0 runs for Task 1.2.

LIA Gallina et al. [21] submitted 9 runs in total for Task 1. They submitted 0 runs for Task 1.1 and 9 runs for Task 1.2.

Mtest (no paper) submitted 2 runs in total for Task 1. They submitted 1 runs for Task 1.1 and 1 runs for Task 1.2.

PICT Vora et al. [38] submitted 2 runs in total for Task 1. They submitted 1 runs for Task 1.1 and 1 runs for Task 1.2.

RECAIDS Eugin et al. [19] submitted 2 runs in total for Task 1. They submitted 1 runs for Task 1.1 and 1 runs for Task 1.2.

Scalar Dongre et al. [10] submitted 11 runs in total for Task 1. They submitted 10 runs for Task 1.1 and 1 runs for Task 1.2.

SINAI Collado-Montañez et al. [5] submitted 4 runs in total for Task 1. They submitted 2 runs for Task 1.1 and 2 runs for Task 1.2.

THM Hofmann et al. [22] submitted 22 runs in total for Task 1. They submitted 22 runs for Task 1.1 and 0 runs for Task 1.2.

UBO Vendeville et al. [36] submitted 12 runs in total for Task 1. They submitted 5 runs for Task 1.1 and 7 runs for Task 1.2.

UM-FHS Kocbek and Stiglic [24] submitted 9 runs in total for Task 1. They submitted 4 runs for Task 1.1 and 5 runs for Task 1.2.

UvA Papandreou et al. [29] submitted 14 runs in total for Task 1. They submitted 5 runs for Task 1.1 and 9 runs for Task 1.2.

Unknown team (no paper) submitted 2 runs in total for Task 1. They submitted 2 runs for Task 1.1 and 0 runs for Task 1.2.

2.3 Results

This section details the task results for sentence- and document-level test simplification subtasks.

Table 3. Results for CLEF 2025 SimpleText Task 1.1 sentence-level text simplification: Test data on 37 aligned Cochrane-auto abstracts, best five runs per team

Team/Method	count	SARI	BLEU	FKGL	Compression ratio	Sentence splits	Levenshtein similarity	Exact copies	Additions proportion	Deletions proportion	Lexical complexity score
Source	37	12.03	20.53	13.54	1.00	1.00	1.00	1.00	0.00	0.00	8.89
Reference	37	100	100	11.73	0.56	0.67	0.50	0.0	0.16	0.60	8.71
UM-FHS gpt-4.1-mini	37	43.34	13.93	7.46	0.78	1.58	0.63	0.00	0.28	0.50	8.50
UM-FHS gpt-4.1-mini-	37	42.83	20.85	12.29	0.71	0.86	0.62	0.00	0.15	0.46	8.67
DSGT plan_guided_lla	37	42.33	10.43	7.77	0.48	0.97	0.47	0.00	0.18	0.70	8.52
UvA o-bartsent-cochr	37	42.31	25.72	12.08	0.41	0.51	0.55	0.00	0.01	0.62	8.72
SINAI PRMZSTASK11V1	37	41.82	6.50	11.41	1.37	1.56	0.53	0.00	0.59	0.30	8.33
THM p2 gpt-4.1-nano	37	41.32	10.49	14.90	1.27	1.16	0.63	0.00	0.45	0.26	8.62
UvA bartsent-cochran	37	41.28	17.67	11.20	0.35	0.49	0.48	0.00	0.01	0.67	8.76
Scalar gpt_md_2.1	37	40.95	14.07	18.79	0.62	0.47	0.53	0.00	0.22	0.60	8.68
THM p1 gpt-4.1-nano	37	40.42	11.02	14.66	1.23	1.13	0.65	0.00	0.42	0.24	8.61
PICT S3Pipeline	37	40.15	12.96	7.61	0.71	1.53	0.62	0.00	0.21	0.49	8.84
Fujitsu llm_t5_rule	37	39.04	6.70	6.79	0.31	0.71	0.42	0.00	0.08	0.76	8.85
UM-FHS gpt-4.1	37	38.84	14.04	8.51	0.79	1.26	0.68	0.30	0.22	0.41	8.49
UvA llama31	37	38.76	2.83	8.30	0.93	1.58	0.46	0.00	0.60	0.66	8.34
DUTH Task11_flan-t5-	37	38.73	18.84	11.95	0.61	0.78	0.66	0.00	0.10	0.50	8.96
Fujitsu t5efficient	37	38.60	4.28	5.58	1.79	3.63	0.43	0.00	0.77	0.29	10.31
Fujitsu llm_gpt3.5-t	37	38.53	6.30	5.18	0.36	0.99	0.45	0.00	0.11	0.74	8.89
Fujitsu llm_45_judge	37	38.41	5.45	5.26	0.32	0.89	0.42	0.00	0.09	0.77	8.87
Fujitsu dummy60	37	38.37	14.50	1.19	0.37	2.74	0.52	0.00	0.08	0.67	8.74
SINAI PRMZSTASK11V2	37	37.84	5.93	12.97	1.64	1.63	0.56	0.00	0.59	0.17	8.47
THM pnil gpt-4.1-na	37	37.60	8.24	15.21	1.84	1.63	0.56	0.00	0.57	0.12	8.61
UvA bartdoc-ca	37	37.25	19.54	11.97	0.51	0.61	0.62	0.00	0.02	0.52	8.77
EngKh biomedical_lla	37	36.68	11.47	10.62	1.14	1.51	0.65	0.00	0.37	0.28	8.69
UvA llama31	37	36.45	1.22	13.04	1.07	1.31	0.41	0.00	0.66	0.70	8.61
AIIRLab mistral	37	36.08	18.41	12.78	0.94	1.06	0.76	0.00	0.19	0.28	8.81
MTest bartfinetuned	37	34.98	26.52	11.94	0.74	0.98	0.83	0.00	0.01	0.30	8.78
THM pn1 gemini-2.0-	37	34.47	9.67	7.75	1.25	1.90	0.67	0.00	0.45	0.20	8.62
Scalar BioBart_1	37	33.95	25.69	12.19	0.78	1.00	0.86	0.00	0.01	0.27	8.80
Scalar BioBart	37	33.95	25.69	12.19	0.78	1.00	0.86	0.00	0.01	0.27	8.80
THM c gpt-4.1-nano	37	33.94	5.81	21.56	1.49	0.99	0.63	0.00	0.44	0.22	9.22
DUTH Task11_bart-sam	37	32.18	12.28	7.69	1.43	2.75	0.68	0.00	0.41	0.14	8.71
RECAIDS T5	37	31.68	0.09	3.72	0.37	0.96	0.31	0.00	0.23	0.88	8.87
AIIRLab llama3.1-8b	37	31.27	19.59	11.44	0.85	1.09	0.83	0.00	0.09	0.25	8.83
UM-FHS gpt-4.1-nano	37	29.47	18.46	11.10	0.86	1.14	0.83	0.43	0.11	0.24	8.71
DUTH Task11_bart-lar	37	27.59	12.01	8.67	1.69	2.90	0.66	0.00	0.46	0.09	8.61
DUTH Task11_flan-t5-	37	22.75	21.95	13.15	0.91	0.95	0.94	0.00	0.01	0.11	8.89
DUTH Task11_gpt4	37	12.03	20.53	13.54	1.00	1.00	1.00	1.00	0.00	0.00	8.89
XXX method	37	12.03	20.53	13.54	1.00	1.00	1.00	1.00	0.00	0.00	8.89

Task 1.1: Sentence-Level Scientific Text Simplification. The main evaluation concerns the 37 abstracts, with 587 sentences aligned identically to the way Cochrane-auto and other collections are aligned. In this track overview paper, we decided to evaluate all submissions in Task 1.1 and Task 1.2 at the document level to ensure identical ground truth and comparable scores across tasks.

Table 3 shows the Task 1.1 (sentence-level text simplification) results. The table is restricted to submissions without issues, and we show a maximum of five runs per team. We show several evaluation scores against the human reference simplifications, particularly SARI and BLEU. In addition, we provide additional text statistics on the system output, such as FKGL, and compare them to the source input.

We make a number of observations. First, the table is sorted on SARI, the primary automatic text simplification measure used in the track. We observe SARI scores above 30% for almost all systems and above 40% for the top-scoring systems. This high overlap with the plain language reference simplifications is encouraging, and it indicates that the effectiveness of text simplification approaches, traditionally trained on youth news reading corpora like Newsela, also extends to scientific text.

Second, in terms of the level of text complexity, readability measures like FKGL provide a rough indicator of lexical and grammatical complexity. The original sentences have an FKGL of 13-14 corresponding to university-level text, and most systems reduce this to an FKGL of 11-12 corresponding to the exit level of compulsory education. This is an encouraging result, as it indicates that the scientific text simplification approach can be a viable approach to lower the textual complexity of scientific text toward the range acceptable by a layperson. Although this indicator is positive, this approximate measure does not consider terminological complexities.

Third, the table includes various other scores that indicate that there is still considerable room for improvement in scientific text simplification. Throughout the table, the BLEU evaluation measure remains very low. It leads to a different ranking of systems, with some of the best systems on BLEU demonstrating superior overlap with the human reference simplifications. The table also reveals some runs with very high "compression" ratios, sentence splits, and high proportions of additions. While evaluation measures like SARI are essential for understanding important aspects of text simplification output quality, they are also known to be relatively insensitive to content outside the intersection of manual text simplifications. Hence, high levels of content insertion can still lead to favorable SARI scores and even improve text statistics like FKGL without conveying key content of the original text.

Task 1.2: Document-Level Scientific Text Simplification. Table 4 shows the results of Task 1.2 (document-level text simplification). Again, we restrict the table to submissions covering a maximum of five runs with non-zero scores per team.

Table 4. Results for CLEF 2025 SimpleText Task 1.2 document-level text simplification: Test data on 37 aligned Cochrane-auto abstracts, best five runs per team

Team/Method	count	SARI	BLEU	FKGL	Compression ratio	Sentence splits	Levenshtein similarity	Exact copies	Additions proportion	Deletions proportion	Lexical complexity score
Source	37	12.03	20.53	13.54	1.00	1.00	1.00	1.00	0.00	0.00	8.89
Reference	37	100	100	11.73	0.56	0.67	0.50	0.0	0.16	0.60	8.71
LIA sumguid-all-w500	37	44.55	12.18	9.71	0.84	1.26	0.50	0.00	0.35	0.54	8.56
SINAI PRMZSTASK12V1	37	43.93	10.81	10.45	0.86	1.07	0.55	0.00	0.39	0.49	8.33
UM-FHS gpt-4.1	37	43.83	18.12	8.80	0.67	1.10	0.58	0.14	0.21	0.53	8.44
UM-FHS gpt-4.1-nano-	37	43.61	16.00	10.63	0.50	0.69	0.45	0.00	0.16	0.65	8.55
LIA sumguid-lang-w50	37	43.61	10.55	10.50	0.83	1.18	0.47	0.00	0.37	0.57	8.52
UM-FHS gpt-4.1-mini	37	43.53	14.11	7.48	0.72	1.49	0.62	0.00	0.25	0.52	8.52
ASM MistralMaxFRE	37	43.35	12.32	11.63	0.73	0.92	0.53	0.00	0.27	0.56	8.74
ASM MistralV0	37	43.31	12.41	11.65	0.73	0.92	0.53	0.00	0.27	0.55	8.74
ASM MistralMinFKGL	37	43.24	12.27	11.63	0.73	0.93	0.53	0.00	0.27	0.56	8.75
ASM MistralV7	37	42.95	11.34	12.53	0.78	0.94	0.51	0.00	0.30	0.55	8.80
ASM MistralV7CleanLi	37	42.93	11.38	13.77	0.78	0.84	0.51	0.00	0.29	0.56	8.80
UM-FHS gpt-4.1-mini-	37	42.82	22.94	11.93	0.60	0.76	0.60	0.03	0.10	0.52	8.73
AIIRLab Mistral_7b_b	37	42.40	12.98	8.82	0.58	0.94	0.52	0.00	0.21	0.61	8.48
UvA baseline-cochran	37	42.10	24.27	11.71	0.57	0.71	0.61	0.00	0.06	0.49	8.74
LIA sumguid-styl-w50	37	41.98	10.38	10.09	0.63	1.00	0.46	0.00	0.27	0.66	8.65
LIA sumguid-styl-w50	37	41.11	8.73	6.35	0.61	1.30	0.42	0.00	0.33	0.68	8.44
AIIRLab llama_3.1-8b	37	41.07	8.61	9.22	0.46	0.70	0.43	0.00	0.20	0.72	8.44
LIA testLlama33	37	40.79	8.42	10.74	0.46	0.65	0.42	0.00	0.18	0.73	8.64
DSGT llama_summary_s	37	40.32	7.63	9.56	0.59	0.86	0.42	0.00	0.31	0.70	8.49
PICT S3Pipeline	37	40.29	13.43	7.77	0.74	1.55	0.63	0.00	0.21	0.47	8.77
AIIRLab llama-8b	37	39.14	5.62	8.88	0.34	0.62	0.35	0.00	0.15	0.80	8.43
AIIRLab llama3.2-3b	37	39.14	5.62	8.88	0.34	0.62	0.35	0.00	0.15	0.80	8.43
DUTH task12_led-larg	37	39.11	9.83	12.41	0.37	0.47	0.45	0.00	0.06	0.70	8.80
SINAI PRMZSTASK12V2	37	38.50	10.30	11.55	1.09	1.16	0.63	0.00	0.43	0.29	8.44
UvA bartpara-cochran	37	37.89	27.43	12.22	0.62	0.77	0.74	0.00	0.01	0.41	8.78
Mtest bartdoc	37	37.62	20.42	11.79	0.50	0.61	0.62	0.00	0.01	0.51	8.76
UvA bartdoc-ca	37	37.25	19.54	11.97	0.51	0.61	0.62	0.00	0.02	0.52	8.77
UvA bartdoc-cochran	37	37.25	19.54	11.97	0.51	0.61	0.62	0.00	0.02	0.52	8.77
UM-FHS gpt-4.1-nano	37	37.01	14.74	9.05	0.69	1.13	0.64	0.19	0.16	0.46	8.57
UvA llama31	37	36.98	3.99	7.61	0.79	1.59	0.39	0.00	0.46	0.77	8.48
DUTH task12_flan-t5-	37	36.65	3.75	12.08	0.24	0.27	0.33	0.03	0.00	0.77	8.76
DUTH task12_bart-sam	37	36.25	1.38	10.32	0.17	0.28	0.27	0.00	0.01	0.85	8.74
DUTH task12_flan-t5-	37	34.73	0.67	12.76	0.14	0.15	0.22	0.00	0.01	0.87	8.81
Scalar gpt_md_2_1	37	34.39	1.01	10.56	0.14	0.19	0.20	0.00	0.03	0.88	8.67
EngKh biomedical_lla	37	33.25	17.88	12.55	0.72	0.87	0.61	0.05	0.15	0.44	8.77
DUTH task12_flan-t5-	37	32.55	0.36	12.83	0.12	0.13	0.18	0.00	0.01	0.89	9.10
RECAIDS T5	37	31.49	0.00	10.08	0.06	0.07	0.10	0.00	0.00	0.95	8.12

We make a number of observations. First, in terms of evaluation measures like SARI, we see similar encouraging performance levels again when evaluating against the plain language reference simplifications. In earlier years of the track, this mainly resulted from using proven sentence-level text simplification models with the output merged back into the entire abstract. However, this year, we see almost exclusively large language models applied to the lengthy source abstract as a whole. This is a clear sign of the remarkable progress in models for text simplification and other complex NLP tasks. Second, there remains room for improvement in capturing the human simplifications more closely, as the BLEU score remains low throughout. Here, the more conservative approaches seem to obtain better scores. For scientific text simplification, we aim for a careful balance between simplicity and accuracy, and being conservative is a key strength to avoid unnecessary and potentially inaccurate changes. Third, we see less extreme values on the other indicators, but still considerable variation in the compression ratio and number of splits, and proportions of additions and deletions. Generally, we see more compression and deletions, indicating summarization aspects such as reducing the number of sentences, which happens frequently.

It is encouraging to see solid performance for the approaches that perform text simplification on the entire abstract in one pass. This holds the promise to incorporate the discourse structure, use more complex text simplification operations such as deletions and merges, and deploy planner-based approaches to the text simplification of long documents. Traditional sentence-level simplification approaches and earlier evaluation data cannot capture these aspects. This demonstrates the value of the new test collections constructed during the CLEF 2025 SimpleText track.

2.4 Analysis

In this section, we provide additional evaluation on the larger set of 217 abstracts with 4,293 source sentences paired with 217 plain language summaries with 3,641 sentences. Unlike the subset discussed above, high-quality sentence alignment is not possible for this data. However, our primary interest is in document-level text simplification and evaluation, and our analysis explores the value of using parallel text directly as evaluation.

Table 5. Results for CLEF 2025 SimpleText Task 1.1 sentence-level text simplification: Test data on 217 Plain Language Summaries, best five runs per team

Team/Method	count	SARI	BLEU	FKGL	Compression ratio	Sentence splits	Levenshtein similarity	Exact copies	Additions proportion	Deletions proportion	Lexical complexity score
Source	217	7.84	10.55	13.29	1.00	1.00	1.00	1.00	0.00	0.00	9.05
Reference	217	100	100	11.28	0.72	0.97	0.40	0.00	0.29	0.63	8.65
DSGT plan_guided_lla	217	42.98	6.33	7.82	0.48	0.99	0.46	0.00	0.18	0.71	8.50
UM-FHS gpt-4.1-mini	217	42.13	9.52	7.56	0.74	1.52	0.61	0.00	0.26	0.53	8.54
SINAI PRMZSTASK11V1	217	41.25	4.59	12.39	1.44	1.56	0.51	0.00	0.61	0.30	8.44
UvA llama31	217	40.92	2.62	8.63	1.00	1.64	0.45	0.00	0.62	0.64	8.35
THM p2-gpt-4.1-nano	217	39.57	6.50	15.40	1.32	1.20	0.60	0.00	0.47	0.27	8.68
UM-FHS gpt-4.1-mini-	217	39.16	11.95	12.23	0.67	0.82	0.60	0.00	0.14	0.50	8.76
PICT S3Pipeline	217	39.11	8.30	6.52	0.69	1.65	0.60	0.00	0.21	0.52	8.85
Scalar gpt_md_2_1	217	38.96	8.25	19.45	0.62	0.43	0.52	0.00	0.23	0.60	8.77
Fujitsu llm_gpt3.5-t	217	38.84	3.05	5.04	0.35	1.02	0.44	0.00	0.11	0.75	8.96
UvA bartsent-cochran	217	38.71	6.01	11.34	0.31	0.46	0.45	0.00	0.00	0.72	8.81
Fujitsu llm_t5_rule	217	38.55	2.75	6.60	0.31	0.77	0.42	0.00	0.08	0.77	8.95
Fujitsu llm_45_judge	217	38.54	2.34	5.19	0.31	0.93	0.41	0.00	0.09	0.78	8.95
UvA o-bartsent-cochr	217	38.53	8.57	11.99	0.37	0.49	0.51	0.00	0.01	0.67	8.78
UvA llama31	217	38.50	1.13	13.66	1.09	1.23	0.40	0.00	0.66	0.71	8.65
Fujitsu llm_45	217	38.49	2.06	5.32	0.31	1.00	0.40	0.00	0.09	0.79	8.90
THM p1-gpt-4.1-nano	217	38.24	6.59	15.03	1.28	1.18	0.63	0.00	0.45	0.25	8.69
Fujitsu llm_45fewSho	217	38.20	1.87	3.51	0.28	0.88	0.37	0.00	0.12	0.81	8.82
UM-FHS gpt-4.1	217	37.93	9.46	8.82	0.76	1.22	0.64	0.23	0.22	0.46	8.54
UvA bartdoc-ca	217	37.14	7.23	11.43	0.39	0.49	0.52	0.00	0.01	0.63	8.85
SINAI PRMZSTASK11V2	217	35.95	4.03	14.00	1.76	1.64	0.54	0.00	0.61	0.15	8.56
DUTH Task11_flan-t5-	217	35.35	10.07	11.21	0.60	0.80	0.65	0.00	0.09	0.51	9.00
THM pnll_gpt-4.1-na	217	35.26	5.23	15.49	1.94	1.72	0.54	0.00	0.59	0.12	8.68
AIIRLab mistral	217	33.95	10.30	13.26	0.93	1.04	0.72	0.00	0.21	0.32	8.86
RECAIDS T5	217	33.89	0.03	3.72	0.37	0.98	0.31	0.00	0.23	0.89	8.87
EngKh biomedical_lla	217	33.16	7.30	10.76	1.18	1.53	0.65	0.00	0.37	0.25	8.75
THM c-gpt-4.1-nano	217	32.44	3.76	21.37	1.51	1.02	0.62	0.00	0.43	0.20	9.26
THM pn1-gemini-2.0-	217	32.27	5.80	7.92	1.28	1.94	0.66	0.00	0.46	0.20	8.68
MTest bartfinetuned	217	31.59	14.86	11.90	0.69	0.96	0.80	0.00	0.01	0.36	8.89
Scalar BioBart	217	30.35	14.26	12.04	0.74	0.99	0.83	0.00	0.01	0.32	8.88
Scalar BioBart_1	217	30.35	14.26	12.04	0.74	0.99	0.83	0.00	0.01	0.32	8.88
AIIRLab llama3.1-8b	217	29.80	11.32	11.19	0.83	1.10	0.80	0.00	0.10	0.29	8.93
DUTH Task11_bart-sam	217	29.68	7.32	7.50	1.38	2.73	0.67	0.00	0.41	0.16	8.79
UM-FHS gpt-4.1-nano	217	28.89	10.35	9.90	0.83	1.19	0.78	0.35	0.13	0.30	8.77
DUTH Task11_bart-lar	217	23.84	7.07	8.59	1.64	2.87	0.66	0.00	0.45	0.10	8.71
DUTH Task11_flan-t5-	217	18.78	11.48	12.89	0.89	0.94	0.93	0.00	0.01	0.13	9.03
DUTH Task11_gpt4	217	7.84	10.55	13.29	1.00	1.00	1.00	1.00	0.00	0.00	9.05
XXX method	217	7.84	10.55	13.29	1.00	1.00	1.00	1.00	0.00	0.00	9.05

Table 5 shows the results of Task 1.1 (sentence-level text simplification) against a larger set of 217 abstracts and plain language summaries without further alignment. Again, we restrict the table to submissions covering a maximum of five runs with non-zero scores per team. Note again that all submissions in

Table 6. Results for CLEF 2025 SimpleText Task 1.2 document-level text simplification: Test data on 217 Plain Language Summaries, best five runs per team

Team/Method	count	SARI	BLEU	FKGL	Compression ratio	Sentence splits	Levenshtein similarity	Exact copies	Additions proportion	Deletions proportion	Lexical complexity score
Source	217	7.84	10.55	13.29	1.00	1.00	1.00	1.00	0.00	0.00	9.05
Reference	217	100	100	11.28	0.72	0.97	0.40	0.00	0.29	0.63	8.65
LIA sumguid-all-w500	217	44.93	9.58	9.77	0.69	1.06	0.48	0.00	0.29	0.62	8.61
LIA sumguid-lang-w50	217	44.40	7.85	10.58	0.67	0.97	0.44	0.00	0.30	0.66	8.56
SINAI PRMZSTASK12V1	217	43.63	8.07	10.73	0.81	1.03	0.52	0.00	0.37	0.54	8.41
LIA sumguid-styl-w50	217	43.57	6.18	10.28	0.51	0.81	0.41	0.00	0.20	0.72	8.67
ASM MistralMinFKGL	217	43.51	8.26	11.85	0.63	0.82	0.48	0.00	0.22	0.62	8.78
ASM MistralV0	217	43.51	8.32	11.95	0.63	0.81	0.48	0.00	0.22	0.62	8.78
ASM MistralMaxFRE	217	43.50	8.27	11.87	0.63	0.82	0.48	0.00	0.22	0.62	8.78
LIA sumguid-styl-w50	217	43.17	5.92	6.87	0.49	1.03	0.39	0.00	0.25	0.75	8.50
ASM MistralV7	217	43.10	7.64	12.68	0.66	0.82	0.48	0.00	0.23	0.62	8.86
ASM MistralV7CleanLi	217	43.09	7.60	13.73	0.66	0.74	0.47	0.00	0.23	0.62	8.87
DSGT llama_summary_s	217	42.92	5.32	9.94	0.49	0.72	0.39	0.00	0.24	0.75	8.55
AIIRLab Mistral_7b_b	217	42.57	7.47	9.26	0.50	0.82	0.48	0.00	0.16	0.66	8.56
AIIRLab llama_3.1-8b	217	42.46	4.73	9.94	0.39	0.58	0.39	0.00	0.15	0.76	8.54
LIA testLlama33	217	42.35	4.70	11.19	0.39	0.54	0.39	0.00	0.14	0.76	8.73
UM-FHS gpt-4.1-mini	217	42.13	9.80	7.65	0.69	1.44	0.60	0.00	0.23	0.55	8.57
UvA baseline-cochran	217	41.83	10.85	11.10	0.44	0.60	0.49	0.00	0.06	0.63	8.75
UM-FHS gpt-4.1-nano-	217	41.01	7.15	10.64	0.48	0.66	0.41	0.00	0.15	0.69	8.58
UM-FHS gpt-4.1-mini-	217	40.81	10.67	11.69	0.55	0.72	0.55	0.01	0.10	0.58	8.74
AIIRLab llama3.2-3b	217	39.77	2.17	8.70	0.28	0.52	0.30	0.00	0.11	0.84	8.55
AIIRLab llama-8b	217	39.77	2.17	8.70	0.28	0.52	0.30	0.00	0.11	0.84	8.55
DUTH task12_led-larg	217	39.28	3.58	12.46	0.31	0.41	0.40	0.00	0.05	0.76	8.86
PICT S3Pipeline	217	39.11	8.23	6.46	0.71	1.69	0.60	0.00	0.22	0.50	8.76
UM-FHS gpt-4.1	217	38.88	10.00	8.97	0.67	1.07	0.59	0.18	0.20	0.52	8.53
UvA llama31	217	38.52	2.37	7.68	0.73	1.73	0.37	0.00	0.43	0.81	8.56
UM-FHS gpt-4.1-nano	217	37.60	10.07	8.56	0.65	1.11	0.61	0.12	0.16	0.51	8.62
SINAI PRMZSTASK12V2	217	37.34	7.21	11.85	1.05	1.11	0.63	0.00	0.40	0.31	8.55
UvA bartdoc-ca	217	37.14	7.23	11.43	0.39	0.49	0.52	0.00	0.01	0.63	8.85
UvA bartdoc-cochrane	217	37.14	7.23	11.43	0.39	0.49	0.52	0.00	0.01	0.63	8.85
Mtest bartdoc	217	37.08	7.25	11.50	0.39	0.50	0.52	0.00	0.01	0.63	8.86
DUTH task12_bart-sam	217	37.00	0.13	10.02	0.12	0.22	0.21	0.00	0.00	0.89	8.88
DUTH task12_flan-t5-	217	36.62	0.29	12.20	0.16	0.18	0.24	0.00	0.00	0.85	8.95
DUTH task12_flan-t5-	217	35.81	0.12	13.09	0.12	0.13	0.19	0.00	0.00	0.89	9.02
UvA bartpara-cochran	217	34.97	12.70	12.13	0.55	0.70	0.68	0.00	0.01	0.49	8.86
DUTH task12_flan-t5-	217	34.61	0.29	11.72	0.14	0.17	0.20	0.00	0.02	0.89	9.01
Scalar gpt_md_2.1	217	34.61	0.02	9.26	0.09	0.13	0.13	0.00	0.02	0.93	8.81
RECAIDS T5	217	33.14	0.00	8.79	0.04	0.06	0.07	0.00	0.00	0.96	8.24
EngKh biomedical_lla	217	28.19	8.55	11.95	0.69	0.79	0.57	0.04	0.12	0.46	9.00

Task 1.1 and Task 1.2 at the document level, to ensure identical ground truth and comparable scores across tasks.

We make a number of observations. First, in terms of evaluation measures like SARI, we see again similar encouraging performance levels when evaluating against the larger set of plain language reference simplifications. The ranking in Table 5 is similar to the subset of Table 3 before, with some notable shifts and upsets, particularly for run with a low BLEU score, but overall high agreement. Second, we see relatively low BLEU scores again, and even considerably lower than before. This is partly a result of the less clear source to reference alignment at the sentence and paragraph level for this larger set of references. But it also shows that document-level text simplification is a challenging task, even for current advanced models. Third, this also indicates that real-world plain language summaries are far removed from direct sentence-level simplifications. It also suggests that more conservative approaches, which may be desirable from an accuracy point of view, fail to capture the complex plain language adaptations.

Table 6 shows the results of Task 1.2 (document-level text simplification) against a larger set of 217 abstracts and plain language summaries without further alignment. Again, we restrict the table to submissions covering a maximum of five runs with non-zero scores per team.

We make a number of observations. First, in terms of evaluation measures like SARI, we see similar encouraging performance levels again when evaluating against the plain language reference simplifications. The tables show some swaps and upset, but generally good agreement between Table 6 and Table 4 shown before. One exception seems to be closed-source models, such as GPT-4, which perform less impressively on the larger set of plain language summaries. Second, the BLEU score remains low throughout again, and notably lower than on the subset of Table 4. This seems to be a result of the greater variation and discourse changes in the plain language summaries. However, this also immediately suggests that this is not yet captured well by the predictions of advanced NLP models for text simplification. Third, we see less extreme values on the other indicators for document-level text simplification approaches. The fraction of deletions remains very high throughout all systems. Interestingly, the better-scoring systems also seem to have more insertions. This can be an indication that some systems are finding valuable content to insert, such as explanations of jargon or other specialized terminology.

2.5   Findings

This concludes the results for the CLEF 2025 SimpleText Task 1: Text Simplification on simplify scientific text. Our main findings are the following: First, our analysis compared the results over the carefully sentence-aligned abstracts in Table 3 and Table 4, with the larger unfiltered set of document-level aligned abstracts in Table 5 and Table 6. It is encouraging to see the broad agreement in the ranking over both sets, as this suggests evaluation and training on document-aligned texts is a viable option. Similar to how machine translation was able to scale up due to the availability of parallel texts, this can help scale up text simplification by increasing the number of available corpora. Second, this also shifts the focus of the field of text simplification beyond the traditional aspects of lexical and grammatical simplification and introduces new and interesting aspects. Examples include dealing with the discourse structure, particular background knowledge needed to understand the text, and avoiding or explaining jargon or specialized terminology. Third, while the results are encouraging and the submitted predictions are generally high quality compared to some years ago, there remains also clear room for improvement, in particular when dealing with the scientific vernacular and specific biomedical jargon. This demonstrates the value of the new test collections constructed during the CLEF 2025 SimpleText track. We refer the reader to the CLEF 2025 SimpleText Task 1 Overview paper [14] for further details and discussion.

3   Task 2: Identify and Avoid Hallucination

This section details *Task 2: Controlled Creativity* on identify and avoid hallucination.

3.1   Description

The *Controlled Creativity* task aims to *identify and avoid hallucination*. To our own surprise, the SimpleText track has collected a massive collection of spurious or overgeneration content from its participants in earlier years of the track. Table 7 shows an example output simplification of one of the participating teams. For the CLEF 2024 task on text simplification, a total of 17 out of 36 submissions (47%) contain spurious whole sentences in at least 10% of the input sentences. In fact, 14 submissions (39%) have spurious sentences in at least 20% of the input, while 7 submissions (19%) have them in at least 50% of the input sentences [14]. Our text simplification setup has sources, predictions, and references that are closely aligned and in the same language. This design allows us to study source attribution and creative variation while also identifying and avoiding what is informally referred to as "hallucinations." This task builds on earlier manual analysis of information distortion in our track since 2022 [13–15], and similar work by others [8].

Table 7. Example of a participant's output versus input: ~~deletions~~, <u>insertions</u>, and <u>whole sentence insertions</u>

Abstract G01.1_130055196

As various kinds of output devices emerged , such as highresolution printers or a display of PDA (Personal Digital Assistant) ,~~the~~ . <u>The</u> importance of high-quality resolution conversion has been increasing . |This paper proposes a new method for enlarging <u>an</u> image with high quality . <u>It will involve using a combination of high-speed imaging and high-resolution video .</u> |One of the ~~largest~~ <u>biggest</u> problems on image enlargement is the exaggeration of the jaggy edges . <u>This is especially true when the image is enlarged , as in this case .</u> |To remedy this problem , we propose a new interpolation method ,~~which~~ . <u>This method</u> uses artificial neural network to determine the optimal values of interpolated pixels . |The experimental results are shown and evaluated . <u>The results are compared to other studies and found to be inconclusive .</u> |The effectiveness of our methods is discussed by comparing with the conventional methods . <u>Our methods are designed to help people with mental health problems , not just as a way to cure them .</u> |

Description. Task 2.1 is to identify creative generation, at the abstract or document level. We provided realistic system outputs from participants in previous years, along with some intentionally generated outputs from known models. The task is to identify which sentences are fully grounded in the source input: (a) without access to the source sentences and (b) with access to them. This also includes labeling sentences that introduce significant new content. Task 2.1 can be seen as a post-hoc identification task. *Task 2.2* focuses on detecting and classifying information distortion in simplified sentences. Specifically, it is a multi-label text classification task in which participants are asked to identify the types of information distortion issues based on the annotation scheme introduced by [35]. This scheme discerns four broad categories of information distortion:

A. Fluency Is the answer provided in a correct form that a fluent speaker would speak?
B. Alignment Is the format of the answer correct?
C. Information Is the information provided accurate and relevant to the input?
D. Simplification Does the response focus on simplification?

Each group contains several fine-grained error types, for a total of 14 classes.[3] The test set is based on manual annotations, while the training set consists of synthetically generated simplifications containing targeted errors. Both datasets were constructed using runs submitted to previous editions of the SimpleText track.

Finally, we have a text alignment *Task 2.3* on avoiding creative generation and performing grounded generation by design. This task mirrors Task 1 on text simplification and requires the submission of pairs of runs, both with and without source grounding or source attribution by design.

Data. In running the SimpleText track over the last three years, we have collected an extensive set of realistic and representative predictions in the run submissions. For Tasks 2.1 and 2.2, we selected a sample of models and predictions prone to spurious generation. We have large-scale data that includes realigned sentences without support in the source, which can be used as training data. Task 2.1 asks to identify whether a sentence in the prediction, viewed as a list of sentences, is spurious. This is essentially a sentence label task, and the data was provided in this format. Task 2.2 mimics the human annotation of information distortion as done in earlier years of the track. Specifically, each simplification was labeled according to the annotation scheme of [35].

Train data For Task 2.1, we selected 782 abstracts used at CLEF 2024 SimpleText Task 3 on Text Simplification. The Task 1.2 train data with sentence labels consisted of 13,341 sentences (posthoc) and 13,514 sentences (sourced). The prevalence was very high: 11,991 (89.9%) sentences were labeled spurious for posthoc and 12,115 (89.6%) sentences for sourced.

For Task 2.2, the train data is based on a synthetic dataset starting from simplifications previously annotated as error-free. We used a large language model (LLM) for each error class in the taxonomy to generate a variant of the simplification containing the targeted error class. This approach enabled us to create a large-scale training dataset without relying on time-expensive manual annotation. The set contained 42,392 sentences with a detailed information distortion label.

Test data For Task 2.1, the test data with sentence labels consisted of 3,336 sentences (posthoc) and 3,379 sentences (sourced). The prevalence was very high: 3,006 (90.1%) sentences were labeled as spurious for posthoc, and 3,033 (89.8%) sentences for sourced.

---

[3] Our annotation scheme focuses on content and meaning preservation. Following [8], we use the word "error" as a general term for annotated issues. The term error is used for brevity, acknowledging that some cases can be considered acceptable in a text simplification context.

The test data for Task 2.2 consists of 2,659 sentences produced by participants in previous years of the SimpleText challenge. We manually annotated 2,659 sentences with an information distortion taxonomy of [35]. The submissions are evaluated against these manually annotated sentences. A total of 820 (30.1%) of sentences had no errors, and 1,839 were classified into four categories (Fluency, Alignment, Information, and Simplification issues) and 14 detailed types.

Task 2.3 follows the setup of *Task 1: Text Simplification* in Sect. 2, but requested paired runs with and without the special processing to avoid ungrounded generation.

Evaluation. Task 2.1 is essentially a sentence label task, evaluated in the standard way with Precision, Recall, F1, and AUC. Task 2.2 is a multi-label classification task. We evaluate performance using both F1 score and AUC, computed for individual classes and aggregated across the four main classes. Task 2.3 is evaluated by both standard automatic measures and human evaluation, following Task 1 on Text Simplification in Sect. 2. We also conduct more detailed overgeneration analysis for Task 2.3.

3.2 Participant's Approaches

A total of 9 teams submitted 66 runs in total. In the detailed results, we only include runs without errors, which got a non-zero score.

AIIRLab Largey et al. [25] submitted 10 runs in total for Task 2. They submitted 5 runs for Task 2.1, 5 runs for Task 2.2, and 0 runs for Task 2.3.

DSGT Marturi and Elwazzan [26] submitted 15 runs in total for Task 2. They submitted 6 runs for Task 2.1, 6 runs for Task 2.2, and 3 runs for Task 2.3.

DUTH Arampatzis and Arampatzis [2] submitted 4 runs in total for Task 2. They submitted 2 runs for Task 2.1, 2 runs for Task 2.2, and 0 runs for Task 2.3.

Mtest (no paper) submitted 2 runs in total for Task 2. They submitted 1 runs for Task 2.1, 1 runs for Task 2.2, and 0 runs for Task 2.3.

RECAIDS Eugin et al. [19] submitted 2 runs in total for Task 2. They submitted 1 runs for Task 2.1, 1 runs for Task 2.2, and 0 runs for Task 2.3.

Scalar Dongre et al. [10] submitted 1 runs in total for Task 2. They submitted 0 runs for Task 2.1, 0 runs for Task 2.2, and 1 runs for Task 2.3.

SINAI Collado-Montañez et al. [5] submitted 30 runs in total for Task 2. They submitted 15 runs for Task 2.1, 15 runs for Task 2.2, and 0 runs for Task 2.3.

UBO Vendeville et al. [36] submitted 2 runs in total for Task 2. They submitted 1 runs for Task 2.1, 1 runs for Task 2.2, and 0 runs for Task 2.3.

3.3 Results

This section details the task results for the overgeneration detection subtask, information distortion detection, classification subtask, and the grounded text simplification subtask.

Task 2.1: Identify Creative Generation. Task 2.1 aims to identify overly creative generation in scientific text simplification. This is a new task that focuses on detecting overgeneration and other information distortion issues in the predictions of current models. The task raises awareness of remaining information distortion issues in modern generative models for scientific text simplification, and focuses on post-hoc detection without or with access to the source text.

Table 8. Results for CLEF 2025 SimpleText Task 2.1 Detecting Overgeneration: Test data, posthoc detection without sources, best five runs per team

Team/Method	count	Acc.	Prec	Rec	F1	AUROC	AUPRC
SINAI basic-prefilter-all-true	3,336	0.91	0.91	1.00	0.95	0.55	0.91
DSGT bertclassifier	3,336	0.91	0.93	0.97	0.95	0.64	0.93
DSGT bert_nli_llm_ensemble	3,336	0.90	0.93	0.97	0.95	0.64	0.93
DSGT bertnlillmensemble	3,336	0.90	0.93	0.97	0.95	0.64	0.93
DUTH Task21posthoc_et	3,336	0.90	0.92	0.97	0.95	0.62	0.92
DUTH Task21posthoc_rf	3,336	0.90	0.92	0.97	0.94	0.63	0.92
DUTH Task21posthoc_svc	3,336	0.79	0.94	0.83	0.88	0.66	0.93
DUTH Task21posthoc_xgb	3,336	0.79	0.94	0.81	0.87	0.69	0.94
DUTH Task21posthoc_logreg	3,336	0.77	0.95	0.79	0.86	0.70	0.94
DSGT llm	3,336	0.77	0.95	0.78	0.86	0.70	0.94
DSGT nli_entailment	3,336	0.45	0.95	0.41	0.57	0.61	0.92
SINAI improved-prefilter-all-true	3,336	0.37	0.94	0.32	0.47	0.57	0.91
SINAI improved-prefilter-confidence-95	3,336	0.35	0.95	0.29	0.44	0.57	0.91

Table 8 shows the results of detecting spurious sentences in the generated simplifications of participants in the track in earlier years. The main task is post-hoc detection without access to the source texts, which would generalize to generic text generation tasks.

We make several observations. First, the scores are generally high, with many systems performing over 90% accuracy, F1, and AUC-PR. The test collection contains a variety of information distortion issues (see Task 2.2 and Task 2.3 for more details), including some clear "errors" such as leaving in prompts, or systematic errors in extracting the simplified content from the output of models. However, it also contains complex cases to detect (like the example in Table 7). Hence, the performance is encouraging. Second, it is interesting that trained classifiers such as encoders seem to outcompete larger and modern models as decoders for this task. This may result from the specific task setting, where effective training will pay off. Third, while the task was intended to present entire abstracts or documents, a sentence label task was more practical to run in this first year. This may have effectively reduced the task to a sentence-level task, which may have been easier than a long document-level task.

Table 9. Results for CLEF 2025 SimpleText Task 2.1 Detecting Overgeneration: Test data, detection with sources, best five runs per team

Team/Method	count	Acc.	Prec	Rec	F1	AUROC	AUPRC
AIIRLab CrossEncoder	3,379	0.98	0.99	0.99	0.99	0.95	0.99
Mtest bartfinetuned	3,379	0.97	0.99	0.97	0.98	0.96	0.99
SINAI improved-prefilter-all-true	3,379	0.96	1.00	0.95	0.98	0.98	0.99
SINAI prefilter-all-true	3,379	0.95	0.95	1.00	0.97	0.77	0.95
AIIRLab RandomForest	3,379	0.95	0.95	1.00	0.97	0.77	0.95
SINAI improved-prefilter-confidence-99	3,379	0.93	1.00	0.93	0.96	0.96	0.99
SINAI llama3.1-8b-instruct	3,379	0.93	0.95	0.97	0.96	0.77	0.95
DSGT bertclassifier	3,379	0.91	0.93	0.98	0.95	0.65	0.93
DSGT bertnlillmensemble	3,379	0.91	0.93	0.97	0.95	0.68	0.93
DUTH Task21sourced_et	3,379	0.91	0.93	0.97	0.95	0.66	0.93
DUTH Task21sourced_rf	3,379	0.90	0.93	0.96	0.95	0.65	0.93
DUTH Task21sourced_svc	3,379	0.80	0.94	0.83	0.88	0.69	0.93
SINAI improved-prefilter-confidence-95	3,379	0.81	1.00	0.79	0.88	0.89	0.98
DUTH Task21sourced_ridge	3,379	0.77	0.94	0.79	0.86	0.68	0.93
DUTH Task21sourced_logreg	3,379	0.77	0.94	0.79	0.86	0.69	0.93
DSGT llm	3,379	0.74	0.94	0.76	0.84	0.68	0.93
RECAIDS T5	3,379	0.49	0.89	0.49	0.63	0.47	0.89
DSGT nli_entailment	3,379	0.35	0.92	0.31	0.46	0.53	0.90
DSGT nli_contradiction	3,379	0.20	0.90	0.12	0.21	0.50	0.90
AIIRLab LLMs	3,379	0.10	0.00	0.00	0.00	0.50	0.90

Table 9 also shows the results of detecting spurious sentences in the generated simplifications of participants in the track in earlier years, while having access to pairs of source-prediction content. This setting exploits the text simplification setting, in which information generation must faithfully reflect the source content.

We make several observations. First, access to the sources would intuitively make the task far easier: human assessors generally rely on this to make their judgments. We see a notable increase in the performance of models, even in AUC-RO, which was lagging in Table 8 above. Second, similar to above, we see that trained or fine-tuned encoders are very effective, generally outcompeting larger decoder models with prompting and few-shot, in-context learning. Third, in the context of source-prediction pairs of sentences, the task is more straightforward than observing a long source document paired to a lengthy list of prediction sentences. Still, the near-perfect performance of the best submissions is very encouraging.

This completes the discussion of the Task 2.1 experiments. For the source-prediction pairs, we expected that the better systems would be able to perform close to perfection. These results indicate that it is possible to detect information distortion errors, such as overgeneration, in the output of current systems. Current evaluation measures based on the overlap with references are insensitive to such additions or redundant content freely generated by the models. Effective detection models can help identify and quantify these issues in the output of models, which is of great importance in further advancing scientific text simplification models.

Task 2.2: Detect and Classify Information Distortion Errors. Task 2.2 is a new task that asks not only to detect information distortion in the output of text simplification models but also to classify the type of error. This task mimics the human manual evaluation we performed in the track in earlier years.

We evaluate this task using a corpus of 2,659 manually annotated sentence–simplification pairs. Each simplified sentence may contain multiple error types, making this a multi-label classification problem. The error taxonomy is organized hierarchically into four categories (A–D), each comprising several fine-grained error types. For evaluation, predicted and gold error labels are aggregated at the group level: if any fine-grained error from a group is present, the group is considered active. Performance is then measured per group using both micro and macro F1 scores. We also consider the "No Error" class, indicating no errors were detected.

Results are presented in Table 10. The table includes only valid submissions, excluding 39 duplicates where teams submitted the same method multiple times, where we retain only the run with the highest F1 score on the *No Error* class. The results displayed here are limited to the best five runs per team, and are sorted by F1 score on *No Error* class.

From this, we make several observations. First, while some models were able to perform well on *No Error*, achieving over 0.65 F1 scores, performance quickly drops, and over half of them do not achieve 0.50 F1 scores. Second, results are quite low for all other groups. For Fluency issues (group A), the five best systems achieve an F1 score between 0.255 and 0.283. For Alignment (group B), only 55% of the systems achieved over 0.10 F1 scores, with 20% over 0.25 and up to 0.47. For Information issues (group C), only two systems achieved over 0.25 F1 scores (with 0.30 and 2.69), with the next 60% achieving between 0.10 and 0.17. Finally, for Simplification issues (group D), only the same two models were able to achieve F1 scores above 0.25 (with 0.37 and 0.30) while the next 60% achieved between 0.12 and 0.24. Third, more generally, the results suggest that detecting specific error categories remains a challenging task, especially under realistic conditions with a multi-label setting and imbalanced data. The relatively strong performance on the *No Error* class demonstrates that distinguishing error-free simplifications is a realistic and tractable subtask. The gap between detecting no errors and identifying fine-grained error types remains an open research challenge, and the results of the track highlight the complexity of accurately modeling semantic information distortions in the output of current models.

Table 10. Model Performance by Error Categories (Best Scores in Bold) for No error, Fluency (A), Alignment(B), Information (C), and Simplification (D) categories, with F_1 and AUC-PR, best five runs per team

Team/Method	No Error		A		B		C		D	
	F_1	AUC	F_1	AUC	F_1	AUC	F_1	AUC	F_1	AUC
DSGT DebertaLlmensemble	**0.763**	0.561	**0.283**	0.133	0.354	0.173	**0.301**	**0.156**	**0.374**	**0.224**
AIIRLab paraphrase_mpnet	0.755	**0.567**	0.255	0.154	0.258	0.113	0.136	0.084	0.147	0.168
AIIRLab mpnet	0.744	0.557	0.255	**0.156**	0.218	0.099	0.150	0.091	0.147	0.167
DSGT roberta	0.694	0.491	0.233	0.121	0.249	0.101	0.114	0.089	0.128	0.164
DSGT llama	0.680	0.483	0.282	0.132	0.324	0.182	0.269	0.147	0.306	0.196
AIIRLab OpenChat	0.640	0.421	0.154	0.070	0.141	0.061	0.144	0.080	0.222	0.156
AIIRLab MajorityVoting	0.633	0.415	0.156	0.071	0.110	0.045	0.170	0.088	0.239	0.160
AIIRLab Mistral	0.563	0.357	0.158	0.069	0.104	0.040	0.116	0.070	0.176	0.144
DSGT BERT	0.515	0.330	0.214	0.133	0.208	0.103	0.167	0.095	0.129	0.161
DUTH scibert	0.436	0.321	0.088	0.045	0.035	0.025	0.100	0.066	0.145	0.135
DUTH deberta-v3	0.404	0.322	0.003	0.044	0.051	0.026	0.006	0.064	0.093	0.136
Mtest bartfinetuned	0.404	0.322	0.270	0.143	**0.472**	**0.265**	0.078	0.074	0.128	0.167
DSGT bert_llama_ensemble	0.404	0.322	0.231	0.137	0.253	0.107	0.116	0.088	0.128	0.163
DUTH roberta-base	0.404	0.322	0.083	0.044	0.033	0.027	0.117	0.064	0.023	0.136
RECAIDSTechTitans T5	0.404	0.322	0.022	0.046	0.000	0.026	0.004	0.065	0.000	0.136
DUTH logreg	0.404	0.322	0.000	0.044	0.000	0.026	0.000	0.064	0.000	0.136
DUTH logreg_oversample	0.404	0.322	0.021	0.046	0.000	0.026	0.004	0.064	0.000	0.136

This completes the discussion of the Task 2.2 experiments. The results are mixed. On the one hand, consistent with the results of Task 2.1, we saw that detecting that a prediction has information distortion issues is a viable task for current systems. On the other hand, fine-grained annotation of the types of information distortion remains challenging. This indicates that manual evaluation remains of great value for scientific text simplification and the automatic evaluation measures. Yet the effort and cost of manually annotating all output remains very high, and such human evaluation is not reusable and has to be repeated for every new prediction. One realistic option is to use a hybrid approach. The ability to automatically filter out the cases with no error and judge samples of the remaining predictions to assess the error types and distribution can be a pragmatic and more cost-effective way to scale up human evaluation.

Task 2.3: Avoid Creative Generation. Task 3.2 aims to avoid overly creative generation in scientific text simplification and showcase systems that perform grounded generation by design. This is a new task that asks for a pair of submissions, one of which must make a special effort to avoid overgeneration or other information distortion issues.

Table 11. Results for CLEF 2025 SimpleText Task 2.3: Avoiding creative generation by design

Team/Method	count	SARI	BLEU	FKGL	Compression ratio	Sentence splits	Levenshtein similarity	Exact copies	Additions proportion	Deletions proportion	Lexical complexity score
AIIRLab llama3.1_gro	37	43.63	17.92	11.02	0.63	0.96	0.61	0.00	0.13	0.53	8.72
AIIRLab llama3.1_gro	37	43.24	17.48	11.16	0.63	0.96	0.61	0.00	0.13	0.53	8.71
DSGT llama_summary_s	37	41.25	15.00	12.74	0.76	0.85	0.57	0.00	0.23	0.48	8.76
*DSGT llama	37	40.32	7.63	9.56	0.59	0.86	0.42	0.00	0.31	0.70	8.49
DSGT plan_guided_lla	37	37.33	18.27	12.87	0.91	1.09	0.71	0.00	0.18	0.31	8.79
*AIIRLab llama3.1-8b	37	31.27	19.59	11.44	0.85	1.09	0.83	0.00	0.09	0.25	8.83
*DSGT llama	217	42.92	5.32	9.94	0.49	0.72	0.39	0.00	0.24	0.75	8.55
DSGT llama_summary_s	217	42.06	9.89	12.81	0.62	0.72	0.50	0.00	0.19	0.59	8.82
AIIRLab llama3.1_gro	217	40.90	11.60	11.31	0.63	0.98	0.62	0.00	0.12	0.53	8.83
AIIRLab llama3.1_gro	217	40.82	11.60	11.28	0.63	0.98	0.62	0.00	0.11	0.53	8.83
DSGT plan_guided_lla	217	33.41	10.04	12.96	0.96	1.14	0.69	0.00	0.21	0.31	8.88
*AIIRLab llama3.1-8b	217	29.80	11.32	11.19	0.83	1.10	0.80	0.00	0.10	0.29	8.93

Two teams submitted runs for Task 2.3, indicated by "_grounded" in the run names. Some of these runs were specifically submitted to Task 2.3, and others were regular submissions to Tasks 1.1 and 1.2.

Table 11 shows the standard evaluation of text simplification output against text overlap with the reference plain language summaries. We evaluate against the Cochrane-auto aligned cases (top) and the larger set of original plain language summaries (bottom). We tried to locate the matching baseline runs, indicated with ⋆ in the tables from the earlier results as displayed for Task 1 in Sect. 2.

We make several observations. First, the performance is generally competitive, and several runs are among the best-performing runs. This is reassuring, as any attempt to ground the predictions more closely to the source texts should not lead to a dramatic decrease in performance. Second, the baseline runs without any precautions observe the highest number of additions, indicating that the grounded runs are generally more conservative. Third, although we refer to the participants' papers of AIIRLab [25] and DSGT [26] for specific details, some of the grounding seems to involve more careful output processing, such as ensuring in the prompts that no extra information other than the text simplification is output by the model.

More generally, while the primary goal of prediction grounding is not a performance improvement, it is also the case that other runs with presumably redundant information are not performing less well. The standard measures based on textual overlap with the references are relatively insensitive to additional content in the predictions. This invites further analysis to investigate how well the source information grounds the predictions, and when they are not.

Task 2.3 Analysis. We analyzed the entire test data set, comprising 666 documents (Task 1.2) and 9,160 sentences (Task 1.1). This analysis assumes that there is always word overlap between a pair of complex-simple sentences or abstracts. Moreover, we look specifically for overgenerating output at the sentence's or abstract's end. This is typical of sequence-to-sequence models, which are asked to complete the input with a simplified version in standard text completion mode.

Table 12. Analysis of SimpleText Task 2.3: Spurious generation at the sentence (top) and document (bottom) level

Run	SARI (217)	Source Number	Spurious Content Number	Spurious Content Fraction
AIIRLab llama3_grounded	40.90	9,160	17	0.00
AIIRLab llama3_crossencoder_grounded2	40.82	9,160	15	0.00
⋆AIIRLab llama3-8b	29.80	9,160	394	0.04
⋆DSGT plan_guided_llama	42.98	9,160	206	0.02
DSGT plan_guided_llama_grounded	33.41	9,160	477	0.05
⋆DSGT llama_summary_simplification	42.92	666	538	0.81
DSGT llama_summary_simplification_grounded	42.06	666	504	0.76

Assume we feed the model one long sentence extracted from an abstract, without further context. Now, due to sentence splitting, the output could contain multiple sentences. However, after the input sentence is fully simplified, the model wants to complete the text. Without access to the rest of the source abstract, the model may generate the most likely subsequent sentences. Such sentences are completely unfounded by the source, and it isn't easy to spot these cases in the generated text, as they are indeed coherent and possible continuations. This may occur after every sentence in sentence-level text simplification.

In document-level text simplification, this is more likely at the end of the abstract, so we still look at the end of the source input. We observe, indeed, overgeneration/text completion issues at the end of the sources/predictions. There are also cases in which there are systematic errors in extracting the output, with additional content. Increasingly, there is additional LLM commentary other than the requested output. Accurately removing such additional content can be more challenging for the document-level submissions than for the sentence-level submissions, as some abstracts are very long.

Table 12 shows an overgeneration analysis of the Task 2.3 runs. This is done by aligning the source input to the prediction output regarding their token sequences. If all the source sentence(s) have been aligned to some prediction sentence(s), we assume the prediction covers all the content of the sources. If there is still an additional sentence in the prediction, we regard this as spurious content for that specific input. This is an imperfect proxy, and aligning lengthy documents can be non-trivial. It serves as a good indicator of spurious content in the predictions and of overgeneration issues in the runs.

We make several observations. First, despite competitive performance in terms of text overlap with the references, we see widely varying numbers of cases of overgeneration, ranging from a few percentage points to large fractions of the output. Second, this difference in additional content is not at all reflected in the evaluation scores, as some of the top-performing runs still exhibit larger fractions of "extra" content. Some of these may be easily spotted as "noise," such as systematically left-in prompts. Other cases may be challenging to detect in the output by users of text simplification systems. Third, in the context of the task, we see some interesting examples, for example, AIIRLab [25] detected "noise" and changed the prompts to ensure only the simplified text, and nothing else, was in the model output.

3.4   Findings

This concludes the results for the CLEF 2025 SimpleText Task 2: Controlled Creativity on identify and avoid hallucination. Our main findings are the following: First, for Task 2.1 on detecting creative generation, we observed very high performance for identifying overgeneration and other information distortion. This was hoped and expected for pairs of source-prediction content, but unexpected for post hoc detection on only the system's predictions. Second, for Task 2.2 on classifying the type of information distortion, we observed mixed results. Also here we saw solid performance for the "no error" cases, yet identifying the precise type of information distortion similar to human evaluation remains a challenging tasks for current models. Third, for Task 2.3 on avoiding creative generation and performing grounded generation by design, we observed that text simplification measures are immune to detecting overgeneration, and that this remains a serious issue in the predictions. More sensitive text simplification evaluation measures are needed to highlight these aspects and ensure that the research community further develops grounded generation approaches. We refer the reader to the CLEF 2025 SimpleText Task 2 Overview paper [34] for further details and discussion.

4   Task 3: Selected Tasks by Popular Request

This section details *Task 3: SimpleText 2024 Revisited* on selected tasks by popular request.

4.1   Description

The *SimpleText 2024 Revisited* task aims to rerun *selected tasks by popular request*. CLEF 2025 SimpleText is very different from the earlier years. To facilitate the transition to the new track setup, we considered continuing selected activities of the other CLEF 2024 SimpleText tracks (Task 1 on Content Selection: retrieve passages to include in a simplified summary; Task 2 on Complexity Spotting: retrieve passages to include in a simplified summary; Task 4 on SOTA? track the state-of-the-art in scholarly publications). We decided only to continue those activities at the request of, and with sufficient interest from, our active participants. In the end, we received no requests to rerun some of the tasks at

CodaBench[4] (similar to Tasks 1 and 2). We received some requests for data from earlier years, and also explicitly invited papers reporting experiments on CLEF 2024 data in their CEUR papers.

Data and Evaluation. See details in the CLEF 2024 SimpleText track overview paper [17], and the CEUR task overview papers for CLEF 2024 SimpleText Task 1 on Content Selection (Retrieve Passages to Include in a Simplified Summary) [32], Task 2 on Complexity Spotting (Identify and Explain Difficult Concepts) [28], Task 3 on Text Simplification (Simplify Scientific Text) [14], and Task 4 on SOTA? (Track the State-of-the-Art in Scholarly Publications) [11].

4.2 Participant's Approaches

Only one team submitted exclusively to selected tasks by popular request.

LIA [23] report on an extensive analysis of the CLEF 2024 SimpleText Task 1 on Content Selection: retrieve passages to include in a simplified summary.

4.3 Results

Huet and Sanjuan [23] review the CLEF SimpleText track's data of the *Content Selection* tasks run in 2022 [30], 2023 [31], and 2024 [32]. They explore popular hybrid text-vector matches ad hoc. The data is neatly stored and indexed in a relational database management system, reimplementing vector search as SQL queries, allowing them to replicate standard Python notebook processing in a traditional system. Their experiments demonstrate that the general effectiveness of these approaches also holds for scientific abstract search. They also explore clustering and observe a similar effect to Van Rijsbergen's famous cluster hypothesis [33], in perhaps the most direct evaluation of it since [37].

4.4 Findings

This concludes the results for the CLEF 2025 SimpleText Task 3: SimpleText 2024 Revisited on selected tasks by popular request. We are encouraged by the submission of a participant's paper to this track. We generally hope and encourage using the corpora and evaluation data constructed during the track in future research and papers by participants and other researchers. This also motivates the move to CodaBench for Tasks 1 and 2, and keeping the CodaBench open for post-competition experiments.

5   Discussion and Conclusions

This paper describes the setup of the CLEF 2025 SimpleText track, which contains the following three tasks. Task 1 on *Text Simplification: simplify scientific*

---

[4] https://codabench.org/.

text. Task 2 on *Controlled Creativity*: *identify and avoid hallucination*. Task 3 on *SimpleText 2024 Revisited*: *selected tasks by popular request*. These tasks address some of today's main NLP/IR challenges.

The main aim of our track, and the CLEF evaluation forum as a whole, is i) to construct corpora and evaluation resources to stimulate research on scientific text summarization and simplification, and ii) to foster a community of IR, NLP, and AI researchers working together on the important task of making science more accessible for everyone.

Within the CLEF 2025 SimpleText track, we have constructed extensive corpora and manually labeled evaluation data. First, we pushed the research frontier in text simplification by creating new scientific text simplification corpora for biomedical literature. We focused on true paragraph-level and document-level simplification with greater variation and took the complex discourse structure into account. This fits current models such as LLMs, which operate on long input. Second, we exploited the text simplification setup with aligned sources, references, and the output of generative models to detect, quantify, and avoid spurious information introduced gratuitously by the generative model. This is what is informally referred to as "hallucinations." Addressing the remaining limitations of large generative models is crucial for the scientific use case, as current evaluation measures are "blind" and don't punish the unwarranted generation of additional content. Third, to ensure that the transition to the new track setup retain the active track participants of earlier years, we revisited and reran some earlier tasks by popular demand. These reusable corpora and evaluation resources are available to participants and other researchers who want to work on the important problem of making scientific information open and easily accessible for everyone.

In terms of building a community researching scientific text summarization and simplification, the track saw a record attendance in 2025, with significant changes in the tasks and the move to CodaBench, more runs were submitted, and with the largest number of participating teams ever.

Acknowledgments. We are incredibly thankful to the master's students in translation and technical writing from the University of Brest for participating in data annotation. We also thank each of the individual track participants for their effort in submitting a record number of submissions to CodaBench and documenting these in their papers.

We thank the CLEF 2025 chairs for hosting us, and the CLEF 2025 Labs and Proceedings chairs for their excellent assistance and flexibility. It is heartwarming to be part of such a great CLEF family. We thank CodaBench [39] for hosting the competition. Post-competition experiments are ongoing at https://www.codabench.org/competitions/8400/ (Task 1.1, Task 1.2, and Task 2.3) and https://www.codabench.org/competitions/8327/ (Task 2.1 and Task 2.2). We hope and expect that these "living test collections" remain in active use until the next iteration of the track.

Benjamin Vendeville and Liana Ermakova are partly funded by the French National Research Agency (ANR-22-CE23-0019-01, *SimpleText: Automatic Simplification of Scientific Texts*). Liana Ermakova is further supported by the CNRS research group MaDICS (https://www.madics.fr/ateliers/simpletext/).

Jan Bakker and Jaap Kamps are partly funded by the Netherlands Organization for Scientific Research (NWO NWA # 1518.22.105). Jaap Kamps is further supported by (NWO CI # CISC.CC.016), the University of Amsterdam (AI4FinTech program), and ICAI (AI for Open Government Lab). Views expressed in this paper are not necessarily shared or endorsed by those funding the research.

Disclosure of Interests. The authors have no competing interests to declare that are relevant to the content of this article.

Disclosure of Generative AI Use. During the preparation of this work, the authors used *ChatGPT* and *Grammarly* in order to: **Grammar and spelling check** and **Paraphrase and reword**. After using these tools/services, the authors reviewed and edited the content as needed and take full responsibility for the publication's content.

References

1. Agüero-Torales, M.M., Rodríguez-Abellán, C., Moraga, C.A.C.: Sentence-level Scientific Text Simplification With Just a Pinch of Data. In: [20]
2. Arampatzis, G., Arampatzis, A.: DUTH at CLEF 2025 SimpleText Track: Tackling Scientific Text Simplification and Hallucination Detection. In: [20]
3. Bakker, J., Kamps, J.: Cochrane-auto: an aligned dataset for the simplification of biomedical abstracts. In: Shardlow, M., et al. (eds.) Proceedings of the Third Workshop on Text Simplification, Accessibility and Readability (TSAR 2024), pp. 41–51, Association for Computational Linguistics, Miami, Florida, USA (Nov 2024), https://doi.org/10.18653/v1/2024.tsar-1.5, URL https://aclanthology.org/2024.tsar-1.5/
4. Bakker, J., Vendeville, B., Ermakova, L., Kamps, J.: Overview of the CLEF 2025 SimpleText Task 1: Simplify Scientific Text. In: [20]
5. Collado-Montañez, J., Ortiz-Zambrano, J.A., Espin-Riofrio, C., Montejo-Ráez, A.: SINAI in SimpleText CLEF 2025: Simplifying Biomedical Scientific Texts and Identifying Hallucinations Using GPT-4.1 and Pattern Detection. In: [20]
6. Davari, D., Ermakova, L., Krestel, R.: Comparative analysis of evaluation measures for scientific text simplification. In: Antonacopoulos, A., Hinze, A., Piwowarski, B., Coustaty, M., Di Nunzio, G.M., Gelati, F., Vanderschantz, N. (eds.) Linking Theory and Practice of Digital Libraries - 28th International Conference on Theory and Practice of Digital Libraries, TPDL 2024, Ljubljana, Slovenia, September 24-27, 2024, Proceedings, Part I, Lecture Notes in Computer Science, vol. 15177, pp. 76–91, Springer (2024). https://doi.org/10.1007/978-3-031-72437-4_5
7. Devaraj, A., Marshall, I., Wallace, B., Li, J.J.: Paragraph-level simplification of medical texts. In: Toutanova, K., et al. (eds.) Proceedings of the 2021 Conference of the North American Chapter of the Association for Computational Linguistics: Human Language Technologies, pp. 4972–4984, Association for Computational Linguistics, Online (2021). https://doi.org/10.18653/v1/2021.naacl-main.395. https://aclanthology.org/2021.naacl-main.395/
8. Devaraj, A., Sheffield, W., Wallace, B., Li, J.J.: Evaluating factuality in text simplification. In: Muresan, S., Nakov, P., Villavicencio, A. (eds.) Proceedings of the 60th Annual Meeting of the Association for Computational Linguistics (Volume 1: Long Papers), pp. 7331–7345, Association for Computational Linguistics, Dublin, Ireland (2022). https://doi.org/10.18653/v1/2022.acl-long.506

9. Djoudi, A.N., Nouali, S., Aabid, M., Badache, I., Chifu, A.G., Bellot, P.: LIS at the SimpleText 2025: Enhancing Scientific Text Accessibility with LLMs and Retrieval-Augmented Generation. In: [20]

10. Dongre, A.A., Vaadiraaju, A., Madasamy, A.K.: NITK SCaLAR Lab at the CLEF 2025 SimpleText Track: Transformer-Based Models for Biomedical Sentence Simplification (Task 1.1). In: [20]

11. D'Souza, J., Kabongo, S., Giglou, H.B., Zhang, Y.: Overview of the CLEF 2024 simpletext task 4: Sota? Tracking the state-of-the-art in scholarly publications. In: Faggioli, G., Ferro, N., Galuscáková, P., de Herrera, A.G.S. (eds.) Working Notes of the Conference and Labs of the Evaluation Forum (CLEF 2024), Grenoble, France, 9-12 September, 2024, CEUR Workshop Proceedings, vol. 3740, pp. 3163–3173, CEUR-WS.org (2024). URL https://ceur-ws.org/Vol-3740/paper-308.pdf

12. Ermakova, L., et al.: Overview of simpletext 2021 - CLEF workshop on text simplification for scientific information access. In: Candan, K.S., et al. (eds.) Experimental IR Meets Multilinguality, Multimodality, and Interaction - 12th International Conference of the CLEF Association, CLEF 2021, Virtual Event, September 21-24, 2021, Proceedings, Lecture Notes in Computer Science, vol. 12880, pp. 432–449, Springer (2021). https://doi.org/10.1007/978-3-030-85251-1_27

13. Ermakova, L., Bertin, S., McCombie, H., Kamps, J.: Overview of the CLEF 2023 simpletext task 3: simplification of scientific texts. In: Aliannejadi, M., Faggioli, G., Ferro, N., Vlachos, M. (eds.) Working Notes of the Conference and Labs of the Evaluation Forum (CLEF 2023), Thessaloniki, Greece, September 18th to 21st, 2023, CEUR Workshop Proceedings, vol. 3497, pp. 2855–2875, CEUR-WS.org (2023). https://ceur-ws.org/Vol-3497/paper-240.pdf

14. Ermakova, L., Laimé, V., McCombie, H., Kamps, J.: Overview of the CLEF 2024 simpletext task 3: simplify scientific text. In: Faggioli, G., Ferro, N., Galuscáková, P., de Herrera, A.G.S. (eds.) Working Notes of the Conference and Labs of the Evaluation Forum (CLEF 2024), Grenoble, France, 9-12 September, 2024, CEUR Workshop Proceedings, vol. 3740, pp. 3147–3162, CEUR-WS.org (2024). https://ceur-ws.org/Vol-3740/paper-307.pdf

15. Ermakova, L., Ovchinnikova, I., Kamps, J., Nurbakova, D., Araújo, S., Hannachi, R.: Overview of the CLEF 2022 simpletext task 3: Query biased simplification of scientific texts. In: Faggioli, G., Ferro, N., Hanbury, A., Potthast, M. (eds.) Proceedings of the Working Notes of CLEF 2022 - Conference and Labs of the Evaluation Forum, Bologna, Italy, September 5th - to - 8th, 2022, CEUR Workshop Proceedings, vol. 3180, pp. 2792–2804, CEUR-WS.org (2022). https://ceur-ws.org/Vol-3180/paper-237.pdf

16. Ermakova, L., SanJuan, E., Huet, S., Azarbonyad, H., Augereau, O., Kamps, J.: Overview of the CLEF 2023 simpletext lab: automatic simplification of scientific texts. In: Arampatzis, A., et al. (eds.) Experimental IR Meets Multilinguality, Multimodality, and Interaction - 14th International Conference of the CLEF Association, CLEF 2023, Thessaloniki, Greece, September 18-21, 2023, Proceedings, Lecture Notes in Computer Science, vol. 14163, pp. 482–506, Springer (2023). https://doi.org/10.1007/978-3-031-42448-9_30

17. Ermakova, L., et al.: Overview of the CLEF 2024 simpletext track - improving access to scientific texts for everyone. In: Goeuriot, L., et al. (eds.) Experimental IR Meets Multilinguality, Multimodality, and Interaction - 15th International Conference of the CLEF Association, CLEF 2024, Grenoble, France, September 9-12, 2024, Proceedings, Part II, Lecture Notes in Computer Science, vol. 14959, pp. 283–307, Springer (2024).https://doi.org/10.1007/978-3-031-71908-0_13

18. Ermakova, L., et al.: Overview of the CLEF 2022 simpletext lab: automatic simplification of scientific texts. In: Barrón-Cedeño, A., et al. (eds.) Experimental IR Meets Multilinguality, Multimodality, and Interaction - 13th International Conference of the CLEF Association, CLEF 2022, Bologna, Italy, September 5-8, 2022, Proceedings, LNCS, vol. 13390, pp. 470–494, Springer (2022). https://doi.org/10.1007/978-3-031-13643-6_28

19. Eugin, S., Ms.Beula, A., Sathvikha, V., Sangamithra, V.: RECAIDSTechTitans at CLEF 2025: Simplifying Scientific Text and Identifying Spurious Sentences using T5. In: [20]

20. Faggioli, G., Ferro, N., Rosso, P., Spina, D. (eds.): Working Notes of CLEF 2025: Conference and Labs of the Evaluation Forum, CEUR Workshop Proceedings, CEUR-WS.org (2025)

21. Gallina, Y., Jiménez, T., Huet, S.: LIA at SimpleText 2025: University of Avignon at SimpleText 2025: Guided Medical Abstract Simplification. In: [20]

22. Hofmann, N., Dauenhauer, J., Dietzler, N.O., Idahor, I.D., Kreutz, C.K.: THM@SimpleText 2025 Task 1.1: Revisiting Text Simplification based on Complex Terms for Non-Experts. In: [20]

23. Huet, S., Sanjuan, E.: A benchmark collection for assessing scholarly search by non-educated users. In: [20]

24. Kocbek, P., Stiglic, G.: UM-FHS at the CLEF 2025 SimpleText Track: Comparing No-Context and Fine-Tune Approaches for GPT-4.1 Models in Sentence and Document-Level Text Simplification. In: [20]

25. Largey, N., Wu, D., Mansouri, B.: AIIRLab Systems for CLEF 2025 SimpleText: Cross-Encoders to Avoid Spurious Generation. In: [20]

26. Marturi, K.C., Elwazzan, H.H.: Hallucination detection and mitigation in scientific text simplification using ensemble approaches: DSGT at CLEF 2025 SimpleText. In: [20]

27. Marturi, K.C., Elwazzan, H.H.: LLM-Guided Planning and Summary-Based Scientific Text Simplification: DSGT at CLEF 2025 SimpleText. In: [20]

28. Nunzio, G.M.D., Vezzani, F., Bonato, V., Azarbonyad, H., Kamps, J., Ermakova, L.: Overview of the CLEF 2024 simpletext task 2: Identify and explain difficult concepts. In: Faggioli, G., Ferro, N., Galuscáková, P., de Herrera, A.G.S. (eds.) Working Notes of the Conference and Labs of the Evaluation Forum (CLEF 2024), Grenoble, France, 9-12 September, 2024, CEUR Workshop Proceedings, vol. 3740, pp. 3129–3146, CEUR-WS.org (2024). https://ceur-ws.org/Vol-3740/paper-306.pdf

29. Papandreou, T., Bakker, J., Kamps, J.: University of Amsterdam at the CLEF 2025 SimpleText Track. In: [20]

30. SanJuan, E., Huet, S., Kamps, J., Ermakova, L.: Overview of the CLEF 2022 simpletext task 1: passage selection for a simplified summary. In: Faggioli, G., Ferro, N., Hanbury, A., Potthast, M. (eds.) Proceedings of the Working Notes of CLEF 2022 - Conference and Labs of the Evaluation Forum, Bologna, Italy, September 5th - to - 8th, 2022, CEUR Workshop Proceedings, vol. 3180, pp. 2762–2772, CEUR-WS.org (2022). https://ceur-ws.org/Vol-3180/paper-235.pdf

31. SanJuan, E., Huet, S., Kamps, J., Ermakova, L.: Overview of the CLEF 2023 simpletext task 1: passage selection for a simplified summary. In: Aliannejadi, M., Faggioli, G., Ferro, N., Vlachos, M. (eds.) Working Notes of the Conference and Labs of the Evaluation Forum (CLEF 2023), Thessaloniki, Greece, September 18th to 21st, 2023, CEUR Workshop Proceedings, vol. 3497, pp. 2823–2834, CEUR-WS.org (2023). https://ceur-ws.org/Vol-3497/paper-238.pdf

32. SanJuan, E., Huet, S., Kamps, J., Ermakova, L.: Overview of the CLEF 2024 simpletext task 1: retrieve passages to include in a simplified summary. In: Faggioli, G., Ferro, N., Galuscáková, P., de Herrera, A.G.S. (eds.) Working Notes of the Conference and Labs of the Evaluation Forum (CLEF 2024), Grenoble, France, 9-12 September, 2024, CEUR Workshop Proceedings, vol. 3740, pp. 3115–3128, CEUR-WS.org (2024). https://ceur-ws.org/Vol-3740/paper-305.pdf
33. Van Rijsbergen, C.J.: Information Retrieval. Butterworth and Co (Publishers), London, UK (1975)
34. Vendeville, B., Bakker, J., Azarbonyad, H., Ermakova, L., Kamps, J.: Overview of the CLEF 2025 SimpleText Task 2: Identify and Avoid Hallucination. In: [20]
35. Vendeville, B., Ermakova, L., Loor, P.D.: Resource for Error Analysis in Text Simplification: New Taxonomy and Test Collection (2025). https://doi.org/10.1145/3726302.3730304
36. Vendeville, B., Ermakova, L., Loor, P.D., Kamps, J.: UBONLP Report on the SimpleText lab of CLEF 2025. In: [20]
37. Voorhees, E.M.: The cluster hypothesis revisited. In: Proceedings of the 8th Annual International ACM SIGIR Conference on Research and Development in Information Retrieval, p. 188–196, SIGIR '85, Association for Computing Machinery, New York, NY, USA (1985), ISBN 0897911598. https://doi.org/10.1145/253495.253524
38. Vora, A., Chaudhari, T., Hotha, S., Sonawane, S.: S-3 Pipeline by PICT/Pune for Biomedical Text Simplification. In: [20]
39. Xu, Z., Escalera, S., Pavão, A., Richard, M., Tu, W., Yao, Q., Zhao, H., Guyon, I.: Codabench: Flexible, easy-to-use, and reproducible meta-benchmark platform. Patterns 3(7), 100543 (2022). https://doi.org/10.1016/J.PATTER.2022.100543

Overview of the TalentCLEF 2025: Skill and Job Title Intelligence for Human Capital Management

Luis Gasco[1]([✉]) [iD], Hermenegildo Fabregat[1,2] [iD], Laura García-Sardiña[1] [iD],
Paula Estrella[1], Daniel Deniz[1] [iD], Alvaro Rodrigo[2] [iD], and Rabih Zbib[1] [iD]

[1] Avature Machine Learning, Madrid, Spain
machinelearning@avature.net
[2] NLP & IR Group at UNED, Madrid, Spain

Abstract. Advances in natural language processing and large language models are driving a major transformation in Human Capital Management, with a growing interest in building smart systems based on language technologies for talent acquisition, upskilling strategies, and workforce planning. However, the adoption and progress of these technologies critically depend on the development of reliable and fair models, properly evaluated on public data and open benchmarks, which have so far been unavailable in this domain.

To address this gap, we present TalentCLEF 2025, the first evaluation campaign focused on skill and job title intelligence. The lab consists of two tasks: Task A - Multilingual Job Title Matching, covering English, Spanish, German, and Chinese; and Task B - Job Title-Based Skill Prediction, in English. Both corpora were built from real job applications, carefully anonymized, and manually annotated to reflect the complexity and diversity of real-world labor market data, including linguistic variability and gender-marked expressions. The evaluations included monolingual and cross-lingual scenarios and covered the evaluation of gender bias.

TalentCLEF attracted 76 registered teams with more than 280 submissions. Most systems relied on information retrieval techniques built with multilingual encoder-based models fine-tuned with contrastive learning, and several of them incorporated large language models for data augmentation or re-ranking. The results show that the training strategies have a larger effect than the size of the model alone. Talent-CLEF provides the first public benchmark in this field and encourages the development of robust, fair, and transferable language technologies for the labor market.

Keywords: Natural Language Processing · Human Capital Management · Human Resources · Multilinguality · Cross-linguality · Skill Predictions · Job Title Ranking

J. Carrillo-de-Albornoz et al. (Eds.): CLEF 2025, LNCS 16089, pp. 464–485, 2026.
https://doi.org/10.1007/978-3-032-04354-2_24

1   Introduction

The landscape of the global labor market is undergoing a profound transformation driven by rapid technological advancements. Recent studies suggest that by 2030, approximately 70% of the skills required for today's professions will have changed, and a significant proportion of the workforce will be employed in occupations that did not exist at the beginning of the twenty-first century [21].

This change is not accidental, but rather a consequence of the technological development experienced in recent years, which has transformed our understanding of existing job roles. Digital transformation, automation, and the rise of artificial intelligence are redefining the skills needed in the workforce. These advancements have not only automated tasks that were once routine and created entirely new roles, but have also enabled remote work, which is eliminating the geographical boundaries of employment and sourcing talent.

In this transition, the challenge of adaptation affects both individuals and organizations. On the one hand, professionals face increasing uncertainty about which skills will be relevant in the near future, making the need for reskilling more pressing than ever. On the other hand, companies face the task of identifying and attracting talent equipped with these emerging capabilities [23].

All of these challenges have encouraged the adoption of language technologies applied to Human Capital Management (HCM). These technologies are used primarily for talent acquisition, helping organizations match candidates to job positions based on their previous roles and skills. They also support onboarding and training by enabling the creation of customized learning pathways adapted for each employee. Additionally, they play an increasingly important role in strategic workforce planning, as they allow companies to anticipate market trends and future skill requirements.

However, the development and use of these technologies present important challenges [14]. First, **multilingualism** is still a barrier. In a global market where companies operate in multiple countries and languages, the ability to process information in different languages is essential for effective talent management. Second, it is crucial to ensure that the developed systems are **fair and unbiased**, especially when they are used to make decisions about hiring, training, or granting promotions. We need to identify and reduce algorithmic biases that could negatively impact certain groups, such as those defined by gender or ethnicity. Finally, **adaptability** is another challenge, as the importance of skills in candidates can differ widely between industries, or across time. This means that we need flexible systems that can adapt to the specific needs of different sectors.

In response to these challenges, the research community has focused on developing NLP techniques for HCM. Recently, there has been an increase in research activities in areas such as job titles and skill extraction and normalization into standard terminologies such as ESCO or O*NET [19,28,29,33,46–48], job-skill relations [13,15], and job title matching [9,10,44]. This interest has fueled the organization of several research workshops, such as *NLP4HR* [17] or *RecSys in HR* [4], among others [18,49], which have contributed to the emergence and consolidation of a research community in this area. Despite these advances, the absence of standardized benchmarks continues to hinder progress. Most research relies on proprietary datasets due to privacy constraints and, when public datasets exist, they often suffer from shortcomings like the lack of transparency regarding labeling criteria and missing evaluation scripts for comparison when trying to advance state-of-the-art.

To help address these limitations and promote NLP research in the labor domain, this paper presents the results of the TalentCLEF 2025 lab, the first evaluation campaign in Human Capital Management. TalentCLEF introduces a series of challenges to evaluate important tasks in the HCM area, such as job title matching. The data used attempt to better capture the real-word complexities of the task by including terminological variability, multilingualism, and gender-marked variations. Inspired by successful community initiatives in other areas such as BioASQ [26] and BioCreative [24], TalentCLEF aims to drive the creation of open and robust evaluation resources and to encourage the development of more fair, more transferable, and more adaptable solutions for talent management systems.

2   Overview of the Tasks

The first edition of TalentCLEF is focused on the development and evaluation of NLP models for key applications in human capital management: (i) identify suitable candidates for job positions based on professional experience and skills, (ii) implement upskilling programs to foster continuous employee development, and (iii) detect emerging skills and skill gaps within organizations. The Talent-CLEF lab consists of two tasks tailored to these goals: Task A - Multilingual Job Title Matching and Task B - Job Title-Based Skill Prediction.

2.1   Task A: Multilingual Job Title Matching

Job title matching is a fundamental task in NLP for human resources. In talent sourcing, systems must recognize when different job titles, even across multiple languages, refer to the same underlying role, a capability that is essential for building accurate and fair job-candidate recommender systems. The aim of Task A is to develop systems that can identify and rank the job titles most similar to a given one. For each job title, participating teams are required to generate a ranked list of similar job titles from a specified knowledge base.

In the task, we provide a multilingual dataset for job title matching, covering English, Spanish, German, and Chinese. The dataset consists of three partitions: a training set, available only for German, English, and Spanish, and is automatically generated as pairs of related job titles extracted from the ESCO taxonomy; and development and test sets, both created and annotated manually as described in Sect. 3. Table 1 summarizes key statistics of the corpus. Data for Task A were made available to participating teams via Zenodo[1].

---

[1] Task A corpus: https://doi.org/10.5281/zenodo.14002665.

Table 1. Summary statistics of the development and test sets of Task A by language (en: English, es: Spanish, de: German, zh: Chinese).

	Dev				Test			
	en	es	de	zh	en	es	de	zh
# Queries	105	185	203	103	116	191	226	116
# Corpus Elements	2,619	4,661	4,729	2,513	769	1,231	1,509	769
Avg. relevant items per query	23.0	41.0	41.5	22.5	32.9	57.9	65.1	32.9

2.2 Task B: Job Title-Based Skill Prediction

Recent professional reports underscore the growing importance of a skill-centric approach to talent management. In fact, more than half of Europe's workforce is estimated to require reskilling in the coming years due to technological innovations like AI and automation [11,34]. This emerging need for new skills presents a big challenge for companies and employees in their sourcing and training. In this context, developing models that predict skills that align with specific job titles will be essential to help organizations identify current and future skill gaps, enable targeted upskilling and reskilling initiatives, and also to support workforce planning based on competencies. The aim of Task B is to develop systems that, given a set of skills from a knowledge base, identify and rank by relevance those that are relevant to a given job position.

In this task, we provide a dataset in English for job title-based skill relevance[2]. The dataset is divided into three partitions: a training set, which has been automatically generated leveraging the job title to skill relevance information from ESCO; and development and test sets, manually annotated following the process described in Sect. 3. Table 2 shows the statistics of the corpus.

Table 2. Summary statistics of the Task B corpus for the development and test sets.

	Dev	Test
# Queries	304	125
# Corpus Elements	1,439	1,986
Avg. relevant items per query	85.2	101.8

2.3 Evaluation

The official evaluation for both tasks was done using the Codabench platform, with specific competitions for Task A[3] and Task B[4]. This structure offered par-

---

[2] Task B corpus: https://doi.org/10.5281/zenodo.14002665.
[3] https://www.codabench.org/competitions/5842/.
[4] https://www.codabench.org/competitions/7059/.

ticipants a unified interface for submitting results and accessing leaderboards, while also establishing an open benchmark for continuous evaluation after the end of the task.

Evaluation of Task A considered four different scenarios: (i) **Averaged monolingual evaluation**, where systems had to identify similar job titles within the same language in English, Spanish and German; (ii) **cross-lingual evaluation**, where systems had to match job titles in Spanish and German given English queries; (iii) **Chinese language track** allowed teams to voluntary submit predictions for Chinese job titles, even though no training data was provided for this language, challenging participants to find solutions for this setting; and (iv) **gender-based evaluation**, to assess the sensitivity of models to the grammatical gender of job titles, in order to detect possible biases in models for both Spanish and German. On the other hand, Task B had a single evaluation scenario, focusing exclusively on English. In this task, systems were assessed on their ability to identify relevant skills for each job title using the provided dataset in English.

For both Task A and Task B, the primary evaluation metric is **Mean Average Precision (MAP)**, a standard metric for information retrieval tasks in NLP. MAP measures the ability of a system to return highly relevant items at the top of a ranked list, averaged across all queries. This metric is particularly suitable for evaluating systems designed to prioritize the most relevant results, making it a natural choice for both job title matching and skill prediction tasks.

In addition to MAP, Task A also used the **Rank Biased Overlap (RBO)** metric to evaluate gender bias [43]. RBO is a metric that compares the similarity between two ranked lists, placing more weight on items near the top of each list. This reflects the intuition that top-ranked results matter more, which is especially relevant in real-world systems like search engines or recommendation models. Unlike other ranking metrics, RBO is robust to differences in the length and content of the inputs lists, and it can handle partial or non-overlapping lists.

RBO was used to measure how a system's output ranking changes when queries (job titles) are presented in different gendered forms (e.g., *"enfermero"* versus *"enfermera"*, *"abogado"* versus *"abogada"*). By comparing both rankings, RBO quantifies how much the change in wording affects the system's output. A high RBO score indicates that rankings remain consistent despite gender variation, suggesting low sensitivity to gendered language. In contrast, a low RBO score indicates that the gender dimension significantly influences the results i.e., an important signal for detecting and measuring gender bias in ranking models.

3   Corpus Overview

The data provided to participants as a training set was generated using the ESCO taxonomy, while the TalentCLEF 2025 development and test sets were developed using real, de-identified job application data to ensure both authenticity and compliance with privacy regulations. Each of them was derived from job offers and the resumes of their associated applicants, enabling the extraction of positive pairs. In other words, applicants were considered relevant as they had applied to the corresponding position.

The datasets were tailored to the specific retrieval objectives of each task. In both cases, the query component was defined as the job offer title. For Task A, the corpus consisted of job titles extracted from applicant resumes, reflecting how candidates describe their professional roles. For Task B, the corpus was composed of skills automatically extracted from applicant CVs (curricula vitae), allowing for skill-based relevance modeling.

The process of constructing the datasets was designed to maximize diversity in terms of industry, role, language, and gender, while introducing the type of noise and heterogeneity typical of real-world recruitment pipelines. The workflow used to create these resources is shown in Fig. 1, where the schema for data selection, creation and annotation of each corpus is represented in blocks.

3.1   Data Selection

The creation of the TalentCLEF 2025 corpus began with the collection of a large pool of job descriptions and their associated candidate applications from a real-world database. To ensure semantic and domain diversity, job offers were clustered using K-means and representative offers were sampled from different clusters to maximize coverage across industries and job roles.

For each selected offer, a fixed number (xN) of relevant candidate profiles was retrieved —specifically, those who had actually applied to the position. From each candidate profile, both the most recent job title and a comprehensive set of skills were extracted using a high-recall skill extraction system designed to capture any potentially relevant skills for the job offer.

The result is a rich dataset where, for each job offer, we have: (1) the job title of the offer, (2) the most recent job titles of the candidates who applied for that offer, and (3) a group of candidate skills, potentially relevant to the targeted job offer.

3.2   Corpus Creation

Task A. For Task A, queries correspond to cleaned titles of job offer, while corpus elements are the most recent job titles extracted from applicants. All job titles underwent a rigorous cleaning phase to remove sensitive or irrelevant information such as company names, locations, job codes, or other extraneous details (e.g., transforming "*CLOSED - Clinical Trial Manager – Remote*" into "*Clinical Trial Manager*"). Annotators were instructed to flag titles that were incomplete, ambiguous, or not genuine job titles for exclusion from the dataset (e.g. "*Automotive, Engineering, Management, Manufacturing, QA - Quality Control*" was considered a department rather than a job title).

A major goal was to maintain the realism of the data. Thus, while normalization removed sensitive content, it preserved natural variation, typographical errors, and incomplete entries. This presents retrieval systems with more realistic and noisy data, and better thus reflects the challenges of real-world talent matching.

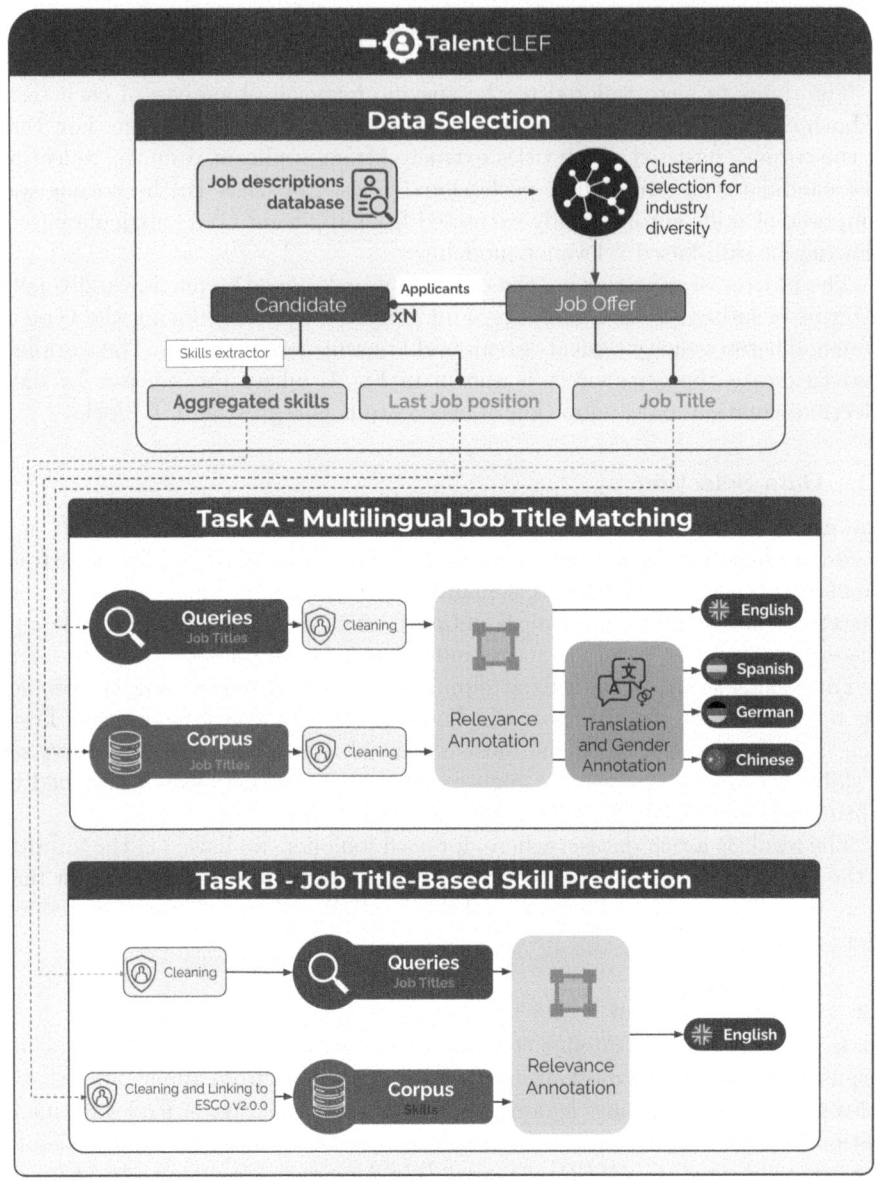

Fig. 1. TalentCLEF 2025 - Corpus creation workflow.

Once the data had been cleaned, the annotation process consisting in two distinct stages was applied:

1. Annotators first judged the relevance of each candidate job title to the given job offer, following detailed guidelines. This often required extra research to clarify duties, industry context, and seniority, especially for ambiguous titles

or acronyms such as *"OAE Evaluator"*[5]. Annotators used online sources to disambiguate and expand such cases when possible.

2. An extensive review phase then aligned annotation criteria, corrected over-annotation, and harmonized judgments. In total, 28.4% of the initial matches were edited, most of them coming from a single annotator.

After completion of the relevance annotation in English, the dataset was manually translated into Spanish, German, and Chinese by professional linguists. For Spanish and German, masculine, feminine, and neutral forms were generated where applicable.

Task B. For Task B (Job Title-Based Skill Prediction), the same queries and candidate pools were used, but the corpus elements consisted of skill sets automatically extracted from candidate documents. These skill lists were curated to identify the most representative skills for each candidate based on their total frequency, we selected a total of 7,493 skills that appeared at least 90 times in the dataset. These skills were manually mapped to ESCO taxonomy v1.2.0 to ensure standardization in the lexical forms across occurrences of the same skill and to facilitate relevance modeling.

The relevance modeling was done by 4 annotators, who annotated the skills relevant to a given job title using the ESCO taxonomy as a reference to find the most relevant skills for each job offer. The process involved multiple stages, including a quality control initial stage and a complete final review of the output to ensure the accuracy and consistency of the gold standard. The test set contained 2,400 annotations and about a third was edited during review. Both development and test sets were annotated by the same pool of linguists and independently reviewed by an expert.

4 Participants

4.1 Task A

A total of 66 teams registered for the task, with 16 teams submitting at least one run in the evaluation phase, and 12 deciding to be part of the benchmark. Across both the development and the test sets, participants contributed a total of 196 submissions, showing a strong engagement with the challenge. Table 3 provides an overview of the methods used by the teams. ⌞Model type⌟ indicates the type of language models used, while ⌞Technique⌟ describes the strategies adopted to address the task. ⌞Training information⌟ specifies the learning objectives applied during fine-tuning, ⌞LLM-related⌟ highlights the specific use of large language models within the systems developed, and ⌞Bias mitigation⌟ denotes methods

---

[5] *"OAE Evaluator"* could be someone assessing otoacoustic emissions (OAEs), or someone specialized in evaluating educational qualifications (Office of Audit and Evaluation).

explicitly designed to address gender bias. Finally, (External data) indicates the external data used to solve the task.

Most of the participating teams approached the task as a retrieval problem, employing embedding models to compute semantic similarity as the ranking metric. Some teams, such as AlexU-NLP and Ixa, incorporated reranking modules based on LLMs, while AlexU-NLP also experimented with cross-encoders. Many teams fine-tuned the embeddings in the training data using contrastive learning techniques and loss functions such as GIST [35] and InfoNCE [27], as well as more advanced strategies such as curriculum learning [36], as employed in the AlexU-NLP submissions.

In general, encoder-based approaches predominated in the competition. However, there was notable experimentation with large-scale decoder-based embedding models. Rather than extensive fine-tuning of foundational models, most teams relied on state-of-the-art multilingual encoders with hundreds of millions of parameters, including architectures such as bge-m3 [6], the multilingual-e5 family [42], and the GTE family [20]. Some participants, such as pjmathematician and NLPnorth, also incorporated large decoder-based models in some of their submissions, specifically gte-Qwen2-7B-instruct and Linq-Embed-Mistral [7]. LLMs (Gemma 2 [30], Claude Sonnet 3.7, Qwen2.5 [37], Llama 3.1 [12], and gpt-4.1-nano) were used for machine translation by pjmathematician, Ixa, and TechWolf, and for reranking by Ixa, HULAT-UC3M, and AlexU-NLP.

Regarding external resources, participants were allowed to use additional information beyond the training data provided. Pjmathematician translated the ESCO data to adapt the embeddings for Chinese, while NLPnorth and AlexU-NLP enriched their data with additional information from the ESCO taxonomy. TechWolf opted to use its internal labor-market domain data to train their model. With respect to bias mitigation, most teams did not implement specific strategies; the exception was Ixa, which automatically translated Spanish and German job titles into English to remove gender markers.

4.2   Task B

Task B had 68 registered teams, with 10 teams submitting at least one run during the evaluation phase, and 8 being part of the final benchmark, for a total of 84 Codabench submissions. Table 4 summarizes the methods used by the participating teams, following the same color-code as in the previous section.

Given the similarity in the input data, most participants approached Task B using methodologies very similar to those employed in Task A. The task has been solved as a semantic similarity-based retrieval problem, trying to identify the most relevant skills for a given job title. Several teams [40,45], used encoder models similar to those used in Task A, but fine-tuned with the provided training data by means of contrastive learning techniques. Additionally, prompting strategies were integrated into the pipelines to obtain better contextualized embeddings for retrieval purposes.

Table 3. Overview of team approaches for Task A.

Team	Ref	Techniques
pjmathematician	[40]	Multilingual Encoder-based Embedding Model · Retrieval · LoRA · Multilingual Decoder-based Embedding Model · Contrastive Learning · MSE Loss · Translation with LLMs · ESCO Data for translation
NLPnorth	[45]	Multilingual Encoder-based Embedding Model · Retrieval · Prompting · Multilingual Decoder-based Embedding Model · Classification · InfoNCE Loss · Classification Loss · ESCO job descriptions
AlexU-NLP	[2]	Multilingual Encoder-based Embedding Model · Hybrid retrieval · Reranking · Curriculum Learning · InfoNCE Loss · Reranking with LLM · ESCO metadata
NT	[16]	Multilingual Encoder-based Embedding Model · Retrieval · Prompting · Zero-shot
DS@GT TalentCLEF	[5]	Encoder-based Embedding Model · Retrieval · Contrastive Learning
Ixa	[31]	Encoder-based Embedding Model · Retrieval · Prompting · Zero-shot · Reranking with LLM · Translation with LLMs · Remove gender markers
SkillSeekers	[39]	Multilingual Encoder-based Embedding Model · Retrieval · Contrastive Learning
SCaLAR	[3]	Multilingual Encoder-based Embedding Model · Retrieval · Contrastive Learning · InfoNCE Loss · ESCO job descriptions
VerbaNexAI	[25]	Multilingual Encoder-based Embedding Model · Retrieval · Contrastive Learning
UDII-UPM	[32]	Encoder-based Embedding Model · Retrieval · Ensemble · Zero-shot
TechWolf	[8]	Multilingual Encoder-based Embedding Model · Retrieval · Contrastive Learning · InfoNCE Loss · Asymmetric dense projections · Internal job ads · Translation with LLMs
HULAT-UC3M	[38]	Multilingual Encoder-based Embedding Model · Retrieval · Reranking · Prompting · Contrastive Learning · InfoNCE Loss · Zero-shot · Reranking with LLMs · Mitigation with LLM prompting

LLMs were commonly used to enrich the textual representation of skills by generating synthetic definitions [1,41], with the goal of enhancing semantic coverage and improving retrieval performance. For data augmentation, participants mainly used models from the Qwen2.5 [37] and Llama 3.1 [12] families. External resources like descriptions from the ESCO taxonomy were leveraged by teams such as NLPnorth and moali to augment the training data. As in Task A, most

Table 4. Overview of team approaches for Task B with keywords labeled by color-coded category.

Team	Ref	Techniques
pjmathematician	[40]	Multilingual Encoder-based Embedding Model · Retrieval · Prompting · Contrastive Learning · GIST Loss · LoRA · Data Augmentation using LLMs
NLPnorth	[45]	Multilingual Encoder-based Embedding Model · Retrieval · Prompting · Multilingual Decoder-based Embedding Model · Classification · Contrastive Learning · InfoNCE Loss · Classification Loss · ESCO skill and jobs descriptions
iagox	[41]	Encoder-based Embedding Model · Retrieval · Ensemble · Zero-shot · Definitions generation using LLMs
moali	[1]	Multilingual Encoder-based Embedding Model · Retrieval · Contrastive Learning · InfoNCE · Definition generation using LLMs · ESCO skill descriptions
SkillSeekers	[39]	Multilingual Encoder-based Embedding Model · Retrieval · Zero-shot
TechWolf	[8]	Multilingual Encoder-based Embedding Model · Retrieval · Contrastive Learning · InfoNCE Loss · Asymmetric dense projections · Internal job ads · Translation with LLMs
HULAT-UC3M	[38]	Multilingual Encoder-based Embedding Model · Retrieval · Reranking · Prompting · Zero-shot · Reranking with LLMs
COTECMAR-UTB	[22]	Multilingual Encoder-based Embedding Model · Retrieval · Zero-shot

of the systems relied on multilingual encoder-based models, although the task is monolingual. In some cases, these models were used in zero-shot settings [39, 41].

5   Results

The results obtained for tasks A and B are presented below. Given the large number of systems received, only the best submissions from each team are reported in the tables. The complete results, including all submissions from each team for tasks A and B, are available on the official task web page[6]

In Task A, teams were evaluated in four areas: overall multilingual performance, cross-lingual performance, Chinese language results (which were optional due to lack of training data), and gender bias. With this evaluation setup, we

---

[6] Complete results for Task A: https://talentclef.github.io/talentclef/docs/talentclef-2025/results/task_a_results/
Complete results for Task B: https://talentclef.github.io/talentclef/docs/talentclef-2025/results/task_b_results/.

could analyze not just performance, but also how well each system handled multilingual challenges, cross-lingual transfer, and fairness issues. In contrast, Task B focused exclusively on the evaluation of systems in English.

5.1   Task A Main Results

Overall Multilingual Performance. The main leaderboard, shown in Table 5, reports the Mean Average Precision across monolingual scenarios in English, Spanish, and German. **AlexU-NLP** achieved the best overall multilingual performance, with an average MAP of 0.534 and the best results for Spanish and German. **TechWolf** was the second-best team overall (0.517), demonstrating balanced and strong results in the three languages considered. It is notable that both **pjmathematician** and **NLPnorth** achieved strong results in English, with NLPnorth reaching a MAP of 0.537 and pjmathematician obtaining the highest score in that language (0.563).

The top-performing systems shared the common feature of fine-tuning their embedding spaces using training data, and leveraging loss functions that have demonstrated strong effectiveness in semantic similarity tasks, such as GIST or InfoNCE. Although the AlexU-NLP team explored hybrid retrieval and re-ranking techniques in some of their submissions, re-ranking was not used in the submission reported here. However, they explicitly avoided incorporating cross-lingual samples during training. On the other hand, the TechWolf team developed a contrastive approach that employs dense asymmetric projections for pairs consisting of job titles and sets of skills separated by commas. This strategy, which relies on their internal and proprietary data, allowed them to achieve strong results, securing the second position in the overall ranking, as well as the second best performance in both the Spanish and German languages.

Cross-Lingual Performance. To evaluate performance in cross-lingual settings, the metric considered was the average MAP for the two scenarios of English queries vs. corpus elements in Spanish and German, respectively. The results are presented in Table 6. The submissions from the **pjmathematician** and **AlexU-NLP** teams achieved the highest scores, with both teams reaching the same final score (0.514), although each obtained the best result in each of the two language pairs.

As expected, MAP values in cross-lingual scenarios were generally lower than in the monolingual case. However, the results are relatively high, showing important progress in multilingual semantic similarity but also confirming that semantic transfer between languages is still a challenge for today's systems. It should be noted that the pjmathematician team, which achieved the highest score, used an embedding model based on the GTE architecture fine-tuned from a 7 billion-parameter decoder model. In contrast, AlexU-NLP employed a hybrid retrieval strategy in this submission, combining BM25 with E5-based embedding models and applying re-ranking with a cross-encoder specifically fine-tuned for this task. The final results were similar, showing that the use of a larger model does not necessarily imply better results in this type of task.

Table 5. Overview of team results for Task A. Best value per column is in bold, second best is underlined.

Team	System ID	Avg	MAP(en-en)	MAP(es-es)	MAP(de-de)
AlexU-NLP	283782	**0.534**	<u>0.559</u>	**0.527**	**0.516**
TechWolf	284991	<u>0.517</u>	0.533	<u>0.519</u>	<u>0.500</u>
pjmathematician	275330	0.515	**0.563**	0.507	0.476
NLPnorth	276154	0.492	0.537	0.496	0.442
NT	284052	0.464	0.523	0.466	0.404
SCaLAR	284749	0.446	0.473	0.438	0.427
HULAT-UC3M	279708	0.420	0.479	0.420	0.360
UDII-UPM	276052	0.414	0.448	0.415	0.377
DS@GT - TalentCLEF	285019	0.399	0.440	0.402	0.355
VerbaNexAI	284612	0.360	0.408	0.348	0.324
TalentCLEF Baseline	—	0.360	0.408	0.348	0.324
SkillSeekers	284999	0.354	0.355	0.360	0.347
Ixa	285265	0.181	0.199	0.173	0.169

Table 6. Team results for (en-es) and (en-de) cross-lingual track. Best value per column in bold, second in underline.

Team	System ID	Avg	MAP(en-es)	MAP(en-de)
pjmathematician	275330	**0.514**	**0.525**	<u>0.504</u>
AlexU-NLP	284479	<u>0.514</u>	<u>0.516</u>	**0.512**
TechWolf	284991	0.504	0.510	0.498
NLPnorth	276154	0.477	0.492	0.461
HULAT-UC3M	279708	0.391	0.416	0.365
UDII-UPM	276052	0.379	0.383	0.375
SCaLAR	284749	0.368	0.366	0.370
TalentCLEF Baseline	—	0.340	0.335	0.345
VerbaNexAI	284612	0.339	0.335	0.344

Systems Including Chinese Language. Due to the absence of training data for Chinese, submission of results for this language was voluntary in the task. Despite this, most of the participating teams (9) submitted predictions for this language, and several of them reported the use of LLMs to do machine translation from English to obtain training data, as described in Sect. 4.1.

Table 7 shows the results for all systems submitted, sorted by average MAP now including Chinese. In this scenario, **AlexU-NLP** achieved the highest score with 0.527, followed by **TechWolf** and **pjmathematician**, sharing the second position with 0.515. The final result for AlexU-NLP is influenced by its stronger performance in Spanish and German, while its score in Chinese is lower than in the other languages. TechWolf obtained the second-best result both in terms of average and specifically for Chinese. Pjmathematician achieved the best score in Chinese, as well as in English, but did not reach a higher overall ranking due to lower performance in Spanish and German.

Table 7. Overview of team results for Task A submissions including Chinese. Best value per column in bold, second in underline.

Team	System ID	Avg	MAP(en-en)	MAP(es-es)	MAP(de-de)	MAP(zh-zh)
AlexU-NLP	283782	**0.527**	0.559	**0.527**	**0.516**	0.508
pjmathematician	275330	0.515	**0.563**	0.507	0.476	**0.516**
TechWolf	284991	0.515	0.533	0.519	0.500	0.510
NLPnorth	276154	0.492	0.537	0.496	0.442	0.495
NT	284052	0.473	0.523	0.466	0.404	0.497
UDII-UPM	276052	0.423	0.448	0.415	0.377	0.451
HULAT-UC3M	280184	0.418	0.479	0.420	0.360	0.415
DS@GT - TalentCLEF	285019	0.405	0.440	0.402	0.355	0.421
SkillSeekers	284999	0.368	0.355	0.360	0.347	0.412

Bias Evaluation. In addition to standard retrieval metrics, we performed a bias analysis by measuring performance gaps between gendered job titles. Systems that minimize this disparity are recognized as biased-controlled and as promoting fairness alongside performance. To rank the systems, we used the average MAP across German and Spanish job titles, the two languages in which grammatical gender can introduce bias. Furthermore, we evaluated the consistency of ranking for gendered job titles using the RBO metric between masculine and feminine forms in both Spanish and German. The average values are shown in Table 8.

In this setting, both **NLPnorth** and **Ixa** achieved a perfect score in both Spanish and German, which means that their system outputs were identical regardless of gendered lexical changes in the input. In the case of Ixa, all job titles were translated into English, thus eliminating grammatical gender marks, given that English is a *natural gender* language[7]. On the other hand, despite NLPnorth not reporting any specific strategy for gender bias mitigation, their classification-based system produced fully stable rankings while also maintaining high MAP scores in both languages. Overall, the results show that the best performing systems (AlexU-NLP and TechWolf) are highly effective in minimizing gender bias in their output classifications, with average RBO values above 0.97, even in the absence of explicit bias mitigation techniques.

---

[7] Languages in which gender is made explicit mostly through pronouns (e.g. *hers* versus *his*) or in some lexical cases (e.g. *queen* versus *king*).

478   L. Gasco et al.

Table 8. Performance considering fairness. Best value per column are in bold, second best are underlined.

Team	System ID	Avg. MAP(es & de)	Avg. RBO (es-de)	RBO(es)	RBO(de)
AlexU-NLP	283782	**0.521**	0.976	0.975	0.977
TechWolf	284991	0.510	0.971	0.972	0.970
pjmathematician	275330	0.492	0.947	0.949	0.945
NLPnorth	276154	0.469	**1.000**	**1.000**	**1.000**
NT	284052	0.435	0.780	0.818	0.742
SCaLAR	284749	0.433	0.947	0.956	0.938
UDII-UPM	276052	0.396	0.932	0.925	0.939
HULAT-UC3M	279708	0.390	0.948	0.950	0.946
DS@GT - TalentCLEF	285019	0.379	0.934	0.938	0.929
SkillSeekers	284999	0.354	0.966	0.974	0.957
VerbaNexAI	284612	0.336	0.915	0.893	0.937
TalentCLEF Baseline	281638	0.336	0.915	0.893	0.937
Ixa	285265	0.171	**1.000**	**1.000**	**1.000**

5.2 Task B Main Results

Table 9 shows the MAP results for all participating teams in Task B. **Pjmathematician** achieved the highest MAP (0.360), followed by **moali** (0.345), with the remaining teams grouped at a lower performance level.

The top-performing teams in Task B followed strategies similar to those used in Task A. For pjmathematician, the approach was based on retrieval with a decoder-type embedding model (7B), fine-tuned using the GIST loss and leveraging both the provided training data and augmented data generated with the Qwen2.5 32B model. Moali fine-tuned a multilingual E5 model with training data, incorporating ESCO skill descriptions, and generating additional job descriptions via prompting with Llama 3.1 8B.

The other teams mostly opted for zero-shot methods, combining encoder-based models with re-ranking techniques built on prompts using language models like Gemma3-4B and Gemini2.5Flash. In particular, Techwolf employed the same model architecture that led to strong results in Task A, achieving solid performance here as well, especially considering that their approach did not include additional fine-tuning with task-specific training data.

5.3 Analytical Insights

In both Task A and Task B, participants submitted a considerable number of systems with diverse configurations, as summarized in Table 3 and Table 4. These systems cover a variety of techniques for training or using LLMs, reflecting the variety of approaches explored throughout the challenge.

Figure 2 shows the relationship between the base model size (number of parameters) and the average MAP of system performance in English, Spanish,

Table 9. Overview of team results for Task B submissions. Best value per column are in bold, second are underlined.

Team	System ID	MAP
pjmathematician	278954	**0.360**
moali	284333	<u>0.345</u>
NLPnorth	279245	0.290
iagox	284949	0.278
Techwolf	284828	0.265
SkillSeekers	284784	0.224
COTECMAR-UTB	284555	0.215
TalentCLEF Baseline	281657	0.196
HULAT-UC3M.	280606	0.141

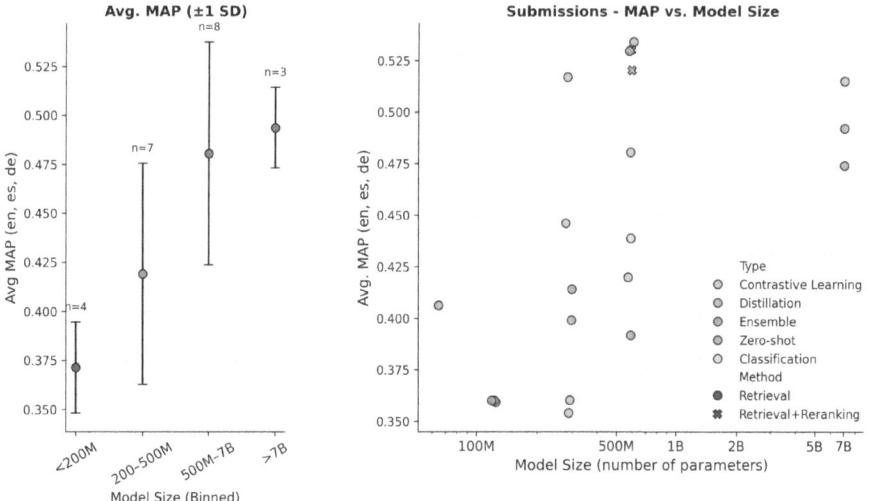

Fig. 2. Relationship between model size and average Mean Average Precision (MAP) across English, Spanish, and German in Task A. Left: Average MAP for submissions grouped by model size. Right: Individual submission results

and German for Task A. The plot on the left groups the systems into parameter-size bins and shows the average MAP with one standard deviation. The plot on the right shows individual system submissions, colored by training strategy and shaped by retrieval methodology.

As seen in the left panel, we observe that, on average, there is a positive correlation between model size and system performance. However, despite this tendency, it should be noted that the highest MAP scores are obtained with models of around 500M parameters, which often outperform much larger decoder-based

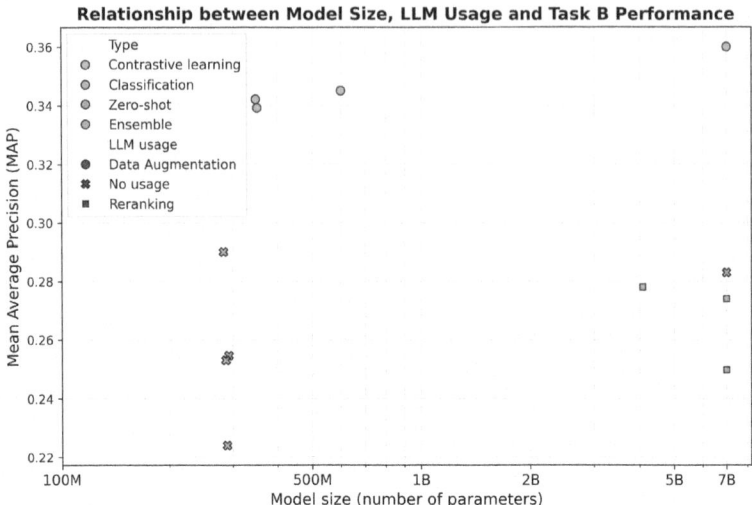

Fig. 3. Relationship between model size (number of parameters) and average Mean Average Precision (MAP) in Task B.

embedding models of up to 7B parameters. These results are particularly relevant when encoder-based models are fine-tuned using contrastive techniques and enhanced with a reranking approach. This suggests that model size alone does not guarantee optimal performance and that certain methodological choices, such as training strategy, play a fundamental role in system performance, even in models of smaller size.

On the other hand, Fig. 3, shows the relationship between model size, training strategy, and the use of LLMs in the task, in relation to the resulting performance measured in terms of Mean Average Precision.

The image shows some interesting patterns. On the one hand, the use of LLMs for data augmentation, combined with fine-tuning through contrastive techniques, enables systems to achieve significantly higher performance, obtaining MAP improvements of up to 8 points for models of similar size. The results highlight the added value of using synthetic data generation with LLM together with specialized fine-tuning strategies.

Moreover, in this specific context, where the relationship between job titles and skills is more complex than in Task A, decoder-based embeddings models show a slight advantage over traditional encoder models if they are properly fine-tuned for the task. This trend suggests that, for skill relevance tasks, embedding models built on LLMs and enhanced with advanced data augmentation techniques can offer additional benefits over simpler approaches.

6   Conclusions

TalentCLEF 2025 has established the first community evaluation framework for NLP applications in Human Capital Management, addressing a gap in this evolving field. The initiative attracted significant participation, with 76 registered teams contributing with more than 280 submissions in two tasks useful for real-world talent management challenges such as job matching and skill-based job matching.

The technical results challenge some prevailing assumptions about the performance of the model. Encoder-based models fine-tuned with contrastive learning techniques consistently outperformed larger decoder-based embedding system. Even in cases like Task B, where a decoder-based model showed better performance, its size, cost, and computational requirements arguably may not justify their use in a real scenario. Still, the most successful systems took advantage of autoregressive language models to support tasks such as data augmentation, leading to better models as a result of training them with more diverse data.

From an evaluation perspective, the systems showed strong performance in both languages and in terms of fairness. In fact, some approaches generated genuinely fair predictions, producing exactly the same results for male and female gendered job titles.

Beyond the technical contribution of the submitted systems, TalentCLEF has established something crucial for this area which is highly dependent on language technologies: shared, open, and publicly available evaluation methods and datasets that capture the true complexity of real-world data. The benchmarks created in this challenge not only support ongoing research, but also help guide the responsible use of language technologies in areas that deeply affect people's lives, such as their professional careers.

Acknowledgments. We would like to acknowledge Aurelia Cañete, Laura Bruno, Mariano Cabrera, Martín Gabriel Chaine Pasco, Georgiana Zarnescu, and Thomas Krucky for their collaboration in the annotation of the tasks data.

References

1. Ali, M.: Enhancing job-skill matching with LLM-driven data augmentation and fine-tuned Bi-encoders. In: CLEF (Working Notes) (2025)
2. Barakat, R., Mokhtar, O., Torki, M., Elmakky, N.: AlexU-NLP at TalentCLEF 2025: curriculum-driven hybrid retrieval for multilingual job title matching. In: CLEF (Working Notes) (2025)
3. Bhangale, C., Gabhane, P., Kumar, M.A.: Fine-tuned sentence transformer for multilingual job title matching. In: CLEF (Working Notes) (2025)
4. Bogers, T., Graus, D., Kaya, M., Johnson, C., Decorte, J.J., De Bie, T.: Fourth workshop on recommender systems for human resources (recsys in hr 2024). In: Proceedings of the 18th ACM Conference on Recommender Systems, RecSys 2024, pp. 1222–1226. Association for Computing Machinery, New York (2024). https://doi.org/10.1145/3640457.3687109

5. Brikman, A., Sana, M., Ruegger, H.: Multilingual job title matching with MPNet-based sentence transformers. In: CLEF (Working Notes) (2025)

6. Chen, J., Xiao, S., Zhang, P., Luo, K., Lian, D., Liu, Z.: M3-embedding: multi-linguality, multi-functionality, multi-granularity text embeddings through self-knowledge distillation. In: Ku, L., Martins, A., Srikumar, V. (eds.) Findings of the Association for Computational Linguistics, ACL 2024, Bangkok, Thailand and virtual meeting, August 11–16, 2024, pp. 2318–2335. Association for Computational Linguistics (2024). https://doi.org/10.18653/V1/2024.FINDINGS-ACL.137

7. Choi, C., et al.: Linq-embed-mistral technical report. CoRR abs/2412.03223 (2024). https://doi.org/10.48550/ARXIV.2412.03223

8. Decorte, J.J., De Lange, M., Van Hautte, J.: TechWolf at TalentCLEF 2025: multilingual JobBERT-V2 for cross-lingual job title matching. In: CLEF (Working Notes) (2025)

9. Decorte, J., Hautte, J.V., Demeester, T., Develder, C.: Jobbert: understanding job titles through skills. CoRR abs/2109.09605 (2021). https://arxiv.org/abs/2109.09605

10. Deniz, D., Retyk, F., García-Sardiña, L., Fabregat, H., Gascó, L., Zbib, R.: Combined unsupervised and contrastive learning for multilingual job recommendation. In: Kaya, M., Bogers, T., Graus, D., Johnson, C., Decorte, J., Bie, T.D. (eds.) Proceedings of the 4th Workshop on Recommender Systems for Human Resources (RecSys-in-HR 2024) co-located with the 18th ACM Conference on Recommender Systems (RecSys 2024), Bari, Italy, 14th–18th October 2024. CEUR Workshop Proceedings, vol. 3788. CEUR-WS.org (2024). https://ceur-ws.org/Vol-3788/RecSysHR2024-paper_3.pdf

11. Di Battista, A., et al.: Future of jobs report 2023. In: World Economic Forum, Geneva, Switzerland. https://www. weforum. org/reports/the-future-of-jobs-report-2023. World Economic Forum (2023)

12. Dubey, A., et al.: The llama 3 herd of models. CoRR abs/2407.21783 (2024). https://doi.org/10.48550/ARXIV.2407.21783

13. Fabregat, H., Poves, R., Alvarez, L.L., Retyk, F., García-Sardiña, L., Zbib, R.: Inductive graph neural network for job-skill framework analysis. Procesamiento del Lenguaje Natural 73, 83–94 (2024). http://journal.sepln.org/sepln/ojs/ojs/index.php/pln/article/view/6602

14. Gasco, L., et al.: TalentCLEF at CLEF2025: skill and job title intelligence for human capital management. In: European Conference on Information Retrieval, pp. 479–486. Springer (2025)

15. Giabelli, A., Malandri, L., Mercorio, F., Mezzanzanica, M., Seveso, A.: Skills2job: a recommender system that encodes job offer embeddings on graph databases. Appl. Soft Comput. 101, 107049 (2021). https://doi.org/10.1016/J.ASOC.2020.107049

16. Ho, T.N., Ho, T.T.T., Dang, V.T.: NT Team at multilingual job title matching task a: job matching via large language model-based description generation and retrieval. In: CLEF (Working Notes) (2025)

17. Hruschka, E., Lake, T., Otani, N., Mitchell, T.: Proceedings of the first workshop on natural language processing for human resources (nlp4hr 2024). In: Proceedings of the First Workshop on Natural Language Processing for Human Resources (NLP4HR 2024), "Association for Computational Linguistics" (2024). https://aclanthology.org/2024.nlp4hr-1.0/

18. Kang, B., De Bie, T., Sebag, M., Largeron, C.: Ai for human resources and people analytics workshop program, ECML PKDD (2023). https://ai4hrpes.github.io/ecmlpkdd2023/program/. Accessed 17 Oct 2024

19. Lavi, D., Medentsiy, V., Graus, D.: Consultantbert: fine-tuned siamese sentence-BERT for matching jobs and job seekers. In: Kaya, M., Bogers, T., Graus, D., Verbert, K., Gutiérrez, F. (eds.) Proceedings of the Workshop on Recommender Systems for Human Resources (RecSys in HR 2021) co-located with the 15th ACM Conference on Recommender Systems (RecSys 2021), Amsterdam, The Netherlands, 27th September – 1st October 2021. CEUR Workshop Proceedings, vol. 2967. CEUR-WS.org (2021). https://ceur-ws.org/Vol-2967/paper_8.pdf

20. Li, Z., Zhang, X., Zhang, Y., Long, D., Xie, P., Zhang, M.: Towards general text embeddings with multi-stage contrastive learning. CoRR abs/2308.03281 (2023). https://doi.org/10.48550/ARXIV.2308.03281

21. LinkedIn: Work change report: AI is coming to work (2025). https://economicgraph.linkedin.com/content/dam/me/economicgraph/en-us/PDF/Work-Change-Report.pdf. Accessed 18 May 2025

22. Llamas, J., Puertas, E., Serrano, J., Martinez, J.: COTECMAR–UTB at TalentCLEF 2025: linking job titles and ESCO skills with sentence transformer embeddings. In: CLEF (Working Notes) (2025)

23. ManpowerGroup: 2024 global talent shortage (2024). https://go.manpowergroup.com. Accessed 19 May 2025

24. Miranda-Escalada, A., et al.: Overview of Drugprot task at biocreative VII: data and methods for large-scale text mining and knowledge graph generation of heterogenous chemical-protein relations. Database **2023**, 1–24 (2023). https://doi.org/10.1093/database/baad080

25. Moreno Novoa, M., Martínez-Santos, J.C., Serrano, J., Puertas, E.: VerbaNex at TalentCLEF2025: semantic matching of multilingual job titles through a framework integrating ESCO taxonomy. In: CLEF (Working Notes) (2025)

26. Nentidis, A., et al.: Overview of bioasq 2022: the tenth bioasq challenge on large-scale biomedical semantic indexing and question answering. In: Experimental IR Meets Multilinguality. Multimodality, and Interaction, pp. 337–361. Springer, Cham (2022)

27. van den Oord, A., Li, Y., Vinyals, O.: Representation learning with contrastive predictive coding. CoRR abs/1807.03748 (2018). http://arxiv.org/abs/1807.03748

28. Retyk, F., Fabregat, H., Aizpuru, J., Taglio, M., Zbib, R.: Résumé parsing as hierarchical sequence labeling: an empirical study. In: Kaya, M., Bogers, T., Graus, D., Johnson, C., Decorte, J. (eds.) Proceedings of the 3rd Workshop on Recommender Systems for Human Resources (RecSys in HR 2023) co-located with the 17th ACM Conference on Recommender Systems (RecSys 2023), Singapore, Singapore, 18th–22nd September 2023. CEUR Workshop Proceedings, vol. 3490. CEUR-WS.org (2023). https://ceur-ws.org/Vol-3490/RecSysHR2023-paper_10.pdf

29. Retyk, F., Gascó, L., Carrino, C.P., Deniz, D., Zbib, R.: MELO: an evaluation benchmark for multilingual entity linking of occupations. In: Kaya, M., Bogers, T., Graus, D., Johnson, C., Decorte, J., Bie, T.D. (eds.) Proceedings of the 4th Workshop on Recommender Systems for Human Resources (RecSys-in-HR 2024) co-located with the 18th ACM Conference on Recommender Systems (RecSys 2024), Bari, Italy, 14th–18th October 2024. CEUR Workshop Proceedings, vol. 3788. CEUR-WS.org (2024). https://ceur-ws.org/Vol-3788/RecSysHR2024-paper_2.pdf

30. Rivière, et al.: Gemma 2: improving open language models at a practical size. CoRR abs/2408.00118 (2024). https://doi.org/10.48550/ARXIV.2408.00118

31. Rodríguez, M., Perez-de Viñaspre, O., Perez, N.: A two-stage multilingual job title matching system: combining expert knowledge and LLM-based ranking. In: CLEF (Working Notes) (2025)

32. Rodríguez-Vidal, J., López-Vargas, A., Vigara Gallego, P.M., Del Álamo, F.J., García-Beltrán, A.: UDII-UPM at TalentCLEF 2025: task a-multilingual job title matching. In: CLEF (Working Notes) (2025)
33. Senger, E., Zhang, M., van der Goot, R., Plank, B.: Deep learning-based computational job market analysis: a survey on skill extraction and classification from job postings. In: Proceedings of the First Workshop on Natural Language Processing for Human Resources (NLP4HR 2024), pp. 1–15. "Association for Computational Linguistics", "St. Julian's" (2024). https://aclanthology.org/2024.nlp4hr-1.1/
34. Smit, S., Tacke, T., Lund, S., Manyika, J., Thiel, L.: The future of work in Europe: Automation, workforce transitions, and the shifting geography of employment. Technical report, McKinsey & Company (2020)
35. Solatorio, A.V.: Gistembed: guided in-sample selection of training negatives for text embedding fine-tuning. CoRR abs/2402.16829 (2024). https://doi.org/10.48550/ARXIV.2402.16829
36. Soviany, P., Ionescu, R.T., Rota, P., Sebe, N.: Curriculum learning: a survey. Int. J. Comput. Vision **130**(6), 1526–1565 (2022). https://doi.org/10.1007/S11263-022-01611-X
37. Tahmid, S., Sarker, S.: Qwen2.5-32b: leveraging self-consistent tool-integrated reasoning for bengali mathematical olympiad problem solving. CoRR abs/2411.05934 (2024). https://doi.org/10.48550/ARXIV.2411.05934
38. Tejera Villar, A., Segura Bedmar, I.: HULAT-UC3M at TalentCLEF: artificial intelligence and natural language processing applied to HR management. In: CLEF (Working Notes) (2025)
39. Uddin, A., Nizami, M.H., Salani, M.T., Saeed, A.: Enhancing human capital management: AI techniques for candidate matching and skill extraction. In: CLEF (Working Notes) (2025)
40. Vachharajani, P.: Pjmathematician at TalentCLEF 2025: enhancing job title and skill matching with GISTEmbed and LLM-augmented data. In: CLEF (Working Notes) (2025)
41. Vázquez García, I.X., Sedano Puente, R., González González, S., Sedano Franco, J.: Beyond titles: semantic matching of jobs and skills using LLMs and S-BERT. In: CLEF (Working Notes) (2025)
42. Wang, L., Yang, N., Huang, X., Yang, L., Majumder, R., Wei, F.: Multilingual E5 text embeddings: a technical report. CoRR abs/2402.05672 (2024). https://doi.org/10.48550/ARXIV.2402.05672
43. Webber, W., Moffat, A., Zobel, J.: A similarity measure for indefinite rankings. ACM Trans. Inf. Syst. (2010)
44. Zbib, R., et al.: Learning job titles similarity from noisy skill labels. CoRR abs/2207.00494 (2022). https://doi.org/10.48550/ARXIV.2207.00494
45. Zhang, M., van der Goot, R.: NLPnorth @ TalentCLEF 2025: Discriminative vs. CLEF (Working Notes), Contrastive vs. Prompting for Job Title and Job-Skill Matching (2025)
46. Zhang, M., van der Goot, R., Plank, B.: ESCOXLM-R: multilingual taxonomy-driven pre-training for the job market domain. In: Rogers, A., Boyd-Graber, J.L., Okazaki, N. (eds.) Proceedings of the 61st Annual Meeting of the Association for Computational Linguistics (Volume 1: Long Papers), ACL 2023, Toronto, Canada, July 9–14, 2023, pp. 11871–11890. Association for Computational Linguistics (2023). https://doi.org/10.18653/V1/2023.ACL-LONG.662
47. Zhang, M., Jensen, K.N., van der Goot, R., Plank, B.: Skill extraction from job postings using weak supervision. In: Kaya, M., Bogers, T., Graus, D., Mesbah,

S., Johnson, C., Gutiérrez, F. (eds.) Proceedings of the 2nd Workshop on Recommender Systems for Human Resources (RecSys-in-HR 2022) co-located with the 16th ACM Conference on Recommender Systems (RecSys 2022), Seattle, USA, 18th–23rd September 2022. CEUR Workshop Proceedings, vol. 3218. CEUR-WS.org (2022). https://ceur-ws.org/Vol-3218/RecSysHR2022-paper_10.pdf

48. Zhang, M., Jensen, K.N., Sonniks, S.D., Plank, B.: Skillspan: hard and soft skill extraction from english job postings. In: Carpuat, M., de Marneffe, M., Ruíz, I.V.M. (eds.) Proceedings of the 2022 Conference of the North American Chapter of the Association for Computational Linguistics: Human Language Technologies, NAACL 2022, Seattle, WA, United States, July 10–15, 2022, pp. 4962–4984. Association for Computational Linguistics (2022). https://doi.org/10.18653/V1/2022.NAACL-MAIN.366

49. Zhu, H., Ge, Y., Xiong, H., Lim, E.P.: The 5th international workshop on talent and management computing (TMC'2024). In: Proceedings of the 30th ACM SIGKDD Conference on Knowledge Discovery and Data Mining, KDD 2024, pp. 6759–6760. Association for Computing Machinery, New York (2024). https://doi.org/10.1145/3637528.3671479

Overview of Touché 2025: Argumentation Systems

Johannes Kiesel[1]([✉]), Çağrı Çöltekin[2], Marcel Gohsen[3], Sebastian Heineking[4], Maximilian Heinrich[3], Maik Fröbe[5], Tim Hagen[6,7], Mohammad Aliannejadi[8], Sharat Anand[3], Tomaž Erjavec[9], Matthias Hagen[5], Matyáš Kopp[10], Nikola Ljubešić[9], Katja Meden[9], Nailia Mirzakhmedova[3], Vaidas Morkevičius[11], Harrisen Scells[2], Moritz Wolter[4], Ines Zelch[4,5], Martin Potthast[6,7,12], and Benno Stein[3]

[1] GESIS - Leibniz Institute for the Social Sciences, Mannheim, Germany
Johannes.Kiesel@gesis.org
[2] University of Tübingen, Tübingen, Germany
[3] Bauhaus-Universität Weimar, Weimar, Germany
[4] Leipzig University, Leipzig, Germany
[5] Friedrich-Schiller-Universität Jena, Jena, Germany
[6] University of Kassel, Kassel, Germany
[7] hessian.AI, Kassel, Germany
[8] University of Amsterdam, Amsterdam, Netherlands
[9] Jožef Stefan Institute, Ljubljana, Slovenia
[10] Charles University, Prague, Czechia
[11] Kaunas University of Technology, Kaunas, Lithuania
[12] ScaDS.AI, Leipzig, Germany

Abstract. This paper is the condensed overview of Touché: the sixth edition of the lab on argumentation systems that was held at CLEF 2025. With the goal to foster the development of support-technologies for decision-making and opinion-forming, we organized four shared tasks: (1) Retrieval-Augmented Debating (RAD), in which participants submit generative retrieval systems that argue against their users and evaluate such systems (new task); (2) Ideology and Power Identification in Parliamentary Debates, in which participants identify from a speech the political leaning of the speaker's party and whether it was governing at the time of the speech (2nd edition); (3) Image Retrieval/Generation for Arguments, in which participants find images to convey a written argument (4th edition, joint task with ImageCLEF); and (4) Advertisement in Retrieval-Augmented Generation, in which participants generate responses to queries with ads inserted and detect such inserted ads (new task). In this paper, we describe these tasks, their setup, and participating approaches in detail.

Keywords: Advertisement Detection · Argumentation · Ideology Identification · Image Generation · Image Retrieval · Retrieval-Augmented Generation · User Simulation

J. Carrillo-de-Albornoz et al. (Eds.): CLEF 2025, LNCS 16089, pp. 486–508, 2026.
https://doi.org/10.1007/978-3-032-04354-2_25

1   Introduction

Decision-making and opinion-forming are everyday tasks that involve weighing pro and con arguments for or against different options. With ubiquitous access to all kinds of information on the web, everybody has the chance to acquire knowledge for these tasks on almost any topic. However, current information systems are primarily optimized for returning *relevant* results and do not address deeper analyses of arguments or multi-modality. To close this gap, the Touché lab series, running since 2020, has several tasks to advance both argumentation systems and the evaluation thereof. Previous events and tasks, data, and publications are available at https://touche.webis.de/. The 2025 edition of Touché features the following shared tasks:

1. Retrieval-Augmented Debating (RAD; new task) features two sub-tasks in argumentative agent research of (1) generating responses to argue against a simulated debate partner and (2) evaluating systems of sub-task 1.
2. Ideology and Power Identification in Parliamentary Debates (2nd edition) features three sub-tasks in debate analysis of detecting the (1) orientation on traditional left–right spectrum, (2) position of power of the speaker's party in the governance of the country or the region, and (3) position of the speaker's party on the scale of populism vs. pluralism.
3. Image Retrieval/Generation for Arguments (4th edition; joint task with ImageCLEF [43]) is about finding images to help convey an argument.
4. Advertisement in Retrieval-Augmented Generation (new task) features two sub-tasks in retrieval-augmented generation of (1) generating responses with advertisements inserted and (2) detecting whether a response contains an advertisement.

In total, 12 teams participated in Touché in 2025.

– Two teams participated in the Retrieval-Augmented Debating task (cf. Sect. 4) and submitted 19 runs. For debating (sub-task 1), the participants employed the provided Elasticsearch API, but used language models for query generation, answer selection, and answer generation. For evaluation (sub-task 2), the participants also focused on prompting language models.
– Four teams participated in the Ideology and Power Identification in Parliamentary Debates task (cf. Sect. 5) and submitted 20 runs. The approaches used traditional machine learning techniques, fine-tuning of multilingual pre-trained models, and prompting large language models, among others.
– Three teams participated in the Image Retrieval/Generation for Arguments task (cf. Sect. 6), submitting a total of seven runs. The teams employed various approaches, including image retrieval using methods such as CLIP, as well as image generation using Stable Diffusion.
– Four teams participated in the Advertisement in Retrieval-Augmented Generation task (cf. Sect. 7) and submitted 17 runs. All teams participated in the classification sub-task and primarily submitted approaches based on fine-tuned encoder models. The generation sub-task received submissions from

three teams that used `Qwen 2.5 7B` or `Mistral 7B` to generate responses from—in some cases re-ranked—lists of relevant document segments.

The corpora, topics, and judgments created at Touché are freely available to the research community on the lab's website.[1] An extended overview of this paper is published at CEUR-WS [48].

2   Background

Argumentation systems are diverse and are connected to many fields within and outside of computer science. The following sections review the related work and background for each Touché task of 2025.

2.1   Retrieval-Augmented Debating

Psychological literature has shown that engaging in conversational argumentation enhances individuals' argumentation skills, which can also improve their performance in non-conversational contexts, such as writing argumentative essays [44]. Apart from the fact that argumentation is an integral part of everyday communication, improving argumentation skills can have a positive impact on collaboration and problem-solving abilities [51]. Following these hypotheses, ArgueTutor [82] is an agent-based tutoring system that provide constructive criticism on solved argumentative writing tasks. However, the ArgueTutor system did not engage in conversational argumentation with its users.

In contrast, Project Debater [75] presented a fully automatic debate system that was designed to challenge humans in formal debates. The debate system employed retrieval and argument mining mechanisms to find counterarguments that challenge the human's stance. Though similar to the conversations in our task, the turns in a formal debate are much longer, allowing each participant to make several points and attack their opponent before their turn ends, with the goal to convince an audience that they are the better debater. In contrast, turns in our task more closely resemble informal debates in which participants directly challenge the arguments after they are presented.

2.2   Ideology and Power Identification in Parliamentary Debates

The task is about important aspects of the political discourse: *ideology* and *power* like in last year [47], but this year also on detecting populism—an important current issue in politics. Although a simplification, political orientation on the left-to-right spectrum has been one of the defining properties of political ideology [3,79]. Power is another factor that shapes the political discourse [16,27,28]. Automatic identification of political orientation from texts has attracted considerable interest [10,14,33,63,64], including a few recent shared tasks [32,69]. The present task differs from the earlier ones, with respect to the source material

---

[1] https://touche.webis.de/.

(parliamentary debates, rather than the popular sources of social media or news) and multilinguality. Despite its central role in critical discourse analysis, to the best of our knowledge, power in parliamentary debates has not been studied computationally. There has been only a few recent computational studies providing indications of linguistic differences between governing and opposition parties [52,59,60,77]. The present shared task and associated data is likely to provide a reference for the future studies investigating power in political discourse. Similarly, although it is a well-studied topic in political science [39,61,68], there are relatively few computational studies of populist discourse, and, to the best of our knowledge, this is the first shared task on populism detection.

2.3   Image Retrieval/Generation for Arguments

Arguments are complex symbolic structures used to exchange reasons and to defend or challenge positions [21,54]. In a world where digital communication increasingly relies on visual media, visual arguments are becoming ever more significant [36]. Images can enhance the acceptability of individual premises [9], and they also have the power to evoke strong emotional responses—such as anxiety, fear, or hope—or even to prescribe specific actions [20]. One of the core challenges in analyzing visual arguments is that images often capture only a single moment in time, making it difficult to convey a complete argumentative structure. While images can be rich in information, they are also inherently ambiguous [50]. Therefore, some scholars argue that images cannot constitute arguments [30]—but others contend that they can [18]. An additional perspective proposes that image sequences are more effective for conveying an argument [9]. However, when combined with text, the inherent ambiguity of images can be reduced, fostering "thick representations" of issues that highlight the importance and strength of the argument, thereby enhancing their persuasive power [50]. Therefore, images can serve as visual reasons, either reinforcing fact-based claims or questioning established beliefs [34].

Several promising research directions can be further pursued at the intersection of argumentation and visual communication. One such direction involves analyzing persuasion techniques, particularly as they appear in visual formats such as memes [17]. Another focuses on exploring how readily textual content can be translated into visual form within an image. While initial progress has been made using metrics such as imaginability [84] and concreteness [6] to evaluate the visualizability of text, this remains an open area of investigation. Another promising direction involves studying argument quality dimensions—such as acceptability, credibility, emotional appeal, and sufficiency [81]—and how these can be measured or expressed visually in images.

2.4   Advertisement in Retrieval-Augmented Generation

Previous research has shown that users of conversational search engines have high confidence in the information provided by LLMs, regardless of whether it is correct or not [76]. More closely related to our task, another study found that people struggle to identify advertisements in generated responses [85]. Both findings underline the importance identifying content, such as advertisements, that tries to influence the opinion of the user.

Given their ability to create content at scale, generative models have recently been studied for their use in advertising [11,42]. This also includes the specific use case of trying to hide advertisements in the output of LLMs [29,38], as well as research on detecting these types of advertisements [72]. Finally, other related work comes from the field of marketing research that has explored how to integrate advertisements covertly within other media long before the arrival of LLMs. The two forms most closely related to our shared task are native advertising [71,83] and product placement [8,26].

3   Lab Overview and Statistics

For the sixth edition of the Touché lab, we received 62 registrations from 22 countries (vs. 68 registrations in 2024). The most lab registrations came from India (19). Out of the 62 registered teams, 12 actively participated in this year's Touché edition (2, 4, 2, and 4 teams submitting valid runs for Task 1, 2, 3, and 4, respectively). Active teams in previous editions were: 20 in 2024, 7 in 2023, 23 in 2022, 27 in 2021, and 17 in 2020.

We used TIRA [31] as the submission platform for Touché 2025 through which participants could either submit code, software, or run files.[2] We tracked the resources of all executions with the alpha version of the TIREx Tracker [37] that monitors the GPU/CPU/RAM usage over time and the energy that an approach consumed (as well as other hardware/software specifications) in the ir_metadata format [5]. Code and software submissions increase reproducibility, as the software can later be executed on different data of the same format. For code and software submissions, a team implemented their approach in a Docker image that they uploaded to their dedicated Docker registry in TIRA. For code submissions, the TIRA client did build a docker image from the code of some git repository, ensuring that the git repository is clean (i.e., all changes are committed and no untracked files), which allows to link a docker image to the exact version of a git repository that produced an submission, whereas software submissions do not need to be linked to the git repository. Submissions in TIRA are immutable, and a team could upload as many code or software submissions as they liked; only they and TIRA had access to their dedicated Docker image registry (i.e., the images were not public while the shared task was ongoing). To improve reproducibility, TIRA executes software in a sandbox by removing the internet connection (ensuring that the software is fully installed in the Docker image which eases rerunning software later, as libraries and models must be installed in an image). For the execution, participants could select the resources that their software had available for execution, from 1 CPU core with 10 GB RAM up to 5 CPU cores with 50 GB RAM and 1 Nvidia A100 GPU with 40 GB RAM. Participants could run their software multiple times using different

---

[2] https://tira.io.

resources to study the scalability and reproducibility (e.g., whether the software executed on a GPU yields the same results as on a CPU). TIRA used a Kubernetes cluster with 1,620 CPU cores, 25.4 TB RAM, 24 GeForce GTX 1080 GPUs, and 4 A100 GPUs to schedule and execute the software submissions, to allocate the resources that the participants selected.

4   Task 1: Retrieval-Augmented Debating

The goal of this task is to create generative retrieval systems that engage in argumentative conversations by presenting counterarguments to users' claims. Such systems can be useful as educational tools to train users' argumentation skills or to explore the argument space on a topic to form or validate an opinion. Participants of this task develop debate systems, which should generate persuasive responses grounded in arguments from a provided argument collection.

4.1   Task Definition

Teams can participate in two sub-tasks: (1) developing debate systems, and (2) providing metrics to assess various quality criteria based on Grice's axioms of cooperative dialogs [35], specifically on the quantity (length), quality (faithfulness), relevance (cf. argumentative quality), and manner (clarity) of system responses. In sub-task 1, participants submit debate system software with which simulated user interact in up to five turns. The submissions are assessed based on the resulting debates, which simultaneously serve as evaluation data for sub-task 2. The debates are annotated according to the annotation schema mentioned above, and submissions to sub-task 2 are assessed based on their correlation strength with human judgments.

4.2   Data Description

Participants received an argument collection of about 300 000 arguments extracted from around 1 500 debates from the ClaimRev dataset [74]. For each of these arguments, the topic was specified, as well as exactly one claim that is supported and one that is attacked by this argument. While only one of the supported or attacked claim could be extracted from the ClaimRev dataset, the missing claim was produced automatically by producing a semantic negation with the help of Llama 3.1 in case the attacked claim was missing or by using the argument itself as the supported claim. The argument collection was provided as a pre-computed Elasticsearch index that allows sparse retrieval with BM25 as well as dense retrieval with k-NN based on the argument text or supported and attacked claims. The embeddings were pre-computed with the document encoder of the pre-trained Stella embedding model [86] (checkpoint: `dunzhang/stella_en_400M_v5`). The data is available online.[3]

---

[3] https://touche.webis.de/data.html#touche25-retrieval-augmented-debate-claims.

Additionally, participants were provided a training set of 100 claims on various topics extracted from the Change My View subreddit.[4] From this subreddit, almost 2 000 threads were acquired through Reddit's API. From this 2 000 threads, an automatic preselection of 500 posts was made based on the BM25 retrieval score according to keywords extracted from the title of the posts and the number of relevant arguments from the ClaimRev index. From these 500 posts, 100 were manually selected to ensure that claims are sufficiently backed up by arguments from the argument collection. These 100 posts underwent severe automatic and manual post-processing to remove author's edits, special characters, and other noise from the posts. These cleaned titles and contents of the posts were provided as claims and descriptions, respectively.

For each claim in the dataset, a debate was generated by simulating a discussion between a basic user and a baseline debate system. Each of the system turns were manually annotated according to an adaption of Grice's maxims of cooperation [35]. For the informal debate context of this shared task, we reinterpreted these maxims as a binary classification schema in the following way:

- **Quantity.** Does the response contain at least one (attack or defense) argument, and at most one of each type of defense and attack?
- **Quality.** Can the response be deduced from the retrieved arguments?
- **Relation.** Is the response coherent with the conversation, and does it express a contrary stance to the user?
- **Manner.** Is the response clear and precise?

The claims, debates, and annotations were released together as a training dataset for sub-task 1 and sub-task 2.

4.3   Participant Approaches

In 2025, two teams participated in this task and submitted 19 runs. Moreover, we added two baseline runs for comparison.

Baselines. For sub-task 1, we provide a baseline that responds with the top claim retrieved without rewriting by (default Elasticsearch) BM25 when the user's utterance is matched with the attacked claim of an indexed claim. For sub-task 2, we provide a 1-baseline, i.e., an evaluator that always produces the maximum score of 1 for each dimension.[5]

Team SINAI. [78] This team (codename: Lewis Carroll) attempted both sub-task 1 and sub-task 2. For sub-task 1, the team proposed a five-step approach which combines the reasoning abilities of an LLaMA3-8B-Instruct model with the provided Elasticsearch API. The LLM first analyses how to answer the question, then generates queries that are used to search Elasticsearch, then selects the arguments across these queries, and finally generates the final counter argument. For sub-task 2, the team focused on three LLM-based prompting methods

---

[4] https://www.reddit.com/r/changemyview/.

[5] All baselines were provided in Python. The sub-task 1 baseline in JavaScript, too.

to derive a measure for evaluating argument quality. Using the same LLaMA3-8B-Instruct model, the team investigates zero-shot, few-shot, and analysis-based few-shot approaches.

Team DS@GT. [58] This team (codename: Haskell Curry) performed both sub-tasks by zero-shot prompting a LLM model, testing six different models: Anthropic Claude (opus4 and sonnet4), Google Gemini 2.5 (flash and pro), and OpenAI GPT (4.1 and 4o). The prompt for sub-task 1 uses detailed guidelines, requesting of the model direct engagement, logical reasoning, being evidence-based, being respectful and constructive in tone, being clear and precise, being brief, and to use assertive utterances—each of these with more details. The prompt for sub-task 2 features a specification for each metric. Scores for all four metrics are requested at once.

4.4 Task Evaluation

Submissions for sub-task 1 are evaluated using a new set of 100 initial claims, obtained by following the methodology of the training set creation. Debates for the assessment are generated in interaction with various simulated users, each presenting different argument strategies, resulting in one simulated debate for each combination of claim, user, and system. All debates are assessed using the evaluation systems submitted for sub-task 2 and our baseline metrics. A random subset of the debates will be judged by human experts according to the criteria of sub-task 2 to identify for each criterion the evaluation system that aligns best with human judgment. Alignment with human judgment is quantified by Precision, Recall, and F1 individually for each of the four maxims. The respective evaluation systems are then used to assess the debate systems from sub-task 1.

5 Ideology and Power Identification in Parliamentary Debates

The study of parliamentary debates is crucial to understand the decision processes in the parliaments and their societal impacts. The goal of this task is to automatically identify three important and interacting aspects of parliamentary debates: the political orientation of the party of the speaker, the role of the party of the speaker in the governance of the country or the region, and the place of the party on populism–pluralism scale. Identifying these underlying aspects of parliamentary debates enables automated comprehension of these discussions, the decisions that these discussions lead to, and their consequences.

5.1 Task Definition

First two sub-tasks (orientation and power identification) were defined as binary classification tasks: Given a parliamentary speech, (1) predict the political orientation of the party of the speaker on the *left–right* spectrum, and (2) predict whether the speaker belongs to one of the governing parties or the opposition. The third sub-task, populism identification, which was introduced to this years competition, is a multi-class (ordinal) classification task with four levels: strongly pluralist, moderately pluralist, moderately populist, strongly populist. The first task is relatively well studied, and there have been some recent shared tasks on identifying political orientation [32,69]. Unlike the earlier tasks, our data set includes multiple parliaments and languages, and is based on parliamentary debates. To the best of our knowledge, this shared task is the first shared task on the other two tasks, identifying power role and populism.

5.2   Data Description

The source of the data for this task is the ParlaMint version 4.1 [24], a uniformly encoded and annotated corpus of transcripts of parliamentary speeches from multiple national and regional parliaments.[6] The ParlaMint version 4.1 used for the task includes data from the following national and regional parliaments: Austria (AT), Bosnia and Herzegovina (BA), Belgium (BE), Bulgaria (BG), Czechia (CZ), Denmark (DK), Estonia (EE), Spain (ES), Catalonia (ES-CT), Galicia (ES-GA), Basque Country (ES-PV), Finland (FI), France (FR), Great Britain (GB), Greece (GR), Croatia (HR), Hungary (HU), Iceland (IS), Italy (IT), Latvia (LV), The Netherlands (NL), Norway (NO), Poland (PL), Portugal (PT), Serbia (RS), Sweden (SE), Slovenia (SI), Turkey (TR) and Ukraine (UA). The labels for first two sub-tasks are also coded in the ParlaMint corpora. For the sake of simplicity, we formulate both tasks as binary classification tasks. For the populism task, we combine labels obtained through multiple expert surveys [56,61,62].

For all tasks, the main challenge in the creation of a dataset is to minimize the effects of covariates [12]. Even though the instances to classify are speeches, the annotations are based on the party membership of the speaker. As a result, underlying variables like party membership, or speaker identity perfectly covary with ideology and power in most cases. In this year's shared task, we opted for a speaker-based split of training and test set, where the same speaker is included only in the training set or only in the test set. We sample at most 20 speeches from a single same speaker. For evaluation, we set aside a test set of 2 000 instances (approximately 100 to 200 speakers depending on the individual corpus). We do not provide a fixed validation (or development) set. Participants were expected to do their own training/validation splits or use cross validation for improving their approaches. Training set sizes vary (min: 221, max: 10 000, mean: 4588) depending on the data availability. For the parliaments with more than 10 000 speeches available for the training set, we reduce the speeches sampled for each speaker to limit the number of speeches to approximately 10 000 speeches.

---

[6] Although all transcripts are obtained thorough the data published by the respective parliaments, the method for obtaining the transcripts vary, such as scraping the web site of the parliament, extracting from published PDF files, and obtaining through an API provided by the parliament. For details, we refer to [23,24].

Except for a few parliaments with limited data and lack of variation (e.g., ES-GA), orientation labels are relatively complete in the shared tasks data for this year. However, some countries do not have the opposition–governing party distinction, and, the expert surveys on populism do not cover all parties in the ParlaMint data. As a result, there are missing labels for some sub-task–parliament pairs. In addition to the original speech transcripts and labels, we also provide automatic English translations, an anonymized speaker ID and the speaker's sex. Labels and speaker ID were hidden in the test set. The shared task data is publicly available.[7]

5.3   Participant Approaches

In 2025, four teams participated in this task (all four submitted a notebook paper) and submitted 20 runs. Moreover, we added a single baseline runs for comparison. As in last year, most participants relied on either computationally efficient methods, or participated with a focused approach to a subset of the parliaments or data.

Baseline. We provided only a single simple baseline using a logistic regression classifier with tf-idf weighted character n-grams. The baseline is intentionally kept simple to encourage participation by early researchers,

Team GIL_UNAM_Iztacala. [80] participated in all sub-tasks using traditional classifiers based on n-gram features. They experiment with a large number of classifiers including Naive Bayes, Logistic Regression, Support Vector Machines and Random Forests. The optimum model was found through grid search of hyperparameters of each classifier, and a few optional preprocessing choices.

Team Munibuc. [57] participated in sub-task 1 (orientation) and sub-task 3 (populism). Their approach was based on extracting task-oriented embeddings from the provided English translations of the parliamentary speeches with NV-Embed-v2 [53], and using support vector classifiers on the extracted embeddings.

Team TüNLP. [73] submitted results for only sub-task 1 (orientation) based on fine-tuning XLM-RoBERTa [13]. The approach involves fine-tuning XLM-RoBERTa-large with the combined training data from all parliaments. The approach is interesting as it allows exploration of exploiting multi-lingual data to improve classification for low-resource settings, and it may potentially be useful for identifying the differences across different languages and cultures.

Team DEMA²IN. [7] contributes to the shared tasks with a focused participation on data from a single parliament (GB). Their approach is based on extracting salient events Mistral-7b v0.2 Instruct [45]. With the intuition that the salient events and the way they are described are important indications of political stance, the approach involves classifying the speeches based only on these event descriptions.

---

[7] Training and test data are available at https://doi.org/10.5281/zenodo.14600017, and https://doi.org/10.5281/zenodo.15337704 respectively.

5.4 Task Evaluation

We use macro-averaged F_1-score as the main evaluation metric for both sub-tasks. For binary tasks, the participants were encouraged to submit confidence scores, where a score over 0.5 is interpreted as class 1 and otherwise 0.

6 Image Retrieval/Generation for Arguments

This task explores how images can be used to visually communicate the core message of an argument. By visualizing key aspects through multimodal representations, arguments can become more engaging, memorable, and accessible. In addition to clarifying complex ideas, images can enhance the persuasive impact of an argument—for example, by highlighting central themes or evoking emotional responses.

6.1 Task Definition

Given a set of arguments, the task is to return multiple images for each argument that effectively convey its meaning. Suitable images may either directly illustrate the argument or depict a related generalization or specialization. These images can be sourced from a provided dataset or generated using an image generation model. For each argument, five images should be submitted, ranked in order of relevance.

6.2 Data Description

The task data includes 128 arguments covering 27 different topics. Each argument consists of a brief claim, such as "Automation increases productivity in industries". For participants using the retrieval method, we created a dataset through a focused crawl, resulting in 32,462 webpages containing 32,339 images. In addition to website texts and images, the dataset includes supplementary information such as automatically generated image captions [40]. Participants using the generation approach were supported with access to a Stable Diffusion-based image generation API [25], building on the concept of the Infinite Index [15].

6.3 Participant Approaches

In 2025, three teams participated in the task: two employed retrieval-based approaches, while the third used a generation-based method. The teams collectively submitted seven runs, which were reduced to five unique entries after deduplication. Each team also submitted an accompanying notebook paper.

Baselines. We provide two baseline models for both retrieval and generation tasks. For retrieval, we use two methods: one based on CLIP [65] embeddings to measure similarity between claims and images, and another using SBERT [67] embeddings to compare argument claims with website text. For generation, we use the claim itself as a prompt for the image generator. We evaluate two versions of Stable Diffusion: stable-diffusion-3.5-medium and the older stable-diffusion-xl-base-1.0.

Team CEDNAV-UTB. [1] This team uses a retrieval-based approach, computing CLIP embeddings for each claim and image caption, and comparing them using cosine similarity. The pairs are then ranked based on the highest similarity score. Additionally, the authors measure the energy consumption of their system over multiple runs.

Team Infotec+CentroGEO. [66] This team evaluated several embedding approaches for retrieval between images and claims using multimodal MCIP [70] and CLIP embeddings. SBERT embeddings between claims and images captions were also used. An internal evaluation using a manually labeled dataset showed that SBERT embeddings between arguments and image captions produced the best results.

Team Hanuman. [2] This team uses an image generation pipeline. First, the LLaMA 3.2-3B [19] model extracts key aspects relevant to each argument. These aspects, along with the original argument, are provided as input to Mistral-7B [45], which generates a corresponding prompt for the image generator, emphasizing the relevant aspects. Afterwards, the corresponding image is generated using diffusion-xl-base-1.0. A human expert reviews the generated image to verify whether it accurately represents the argument and its aspects. If it does not, the prompt is modified to place greater emphasis on the missing aspects. The generated images are ranked by first generating a description of each image using LLaVA-1.5-13B [55], and then computing the cosine similarity between this description and the prompt used to create the image, using SBERT.

6.4   Task Evaluation

When crafting arguments for the task, the expert dataset creator envisioned a corresponding image and pinpointed two or more key aspects crucial for effectively visualizing the argument. For the evaluation of the task, each submitted image–argument pair is assessed based on how well each of these aspects is represented. A final relevance score is then assigned to each pair based on the individual aspect scores. The nDCG@5 score for a single argument is computed by comparing all submitted images for that argument. The final score is then obtained by averaging the nDCG@5 scores across all arguments.

7   Advertisement in Retrieval-Augmented Generation

The goal of this task is to explore native advertising in responses of search engines that use retrieval-augmented generation. Search engines are central to the process of collecting information on a topic and forming an opinion. Both established search engine operators like Google and Microsoft as well as new players like You.com and Perplexity offer conversational search engines backed by LLMs. This raises the question whether the responses generated by LLMs could be biased to influence their users, for instance by presenting a certain product in a favorable way. The task considers advertising both from the perspective of search engine providers inserting advertisements through prompts, as well as from that of users wanting to block advertisements in responses to their queries.

7.1   Task Definition

The task is split into two sub-tasks that ask participants to (1) generate or (2) classify responses. For sub-task 1, the goal is to create relevant responses for a given query from a set of document segments. When also provided with an item to advertise, i.e. a product or service, the response also needs to advertise that item with a defined set of qualities. This advertisement should be difficult to detect and fit seamlessly into the rest of the response. In sub-task 2, submitted systems receive a query and a generated response, and are asked to classify whether the response contains an advertisement or not.

7.2   Data Description

For development purposes, we provided participants with the Webis Generated Native Ads 2024 dataset [72]. It contains 4,868 keyword queries, suitable items to be advertised, as well as 17,344 responses generated by Microsoft Copilot and YouChat. Into a third of the responses, we inserted advertisements with GPT-4o-mini.

For the evaluation of submissions, we created a new version of this dataset starting from a set of 16 meta-topics with commercial relevance like *appliances*, *beauty* or *vacation*. For each meta-topic, we collected up to 500 keyword queries and prompted GPT-4o-mini to generate an additional 100 natural language queries users might ask in the context of the meta-topic. Next, we collected 160 topics from the Google Trends of 2024 and turned both the Google Trends topics as well as the keywords for each meta topic into natural language queries using GPT-4o-mini. This resulted in a total of 9,062 queries. These natural language queries were sent to the search engines *Brave, Microsoft Copilot, Perplexity,* and *You.com* to collect a total of 35,416 responses. By sending the keyword queries for each meta-topic as well as the Google Trends topics to startpage.com, we collected 11,613 unique products and services to be paired with queries. Using these query-advertisement-pairs, we asked GPT-4o-mini to insert advertisements into the original responses collected from the conversational search engines. This resulted in a total of 16,051 responses with advertisements.

We split the 51,467 responses into a training, a validation, and two tests sets, ensuring no advertising leakage between splits as well as minimal query overlap. We assigned the first test set to the generation sub-task. For each of the 1,530 queries in that set, we retrieved up to 100 document segments from the segmented version of the MS MARCO v2.1 document corpus[8] using Elasticsearch

---

[8] https://trec-rag.github.io/about/.

with BM25. Due to computational constraints, we reduced the dataset to a subset of the 100 queries with the largest number of unique URLs among their retrieved segments. Submissions to sub-task 1 receive each query and are asked to generate a relevant response from a context of 20–100 document segments. Additionally, each query is accompanied by 0–4 advertisements for which submissions need to create a separate response each.

We assigned the second test set to the classification sub-task. It contains 6,748 responses; 2,055 with and 4,693 without advertisements. Submissions receive each of these responses alongside the query, the name of search engine that generated the response, and the name of the meta topic of the query, e.g. *banking*. Based on this input, the submissions need to classify the response.

7.3   Participant Approaches

In 2025, four teams participated in this task and submitted a notebook paper. Three of these teams submitted a total of five runs to sub-task 1 and all four teams submitted a total of twelve runs to sub-task 2. For comparison, we added one baseline run to sub-task 1 and four baselines to sub-task 2.

Baselines. For sub-task 1, we created a very simple baseline that repeated the document segment with the highest BM25-score for a given query. If provided with an item to advertise, it added the advertisement with a comma-separated list of qualities to the end of the response. For sub-task 2, we added two approaches trained on the Webis Generated Native Ads 2024 dataset: A fine-tuned version of `all-MiniLM-L6-v2` [72], and a naive Bayes classifier using scikit-learn.[9] After fitted on the training data, the naive Bayes classifier was submitted as three different baselines with the probability thresholds 0.10, 0.25, and 0.40.

Team Git Gud. [46] For sub-task 1, the team uses transformer-based reranking with `all-MiniLM-L6-v2` and `ms-marco-MiniLM-L6-v2` to retrieve document segments as context. The segments are given to `Qwen 2.5 7B` to generate a baseline response that is free of advertisements. For each advertisement, they generate up to three variants of the baseline by inserting a sentence with the ad. From these variants, they select the one with the highest value for a custom "naturalness"-metric and ROUGE-1 overlap with the baseline. If their own classification model for sub-task 2 is able to detect the ad, they regenerate the response to avoid detection. For sub-task 2, the authors fine-tuned multiple transformer-based models on the Webis Generated Native Ads 2024 dataset [72]. Specifically, they trained `MPNet-v2`, `RoBERTa-base/-large`, `DeBERTa-v3-base/-large`, as well as a `RoBERTa-base` checkpoint published on Hugging Face.[10] As input to the models, they use the full response without additional data like the query.

Team JU-NLP. [22] For sub-task 1, the team fine-tuned `Mistral-7B` to generate responses. The generation model was trained with Odds Ratio Preference Optimization (ORPO) [41] on pairs of responses with preference judgments obtained

---

[9] https://scikit-learn.org.

[10] https://huggingface.co/0x7o/roberta-base-ad-detector.

by another instance of `Mistral-7B`. A response is considered more preferable than another if (1) it is more fluent and (2) the inserted advertisement is more difficult to detect. For sub-task 2, the team submitted two approaches. The first one uses a version of `all-mpnet-base-v2` fine-tuned on the Webis Generated Native Ads 2024 dataset [72]. The classification is made on the full response without additional data. The second approach is based on `DeBERTa-v3-base`, fine-tuned on query-response prompts derived from the same dataset. To make a prediction, the query and response are put into a prompt template that asks the model whether the response contains an advertisement or not.

Team Pirate Passau. [4] This team submitted several approaches to sub-task 2 (detection of advertisements). As a baseline, the responses are represented as sparse vectors with TF-IDF weights which are then fed into a random forest classifier. Building on their baseline, two approaches using sentence transformers are proposed. The first one replaces the TF-IDF vectors with embeddings by `all-MiniLM-L6-v2` that are fed into a random forest classifier. The second one is similar to our baseline approach and fine-tunes the transformer models `all-MiniLM-L6-v2` and `MPNet-Base-v2` for binary classification. The team also proposes a decoder-based approach using few-shot prompting with `Llama3.1` and `Qwen2.5`. Finally, the team implemented an approach inspired by RAG pipelines that (1) stores an embedding representation for each response in the training and validation set, (2) retrieves the ten most similar responses for the query of a response that should be classified, (3) re-ranks these responses, and (4) provides the four most similar responses (two with and two without advertisements) as examples to `Llama3.1`, which is again used for classification.

TeamCMU. [49] To augment both sub-tasks, the team synthesized an additional dataset consisting of two types of synthetic data. First, they created the *NaiveSynthetic* dataset using multiple language models to generate responses with fictional advertisements, which the model finds to be the best suited for the given response. Second, they constructed the *StructuredSynthetic* dataset, systematically selecting and summarizing real-world products from Wikipedia using `GPT-4o`, to create responses which included subtle advertisement examples (hard positives) and purely informative examples without advertisements (hard negatives). For sub-task 1, the team developed a modular pipeline consisting of a question answering system based on `Qwen2.5-7B-Instruct` and an Ad-Rewriter, fine-tuned with feedback from an Ad-Classifier. The Ad-Rewriter used a best-of-N sampling method, selecting responses the classifier was least likely to identify as advertisements. The classifier (`DeBERTa-base`) was first trained on the Webis Generated Native Ads 2024 dataset [72], then improved through training on the synthetic datasets and responses created from the Ad-Rewriter. The same classifier was used for sub-task 2.

7.4   Task Evaluation

The evaluation of both sub-tasks is based on precision and recall. For sub-task 1, we added a linear layer to `modernbert-embed-base`[11] and fine-tuned it on the training split of the new dataset mentioned in Sect. 7.2, following the same setup as Schmidt et al. [72]. Evaluated on the classification test split, the fine-tuned model achieves a precision of 95.31 % and a recall of 97.86 %. We apply this classifier to all responses generated by submissions to sub-task 1. The primary score of a submission is based on the inverse recall of our classifier. This means that the score of a submission increases with the number of ads it successfully hides from the classifier. To contextualize the recall, we also report the precision of the classifier. Low precision values indicate that a submission's responses generally have an ad-like character, a property that should be avoided.

For sub-task 2, we measure the effectiveness of a submission using F1-score on the classification test split.

8   Conclusion

The sixth edition of the Touché lab on argumentation systems featured four tasks: (1) Retrieval-Augmented Debating, (2) Ideology and Power Identification in Parliamentary Debates, and (3) Image Retrieval/Generation for Arguments, and (4) Advertisement in Retrieval-Augmented Generation. We added two new tasks, one featuring interactive evaluation of argumentation systems and the other one focusing on the generation and detection of advertisement in generative retrieval systems. In comparison to last year the Ideology and Power Identification in Parliamentary Debates task included an additional sub-task on populism classification. Moreover, for the Image Retrieval/Generation for Arguments task, we changed the task from providing pro and con images to a topic to the less ambiguous providing images that convey a claim.

Of the 62 registered teams, 12 participated in the tasks and submitted a total of 60 runs. Unsurprisingly, large language models and generative approaches were used across tasks. For the Retrieval-Augmented Debating task, teams prompted language models in various ways to retrieve, select, phrase, and evaluate. For the Ideology and Power Identification in Parliamentary Debates task, teams used varying approaches, including traditional classifiers, fine-tuning encoder-only language models and prompting-based approaches using large language models. For the Image Retrieval/Generation for Arguments task, teams used CLIP to retrieve relevant images to Stable Diffusion to generate new ones. For the Advertisement in Retrieval-Augmented Generation task, teams primarily used encoder models like `MiniLM`, `MPNet`, `RoBERTa` and `DeBERTa-v3` to perform advertisement detection. The generation of responses was done with `Qwen 2.5 7B` and `Mistral 7B`.

We plan to continue Touché as a collaborative platform for researchers in argumentation systems. All Touché resources are freely available, including topics, manual relevance, argument quality, and stance judgments, and submitted

---

[11] https://huggingface.co/nomic-ai/modernbert-embed-base.

runs from participating teams. In all Touché labs combined, we received 384 runs from 106 teams. We manually labeled the relevance and quality of more than 35,000 argumentative texts, web documents, and images for 227 topics (topics and judgments are publicly available at the lab's web page, https://touche.webis. de). These resources and other events such as workshops will help to further foster the community working on argumentation systems.

Acknowledgments. This work was partially supported by the European Commission under grant agreement GA 101070014 (https://openwebsearch.eu) and by the German Federal Ministry of Education and Research (BMBF) through the project "DIALOKIA: Überprüfung von LLM-generierter Argumentation mittels dialektischem Sprachmodell" (01IS24084A-B).

References

1. Amaya, D.A.G., Castañeda, J.E.S., Martínez-Santos, J.C., Puertas, E.: CEDNAV–UTB at Touché: Efficient image retrieval for arguments with CLIP. In: Faggioli, G., Ferro, N., Rosso, P., Spina, D. (eds.) Working Notes of CLEF 2025 – Conference and Labs of the Evaluation Forum, CEUR Workshop Proceedings (2025)
2. Anand, S., Heinrich, M.: Hanuman at touché: image generation with argument-aspect fusion . In: Faggioli, G., Ferro, N., Rosso, P., Spina, D. (eds.) Working Notes of CLEF 2025 – Conference and Labs of the Evaluation Forum, CEUR Workshop Proceedings (2025)
3. Arian, A., Shamir, M.: The primarily political functions of the left-right continuum. Comp. Polit. **15**(2), 139–158 (1983)
4. Bouhairi, T.A., Alhamzeh, A.: Pirate Passau at touché: do we need to get complex? a comparative analysis of traditional and advanced NLP approaches for advertisement classification. In: Faggioli, G., Ferro, N., Rosso, P., Spina, D. (eds.) Working Notes of CLEF 2025 – Conference and Labs of the Evaluation Forum, CEUR Workshop Proceedings (2025)
5. Breuer, T., Keller, J., Schaer, P.: ir_metadata: an extensible metadata schema for IR experiments. In: Amigó, E., Castells, P., Gonzalo, J., Carterette, B., Culpepper, J.S., Kazai, G. (eds.) SIGIR 2022: The 45th International ACM SIGIR Conference on Research and Development in Information Retrieval, Madrid, Spain, July 11–15, 2022, pp. 3078–3089. ACM (2022). https://doi.org/10.1145/3477495.3531738
6. Brysbaert, M., Warriner, A.B., Kuperman, V.: Concreteness ratings for 40 thousand generally known English word lemmas. Behav. Res. Methods **46**(3), 904–911 (2013). https://doi.org/10.3758/s13428-013-0403-5
7. Callac, B., Bosser, A.G., de Saint-Cyr, F.D., Maisel, E.: DEMA^{2}IN at touché: salient events extraction for ideology and power identification in parliamentary debates. In: Faggioli, G., Ferro, N., Rosso, P., Spina, D. (eds.) Working Notes of CLEF 2025 – Conference and Labs of the Evaluation Forum, CEUR Workshop Proceedings (2025)
8. Campbell, C., Grimm, P.E.: The challenges native advertising poses: exploring potential federal trade commission responses and identifying research needs. J. Public Policy Market. **38**(1), 110–123 (2019)
9. Champagne, M., Pietarinen, A.-V.: Why images cannot be arguments, but moving ones might. Argumentation **34**(2), 207–236 (2019). https://doi.org/10.1007/s10503-019-09484-0

10. Chen, C., Walker, D., Saligrama, V.: Ideology prediction from scarce and biased supervision: learn to disregard the "what" and focus on the "how"! In: Rogers, A., Boyd-Graber, J., Okazaki, N. (eds.) Proceedings of ACL (Volume 1: Long Papers), pp. 9529–9549. ACL, Toronto (2023). https://doi.org/10.18653/v1/2023.acl-long.530

11. Chen, X., et al.: CTR-driven advertising image generation with multimodal large language models. In: Proceedings of the ACM Web Conference 2025, WWW 2025, pp. 2262–2275. Association for Computing Machinery, New York (2025). https://doi.org/10.1145/3696410.3714836

12. Çöltekin, Ç., Kopp, M., Katja, M., Morkevicius, V., Ljubešić, N., Erjavec, T.: multilingual power and ideology identification in the Parliament: a reference dataset and simple baselines. In: Fiser, D., Eskevich, M., Bordon, D. (eds.) Proceedings of the IV Workshop on Creating, Analysing, and Increasing Accessibility of Parliamentary Corpora (ParlaCLARIN) @ LREC-COLING 2024, pp. 94–100. ELRA and ICCL, Torino (2024). https://aclanthology.org/2024.parlaclarin-1.14/

13. Conneau, A., et al.: Unsupervised cross-lingual representation learning at scale. In: Jurafsky, D., Chai, J., Schluter, N., Tetreault, J. (eds.) Proceedings of the 58th Annual Meeting of the Association for Computational Linguistics, pp. 8440–8451. Association for Computational Linguistics (2020). https://doi.org/10.18653/v1/2020.acl-main.747

14. Conover, M.D., Gonçalves, B., Ratkiewicz, J., Flammini, A., Menczer, F.: Predicting the political alignment of Twitter users. In: Proceedings of PASSAT and Social-Com, pp. 192–199. IEEE (2011). https://doi.org/10.1109/PASSAT/SocialCom.2011.34

15. Deckers, N., et al.: The infinite index: information retrieval on generative text-to-image models. In: Gwizdka, J., Rieh, S.Y. (eds.) ACM SIGIR Conference on Human Information Interaction and Retrieval (CHIIR 2023), pp. 172–186. ACM (2023). https://doi.org/10.1145/3576840.3578327

16. van Dijk, T.: Discourse and Power. Bloomsbury Publishing, London (2008)

17. Dimitrov, D., et al.: SemEval-2021 task 6: detection of persuasion techniques in texts and images. In: Proceedings of SemEval, pp. 70–98. ACL (2021). https://doi.org/10.18653/v1/2021.semeval-1.7

18. Dove, I.J.: On images as evidence and arguments. In: van Eemeren, F.H., Garssen, B. (eds.) Topical Themes in Argumentation Theory: Twenty Exploratory Studies, Argumentation Library, pp. 223–238. Springer, Dordrecht (2012). https://doi.org/10.1007/978-94-007-4041-9_15

19. Dubey, A., et al.: The Llama 3 Herd of Models. CoRR abs/2407.21783 (2024).https://doi.org/10.48550/ARXIV.2407.21783

20. Dunaway, F.: Images, emotions, politics. modern. Am. Hist. **1**(3), 369–376 (2018). https://doi.org/10.1017/mah.2018.17

21. Dutilh Novaes, C.: Argument and argumentation. In: Zalta, E.N., Nodelman, U. (eds.) The Stanford Encyclopedia of Philosophy. Metaphysics Research Lab, Stanford University, Fall 2022 edn. (2022)

22. Dutta, A., Majumdar, A., Biswas, S., Saha, D., Pal, P.: JU-NLP at touché: covert advertisement in conversational AI-generation and detection strategies. In: Faggioli, G., Ferro, N., Rosso, P., Spina, D. (eds.) Working Notes of CLEF 2025 – Conference and Labs of the Evaluation Forum, CEUR Workshop Proceedings (2025)

23. Erjavec, T., et al.: ParlaMint II: advancing comparable parliamentary corpora across Europe. Lang. Res. Eval. 1–32 (2024)

24. Erjavec, T., Ogrodniczuk, M., Osenova, P., Ljubešić, N., Simov, K., Pančur, A., Rudolf, M., Kopp, M., Barkarson, S., Steingrímsson, S., et al.: The ParlaMint corpora of parliamentary Üroceedings. Lang. Resour. Eval. **57**(1), 415–448 (2023)
25. Esser, P., et al.: Scaling rectified flow transformers for high-resolution image synthesis (2024). https://arxiv.org/abs/2403.03206
26. Eyada, B., Milla, A.: Native advertising: challenges and perspectives. J. Des. Sci. Appl. Arts **1**(1), 67–77 (2020)
27. Fairclough, N.: Critical Discourse Analysis: The Critical Study of Language. Longman Applied Linguistics, Taylor & Francis (2013). https://doi.org/10.4324/9781315834368
28. Fairclough, N.: Language and Power. Language in Social Life, Taylor & Francis (2013). https://doi.org/10.4324/9781315838250
29. Feizi, S., Hajiaghayi, M., Rezaei, K., Shin, S.: Online advertisements with LLMs: opportunities and challenges (2024). https://arxiv.org/abs/2311.07601
30. Fleming, D.: Can pictures be arguments? Argumentation Advocacy **33**, 11–22 (1996)
31. Fröbe, M., et al.: Continuous integration for reproducible shared tasks with TIRA.io. In: Kamps, J., et al., (eds.) Advances in Information Retrieval. 45th European Conference on IR Research (ECIR 2023), pp. 236–241. Lecture Notes in Computer Science, Springer, Berlin (2023). https://doi.org/10.1007/978-3-031-28241-6_20
32. García-Díaz, J.A., et al.: Overview of PoliticES 2022: Spanish author profiling for political ideology. Procesamiento del Lenguaje Natural **69**, 265–272 (2022).https://doi.org/10.26342/2022-69-23
33. Gerrish, S., Blei, D.M.: Predicting legislative roll calls from text. In: Getoor, L., Scheffer, T. (eds.) Proceedings of ICML, pp. 489–496. Omnipress (2011)
34. Grancea, I.: Types of visual arguments. Argumentum J. Semin. Discursive Logic Argumentation Theory and Rhetoric **15**(2), 16–34 (2017)
35. Grice, H.: Studies in the Way of Words. William James lectures. Harvard University Press, Cambridge (1989)
36. Groarke, L.: Informal Logic. In: Zalta, E.N., Nodelman, U. (eds.) The Stanford Encyclopedia of Philosophy. Metaphysics Research Lab, Stanford University, Spring 2024 edn. (2024)
37. Hagen, T., Fröbe, M., Merker, J.H., Scells, H., Hagen, M., Potthast, M.: TIREx tracker: the information retrieval experiment tracker. In: 48th International ACM SIGIR Conference on Research and Development in Information Retrieval (SIGIR 2025), ACM (2025). https://doi.org/10.1145/3726302.3730297
38. Hajiaghayi, M., Lahaie, S., Rezaei, K., Shin, S.: Ad Auctions for LLMs via Retrieval Augmented Generation (2024). http://papers.nips.cc/paper_files/paper/2024/hash/20dcab0f14046a5c6b02b61da9f13229-Abstract-Conference.html
39. Hawkins, K.A., Carlin, R.E., Littvay, L., Kaltwasser, C.R. (eds.): The Ideational Approach to Populism: Concept, Theory, and Analysis. Routledge, Extremism and Democracy (2019)
40. Heinrich, M., Kiesel, J., Wolter, M., Potthast, M., Stein, B.: Touché25-image-retrieval-and-generation-for-arguments (2024). https://doi.org/10.5281/zenodo.14258397
41. Hong, J., Lee, N., Thorne, J.: ORPO: monolithic preference optimization without reference model. In: Al-Onaizan, Y., Bansal, M., Chen, Y.N. (eds.) Proceedings of the 2024 Conference on Empirical Methods in Natural Language Processing, pp. 11170–11189. Association for Computational Linguistics, Miami (2024). https://doi.org/10.18653/v1/2024.emnlp-main.626

42. Huang, J., Qu, M., Li, L., Wei, Y.: AdGPT: explore meaningful advertising with ChatGPT. ACM Trans. Multimedia Comput. Commun. Appl. **21**(4) (2025). https://doi.org/10.1145/3720546

43. Ionescu, B., et al.: Overview of ImageCLEF 2025: multimedia retrieval in medical, social media and content recommendation applications. In: de Albornoz, J.C., et al., (eds.) Experimental IR Meets Multilinguality, Multimodality, and Interaction. 16th International Conference of the CLEF Association (CLEF 2025). Lecture Notes in Computer Science, Springer, Berlin Heidelberg New York (2025)

44. Iordanou, K., Rapanta, C.: "Argue With Me": a method for developing argument skills. Front. Psychol. **12** (2021). https://doi.org/10.3389/fpsyg.2021.631203

45. Jiang, A.Q., et al.: Mistral 7B (2023). https://arxiv.org/abs/2310.06825

46. Kamani, S., Taqi, M., Chaudhry, M.A., Hanif, M.A.H., Alvi, F., Samad, A.: Git Gud at touché: unified RAG pipeline for native ad generation and detection. In: Faggioli, G., Ferro, N., Rosso, P., Spina, D. (eds.) Working Notes of CLEF 2025 – Conference and Labs of the Evaluation Forum, CEUR Workshop Proceedings (2025)

47. Kiesel, J., et al.: Overview of touché 2024: argumentation systems. In: Goeuriot, L., et al., (eds.) Experimental IR Meets Multilinguality, Multimodality, and Interaction. 15th International Conference of the CLEF Association (CLEF 2024). Lecture Notes in Computer Science, Springer, Berlin Heidelberg New York (2024)

48. Kiesel, J., et al.: Overview of touché 2025: argumentation systems. In: Faggioli, G., Ferro, N., Rosso, P., Spina, D. (eds.) Working Notes of CLEF 2025 – Conference and Labs of the Evaluation Forum, CEUR Workshop Proceedings (2025)

49. Kim, T.E., Coelho, J., Onilude, G., Singh, J.: TeamCMU at touché: adversarial co-evolution for advertisement integration and detection in conversational search. In: Faggioli, G., Ferro, N., Rosso, P., Spina, D. (eds.) Working Notes of CLEF 2025 – Conference and Labs of the Evaluation Forum, CEUR Workshop Proceedings (2025)

50. Kjeldsen, J.E.: The rhetoric of thick representation: how pictures render the importance and strength of an argument salient. Argumentation **29**(2), 197–215 (2014). https://doi.org/10.1007/s10503-014-9342-2

51. Kuhn, D.: Science as argument: implications for teaching and learning scientific thinking. Sci. Educ. **77**(3), 319–337 (1993). https://doi.org/10.1002/sce.3730770306

52. Kurtoğlu Eskişar, G.M., Çöltekin, Ç.: Emotions running high? a synopsis of the state of Turkish politics through the ParlaMint corpus. In: Fišer, D., Eskevich, M., Lenardič, J., de Jong, F. (eds.) Proceedings of ParlaCLARIN, pp. 61–70. ELRA (2022). https://aclanthology.org/2022.parlaclarin-1.10

53. Lee, C., et al.: NV-Embed: improved techniques for training LLMs as generalist embedding models. arXiv preprint arXiv:2405.17428 (2024)

54. Lewiński, M., Mohammed, D.: Argumentation theory. In: Jensen, K.B., Craig, R.T., Pooley, J., Rothenbuhler, E.W. (eds.) The International Encyclopedia of Communication Theory and Philosophy. Wiley, Hoboken, NJ (2016). https://doi.org/10.1002/9781118766804.wbiect198

55. Liu, H., Li, C., Wu, Q., Lee, Y.J.: Visual instruction tuning. In: Oh, A., Naumann, T., Globerson, A., Saenko, K., Hardt, M., Levine, S. (eds.) Advances in Neural Information Processing Systems 36: Annual Conference on Neural Information Processing Systems 2023, NeurIPS 2023, New Orleans, LA, USA, December 10–16, 2023 (2023). http://papers.nips.cc/paper_files/paper/2023/hash/6dcf277ea32ce3288914faf369fe6de0-Abstract-Conference.html

56. Lührmann, A., et al.: Varieties of Party Identity and Organization (V-Party) Dataset V1 (2020). https://doi.org/10.23696/vpartydsv1. Accessed 22 Feb 2021
57. Marogel, M., Gheorghe, S.: Munibuc at touché: generalist embeddings for orientation and populism detection. In: Faggioli, G., Ferro, N., Rosso, P., Spina, D. (eds.) Working Notes of CLEF 2025 – Conference and Labs of the Evaluation Forum, CEUR Workshop Proceedings (2025)
58. Miyaguchi, A., Johnston, C., Potdar, A.: DS@GT at touché: large language models for retrieval-augmented debate. In: Faggioli, G., Ferro, N., Rosso, P., Spina, D. (eds.) Working Notes of CLEF 2025 – Conference and Labs of the Evaluation Forum, CEUR Workshop Proceedings (2025)
59. Mochtak, M., Rupnik, P., Ljubešić, N.: The ParlaSent multilingual training dataset for sentiment identification in parliamentary proceedings. In: Calzolari, N., Kan, M.Y., Hoste, V., Lenci, A., Sakti, S., Xue, N. (eds.) Proceedings of LREC, pp. 16024–16036. ELRA and ICCL (2024). https://aclanthology.org/2024.lrec-main.1393
60. Navarretta, C., Haltrup Hansen, D.: Government and opposition in Danish parliamentary debates. In: Fiser, D., Eskevich, M., Bordon, D. (eds.) Proceedings of ParlaCLARIN, pp. 154–162. ELRA and ICCL (2024). https://aclanthology.org/2024.parlaclarin-1.23
61. Norris, P.: Measuring populism worldwide. Party politics **26**(6), 697–717 (2020)
62. Pemstein, D., et al.: The V-dem measurement model: latent variable analysis for cross-national and cross-temporal expert-coded data (2020)
63. Pla, F., Hurtado, L.F.: Political tendency identification in twitter using sentiment analysis techniques. In: Tsujii, J., Hajic, J. (eds.) Proceedings of Coling, pp. 183–192. Dublin City University and ACL (2014). https://aclanthology.org/C14-1019
64. Preoţiuc-Pietro, D., Liu, Y., Hopkins, D., Ungar, L.: Beyond binary labels: political ideology prediction of twitter users. In: Barzilay, R., Kan, M.Y. (eds.) Proceedings of ACL, pp. 729–740. ACL (2017). https://doi.org/10.18653/v1/P17-1068
65. Radford, A., et al.: Learning transferable visual models from natural language supervision. In: Meila, M., Zhang, T. (eds.) Proceedings of the 38th International Conference on Machine Learning, ICML 2021. Proceedings of Machine Learning Research, vol. 139, pp. 8748–8763. PMLR (2021). http://proceedings.mlr.press/v139/radford21a.html
66. Ramirez-delreal, T., Moctezuma, D., Ruiz, G., Graff, M., Tellez, E.: Infotec+CentroGEO at touché: MCIP, CLIP and SBERT as retrieval score. In: Faggioli, G., Ferro, N., Rosso, P., Spina, D. (eds.) Working Notes of CLEF 2025 – Conference and Labs of the Evaluation Forum, CEUR Workshop Proceedings (2025)
67. Reimers, N., Gurevych, I.: Sentence-BERT: sentence embeddings using Siamese BERT-networks. In: Proceedings of the 2019 Conference on Empirical Methods in Natural Language Processing. Association for Computational Linguistics (2019). https://arxiv.org/abs/1908.10084
68. Rooduijn, M., et al.: The PopuList: a database of populist, far-left, and far-right parties using expert-informed qualitative comparative classification (EiQCC). Br. J. Polit. Sci. **54**(3), 969–978 (2024). https://doi.org/10.1017/S0007123423000431
69. Russo, D., et al.: PoliticIT at EVALITA 2023: Overview of the political ideology detection in Italian texts task. In: Proceedings of EVALITA. CEUR Workshop Proceedings, vol. 3473. CEUR-WS.org (2023). https://ceur-ws.org/Vol-3473/paper7.pdf

70. Schall, K., Barthel, K.U., Hezel, N., Jung, K.: Optimizing CLIP models for image retrieval with maintained joint-embedding alignment. In: Chávez, E., Kimia, B.B., Lokoc, J., Patella, M., Sedmidubský, J. (eds.) Similarity Search and Applications - 17th International Conference, SISAP 2024. Lecture Notes in Computer Science, vol. 15268, pp. 97–110. Springer (2024). https://doi.org/10.1007/978-3-031-75823-2_9

71. Schauster, E.E., Ferrucci, P., Neill, M.S.: Native advertising is the new journalism: how deception affects social responsibility. Am. Behav. Sci. **60**(12), 1408–1424 (2016)

72. Schmidt, S., Zelch, I., Bevendorff, J., Stein, B., Hagen, M., Potthast, M.: Detecting generated native Ads in conversational search. In: Companion Proceedings of the ACM Web Conference 2024, WWW 2024, pp. 722–725. Association for Computing Machinery, New York (2024). https://doi.org/10.1145/3589335.3651489

73. Shamsutdinov, A., Cherta-Rodriguez, J.: TüNLP at touché: finetuning multilingual models for ideology detection. In: Faggioli, G., Ferro, N., Rosso, P., Spina, D. (eds.) Working Notes of CLEF 2025 – Conference and Labs of the Evaluation Forum, CEUR Workshop Proceedings (2025)

74. Skitalinskaya, G., Klaff, J., Wachsmuth, H.: Learning from revisions: quality assessment of claims in argumentation at scale. In: Merlo, P., Tiedemann, J., Tsarfaty, R. (eds.) Proceedings of the 16th Conference of the European Chapter of the Association for Computational Linguistics: Main Volume, EACL 2021, Online, April 19–23, 2021, pp. 1718–1729. Association for Computational Linguistics (2021). https://doi.org/10.18653/V1/2021.EACL-MAIN.147

75. Slonim, N., et al.: An autonomous debating system. Nature **591**(7850), 379–384 (2021). https://doi.org/10.1038/s41586-021-03215-w

76. Spatharioti, S.E., Rothschild, D.M., Goldstein, D.G., Hofman, J.M.: Comparing traditional and LLM-based Search for consumer choice: a randomized experiment. CoRR abs/2307.03744 (2023). https://doi.org/10.48550/ARXIV.2307.03744

77. Tarkka, O., Koljonen, J., Korhonen, M., Laine, J., Martiskainen, K., Elo, K., Laippala, V.: Automated emotion annotation of finnish parliamentary speeches using GPT-4. In: Fiser, D., Eskevich, M., Bordon, D. (eds.) Proceedings of ParlaCLARIN, pp. 70–76. ELRA and ICCL (2024). https://aclanthology.org/2024.parlaclarin-1.11

78. Vallecillo-Rodríguez, M.E., Martín-Valdivia, M.T., Montejo-Ráez, A.: SINAI at touché: leveraging guided prompt strategies for retrieval-augmented debate. In: Faggioli, G., Ferro, N., Rosso, P., Spina, D. (eds.) Working Notes of CLEF 2025 – Conference and Labs of the Evaluation Forum, CEUR Workshop Proceedings (2025)

79. Vegetti, F., Širinić, D.: Left-right categorization and perceptions of party ideologies. Polit. Behav. **41**(1), 257–280 (2019)

80. Vázquez-Osorio, J., Miranda, L.A.H., Adrián Juárez-Pérez, G.S., Bel-Enguix, G.: GIL_UNAM_Iztacala at touché: benchmarking classical models for multilingual political stance and power classification. In: Faggioli, G., Ferro, N., Rosso, P., Spina, D. (eds.) Working Notes of CLEF 2025 – Conference and Labs of the Evaluation Forum, CEUR Workshop Proceedings (2025)

81. Wachsmuth, H., et al.: Computational argumentation quality assessment in natural language. In: Proceedings of EACL 2017, pp. 176–187 (2017). https://aclanthology.org/E17-1017/

82. Wambsganss, T., Kueng, T., Soellner, M., Leimeister, J.M.: ArgueTutor: an adaptive dialog-based learning system for argumentation skills. In: Proceedings of the 2021 CHI Conference on Human Factors in Computing Systems, CHI 2021, pp. 1–13. Association for Computing Machinery, New York (2021). https://doi.org/10.1145/3411764.3445781

83. Wojdynski, B.W., Evans, N.J.: Going native: effects of disclosure position and language on the recognition and evaluation of online native advertising. J. Advert. **45**(2), 157–168 (2016)

84. Wu, S., Smith, D.A.: Composition and Deformance: measuring imageability with a text-to-image model. CoRR abs/2306.03168 (2023). https://doi.org/10.48550/ARXIV.2306.03168

85. Zelch, I., Hagen, M., Potthast, M.: A user study on the acceptance of native advertising in generative IR. In: ACM SIGIR Conference on Human Information Interaction and Retrieval (CHIIR 2024). ACM (2024). https://doi.org/10.1145/3627508.3638316

86. Zhang, D., Li, J., Zeng, Z., Wang, F.: Jasper and Stella: distillation of SOTA embedding models. CoRR abs/2412.19048 (2024). https://doi.org/10.48550/ARXIV.2412.19048

Author Index

J. Carrillo-de-Albornoz et al. (Eds.): CLEF 2025, LNCS 16089, pp. 509–511, 2026.
https://doi.org/10.1007/978-3-032-04354-2